Complex Analysis and Spectral Theory

CONTEMPORARY MATHEMATICS

743

Centre de Recherches Mathématiques Proceedings

Complex Analysis and Spectral Theory

Conference in Celebration of
Thomas Ransford's 60th Birthday
Complex Analysis and Spectral Theory
May 21–25, 2018
Laval University, Québec, Canada

H. Garth Dales
Dmitry Khavinson
Javad Mashreghi
Editors

AMS AMERICAN MATHEMATICAL SOCIETY

2010 *Mathematics Subject Classification.* Primary 30Cxx, 30Exx, 30Hxx, 30Dxx, 44Axx, 46Fxx, 46Jxx, 47Axx, 47Bxx, 47Dxx.

Library of Congress Cataloging-in-Publication Data

Names: Dales, H. G. (Harold G.), 1944- editor. | Khavinson, Dmitry, 1956–editor. | Mashreghi, Javad, editor.

Title: Complex analysis and spectral theory : a conference in celebration of Thomas Ransford's 60th birthday : May 21–25, 2018, Laval University, Québec, Canada / H. Garth Dales, Dmitry Khavinson, Javad Mashreghi, editors.

Description: Providence, Rhode Island : American Mathematical Society ; Montreal, Québec, Canada : Centre de Recherches Mathematiques, [2020] | Series: Contemporary mathematics. Centre de recherches mathématiques proceedings | Includes bibliographical references.

Identifiers: LCCN 2019040121 | ISBN 9781470446925 (paperback) | ISBN 9781470454531 (ebook)

Subjects: LCSH: Ransford, Thomas, honoree. | Functions of complex variables–Congresses. | Analytic functions–Congresses. | Spectral theory (Mathematics)–Congresses. | Festschriften. | AMS: Functions of a complex variable {For analysis on manifolds, see 58-XX} – Geometric function theory. | Functions of a complex variable {For analysis on manifolds, see 58-XX} – Miscellaneous topics of analysis in the complex domain. | Functions of a complex variable {For analysis on manifolds, see 58-XX} – Spaces and algebras of analytic functions. | Functions of a complex variable {For analysis on manifolds, see 58-XX} – Entire and meromorphic functions, and related topics. | Integral transforms, operational calculus {For fractional derivatives and integrals, see 26A33. For Fourier transforms, see 42A38, 42B10. For integral transforms in distribution spaces, see 46F12. For | Functional analysis {For manifolds modeled on topological linear spaces, see 57Nxx, 58Bxx} – Distributions, generalized functions, distribution spaces [See also 46T30]. | Functional analysis {For manifolds modeled on topological linear spaces, see 57Nxx, 58Bxx} – Commutative Banach algebras and commutative topological algebras [See also 46E25]. | Operator theory – General theory of linear operators. | Operator theory – Special classes of linear operators. | Operator theory – Groups and semigroups of linear operators, their generalizations and applications.

Classification: LCC QA331.7 .C67178 2020 | DDC 515/.98–dc23

LC record available at https://lccn.loc.gov/2019040121

Contemporary Mathematics ISSN: 0271-4132 (print); ISSN: 1098-3627 (online)

DOI: https://doi.org/10.1090/conm/743

∞ The paper used in this book is acid-free and falls within the guidelines
established to ensure permanence and durability.
Visit the AMS home page at `https://www.ams.org/`

10 9 8 7 6 5 4 3 2 1   25 24 23 22 21 20

Contents

Preface

Spectral theory is the branch of mathematics devoted to the study of matrices and their eigenvalues, as well as their infinite-dimensional counterparts, linear operators and their spectra. Spectral theory is ubiquitous in science and engineering because so many physical phenomena, being essentially linear in nature, can be modelled using linear operators. It draws upon techniques from a variety of other areas of mathematics and leads to problems in these areas that are of interest in their own right. *Complex analysis* is the calculus of functions of a complex variable. The roots of the subject go back to the early 19th century and are associated with the names of Euler, Gauss, Cauchy, Riemann, and Weierstrass. Of particular importance are the differentiable functions, usually called analytic or holomorphic functions. They are widely used in mathematics (for example, in Fourier analysis, analytic number theory, and complex dynamics), in physics (potential theory, string theory) and in engineering (fluid dynamics, control theory and the theory of communication).

Both topics are related to numerous other domains in mathematics as well as other branches of science and engineering. For example, analytic function spaces arise in various different branches of mathematics and science. The list includes, but is not restricted to, analytical mechanics, physics, astronomy (celestial mechanics), geology (weather modeling), chemistry (reaction rates), biology, population modeling, economics (stock trends, interest rates and the market equilibrium price changes). As a matter of fact, it is hard to find a branch of analysis or applied sciences in which function spaces do not appear. Many mathematicians have studied this domain and contributed to the field and it is rather impossible to provide a list.

As another manifestation, *functional analysis* is the branch of mathematics concerned with the study of vector spaces and linear mappings acting upon them. The word "functional" refers to an operation whose argument is a function, integration, for example. Two of the most important names are Hilbert and Banach, and the central notions of the subject are named in their honor: Banach spaces (complete normed vector spaces) and Hilbert spaces (Banach spaces where the norm arises from an inner product). Hilbert spaces, which generalize the notion of Euclidean space to infinite dimensions, are of fundamental importance in many areas, including partial differential equations, quantum mechanics and signal processing. From the earliest days, researchers in functional analysis recognized the importance of studying spaces of functions, as opposed to considering just one function at a time. Together with the development of the Lebesgue integral, this led to new techniques, for example, for analyzing the behavior of analytic functions at the boundary of their domain and for proving the existence of analytic functions with certain properties, hitherto difficult or impossible to construct. In turn, complex analysis repaid its debt to functional analysis by providing methods for defining functions of operators, for example, the square root or the logarithm of an operator or a matrix.

There are many other connections, and in the century that has followed this has become a vast domain of research. In recent years, there has been a tremendous amount of work on reproducing kernel Hilbert spaces of analytic functions, on the operators acting on them, as well as on applications in physics and engineering which arise from pure topics like interpolation and sampling.

In this conference, more than thirty analysts, some up-and-coming, others well-established, and from Europe and North America, were invited. Many different topics in complex analysis—operator theory, matrix analysis, spectral theory, functional analysis, and approximation theory—were discussed during the invited talks. This lively meeting certainly strengthened our understanding of the subjects, how far the applications range, how much is known, and how much is still unknown. The goal of our gathering was to discuss a number of fundamental open problems on Hilbert and Banach spaces of analytic functions and the new ideas that have been developed as well as the recent progress that has been made. We believe this event was worthwhile, since the ideas involved were of widespread interest in the mathematical analysis community.

In this conference, we also celebrated the 60th birthday of Thomas Ransford. Thomas Ransford is Professor in the Département de mathématiques et de statistique of Université Laval and Canada Research Chair in Spectral Theory and Complex Analysis. He obtained his Ph.D. at Cambridge in 1984, as a student of the late Graham Allan, and was awarded an Sc.D. by Cambridge in 1999. Before coming to Québec in 1993, he held teaching positions at Leeds and Cambridge. He has also held visiting positions at Ann Arbor, Bordeaux, Brown, Lille, Marseille, Oxford, and UCLA.

Ransford's research is primarily in complex analysis and spectral theory, though he has also worked in potential theory, dynamical systems and probability. He has over a hundred research publications to his name, many written in collaboration with researchers from around the world (over sixty at last count). He is also the author of two books, both published by Cambridge University Press: *Potential Theory in the Complex Plane* (written sole) and *A Primer on the Dirichlet Space* (co-authored with O. El-Fallah, K. Kellay, and J. Mashreghi). Ransford has presented over eighty invited talks at national and international conferences. Ransford is or has been a member of the editorial boards of ten different journals, including the CMS and LMS journals. He has also served on numerous selection committees for grants, fellowships, and prizes, including grant selection committees of NSERC and FRQNT.

A major aspect of Ransford's research career has been the direction of students at all levels. He has supervised 10 doctoral students, 25 master's students, and 14 postdoctoral fellows, in addition to co-directing 10 other graduate students. Many of these students have continued in academia: 22 of them now hold permanent positions at universities, and five have postdoctoral positions. Of the others, eight are now college instructors, eight others work in the financial sector, and four are computer programmers. In addition to graduate direction, Ransford has also supervised 24 summer undergraduate research projects, eight of which have given rise to publications. In the course of his career, Ransford has taught undergraduate and graduate courses in over twenty different subjects. He has received the teaching award "professeur étoile" from the Faculté des sciences et de génie of Université Laval ten times and was voted "professeur méritant en mathématiques et statistique" at the Gala du Mérite Étudiant four times.

Mathematics runs in the family. Ransford's wife, Line Baribeau, is a mathematician, and they have published six papers together. His eldest son, Julian, is also a mathematician.

H. Garth Dales
Dmitry Khavinson
Javad Mashreghi

List of Invited Speakers

(1) W. Arendt, Ulm University, Germany
(2) R. M. Aron, Kent State University, USA
(3) C. Bénéteau, University of South Florida, USA
(4) A. Bourhim, University of Syracuse, USA
(5) I. Chalendar, Université Paris-Est Marne-la-Vallée, France
(6) J. Cima, University of North Carolina, USA
(7) C. Costara, Ovidius University, Romania
(8) H. G. Dales, Lancaster University, UK
(9) O. El-Fallah, Université Mohammed V, Morocco
(10) J. Esterle, Université de Bordeaux, France
(11) B. Forrest, University of Waterloo, Canada
(12) M. Fortier Bourque, University of Toronto, Canada
(13) R. Fournier, Université de Montréal, Canada
(14) E. Fricain, Université Lille 1, France
(15) J. Galé, University of Zaragoza, Spain
(16) P. M. Gauthier, Université de Montréal, Canada
(17) P. Gorkin, Bucknell University, USA
(18) D. Guillot, University of Delaware, USA
(19) D. Jakobson, McGill University, Canada
(20) K. Kellay, Université de Bordeaux, France
(21) D. Khavinson, University of South Florida, USA
(22) D. Kinzebulatov, Université Laval, Canada
(23) L. Kosinski, Jagiellonian University, Poland
(24) N. Levenberg, Indiana University, USA
(25) L. Marcoux, University of Waterloo, Canada
(26) J. Mashreghi, Université Laval, Canada
(27) M. Mbekhta, Université Lille 1, France
(28) A. G. O'Farrell, Maynooth University, Ireland
(29) S. Pouliasis, Aristotle University of Thessaloniki, Greece
(30) W. T. Ross, University of Richmond, USA
(31) M. Roy, York University, Canada
(32) E. Strouse, Université de Bordeaux, France
(33) M. C. White, Newcastle University, UK
(34) M. Younsi, University of Hawaii at Manoa, USA
(35) N. Zorboska, University of Manitoba, Canada
(36) W. Zwonek, Jagiellonian University, Poland

Contemporary Mathematics
Volume **743**, 2020
https://doi.org/10.1090/conm/743/14954

Additive maps preserving matrices of inner local spectral radius zero at some fixed vector

Constantin Costara

ABSTRACT. We characterize surjective additive maps on the space of complex $n \times n$ matrices which preserve matrices of inner local spectral radius zero at some fixed nonzero vector.

1. Introduction and statement of results

For a Banach space X over the complex field $\mathbb{C}$, let us denote by $\mathcal{L}(X)$ the algebra of all linear bounded operators on it. Fix then a point $x_0 \in X$. Given $T \in \mathcal{L}(X)$, its local resolvent set $\rho_T(x_0)$ at x_0 is defined as the union of all open subsets $U \subseteq \mathbb{C}$ for which there exists an analytic function $f : U \to X$ such that $(T-\lambda)f(\lambda) = x_0$ for every $\lambda \in U$. The local spectrum $\sigma_T(x_0)$ of T at x_0 is defined as $\sigma_T(x_0) = \mathbb{C} \backslash \rho_T(x_0)$, and is always a closed subset of the classical spectrum $\sigma(T)$ of T. Unlike the classical spectrum, which is always non-empty, we may have $\sigma_T(x_0) = \emptyset$. If, for example, T has the single-valued extension property (SVEP) then $x_0 \neq 0$ implies $\sigma_T(x_0) \neq \emptyset$.

For a closed subset $F \subseteq \mathbb{C}$ and an operator $T \in \mathcal{L}(X)$, the glocal spectral subspace of T associated with F is defined as

$$\chi_T(F) = \{x \in X : (T-\lambda)f(\lambda) = x \text{ has an analytic solution on } \mathbb{C}\backslash F\}.$$

Then the inner local spectral radius of T at x_0 is defined by

$$i_T(x_0) = \sup\{r \geq 0 : x_0 \in \chi_T(\{|z| \geq r\})\},$$

and coincides with the minimum modulus of $\sigma_T(x_0)$, provided that T has SVEP. Even if T does not have SVEP, we still have that $i_T(x_0) = 0$ if and only if $0 \in \sigma_T(x_0)$; see, for example, [**19**].

We refer to the books [**1**] and [**18**] for more background information on general local spectra theory. In this paper, we shall mainly work in the particular case when X is of finite dimension. So fix a natural number $n \geq 1$, and denote by $\mathcal{M}_n$ the algebra of all $n \times n$ matrices over $\mathbb{C}$. Then for $T \in \mathcal{M}_n$ we have that $\sigma(T)$ is the set of all its eigenvalues without taking into account multiplicities. Fixing a nonzero vector $x_0 \in \mathbb{C}^n$, in this case we have the following nice characterization of the local spectrum $\sigma_T(x_0)$ of the matrix T at x_0; see, for example, the articles

---

2010 *Mathematics Subject Classification.* Primary 47B49; Secondary 47A10, 47A11.
Key words and phrases. Additive map; Preserver; Local spectrum; Inner local spectral radius.

[**6**] and [**20**]. Let $\lambda_1, ..., \lambda_k$ be the distinct eigenvalues of $T \in \mathcal{M}_n$ and denote by $N_1, ..., N_k$ the corresponding root spaces. Then

$$\sigma_T(x_0) = \{\lambda_j : 1 \le j \le k,\ P_j(x_0) \ne 0\}, \tag{1.1}$$

where for $j = 1, ..., k$, the maps $P_j : \mathbb{C}^n \to N_j \subseteq \mathbb{C}^n$ are the associated canonical projections. Let us observe that in the case of matrices, $\sigma_T(x_0) \subseteq \sigma(T)$ is always a non-empty subset: this comes from the fact that $P_1 + \cdots + P_k = I$, the identity of $\mathbb{C}^n$, so that $P_j(x_0)$, $j = 1, ..., k$, cannot all be zero at the same time. In fact, any operator whose point spectrum has empty interior has SVEP, and in particular any finite rank operator has SVEP. Thus any matrix has SVEP, and therefore the inner local spectral radius of the matrix T at x_0 is

$$i_T(x_0) = \min\{|\lambda| : \lambda \in \sigma_T(x_0)\}.$$

Then (1.1) shows that

$$i_T(x_0) = \min\{|\lambda_j| : 1 \le j \le k,\ P_j(x_0) \ne 0\}. \tag{1.2}$$

Therefore, for a non-invertible matrix $T \in \mathcal{M}_n$, we have that $i_T(x_0)$ equals 0 if and only if $P(x_0) \ne 0$, where $N \subseteq \mathbb{C}^n$ is the root space corresponding to the eigenvalue 0 and $P : \mathbb{C}^n \to N \subseteq \mathbb{C}^n$ is the associated canonical projection.

Over the last years, the study of linear/additive/nonlinear local spectra preserver problems on matrix/operator spaces has been a very active field of research. The first ones to consider this type of problems were Bourhim and Ransford in 2006. They proved in [**5**] that for a complex Banach space X, if $\varphi : \mathcal{L}(X) \to \mathcal{L}(X)$ is an additive map such that

$$\sigma_{\varphi(T)}(x) = \sigma_T(x)$$

for each $T \in \mathcal{L}(X)$ and each $x \in X$, then φ is the identity of $\mathcal{L}(X)$. Afterwards, many different preserver problems stated in terms of the local spectrum/local spectral radius/inner local spectral radius have been considered; see, for example, the last section of the survey article [**8**] and the references therein.

Linear surjective maps on $\mathcal{L}(X)$ preserving operators of local spectral radius zero at vectors $x \in X$ have been characterized by the author in [**10**]. This has been generalized by Bourhim and Mashreghi in [**7**] and further by Elhodaibi and Jaatit in [**13**]. In the case when X is of finite dimension, Bourhim and the author characterized in [**9**] linear maps which preserve matrices of local spectral radius zero at some fixed nonzero vector. It was left as an open question what happens in the case when we work with the inner local spectral radius [**9**, Problem 5].

Bendaoud, Jabbar and Sarih in [**4**, Theorem 1.6] and El Kettani and Benbouziane in [**15**, Corollary 3.2] proved that if $\varphi : \mathcal{L}(X) \to \mathcal{L}(X)$ is a surjective additive map such that

$$i_{\varphi(T)}(x) = 0 \Longleftrightarrow i_T(x) = 0$$

for every $T \in \mathcal{L}(X)$ and every $x \in X$, then there exists a nonzero complex number c such that $\varphi(T) = cT$ for every $T \in \mathcal{L}(X)$. This was further generalized by Jari in [**17**, Theorem 3.1] (keeping the surjectivity assumption on the map φ), and then by Elhodaibi and Jaatit in [**14**, Theorem 3.7], who proved that if $\varphi : \mathcal{L}(X) \to \mathcal{L}(X)$ is a map such that

$$i_{\varphi(T_1)-\varphi(T_2)}(x) = 0 \Longleftrightarrow i_{T_1-T_2}(x) = 0$$

for every $T_1, T_2 \in \mathcal{L}(X)$ and every $x \in X$, then there exists a nonzero complex number c such that $\varphi(T) = cT + \varphi(0)$ for every $T \in \mathcal{L}(X)$. They also proved [**14**, Corollary 4.2] that if $\varphi : \mathcal{L}(X) \to \mathcal{L}(X)$ is a linear surjective map such that $\varphi(I)$ is invertible and either

$$i_T(x) = 0 \Longrightarrow i_{\varphi(T)}(x) = 0$$

for every $T \in \mathcal{L}(X)$ and every $x \in X$, or

$$i_{\varphi(T)}(x) = 0 \Longrightarrow i_T(x) = 0$$

for every $T \in \mathcal{L}(X)$ and every $x \in X$, then there exists a nonzero complex number c such that $\varphi(T) = cT$ for every $T \in \mathcal{L}(X)$.

All the above stated results for maps preserving operators of inner local spectral radius zero have been obtained in the finite or the infinite-dimensional setting, with a preserving property supposed true for *every* $x \in X$. In this paper, we shall consider the corresponding problems for additive maps in the finite-dimensional setting, with the preserving property stated at some *fixed* nonzero vector of $\mathbb{C}^n$. This type of preserver problem, in the linear case, has also been considered in [**3**, Theorem 3.1].

Before stating the main result of this paper, let us make some notations. If $\eta : \mathbb{C} \to \mathbb{C}$ is a field automorphism, then for a matrix $T \in \mathcal{M}_n$ or a vector $x \in \mathbb{C}^n$, by T^η (respectively, x^η) we shall denote the matrix (respectively, the vector) obtained by applying to each entry the map $\eta : \mathbb{C} \to \mathbb{C}$. Also, by T^t (respectively, x^t) we shall denote the transpose of the matrix (respectively, the vector). Let us also observe that for $f \in \mathbb{C}^n$, we have that $x_0 f^t \in \mathcal{M}_n$ is the rank one matrix sending $x \in \mathbb{C}^n$ into $(f^t x)x_0 \in \mathbb{C}^n$; throughout this paper, the elements of $\mathbb{C}^n$ are considered as column vectors.

THEOREM 1.1. *Let $n \geq 2$ be a natural number. Let $x_0 \in \mathbb{C}^n$ be a fixed nonzero vector and let $\varphi : \mathcal{M}_n \to \mathcal{M}_n$ be a surjective additive map. Then*

$$i_T(x_0) = 0 \Longrightarrow i_{\varphi(T)}(x_0) = 0, \qquad (T \in \mathcal{M}_n) \tag{1.3}$$

if and only if there exist a nonzero complex number c, a field automorphism $\eta : \mathbb{C} \to \mathbb{C}$, an invertible matrix $A \in \mathcal{M}_n$ satisfying $A(x_0^\eta) = x_0$ and a vector $f \in \mathbb{C}^n$ satisfying $f^t x_0 \neq 1$ such that

$$\varphi(T) = cA(T - x_0 f^t T)^\eta A^{-1}, \qquad (T \in \mathcal{M}_n). \tag{1.4}$$

We arrive at the same conclusion by supposing

$$i_{\varphi(T)}(x_0) = 0 \Longrightarrow i_T(x_0) = 0, \qquad (T \in \mathcal{M}_n) \tag{1.5}$$

instead of (1.3).

The surjectivity assumption cannot be removed from the statement of Theorem 1.1, even in the case when φ is supposed to be linear. For example, consider a linear map $\gamma : \mathcal{M}_2 \to \mathbb{C}$, fix $x_0 = (1,0)^t \in \mathbb{C}^2$ and define $\varphi : \mathcal{M}_2 \to \mathcal{M}_2$ by putting

$$\varphi(T) = \begin{bmatrix} 0 & \gamma(T) \\ 0 & 0 \end{bmatrix}, \qquad (T \in \mathcal{M}_2).$$

Then φ is a linear non-surjective map such that $\varphi(T)$ is nilpotent for each T, and in particular $i_{\varphi(T)}(x_0) = 0$ for every T. Thus (1.3) holds and φ is not of the form (1.4).

For the same $x_0 \in \mathbb{C}^2$, consider now $\varphi : \mathcal{M}_2 \to \mathcal{M}_2$ given by

$$\varphi\left(\left[\begin{array}{cc} a & b \\ c & d \end{array}\right]\right) = \left[\begin{array}{cc} a & c \\ c & 0 \end{array}\right].$$

Once more, the map φ is linear and non-surjective. If $i_{\varphi(T)}(x_0) = 0$, then $\varphi(T)$ is non-invertible, and therefore c must be zero. Thus

$$\varphi(T) = \left[\begin{array}{cc} a & 0 \\ 0 & 0 \end{array}\right],$$

and since $\varphi(T)x_0 = ax_0$ then $\sigma_{\varphi(T)}(x_0) = \{a\}$. Then $i_{\varphi(T)}(x_0) = 0$ implies $a = 0$. Thus $i_{\varphi(T)}(x_0) = 0$ implies

$$T = \left[\begin{array}{cc} 0 & b \\ 0 & d \end{array}\right].$$

In particular, $Tx_0 = 0$, which gives $\sigma_T(x_0) = \{0\}$. Thus (1.5) holds. Still, the map φ is not of the form (1.4).

Let us also observe that for $n = 1$ in the statement of Theorem 1.1, any surjective additive map $\varphi : \mathbb{C} \to \mathbb{C}$ satisfies (1.3), while surjective additive maps $\varphi : \mathbb{C} \to \mathbb{C}$ satisfying (1.5) are exactly those which are additive and bijective.

2. Preliminary results

It is well-known the fact that the spectrum function is continuous on matrices; see, for example, [**2**, Corollary 3.4.5]. In particular, if $(T_j)_j \subseteq \mathcal{M}_n$ is a sequence such that $T_j \to T \in \mathcal{M}_n$ and $\lambda_j \in \sigma(T_j)$ for each j, then $\lambda_j \to \lambda \in \mathbb{C}$ implies $\lambda \in \sigma(T)$. The local spectrum is not continuous on $\mathcal{M}_n$, being only lower semi-continuous; see, for example, [**12**, Corollary 2.3]. For $x_0 = (1,0)^t \in \mathbb{C}^2$ and

$$T_j = \left[\begin{array}{cc} 1 & 0 \\ 1/j & 0 \end{array}\right] \in \mathcal{M}_2, \qquad (j \geq 1),$$

one can easily check that $0 \in \sigma_{T_j}(x_0)$ for each j, while

$$T_j \to T = \left[\begin{array}{cc} 1 & 0 \\ 0 & 0 \end{array}\right] \in \mathcal{M}_2$$

and $0 \notin \sigma_T(x_0)$.

Of course, if $(T_j)_j \subseteq \mathcal{M}_n$ converges to $T \in \mathcal{M}_n$ and $0 \in \sigma_{T_j}(x_0)$ for each j, we have $0 \in \sigma(T_j)$ for each j and then necessarily $0 \in \sigma(T)$. The next result gives a converse to this observation.

THEOREM 2.1. *Let $x_0 \in \mathbb{C}^n$ be a fixed nonzero vector, and let $T \in \mathcal{M}_n$ be a non-invertible matrix. There exist then $A, B \in \mathcal{M}_n$ such that*

$$0 \in \sigma_{T+A/j+B/j^2}(x_0), \qquad (j \geq 1). \tag{2.1}$$

Proof. If $\sigma(T) = \{0\}$, then $\sigma_T(x_0) = \{0\}$ and one may take $A = B = 0$. So suppose for the remaining of this proof that $k \geq 2$, $0 = \lambda_1, \lambda_2, ..., \lambda_k$ are the distinct eigenvalues of T and $N_1, ..., N_k$ the corresponding root spaces, and denote once more by $P_1, ..., P_k$ the corresponding canonical projections. If $P_{N_1}(x_0) \neq 0$ then $0 \in \sigma_T(x_0)$ and we may take once again $A = B = 0$.

Suppose now that $P_{N_1}(x_0) = 0$. Then $x_0 \in N$, where

$$N := \bigoplus\nolimits_{j=2}^{k} N_j,$$

the direct sum being an algebraic one. Let $1 \leq s < n$ and $\{y_1, ..., y_s\}$ an algebraic basis of N_1 and $\{y_{s+1}, ..., y_n\}$ an algebraic basis of N such that $y_{s+1} = x_0$. Then $\{y_1, ..., y_n\}$ is an algebraic basis of $\mathbb{C}^n$, and consider $f \in \mathbb{C}^n$ such that $f^t y_{s+1} = 1$ while $f^t y_j = 0$ for $j \neq s+1$. Put

$$R = y_1 f^t \in \mathcal{M}_n,$$

that is $R : \mathbb{C}^n \to \mathbb{C}^n$ sends $x \in \mathbb{C}^n$ into $(f^t x)\, y_1 \in \mathbb{C}^n$. By construction, R is a rank one matrix satisfying $R^2 = 0$. Let

$$U_j = I + R/j, \qquad (j \geq 1).$$

Then $U_j^{-1} = I - R/j$ for $j \geq 1$, and by our construction U_j^{-1} sends y_t to y_t for $t \neq s+1$, and $y_{s+1} + y_1/j$ to y_{s+1}. For each j, put

$$\begin{aligned} T_j &= U_j T U_j^{-1} = (I + R/j)T(I - R/j) \\ &= T + (RT - TR)/j - RTR/j^2, \end{aligned}$$

and denote

$$A = RT - TR \in \mathcal{M}_n$$

and

$$B = -RTR \in \mathcal{M}_n.$$

For each j we have

$$\begin{aligned} U_j^{-1}(x_0) &= U_j^{-1}(y_{s+1}) = U_j^{-1}((y_{s+1} + y_1/j) - y_1/j) = y_{s+1} - y_1/j \\ &= x_0 - y_1/j. \end{aligned}$$

Then

$$\sigma_{T_j}(x_0) = \sigma_{U_j T U_j^{-1}}(x_0) = \sigma_T(U_j^{-1} x_0) = \sigma_T(x_0 - y_1/j),$$

and since $P_{N_1}(x_0 - y_1/j) = -y_1/j \neq 0$, by (1.1) we have that $0 \in \sigma_T(x_0 - y_1/j)$, and therefore $0 \in \sigma_{T_j}(x_0)$. Thus

$$0 \in \sigma_{T + A/j + B/j^2}(x_0), \qquad (j \geq 1),$$

that is (2.1) holds. $\square$

By a result of Bračič and Müller from [**11**], given any $T \in \mathcal{M}_n$ and $\lambda \in \sigma(T)$, there exists a sequence $(T_j)_j \subseteq \mathcal{M}_n$ such that $T_j \to T$ and $\lambda \in \sigma_{T_j}(x_0)$ for each j. (In fact, the result holds even in the infinite-dimensional setting if λ is taken from the surjectivity spectrum of T!) Since no continuity assumption is made of the map φ in the statement of Theorem 1.1, we cannot use this result in order to obtain spectrum-preserving properties for the map φ. The sequence $T_j = T + A/j + B/j^2$ which appears in (2.1), besides the fact that it converges to T, has the supplementary property that it allows us to use the additivity of the map φ to see that

$$\varphi(T_j) = \varphi(T) + \varphi(A)/j + \varphi(B)/j^2 \to \varphi(T).$$

This will allow us to obtain invertibility/singularity preserving properties on the additive map φ, so that we may use the following results of Fošner and Šemrl [**16**].

THEOREM 2.2. *(*[**16**, Theorem 1.2 and Corollary 1.3]*) Let $n \geq 2$ be a natural number. Let $\varphi : \mathcal{M}_n \to \mathcal{M}_n$ be a surjective additive map which preserve singularity. There exist then a field automorphism $\eta : \mathbb{C} \to \mathbb{C}$ and invertible matrices $A, B \in \mathcal{M}_n$ such that either*

$$\text{(2.2)} \qquad \varphi(T) = A T^{\eta} B, \qquad (T \in \mathcal{M}_n),$$

or

$$\varphi(T) = A(T^\eta)^t B, \qquad (T \in \mathcal{M}_n). \tag{2.3}$$

We arrive at the same conclusion by supposing that φ is additive, surjective and preserves invertibility.

Before proving our main result, let us also make the following observation. For an automorphism $\eta : \mathbb{C} \to \mathbb{C}$, a nonzero vector $x_0 \in \mathbb{C}^n$ and a matrix $T \in \mathcal{M}_n$, we have that

$$\sigma_{T^\eta}(x_0^\eta) = \eta(\sigma_T(x_0)). \tag{2.4}$$

Indeed, let $\lambda_1, ..., \lambda_k$ be the distinct eigenvalues of $T \in \mathcal{M}_n$ and let once more $N_1, ..., N_k$ denote the corresponding root spaces. For $j = 1, ..., k$, we have $x \in N_j$ if and only if $(T - \lambda_j I)^n x = 0$, which is equivalent to $(T^\eta - \eta(\lambda_j) I)^n x^\eta = 0$. Therefore, the matrix T^η has the distinct eigenvalues $\eta(\lambda_1), ..., \eta(\lambda_k)$, with corresponding root spaces $\eta(N_1), ..., \eta(N_k)$. Now (2.4) follows from (1.1).

3. Proof of the main result

We are now ready for the proof of Theorem 1.1.

Step 1. Let us start by showing that any map φ of the form (1.4) satisfies (1.3) and (1.5). Denoting $S = I - x_0 f^t \in \mathcal{M}_n$, then S is invertible,

$$\varphi(T) = cA(ST)^\eta A^{-1}, \qquad (T \in \mathcal{M}_n)$$

and

$$\begin{aligned} i_{\varphi(T)}(x_0) = 0 &\iff i_{(ST)^\eta}(A^{-1}x_0) = 0 \\ &\iff i_{(ST)^\eta}(x_0^\eta) = 0 \\ &\iff i_{ST}(x_0) = 0. \end{aligned}$$

So suppose, for a contradiction, that $i_T(x_0) = 0$ and $i_{ST}(x_0) \neq 0$ for some matrix T. Then T is not invertible, and therefore ST is not invertible. Since $i_{ST}(x_0)$ is supposed to be nonzero, the matrix ST is not nilpotent. So let $0 = \lambda_1, \lambda_2, ..., \lambda_k$ be the distinct eigenvalues of $ST \in \mathcal{M}_n$ and let $N_1, ..., N_k$ denote the corresponding root spaces. Put $N = \bigoplus_{j=2}^k N_j$. Then ST sends N into N, and $(ST)|_N : N \to N$ is invertible. That $i_{ST}(x_0) \neq 0$ gives $P_{N_1}(x_0) = 0$, and therefore $x_0 \in N$. Then $Tx - (f^t Tx)x_0 = STx \in N$ for each $x \in N$ implies that T also sends N into N. That $(ST)|_N : N \to N$ is invertible implies that $T|_N : N \to N$ is also invertible. So let U be an open disc centered at $0 \in \mathbb{C}$ disjoint from the spectrum of $T|_N$, and define the analytic function $h : U \to N \subseteq \mathbb{C}^n$ by putting

$$h(\lambda) = (T|_N - \lambda I|_N)^{-1} x_0, \qquad (\lambda \in U).$$

Then $(T - \lambda I)h(\lambda) = ((T - \lambda I)|_N)h(\lambda) = x_0$ for every $\lambda \in U$, and therefore $0 \notin \sigma_T(x_0)$, arriving to a contradiction. Thus (1.3) holds.

Since $(x_0 f^t)^2 = (f^t x_0)(x_0 f^t)$ and $f^t x_0 \neq 1$, then $S^{-1} = I - x_0 g^t \in \mathcal{M}_n$ where $g = f/(f^t x_0 - 1) \in \mathbb{C}^n$ satisfies $g^t x_0 \neq 1$. With the same argument as above we obtain that $i_T(x_0) = 0$ implies $i_{S^{-1}T}(x_0) = 0$. Thus $i_{ST}(x_0) = 0$ implies $i_T(x_0) = 0$, that is (1.5) holds.

Step 2. Suppose first that (1.3) holds. Let $T \in \mathcal{M}_n$ be an arbitrary singular matrix. By Theorem 2.1, there exist $A, B \in \mathcal{M}_n$ such that $i_{T+A/j+B/j^2}(x_0) = 0$

for $j \geq 1$, and then (1.3) together with the fact that φ is additive imply that $i_{\varphi(T)+\varphi(A)/j+\varphi(B)/j^2}(x_0) = 0$ for $j \geq 1$. In particular $\varphi(T) + \varphi(A)/j + \varphi(B)/j^2 \in \mathcal{M}_n$ is not invertible for $j \geq 1$. Letting $j \to +\infty$, we obtain that $\varphi(T)$ is singular. Thus the surjective additive map φ sends singular matrices into singular ones. By Theorem 2.2, there exist a field automorphism $\eta : \mathbb{C} \to \mathbb{C}$ and invertible matrices $A, B \in \mathcal{M}_n$ such that φ is either of the form (2.2) or (2.3).

Suppose first that $\varphi(T) = AT^{\eta}B$ for every $T \in \mathcal{M}_n$. If $\{x_0^{\eta}, Bx_0\}$ were linearly independent, then $\{x_0, (Bx_0)^{\eta^{-1}}\}$ will also be linearly independent, and consider then $T \in \mathcal{M}_n$ such that $Tx_0 = 0$ and $T((Bx_0)^{\eta^{-1}}) = (A^{-1}x_0)^{\eta^{-1}}$. That $Tx_0 = 0$ gives $i_T(x_0) = 0$. Also, that $T^{\eta}(Bx_0) = A^{-1}x_0$ gives $\varphi(T)x_0 = x_0$, and therefore $i_{\varphi(T)}(x_0) = 1$, contradicting (1.3). Thus $Bx_0 = \delta x_0^{\eta}$ for some nonzero complex number δ. This gives

$$\begin{aligned} i_{\varphi(T)}(x_0) &= i_{B^{-1}(BAT^{\eta})B}(x_0) = i_{BAT^{\eta}}(Bx_0) \\ &= i_{BAT^{\eta}}(x_0^{\eta}) \end{aligned}$$

for every $T \in \mathcal{M}_n$. Thus $i_T(x_0) = 0$ implies $i_{BAT^{\eta}}(x_0^{\eta}) = 0$, which, by (2.4), gives $i_{(BA)^{\eta^{-1}}T}(x_0) = 0$. Denoting $S = (BA)^{\eta^{-1}} \in \mathcal{M}_n$, then S is invertible and

$$i_T(x_0) = 0 \Longrightarrow i_{ST}(x_0) = 0, \qquad (T \in \mathcal{M}_n). \tag{3.1}$$

Suppose, for a contradiction, that $\{x_0, Sx_0\}$ is linearly independent. Then $\{x_0, S^{-1}x_0\} \subseteq \mathbb{C}^n$ is also a linearly independent system, and consider a $T \in \mathcal{M}_n$ sending x_0 to $S^{-1}x_0$ and $S^{-1}x_0$ into $0 \in \mathbb{C}^n$. That $T^2x_0 = 0$ gives $\sigma_T(x_0) = \{0\}$, and that $ST(x_0) = x_0$ gives $\sigma_{ST}(x_0) = \{1\}$, contradicting (3.1). Thus, there exists a nonzero complex constant α such that $Sx_0 = \alpha x_0$.

Suppose, for a contradiction, that for some $x \in \mathbb{C}^n$ we have that $\{x_0, x, Sx\} \subseteq \mathbb{C}^n$ is linearly independent. Consider then $T \in \mathcal{M}_n$ sending x_0 to x, the vector x into $0 \in \mathbb{C}^n$ and Sx to $S^{-1}x_0 = x_0/\alpha$. Once more, that $T^2x_0 = 0$ gives $\sigma_T(x_0) = \{0\}$. Also, by our construction, $(ST)^2(x_0) = x_0$. Then $\sigma_{(ST)^2}(x_0) = \{1\}$ and, directly from the definition of the local spectrum, we also have $(\sigma_{ST}(x_0))^2 = \sigma_{(ST)^2}(x_0)$. Therefore $\sigma_{ST}(x_0) \subseteq \{\pm 1\}$, contradicting (3.1). Thus, x_0, x and Sx are always linearly dependent.

Let $Y \subseteq \mathbb{C}^n$ be a subspace such that $\mathbb{C}^n = Y \bigoplus (\mathbb{C}x_0)$, an algebraic direct sum. Let $P : \mathbb{C}^n \to Y$ be the corresponding projection. For each $y \in Y$ we have that $\{x_0, y, Sy\} \subseteq \mathbb{C}^n$ is linearly dependent, and by applying P we see that $\{y, PSy\} \subseteq Y$ is linearly dependent. For $I|_Y$ and $PS|_Y$ in $\mathcal{L}(Y)$ this implies the existence of $\beta \in \mathbb{C}$ such that $PS|_Y = \beta I|_Y$ in $\mathcal{L}(Y)$; see, for example, [**2**, Theorem 4.2.7]. Thus $(S - \beta I)(Y) \subseteq (\mathbb{C}x_0)$, and since we also know that $Sx_0 \in (\mathbb{C}x_0)$, then $(S - \beta I)(\mathbb{C}^n) \subseteq (\mathbb{C}x_0)$. This implies the existence of $g \in \mathbb{C}^n$ such that $S = \beta I + x_0 g^t$. That S is invertible implies $\beta \neq 0$, and putting $f = -g/\beta \in \mathbb{C}^n$ we have $S = \beta(I - x_0 f^t)$. That S is invertible also implies $f^t x_0 \neq 1$. Therefore $BA = \eta(\beta)(I - x_0 f^t)^{\eta}$, and therefore φ is indeed of the form (1.4).

Step 3. Let us prove now that (2.3) cannot occur. Indeed, if $\varphi(T) = A(T^{\eta})^t B$ for each T, then $i_{\varphi(T)}(x_0) = i_{BA(T^{\eta})^t}(Bx_0)$ for each T. Denoting $S = (BA)^{\eta^{-1}} \in \mathcal{M}_n$ and $y_0 = (Bx_0)^{\eta^{-1}} \in \mathbb{C}^n$, then S is invertible, y_0 is a nonzero vector and

$$i_T(x_0) = 0 \Longrightarrow i_{ST^t}(y_0) = 0, \qquad (T \in \mathcal{M}_n). \tag{3.2}$$

Let $g \in \mathbb{C}^n$ such that $\{g, x_0\} \subseteq \mathbb{C}^n$ is a linearly independent system and $g^t y_0 \neq 0$. Put

$$T = \frac{g(S^{-1}y_0)^t}{g^t y_0} \in \mathcal{M}_n.$$

Then $T^t y_0 = S^{-1}y_0$, thus $ST^t y_0 = y_0$, which gives $\sigma_{ST^t}(y_0) = \{1\}$. Observe also that T is a rank one matrix, its image in $\mathbb{C}^n$ being generated by $g \in \mathbb{C}^n$. If $0 \notin \sigma_T(x_0)$, then necessarily g and x_0 are linearly dependent, and $(S^{-1}y_0)^t g \neq 0$. Thus by our choice of g we have $i_T(x_0) = 0$, and we contradict (3.2).

Step 4. Suppose that (1.5) holds. Using the fact that φ is surjective exactly as in the first part of Step 2 we obtain that if $\varphi(T)$ is singular, then T is singular. Thus T invertible implies $\varphi(T)$ invertible, and using once more Theorem 2.2 we see that φ is either of the form (2.2) or (2.3). In particular, $\varphi : \mathcal{M}_n \to \mathcal{M}_n$ is bijective, and $\varphi^{-1} : \mathcal{M}_n \to \mathcal{M}_n$ has the property that $i_T(x_0) = 0$ implies $i_{\varphi^{-1}(T)}(x_0) = 0$. Thus φ^{-1} is of the form (1.4), and let us prove now that φ must be of the same form. Indeed, denote $S = I - x_0 f^t$ and let $\tau = \eta^{-1} : \mathbb{C} \to \mathbb{C}$. Then

$$\varphi(T) = \frac{1}{\tau(c)} S^{-1}(A^\tau)^{-1} T^\tau A^\tau, \qquad (T \in \mathcal{M}_n).$$

Denoting $d = 1/\tau(c) \in \mathbb{C}$, $B = (A^\tau)^{-1} \in \mathcal{M}_n$ and $R = A(S^\eta)^{-1}A^{-1} \in \mathcal{M}_n$, then

$$\varphi(T) = dB(RT)^\tau B^{-1}, \qquad (T \in \mathcal{M}_n),$$

where the invertible matrix B satisfies

$$Bx_0^\tau = (A^{-1}x_0)^\tau = (x_0^\eta)^\tau = x_0$$

and

$$\begin{aligned} R &= A\left(I - \frac{x_0 f^t}{f^t x_0 - 1}\right)^\eta A^{-1} = I - \frac{(Ax_0^\eta)((A^{-1})^t f^\eta)^t}{\eta(f^t x_0) - 1} \\ &= I - x_0 g^t, \end{aligned}$$

where $g := (A^{-1})^t f^\eta / (\eta(f^t x_0) - 1) \in \mathbb{C}^n$ satisfies

$$g^t x_0 = \frac{(f^\eta)^t x_0^\eta}{\eta(f^t x_0) - 1} = \frac{\eta(f^t x_0)}{\eta(f^t x_0) - 1} \neq 1.$$

References

[1] P. Aiena, *Fredholm and local spectral theory, with applications to multipliers*, Kluwer Academic Publishers, Dordrecht, 2004. MR2070395

[2] B. Aupetit, *A primer on spectral theory*, Universitext, Springer-Verlag, New York, 1991. MR1083349

[3] H. Benbouziane, M.E. El Kettani and A.M. Vadel, *Linear maps on $\mathcal{M}_n(\mathbb{C})$ preserving inner local spectral radius zero*, Italian Journal of Pure and Applied Mathematics 39 (2018), 757–763.

[4] M. Bendaoud, M. Jabbar, and M. Sarih, *Additive local invertibility preservers*, Publ. Math. Debrecen **85** (2014), no. 3-4, 467–480, DOI 10.5486/PMD.2014.6073. MR3291843

[5] A. Bourhim and T. Ransford, *Additive maps preserving local spectrum*, Integral Equations Operator Theory **55** (2006), no. 3, 377–385, DOI 10.1007/s00020-005-1392-2. MR2244195

[6] A. Bourhim and V. G. Miller, *Linear maps on $M_n(\mathbb{C})$ preserving the local spectral radius*, Studia Math. **188** (2008), no. 1, 67–75, DOI 10.4064/sm188-1-4. MR2430550

[7] A. Bourhim and J. Mashreghi, *Local spectral radius preservers*, Integral Equations Operator Theory **76** (2013), no. 1, 95–104, DOI 10.1007/s00020-013-2041-9. MR3041723

[8] A. Bourhim and J. Mashreghi, *A survey on preservers of spectra and local spectra*, Invariant subspaces of the shift operator, Contemp. Math., vol. 638, Amer. Math. Soc., Providence, RI, 2015, pp. 45–98, DOI 10.1090/conm/638/12810. MR3309349

[9] A. Bourhim and C. Costara, *Linear maps preserving matrices of local spectral radius zero at a fixed vector*, Online first in Canadian Journal of Mathematics, Canadian J. Math, in **71** (2019), no. 4, 749–771. `https://doi.org/10.4153/CJM-2018-017-0`.

[10] C. Costara, *Linear maps preserving operators of local spectral radius zero*, Integral Equations Operator Theory **73** (2012), no. 1, 7–16, DOI 10.1007/s00020-012-1953-0. MR2913657

[11] J. Bračič and V. Müller, *Local spectrum and local spectral radius of an operator at a fixed vector*, Studia Math. **194** (2009), no. 2, 155–162, DOI 10.4064/sm194-2-3. MR2534182

[12] M. B. Dollinger and K. K. Oberai, *Variation of local spectra*, J. Math. Anal. Appl. **39** (1972), 324–337, DOI 10.1016/0022-247X(72)90205-3. MR0318950

[13] M. Elhodaibi and A. Jaatit, *On maps preserving operators of local spectral radius zero*, Linear Algebra Appl. **512** (2017), 191–201, DOI 10.1016/j.laa.2016.10.001. MR3567521

[14] M. Elhodaibi and A. Jaatit, *Inner local spectral radius preservers*, Rend. Circ. Mat. Palermo (2) **67** (2018), no. 2, 215–225, DOI 10.1007/s12215-017-0308-8. MR3833003

[15] M. Ech-Cherif El Kettani and H. Benbouziane, *Additive maps preserving operators of inner local spectral radius zero*, Rend. Circ. Mat. Palermo (2) **63** (2014), no. 2, 311–316, DOI 10.1007/s12215-014-0160-z. MR3274066

[16] A. Fošner and P. Šemrl, *Additive maps on matrix algebras preserving invertibility or singularity*, Acta Math. Sin. (Engl. Ser.) **21** (2005), no. 4, 681–684, DOI 10.1007/s10114-005-0519-3. MR2156942

[17] T. Jari, *Nonlinear maps preserving the inner local spectral radius*, Rend. Circ. Mat. Palermo (2) **64** (2015), no. 1, 67–76, DOI 10.1007/s12215-014-0181-7. MR3324374

[18] K. B. Laursen and M. M. Neumann, *An introduction to local spectral theory*, London Mathematical Society Monographs. New Series, vol. 20, The Clarendon Press, Oxford University Press, New York, 2000. MR1747914

[19] T. L. Miller, V. G. Miller, and M. M. Neumann, *Local spectral properties of weighted shifts*, J. Operator Theory **51** (2004), no. 1, 71–88. MR2055805

[20] A. Torgašev, *On operators with the same local spectra*, Czechoslovak Math. J. 48 (1998), 77–83.

Faculty of Mathematics and Informatics, Ovidius University of Constanţa, Mamaia Boul. 124, Constanţa, Romania, 900527

Email address: `cdcostara@univ-ovidius.ro`

Contemporary Mathematics
Volume **743**, 2020
https://doi.org/10.1090/conm/743/14955

A global domination principle for $P-$pluripotential theory

Norm Levenberg and Menuja Perera

In honor of 60 years of Tom Ransford

ABSTRACT. We prove a global domination principle in the setting of $P-$pluripotential theory. This has many applications including a general product property for $P-$extremal functions. The key ingredient is the proof of the existence of a strictly plurisubharmonic $P-$potential.

1. Introduction

Following [**1**], in [**2**] and [**4**] a pluripotential theory associated to plurisubharmonic (psh) functions on $\mathbb{C}^d$ having growth at infinity specified by $H_P(z) := \phi_P(\log|z_1|, ..., \log|z_d|)$ where

$$\phi_P(x_1, ..., x_d) := \sup_{(y_1,...,y_d)\in P} (x_1y_1 + \cdots + x_dy_d)$$

is the indicator function of a convex body $P \subset (\mathbb{R}^+)^d$ was developed. Given P, the classes

$$L_P = L_P(\mathbb{C}^d) := \{u \in PSH(\mathbb{C}^d) : u(z) - H_P(z) = 0(1), \ |z| \to \infty\}$$

and

$$L_P^+ = L_P^+(\mathbb{C}^d) := \{u \in L_P : u(z) \geq H_P(z) + c_u\}$$

are of fundamental importance. These are generalizations of the standard Lelong classes $L(\mathbb{C}^d)$, the set of all plurisubharmonic (psh) functions u on $\mathbb{C}^d$ with $u(z) - \max[\log|z_1|, ..., \log|z_d|] = 0(1), \ |z| \to \infty$, and

$$L^+(\mathbb{C}^d) = \{u \in L(\mathbb{C}^d) : u(z) \geq \max[0, \log|z_1|, ..., \log|z_d|] + C_u\}$$

which correspond to $P = \Sigma$ where

$$\Sigma := \{(x_1, ..., x_d) \in \mathbb{R}^d : x_1, ..., x_d \geq 0, \ x_1 + \cdots + x_d \leq 1\}.$$

For more on standard pluripotential theory, cf., [**7**].

Given $E \subset \mathbb{C}^d$, the $P-$*extremal function of* E is defined as

$$V_{P,E}^*(z) := \limsup_{\zeta \to z} V_{P,E}(\zeta)$$

where

$$V_{P,E}(z) := \sup\{u(z) : u \in L_P(\mathbb{C}^d), \ u \leq 0 \text{ on } E\}.$$

2010 *Mathematics Subject Classification.* 32U15, 32U20, 31C15.

The first author was supported by Simons Foundation grant No. 354549.

For $P = \Sigma$, we write $V_E := V_{\Sigma,E}$. For E bounded and nonpluripolar, $V_E^* \in L^+(\mathbb{C}^d)$; $V_E^* = 0$ q.e. on E (i.e., on all of E except perhaps a pluripolar set); and $(dd^cV_E^*)^d = 0$ outside of $\bar{E}$ where $(dd^cV_E^*)^d$ is the complex Monge-Ampère measure of V_E^* (see section 2). A key ingredient in verifying a candidate function $v \in L^+(\mathbb{C}^d)$ is equal to V_E^* is the following global domination principle of Bedford and Taylor:

PROPOSITION 1. **[3]** *Let $u \in L(\mathbb{C}^d)$ and $v \in L^+(\mathbb{C}^d)$ and suppose $u \leq v$ a.e.-$(dd^cv)^d$. Then $u \leq v$ on $\mathbb{C}^d$.*

Thus if one finds $v \in L^+(\mathbb{C}^d)$ with $v = 0$ a.e.-$(dd^cv)^d$ on $\bar{E}$ and $(dd^cv)^d = 0$ outside of $\bar{E}$ then $v = V_E^*$. For the proof of Proposition 1 in **[3]** the fact that in the definition of the Lelong classes $\max[\log|z_1|, ..., \log|z_d|]$ and $\max[0, \log|z_1|, ..., \log|z_d|]$ can be replaced by the Kähler potential

$$u_0(z) := \frac{1}{2}\log(1 + |z|^2) := \frac{1}{2}\log(1 + \sum_{j=1}^d |z_j|^2)$$

is crucial; this latter function is strictly psh and $(dd^cu_0)^d > 0$ on $\mathbb{C}^d$.

We prove a version of the global domination principle for very general L_P and L_P^+ classes. We consider convex bodies $P \subset (\mathbb{R}^+)^d$ satisfying

$$\Sigma \subset kP \text{ for some } k \in \mathbb{Z}^+. \tag{1.1}$$

PROPOSITION 2. *For $P \subset (\mathbb{R}^+)^d$ satisfying (1.1), let $u \in L_P$ and $v \in L_P^+$ with $u \leq v$ a.e.-$(dd^cv)^d$. Then $u \leq v$ in $\mathbb{C}^d$.*

As a corollary, we obtain a generalization of Proposition 2.4 of **[4]** on $P-$extremal functions:

PROPOSITION 3. *Given $P \subset (\mathbb{R}^+)^d$ satisfying (1.1), let $E_1, ..., E_d \subset \mathbb{C}$ be compact and nonpolar. Then*

$$V_{P,E_1\times\cdots\times E_d}^*(z_1, ..., z_d) = \phi_P(V_{E_1}^*(z_1), ..., V_{E_d}^*(z_d)). \tag{1.2}$$

The main issue in proving our version of the global domination principle (restated as Proposition 4 below) is the construction of a strictly psh $P-$potential u_P which can replace the logarithmic indicator function $H_P(z)$ used to define L_P and L_P^+. To do this, we utilize a classical result on subharmonic functions in the complex plane which we learned in Tom Ransford's beautiful book **[8]**; thus it is fitting that this article is written in his honor.

2. The global $P-$domination principle

Following **[2]** and **[4]**, we fix a convex body $P \subset (\mathbb{R}^+)^d$; i.e., a compact, convex set in $(\mathbb{R}^+)^d$ with non-empty interior P^o. The most important example is the case where P is the convex hull of a finite subset of $(\mathbb{Z}^+)^d$ in $(\mathbb{R}^+)^d$ with $P^o \neq \emptyset$ (P is a non-degenerate convex polytope). Another interesting class consists of the $(\mathbb{R}^+)^d$ portion of an ℓ^q ball for $1 \leq q \leq \infty$; see (4.2). Recall that $H_P(z) := \phi_P(\log|z_1|, ..., \log|z_d|)$ where ϕ_P is the indicator function of P.

A C^2-function u on $D \subset \mathbb{C}^d$ is *strictly psh* on D if the complex Hessian $H(u) := [\frac{\partial^2 u}{\partial z_j \partial \bar{z}_k}]_{j,k=1,...,d}$ is positive definite on D. We define

$$dd^cu := 2i\sum_{j,k=1}^d \frac{\partial^2 u}{\partial z_j \partial \bar{z}_k} dz_j \wedge d\bar{z}_k$$

and

$$(dd^c u)^d = dd^c u \wedge \cdots \wedge dd^c u = c_d \det H(u) dV$$

where $dV = (\frac{i}{2})^d \sum_{j=1}^d dz_j \wedge d\bar{z}_j$ is the volume form on $\mathbb{C}^d$ and c_d is a dimensional constant. Thus u strictly psh on D implies that $(dd^c u)^d = f dV$ on D where $f > 0$. We remark that if u is a locally bounded psh function then $(dd^c u)^d$ is well-defined as a positive measure, the *complex Monge-Ampère measure* of u; this is the case, e.g., for functions $u \in L_P^+$.

DEFINITION 2.1. We say that u_P is a *strictly psh P−potential* if

(1) $u_P \in L_P^+$ is strictly psh on $\mathbb{C}^d$ and
(2) there exists a constant C such that $|u_P(z) - H_P(z)| \leq C$ for all $z \in \mathbb{C}^d$.

This property implies that u_P can replace H_P in defining the L_P and L_P^+ classes:

$$L_P = \{u \in PSH(\mathbb{C}^d) : u(z) - u_P(z) = 0(1), \ |z| \to \infty\}$$

and

$$L_P^+ = \{u \in L_P : u(z) \geq u_P(z) + c_u\}.$$

Given the existence of a strictly psh P−potential, we can follow the proof of Proposition 1 in [**3**] to prove:

PROPOSITION 4. *For $P \subset (\mathbb{R}^+)^d$ satisfying (1.1), let $u \in L_P$ and $v \in L_P^+$ with $u \leq v$ a.e. $(dd^c v)^d$. Then $u \leq v$ in $\mathbb{C}^d$.*

PROOF. Suppose the result is false; i.e., there exists $z_0 \in \mathbb{C}^d$ with $u(z_0) > v(z_0)$. Since $v \in L_P^+$, by adding a constant to v we may assume $v(z) \geq u_P(z)$ in $\mathbb{C}^d$. Note that $(dd^c u_P)^d > 0$ on $\mathbb{C}^d$. Fix $\delta, \ \epsilon > 0$ with $\delta < \epsilon/2$ in such a way that the set

$$S := \{z \in \mathbb{C} : u(z) + \delta u_P(z) > (1+\epsilon)v(z)\}$$

contains z_0. Then S has positive Lebesgue measure. Moreover, since $\delta < \epsilon$ and $v \geq u_P$, S is bounded. By the comparison principle (cf., Theorem 3.7.1 [**7**]), we conclude that

$$\int_S (dd^c[u + \delta u_P])^d \leq \int_S (dd^c(1+\epsilon)v)^d.$$

But $\int_S (dd^c \delta u_P)^d > 0$ since S has positive Lebesgue measure, so

$$(1+\epsilon)^d \int_S (dd^c v)^d > 0.$$

By hypothesis, for a.e.-$(dd^c v)^d$ points in $\text{supp}(dd^c v)^d \cap S$ (which is not empty since $\int_S (dd^c v)^d > 0$), we have

$$(1+\epsilon)v(z) < u(z) + \delta u_P(z) \leq v(z) + \delta u_P(z),$$

i.e., $v(z) < \frac{1}{2} u_P(z)$ since $\delta < \epsilon/2$. This contradicts the normalization $v(z) \geq u_P(z)$ in $\mathbb{C}^d$. □

In the next section, we show how to construct u_P in Definition 2.1 for a convex body in $(\mathbb{R}^+)^d$ satisfying (1.1).

3. Existence of strictly psh P–potential

For the P we consider, $\phi_P \geq 0$ on $(\mathbb{R}^+)^d$ with $\phi_P(0) = 0$. We write $z^J = z_1^{j_1} \cdots z_d^{j_d}$ where $J = (j_1, ..., j_d) \in P$ (the components j_k need not be integers) so that

$$H_P(z) := \sup_{J \in P} \log |z^J| := \phi_P(\log^+ |z_1|, ..., \log^+ |z_d|)$$

with $|z^J| := |z_1|^{j_1} \cdots |z_d|^{j_d}$. To construct a strictly psh P–potential u_P, we first assume P is a convex polytope in $(\mathbb{R}^+)^d$ satisfying (1.1). Thus

$$(a_1, 0, ..., 0), \ldots, (0, ..., 0, a_d) \in \partial P$$

for some $a_1, ..., a_d > 0$. A calculation shows that

$$\log(1 + |z_1|^{2a_1} + \cdots + |z_d|^{2a_d})$$

is strictly psh in $\mathbb{C}^d$.

We claim then that

$$u_P(z) := \frac{1}{2} \log(1 + \sum_{J \in Extr(P)} |z^J|^2) \tag{3.1}$$

is strictly psh in $\mathbb{C}^d$ and the L_P, L_P^+ classes can be defined using u_P instead of H_P; i.e., u_P satisfies (1) and (2) of Definition 2.1. Here, $Extr(P)$ denotes the extreme points of P but we omit the origin $\mathbf{0}$. Note that $(a_1, 0, ..., 0), ..., (0, ..., 0, a_d) \in Extr(P)$.

Indeed, in this case,

$$H_P(z) = \sup_{J \in P} \log |z^J| = \max[0, \max_{J \in Extr(P)} \log |z^J|]$$

so clearly for $|z|$ large, $|u_P(z) - H_P(z)| = 0(1)$ while on any compact set K,

$$\sup_{z \in K} |u_P(z) - H_P(z)| \leq C = C(K)$$

which gives (2) (and therefore that $u_P \in L_P^+$).

It remains to verify the strict psh of u_P in (3.1). We use reasoning based on a classical univariate result which is exercise 4 in section 2.6 of [**8**]: if u, v are nonnegative with $\log u$ and $\log v$ subharmonic (shm) – hence u, v are shm – then $\log(u + v)$ is shm. The usual proof is to show $(u + v)^a$ is shm for any $a > 0$ – which is exercise 3 in section 2.6 of [**8**] – which trivially follows since u, v are shm and $a > 0$. However, assume u, v are smooth and compute the Laplacian $\Delta \log(u + v)$ on $\{u, v > 0\}$:

$$\begin{aligned} \left(\log(u + v)\right)_{z\bar{z}} &= \frac{(u + v)(u_{z\bar{z}} + v_{z\bar{z}}) - (u_z + v_z)(u_{\bar{z}} + v_{\bar{z}})}{(u + v)^2} \\ &= \frac{[uu_{z\bar{z}} - |u_z|^2 + vv_{z\bar{z}} - |v_z|^2] + [uv_{z\bar{z}} + vu_{z\bar{z}} - 2\Re(u_z v_{\bar{z}})]}{(u + v)^2}. \end{aligned}$$

Now $\log u$, $\log v$ shm implies $uu_{z\bar{z}} - |u_z|^2 \geq 0$ and $vv_{z\bar{z}} - |v_z|^2 \geq 0$ with strict inequality in case of strict shm. Since $\log(u + v)$ is shm, the entire numerator is nonnegative:

$$[uu_{z\bar{z}} - |u_z|^2 + vv_{z\bar{z}} - |v_z|^2] + [uv_{z\bar{z}} + vu_{z\bar{z}} - 2\Re(u_z v_{\bar{z}})] \geq 0$$

so that the "extra term"

$$uv_{z\bar{z}} + vu_{z\bar{z}} - 2\Re(u_z v_{\bar{z}})$$

is nonnegative whenever

$$(\log u)_{z\bar{z}} + (\log v)_{z\bar{z}} = uu_{z\bar{z}} - |u_z|^2 + vv_{z\bar{z}} - |v_z|^2 = 0.$$

We show $\Delta \log(u+v)$ is strictly positive on $\{u, v > 0\}$ if one of $\log u$ or $\log v$ is strictly shm.

PROPOSITION 5. *Let $u, v \geq 0$ with $\log u$ and $\log v$ shm. If one of $\log u$ or $\log v$ is strictly shm, e.g., $\Delta \log u > 0$, then $\Delta \log(u+v) > 0$ on $\{u, v > 0\}$.*

PROOF. We have $u, v \geq 0$, $u_{z\bar{z}}, v_{z\bar{z}} \geq 0$, $vv_{z\bar{z}} - |v_z|^2 \geq 0$ and $uu_{z\bar{z}} - |u_z|^2 > 0$ if $u > 0$. We want to show that

$$uv_{z\bar{z}} + vu_{z\bar{z}} - 2\Re(u_z v_{\bar{z}}) = uv_{z\bar{z}} + vu_{z\bar{z}} - (u_z v_{\bar{z}} + v_z u_{\bar{z}}) > 0$$

on $\{u, v > 0\}$. We start with the identity

$$(uv_z - vu_z)(uv_{\bar{z}} - vu_{\bar{z}}) = u^2 v_z v_{\bar{z}} + v^2 u_z u_{\bar{z}} - uv(u_z v_{\bar{z}} + v_z u_{\bar{z}}) \geq 0. \tag{3.2}$$

Since $uu_{z\bar{z}} - |u_z|^2 > 0$ and $vv_{z\bar{z}} - |v_z|^2 \geq 0$,

$$uu_{z\bar{z}} > u_z u_{\bar{z}}, \quad vv_{z\bar{z}} \geq v_z v_{\bar{z}}.$$

Thus

$$uv_{z\bar{z}} + vu_{z\bar{z}} = \frac{u}{v} vv_{z\bar{z}} + \frac{v}{u} uu_{z\bar{z}} > \frac{u}{v} v_z v_{\bar{z}} + \frac{v}{u} u_z u_{\bar{z}}.$$

Thus it suffices to show

$$\frac{u}{v} v_z v_{\bar{z}} + \frac{v}{u} u_z u_{\bar{z}} \geq u_z v_{\bar{z}} + u_{\bar{z}} v_z.$$

Multiplying both sides by uv, this becomes

$$u^2 v_z v_{\bar{z}} + v^2 u_z u_{\bar{z}} \geq uv(u_z v_{\bar{z}} + u_{\bar{z}} v_z).$$

This is (3.2). □

This proof actually shows that

$$uv_{z\bar{z}} + vu_{z\bar{z}} - 2\Re(u_z v_{\bar{z}}) > 0$$

under the hypotheses of the proposition.

REMARK 3.1. To be precise, this shows strict shm only on $\{u, v > 0\}$. In the multivariate case, this shows the restriction of $\log(u+v)$ to the intersection of a complex line and $\{u, v > 0\}$ is strictly shm if one of $\log u, \log v$ is strictly psh so that $\log(u+v)$ is strictly psh on $\{u, v > 0\}$.

Now with u_P in (3.1) we may write

$$u_P(z) = \log(u+v)$$

where

$$u(z) = 1 + |z_1|^{2a_1} + \cdots + |z_d|^{2a_d} \tag{3.3}$$

– so that $\log u$ is strictly psh in $\mathbb{C}^d$ – and

$$v(z) = \sum_{J \in Extr(P)} |z^J|^2 - |z_1|^{2a_1} - \cdots - |z_d|^{2a_d}.$$

If $v \equiv 0$ (e.g., if $P = \Sigma$) we are done. Otherwise $v \geq 0$ (being a sum of non-negative terms) and $\log v$ is psh (being the logarithm of a sum of moduli squared of holomorphic functions) showing that $u_P(z) := \frac{1}{2}\log(1 + \sum_{J \in Extr(P)} |z^J|^2)$ is strictly psh where $v > 0$. There remains an issue at points where $v = 0$ (coordinate axes). However, if we simply replace the decomposition $u_P(z) = \log(u + v)$ by $u_P(z) = \log(u_\epsilon + v_\epsilon)$ where

$$u_\epsilon := 1 + (1-\epsilon)(|z_1^a|^2 + ... + |z_d^a|^2) \text{ and}$$
$$v_\epsilon := \sum_{J \in ExtrP} |z^J|^2 - (1-\epsilon)(|z_1^a|^2 + ... + |z_d^a|^2)$$

for $\epsilon > 0$ sufficiently small, then the result holds everywhere. We thank F. Piazzon for this last observation.

If $P \subset (\mathbb{R}^+)^d$ is a convex body satisfying (1.1), we can approximate P by a monotone decreasing sequence of convex polytopes P_n satisfying the same property. Since $P_{n+1} \subset P_n$ and $\cap_n P_n = P$, the sequence $\{u_{P_n}\}$ decreases to the function $u_P \in L_P^+$. Since each u_{P_n} is of the form

$$u_{P_n}(z) = \log(u_n + v_n)$$

where $u_n(z) = 1 + |z_1|^{2a_{n1}} + \cdots + |z_d|^{2a_{nd}}$ and $a_{nj} \geq a_j$ for all n and each $j = 1, ..., d$ in (3.3), it follows that u_P is strictly psh and hence satisfies Definition 2.1. This concludes the proof of Proposition 4.

REMARK 3.2. Another construction of a strictly psh $P-$potential as in Definition 2.1 which is based on solving a real Monge-Ampère equation and which works in more general situations was recently given by C. H. Lu [**5**]. Indeed, his construction, combined with Corollary 3.10 of [**6**], yields a new proof of the global domination principle, Proposition 4.

4. The product property

In this section, we prove the product property stated in the introduction:

PROPOSITION 6. *For $P \subset (\mathbb{R}^+)^d$ satisfying (1.1), let $E_1, ..., E_d \subset \mathbb{C}$ be compact and nonpolar. Then*

$$V^*_{P, E_1 \times \cdots \times E_d}(z_1, ..., z_d) = \phi_P(V^*_{E_1}(z_1), ..., V^*_{E_d}(z_d)). \tag{4.1}$$

REMARK 4.1. One can verify the formula

$$V_{P,T^d}(z) = H_P(z) = \sup_{J \in P} \log |z^J|$$

for the $P-$extremal function of the torus

$$T^d := \{(z_1, ..., z_d) : |z_j| = 1, \ j = 1, ..., d\}$$

for a general convex body by modifying the argument in [**7**] for the standard extremal function of a ball in a complex norm. Indeed, let $u \in L_P$ with $u \leq 0$ on T^d. For $w = (w_1, ..., w_d) \notin T^d$ and $w_j \neq 0$, we consider

$$v(\zeta_1, ..., \zeta_d) := u(w_1/\zeta_1, ..., w_d/\zeta_d) - H_P(w_1/\zeta_1, ..., w_d/\zeta_d).$$

This is psh on $0 < |\zeta_j| < |w_j|$, $j = 1, ..., d$. Since $u \in L_P$, v is bounded above near the pluripolar set given by the union of the coordinate planes in this polydisk and hence extends to the full polydisk. On the boundary $|\zeta_j| = |w_j|$, $v \leq 0$ so at $(1, 1, ..., 1)$ we get $u(w_1, ..., w_d) \leq H_P(w_1, ..., w_d)$. Note

$$H_P(z) = \sup_{J \in P} \log |z^J| = \phi_P(\log^+ |z_1|, ..., \log^+ |z_d|)$$

and $V_{T^1}(\zeta) = \log^+ |\zeta|$ so this is a special case of Proposition 6.

PROOF. For simplicity we consider the case $d = 2$ with variables (z, w) on $\mathbb{C}^2$. As in [**4**], we may assume V_E and V_F are continuous. Also, by approximation we may assume ϕ_P is smooth. We write

$$v(z,w) := \phi_P(V_E(z), V_F(w)).$$

An important remark is that, since $P \subset (\mathbb{R}^+)^2$, P is convex, and P contains $k\Sigma$ for some $k > 0$, the function ϕ_P on $(\mathbb{R}^+)^2$ satisfies

(1) $\phi_P \geq 0$ and $\phi_P(x,y) = 0$ only for $x = y = 0$;
(2) ϕ_P is nondecreasing in each variable; i.e., $(\phi_P)_x, (\phi_P)_y \geq 0$;
(3) ϕ_P is convex; i.e., the real Hessian $H_{\mathbb{R}}(\phi_P)$ of ϕ_P is positive semidefinite; and, more precisely, by the homogenity of ϕ_P; i.e., $\phi_P(tx,ty) = t\phi_P(x,y)$,

$$\det H_{\mathbb{R}}(\phi_P) = 0 \text{ away from the origin.}$$

As in [**4**], to see that

$$v(z,w) \leq V_{P,E\times F}(z,w),$$

since $\phi_P(0,0) = 0$, it suffices to show that $\phi_P(V_E(z), V_F(w)) \in L_P(\mathbb{C}^2)$. From the definition of ϕ_P,

$$\phi_P(V_E(z), V_F(w)) = \sup_{(x,y)\in P} [xV_E(z) + yV_F(w)]$$

which is a locally bounded above upper envelope of plurisubharmonic functions. As ϕ_P is convex and V_E, V_F are continuous, $\phi_P(V_E(z), V_F(w))$ is continuous. Since $V_E(z) = \log|z| + 0(1)$ as $|z| \to \infty$ and $V_F(w) = \log|w| + 0(1)$ as $|w| \to \infty$, it follows that $\phi_P(V_E(z), V_F(w)) \in L_P(\mathbb{C}^2)$.

By Proposition 4, it remains to show $(dd^c v)^2 = 0$ outside of $E \times F$. Since we can approximate v from above uniformly by a decreasing sequence of smooth psh functions by convolving v with a smooth bump function, we assume v is smooth and compute the following derivatives:

$$\begin{aligned} v_z &= (\phi_P)_x(V_E)_z, \\ v_w &= (\phi_P)_y(V_F)_w; \\ v_{z\bar{z}} &= (\phi_P)_{xx}|(V_E)_z|^2 + (\phi_P)_x(V_E)_{z\bar{z}}; \\ v_{z\bar{w}} &= (\phi_P)_{xy}(V_E)_z(V_F)_{\bar{w}}; \\ v_{w\bar{w}} &= (\phi_P)_{yy}|(V_F)_w|^2 + (\phi_P)_y(V_F)_{w\bar{w}}. \end{aligned}$$

It follows from (2) that $v_{z\bar{z}}, v_{w\bar{w}} \geq 0$. Next, we compute the determinant of the complex Hessian of v on $(\mathbb{C} \setminus E) \times (\mathbb{C} \setminus F)$ (so $(V_E)_{z\bar{z}} = (V_F)_{w\bar{w}} = 0$):

$$\begin{aligned} v_{z\bar{z}}v_{w\bar{w}} - |v_{z\bar{w}}|^2 &= (\phi_P)_{xx}|(V_E)_z|^2(\phi_P)_{yy}|(V_F)_w|^2 - [(\phi_P)_{xy}]^2|(V_E)_z|^2|(V_F)_w|^2 \\ &= |(V_E)_z|^2|(V_F)_w|^2[(\phi_P)_{xx}(\phi_P)_{yy} - (\phi_P)_{xy}]^2]. \end{aligned}$$

This is nonnegative by the convexity of ϕ_P and, indeed, it vanishes on $(\mathbb{C} \setminus E) \times (\mathbb{C} \setminus F)$ by (3). The general formula for the determinant of the complex Hessian of v is

$$\begin{aligned} & v_{z\bar{z}}v_{w\bar{w}} - |v_{z\bar{w}}|^2 \\ = \ & |(V_E)_z|^2|(V_F)_w|^2[(\phi_P)_{xx}(\phi_P)_{yy} - (\phi_P)_{xy}]^2] + (\phi_P)_{xx}|(V_E)_z|^2(\phi_P)_y(V_F)_{w\bar{w}} \\ & + (\phi_P)_{yy}|(V_F)_w|^2(\phi_P)_x(V_E)_{z\bar{z}} + (\phi_P)_x(V_E)_{z\bar{z}}(\phi_P)_y(V_F)_{w\bar{w}}. \end{aligned}$$

If, e.g., $z \in E$ and $w \in (\mathbb{C} \setminus F)$,

$$|(V_E)_z|^2 |(V_F)_w|^2 [(\phi_P)_{xx}(\phi_P)_{yy} - (\phi_P)_{xy}]^2] = 0$$

by (3) (since $(V_E(z), V_F(w)) = (0, a) \neq (0, 0)$) and $(V_F)_{w\bar{w}} = 0$ so

$$v_{z\bar{z}} v_{w\bar{w}} - |v_{z\bar{w}}|^2 = (\phi_P)_{yy} |(V_F)_w|^2 (\phi_P)_x (V_E)_{z\bar{z}}.$$

However, we claim that

$$(\phi_P)_{yy}(0, a) = 0 \text{ if } a > 0$$

since we have $\phi_P(0, ty) = t\phi_P(0, y)$. Hence

$$v_{z\bar{z}} v_{w\bar{w}} - |v_{z\bar{w}}|^2 = 0$$

if $z \in E$ and $w \in (\mathbb{C} \setminus F)$. Similarly,

$$(\phi_P)_{xx}(a, 0) = 0 \text{ if } a > 0$$

so that

$$v_{z\bar{z}} v_{w\bar{w}} - |v_{z\bar{w}}|^2 = 0$$

if $z \in (\mathbb{C} \setminus E)$ and $w \in F$.

□

REMARK 4.2. In [**4**], a (much different) proof of Proposition 6 was given under the additional hypothesis that $P \subset (\mathbb{R}^+)^d$ be a *lower set*: for each $n = 1, 2, ...,$ whenever $(j_1, ..., j_d) \in nP \cap (\mathbb{Z}^+)^d$ we have $(k_1, ..., k_d) \in nP \cap (\mathbb{Z}^+)^d$ for all $k_l \leq j_l$, $l = 1, ..., d$.

Finally, although computation of the $P-$extremal function of a product set is now rather straightforward, even qualitative properties of the corresponding Monge-Ampère measure are less clear. To be concrete, for $q \geq 1$, let

$$P_q := \{(x_1, ..., x_d) : x_1, ..., x_d \geq 0, \ x_1^q + \cdots + x_d^q \leq 1\} \tag{4.2}$$

be the $(\mathbb{R}^+)^d$ portion of an ℓ^q ball. Then for $1/q' + 1/q = 1$ we have $\phi_{P_q}(x) = ||x||_{\ell^{q'}}$ (for $q = \infty$ we take $q' = 1$ and vice-versa). Hence if $E_1, ..., E_d \subset \mathbb{C}$,

$$\begin{aligned} V^*_{P_q, E_1 \times \cdots \times E_d}(z_1, ..., z_d) &= ||[V^*_{E_1}(z_1), V^*_{E_2}(z_2), \ldots, V^*_{E_d}(z_d)]||_{\ell^{q'}} \\ &= [V^*_{E_1}(z_1)^{q'} + \cdots + V^*_{E_d}(z_d)^{q'}]^{1/q'}. \end{aligned}$$

In the standard case $q = 1$, $P_1 = \Sigma$ and we have the well-known result that

$$V^*_{E_1 \times \cdots \times E_d}(z_1, ..., z_d) = \max[V^*_{E_1}(z_1), V^*_{E_2}(z_2), \ldots, V^*_{E_d}(z_d)].$$

Then if none of the sets E_j are polar,

$$(dd^c V^*_{E_1 \times \cdots \times E_d})^d = \mu_{E_1} \times \cdots \times \mu_{E_d}$$

where $\mu_{E_j} = \Delta V^*_{E_j}$ is the classical equilibrium measure of E_j.

QUESTION 7. What can one say about $\text{supp}(dd^c V^*_{P_q, E_1 \times \cdots \times E_d})^d$ in the case when $q > 1$?

As examples, for $T^d = \{(z_1, ..., z_d) : |z_j| = 1, \ j = 1, ..., d\}$ we have $V_T(z_j) = \log^+ |z_j|$ and hence for $q \geq 1$

$$V_{P_q, T^d}(z) = \phi_{P_q}(\log^+ |z_1|, ..., \log^+ |z_d|) = [\sum_{j=1}^d (\log^+ |z_j|)^{q'}]^{1/q'}.$$

The measure $(dd^cV_{P_q,T^d})^d$ is easily seen to be invariant under the torus action and hence is a positive constant times Haar measure on T^d. Thus in this case $\operatorname{supp}(dd^cV_{P_q,T^d})^d = T^d$ for $q \geq 1$.

For the set $[-1,1]^d$ we have $V_{[-1,1]}(z_j) = \log|z_j + \sqrt{z_j^2 - 1}|$ and hence for $q \geq 1$

$$V_{P_q,[-1,1]^d}(z_1,...,z_d) = \left\{ \sum_{j=1}^{d} \left(\log \left| z_j + \sqrt{z_j^2 - 1} \right| \right)^{q'} \right\}^{1/q'}.$$

In this case, it is not clear for $q > 1$ whether $\operatorname{supp}\,(dd^cV_{P_q,[-1,1]^d})^d = [-1,1]^d$.

References

[1] T. Bayraktar, *Zero distribution of random sparse polynomials*, Michigan Math. J. **66** (2017), no. 2, 389–419, DOI 10.1307/mmj/1490639822. MR3657224

[2] T. Baĭraktar, T. Blum, and N. Levenberg, *Pluripotential theory and convex bodies* (Russian, with Russian summary), Mat. Sb. **209** (2018), no. 3, 67–101, DOI 10.4213/sm8893; English transl., Sb. Math. **209** (2018), no. 3, 352–384. MR3769215

[3] E. Bedford and B. A. Taylor, *Plurisubharmonic functions with logarithmic singularities* (English, with French summary), Ann. Inst. Fourier (Grenoble) **38** (1988), no. 4, 133–171. MR978244

[4] L. Bos and N. Levenberg, *Bernstein-Walsh theory associated to convex bodies and applications to multivariate approximation theory*, Comput. Methods Funct. Theory **18** (2018), no. 2, 361–388, DOI 10.1007/s40315-017-0220-4. MR3806552

[5] C. H. Lu, private communication.

[6] T. Darvas, E. Di Nezza, and C. H. Lu, *Monotonicity of nonpluripolar products and complex Monge-Ampère equations with prescribed singularity*, Anal. PDE **11** (2018), no. 8, 2049–2087, DOI 10.2140/apde.2018.11.2049. MR3812864

[7] M. Klimek, *Pluripotential theory*, London Mathematical Society Monographs. New Series, vol. 6, The Clarendon Press, Oxford University Press, New York, 1991. Oxford Science Publications. MR1150978

[8] T. Ransford, *Potential theory in the complex plane*, London Mathematical Society Student Texts, vol. 28, Cambridge University Press, Cambridge, 1995. MR1334766

[9] E. B. Saff and V. Totik, *Logarithmic potentials with external fields*, Grundlehren der Mathematischen Wissenschaften [Fundamental Principles of Mathematical Sciences], vol. 316, Springer-Verlag, Berlin, 1997. Appendix B by Thomas Bloom. MR1485778

Department of Mathematics, Indiana University, Bloomington, IN 47405
Email address: nlevenbe@indiana.edu

Department of Mathematics, Indiana University, Bloomington, IN 47405
Email address: kmperera@iu.edu

Contemporary Mathematics
Volume **743**, 2020
https://doi.org/10.1090/conm/743/14956

A holomorphic functional calculus for finite families of commuting semigroups

Jean Esterle

ABSTRACT. Let $\mathcal{A}$ be a commutative Banach algebra such that $u\mathcal{A} \neq \{0\}$ for $u \in \mathcal{A} \setminus \{0\}$ which possesses dense principal ideals. The purpose of the paper is to give a general framework to define $F(-\lambda_1\Delta_{T_1}, \dots, -\lambda_k\Delta_{T_k})$ where F belongs to a natural class of holomorphic functions defined on suitable open subsets of $\mathbb{C}^k$ containing the "Arveson spectrum" of the family $(-\lambda_j\Delta_{T_j})_{1\le j\le k}$, where $\Delta_{T_1}, \dots, \Delta_{T_k}$ are the infinitesimal generators of commuting one-parameter semigroups of multipliers on $\mathcal{A}$ belonging to one of the following classes:

(1) The class of strongly continuous semigroups $T = (T(te^{ia}))_{t>0}$ such that $\cup_{t>0}T(te^{ia})\mathcal{A}$ is dense in $\mathcal{A}$, where $a \in \mathbb{R}$.

(2) The class of semigroups $T = (T(\zeta))_{\zeta\in S_{a,b}}$ holomorphic on an open sector $S_{a,b}$ such that $T(\zeta)\mathcal{A}$ is dense in $\mathcal{A}$ for some, or equivalently for all, $\zeta \in S_{a,b}$.

We use the notion of quasimultiplier, introduced in 1981 by the author at the Long Beach Conference on Banach algebras: the generators of the semigroups under consideration will be defined as quasimultipliers on $\mathcal{A}$, and for ζ in the Arveson resolvent set of Δ_T the resolvent $(\Delta_T - \zeta I)^{-1}$ will be defined as a regular quasimultiplier on $\mathcal{A}$, i.e. a quasimultiplier S on $\mathcal{A}$ such that $\sup_{n\ge 1} \lambda^n\|S^n u\| < +\infty$ for some $\lambda > 0$ and some u generating a dense ideal of $\mathcal{A}$ and belonging to the intersection of the domains of S^n, $n \ge 1$. The algebra of quasimultipliers (resp. regular quasimultipliers) on $\mathcal{A}$ will be denoted $\mathcal{QM}(\mathcal{A})$ (resp. $\mathcal{QM}_r(\mathcal{A})$).

The first step consists in "normalizing" the Banach algebra $\mathcal{A}$, i.e. continuously embedding $\mathcal{A}$ in a Banach algebra $\mathcal{B}$ having the same quasimultiplier algebra as $\mathcal{A}$ but for which $\limsup_{t\to 0^+} \|T(te^{ia})\|_{\mathcal{M}(\mathcal{B})} < +\infty$ if T belongs to the class (1), and for which $\limsup_{\substack{\zeta\to 0\\ \zeta\in S_{\alpha,\beta}}} \|T(\zeta)\| < +\infty$ for all pairs (α, β) such that $a < \alpha < \beta < b$ if T belongs to the class (2). Iterating this procedure this allows to consider $(\lambda_j\Delta_{T_j} + \zeta I)^{-1}$ as an element of $\mathcal{M}(\mathcal{B})$ for $\zeta \in Res_{ar}(-\lambda_j\Delta_{T_j})$, the "Arveson resolvent set " of $-\lambda_j\Delta_{T_j}$, and to use the standard integral 'resolvent formula' even if the given semigroups are not bounded near the origin.

A first approach to the functional calculus involves the dual of an algebra of fast decreasing functions, described in Appendix 2. The action of elements of this dual, which is an algebra with respect to convolution can also be implemented via representing measures, Cauchy transforms and Fourier–Borel transforms introduced in Appendix 1.

The second approach to the functional calculus is based on Cauchy's formula. We introduce a family $M_{a,b}$ of products of singletons and/or pair of real numbers associated to the semigroups $(T_j)_{1\le j\le k}$. For $(\alpha, \beta) \in M_{a,b}$, we introduce the family of open subsets U of $\mathbb{C}^k$ which are "admissible" with respect to (α,β), which implies in particular that $U+\epsilon\subset U$ for every $\epsilon\in \overline{S}^*_{\alpha,\beta}$, where $S^*_{\alpha,\beta}=\Pi_{1\le j\le k}S_{-\pi/2-\alpha_j,\pi/2+\beta_j}$. Standard properties of the class $H^{(1)}(U)$ of all holomorphic functions F on U such that

$$\|F\|_{H^{(1)}(U)}:=\sup_{\epsilon\in S^*_{\alpha,\beta}}\int_{\epsilon+\tilde{\partial}U}\|F(\sigma\||d\sigma|<+\infty$$

are given in Appendix 3.

The results of Appendix 3 allow, when an open set $U\subset\mathbb{C}^k$ admissible with respect to $(\alpha,\beta)\in M_{a,b}$ satisfies some more suitable admissibility conditions with respect to $T=(T_1,\dots,T_k)$ and $\lambda=(\lambda_1,\dots,\lambda_k)$, to define $F(-\lambda_1\Delta_{T_1},\dots,-\lambda_k\Delta_{T_k})$ for $F\in H^{(1)}(U)$ by using the formula

$$\begin{aligned}&F(-\lambda_1\Delta_{T_1},\dots,-\lambda_k\Delta_{T_k})\\ =\;&\frac{1}{(2\pi i)^k}\int_{\epsilon+\overline{\partial}U}F(\zeta_1,\dots,\zeta_k)\prod_{j=1}^k(\lambda_j\Delta_{T_j}+\zeta_jI)^{-1}d\zeta_1\dots d\zeta_k,\end{aligned}$$

where $\tilde{\partial}U$ denotes the "distinguished boundary" of U and where $\epsilon\in S^*_{\alpha,\beta}$ is chosen so that $\epsilon+U$ still satisfies the required admissibility conditions with respect to T and λ. Given T and λ, this gives a family $\mathcal{W}_{T,\lambda}$ of open sets stable under finite intersections and an algebra homomorphism $F\to F(-\lambda_1\Delta_{T_1},\dots,-\lambda_k\Delta_{T_k})$ from $\cup_{U\in\mathcal{W}_{T,\lambda}}H^{(1)}(U)$ into the multiplier algebra $\mathcal{M}(\mathcal{B})\subset\mathcal{QM}_r(\mathcal{A})$. This homomorphism extends in a natural way to a bounded algebra homomorphism from $\cup_{U\in\mathcal{W}_{T,\lambda}}H^\infty(U)$ into $\mathcal{QM}_r(\mathcal{B})=\mathcal{QM}_r(\mathcal{A})$, and to a bounded algebra homomorphism from $\cup_{U\in\mathcal{W}_{T,\lambda}}\mathcal{S}(U)$ into $\mathcal{QM}(\mathcal{B})=\mathcal{QM}(\mathcal{A})$, where $\mathcal{S}(U)$ denotes the class of those functions F holomorphic on U such that $FG\in H^\infty(U)$ for some "strongly outer" function $G\in H^\infty(U)$. If $F_j:\zeta\to\zeta_j$ is the j-th coordinate projection, then of course $F_j(-\lambda_1\Delta_{T_1},\dots,-\lambda_k\Delta_{T_k})=-\lambda_j\Delta_{T_j}$.

Keywords: analytic semigroup, infinitesimal generator, resolvent, Cauchy transform, Fourier–Borel transform, Laplace transform, holomorphic functional calculus, Cauchy theorem, Cauchy formula.

AMS Classifications: Primary 47D03; Secondary 46J15, 44A10

1. Introduction

The author observed in **[15]** that if a Banach algebra $\mathcal{A}$ does not possess any nonzero idempotent then $\inf\{\|x^2-x\|: x\in\mathcal{A}, \|x\|\ge 1/2\}\ge 1/4$. If x is quasinilpotent, and if $\|x\|\ge 1/2$, then $\|x\|>1/4$. Concerning (nonzero) strongly continuous semigroups $T=(T(t))_{t>0}$ of bounded operators on a Banach space X, these elementary considerations lead to the following results, obtained in 1987 by Mokhtari **[26]**.

(1) If $\limsup_{t\to 0^+}\|T(t)-T(2t)\|<1/4$, then the generator of the semigroup is bounded, and so $\limsup_{t\to0^+}\|T(t)-T(2t)\|=0$.
(2) If the semigroup is quasinilpotent, then $\|T(t)-T(2t)\|>1/4$ when t is sufficiently small.

If the semigroup is norm continuous, and if there exists a sequence $(t_n)_{n\ge1}$ of positive real numbers such that $\lim_{n\to+\infty}t_n=0$ and $\|T(t_n)-T(2t_n)\|<1/4$, then the closed subalgebra $\mathcal{A}_T$ of $\mathcal{B}(X)$ generated by the semigroup possesses an exhaustive sequence of idempotents, i.e. there exists a sequence $(P_n)_{n\ge1}$ of idempotents of $\mathcal{A}_T$ such that for every compact subset K of the character space of $\mathcal{A}_T$ there exists $n_K>0$ satisfying $\chi(P_n)=1$ for $\chi\in K$, $n\ge n_K$.

More sophisticated arguments allowed A. Mokhtari and the author to obtain later in [**19**] more general results valid for every integer $p \geq 1$.

These results led the author to consider in [**16**] the behaviour of the distance $\|T(s) - T(t)\|$ for $s > t$ near 0. The following results were obtained in [**16**].

(1) If there exist for some $\delta > 0$ two continuous functions $r \to t(r)$ and $r \to s(r)$ on $[0, \delta]$, such that $s(0) = 0$ and such that $0 < t(r) < s(r)$ and $\|T(t(r)) - T(s(r))\| < (s(r) - t(r))s(r)^{s(r)/(s(r)-t(r))}t(r)^{-t(r)/(s(r)-t(r))}$ for $r \in (0, \delta]$, then the generator of the semigroup is bounded, and so $\|T(t) - T(s)\| \to 0$ as $0 < t < s, s \to 0^+$.

(2) If the semigroup is quasinilpotent, there exists $\delta > 0$ such that

$$\|T(t) - T(s)\| > (s-t)s^{\frac{s}{s-t}}t^{-\frac{t}{s-t}}$$

for $0 < t < s \leq \delta$.

(3) If the semigroup is norm continuous, and if there exists two sequences of positive real numbers such that $0 < t_n < s_n$, $\lim_{n \to +\infty} s_n = 0$, and such that $\|T(t_n) - T(s_n)\| < (s_n - t_n)s_n^{s_n/(s_n-t_n)}t_n^{-t_n/(s_n-t_n)}$, then the closed subalgebra $\mathcal{A}_T$ of $\mathcal{B}(X)$ generated by the semigroup possesses an exhaustive sequence of idempotents.

The quantities appearing in these statements are not mysterious: consider the Hilbert space $L^2([0,1])$, and for $t > 0$ define $T_0(t) : L^2([0,1] \to L^2([0,1]$ by the formula $T_0(t)(f)(x) = x^t f(x) \quad (0 < x \leq 1)$. Then

$$\|T_0(t) - T_0(s)\| = (s-t)s^{s/(s-t)}t^{-t/(s-t)}.$$

This remark also shows that assertions (1) and (3) in these statements are sharp, and examples show that assertion (2) is also sharp.

One can consider $T(t)$ as defined by the formula $\int_0^{+\infty} T(x)d\delta_t(x)$, where δ_t denotes the Dirac measure at t. Heuristically, $T(t) = e^{t\Delta_T}$, where Δ_T denotes the generator of the semigroup, and since the Laplace transform of δ_t is defined by the formula $\mathcal{L}(\delta_t)(z) = \int_0^{+\infty} e^{-zx}d\delta_t(x) = e^{-zt}$, it is natural to write $\mathcal{L}(\delta_t)(-\Delta_T) = T(t)$. More generally, if an entire function F has the form $F = \mathcal{L}(\mu)$, where μ is a measure supported by $[a, b]$, with $0 < a < b < +\infty$, we can set

$$F(-\Delta_T) = \int_0^{+\infty} T(x)d\mu(x),$$

and consider the behaviour of the semigroup near 0 in this context.

I. Chalendar, J.R. Partington and the author used this point of view in [**9**]. Denote by $\mathcal{M}_c(0, +\infty)$ the set of all measures μ supported by some interval $[a, b]$, where $0 < a < b < +\infty$. For the sake of simplicity we restrict attention to statements analogous to assertion 2. The following result is proved in [**9**].

Theorem: *Let* $\mu \in \mathcal{M}_c(0, +\infty)$ *be a real measure such that* $\int_0^{+\infty} d\mu(t) = 0$ *and let* $T = (T(t))_{t>0}$ *be a quasinilpotent semigroup of bounded operators.If* $\mu \neq 0$, *then there exists* $\delta > 0$ *such that* $\|F(-s\Delta_T)\| > \max_{x \geq 0} |F(x)|$ *for* $0 < s \leq \delta$.

When $\mu = \delta_1 - \delta_2$ this gives assertion 3 of Mokhtari's result, and when $\mu = \delta_1 - \delta_{p+1}$ this gives assertion 3 of the extension of Mokhtari's result given in [**19**] (but several variables extensions of this functional calculus would be needed in order to obtain extensions of the results of [**16**]).

This theory applies, for example, to $\|T(t) - 2T(2t) + T(3t)\|$, or to the Bochner integrals $\| \int_1^2 T(tx)dx - \int_2^3 T(tx)dx\|$, which are not accessible by the methods of

[26] or **[19]**. Preliminary results concerning semigroups holomorphic in a sector were obtained by I. Chalendar, J.R. Partington and the author in **[10]**.

More generally it would be interesting to obtain lower estimates for quantities of the form $F(-\lambda_1\Delta_T,\dots,-\lambda_k\Delta_T)$ as $(\lambda_1,\dots,\lambda_k)\to(0,\dots,0)$ when the generator Δ_T of the semigroup is unbounded, and when F is an analytic function of several complex variables defined and satisfying natural growth conditions on a suitable neighbourhood of $\sigma_{ar}(\Delta_T)$, where $\sigma_{ar}(\Delta_T)$ denotes the "Arveson spectrum" of the infinitesimal generator Δ_T of T. The purpose of the present paper is to pave the way to such a program by defining more generally $F(-\lambda_1\Delta_{T_1},\dots,-\lambda_k\Delta_{T_k})$ when F belongs to a suitable class of holomorphic functions on some element of a family $\mathcal{W}_{T_1,\dots,T_k,\lambda}$ of open sets, and where $(T_1,\dots,T_k)$ denotes a finite family of commuting semigroups.

More precisely consider $a=(a_1,\dots,a_k)\in\mathbb{R}^k, b=(b_1,\dots,b_k)\in\mathbb{R}^k$ satisfying $a_j\le b_j\le a_j+\pi$ for $1\le j\le k$, and consider a commutative Banach algebra $\mathcal{A}$ such that $u\mathcal{A}$ is dense in $\mathcal{A}$ for some $u\in\mathcal{A}$ and such that $u\mathcal{A}\neq\{0\}$ for $u\in\mathcal{A}\setminus\{0\}$. This allows to consider the algebra $\mathcal{QM}(\mathcal{A})$ of all quasimultipliers on $\mathcal{A}$ and the algebra $\mathcal{QM}_r(\mathcal{A})$ of all regular quasimultipliers on $\mathcal{A}$ introduced by the author in **[15]**, see Section 2, and the usual algebra $\mathcal{M}(\mathcal{A})$ of all multipliers on $\mathcal{A}$ can be identified to the algebra of all quasimultipliers on $\mathcal{A}$ of domain equal to the whole of $\mathcal{A}$. We will be interested here in finite families $(T_1,\dots,T_k)$ of commuting semigroups of multipliers on $\mathcal{A}$ satisfying the following conditions.

- the semigroup T_j is strongly continuous on $e^{ia_j}.(0,+\infty)$, and $\cup_{t>0}T_j(e^{ia_j}t)\mathcal{A}$ is dense in $\mathcal{A}$ if $a_j=b_j$,
- the semigroup T_j is holomorphic on S_{a_j,b_j} and $T_j(\zeta)\mathcal{A}$ is dense in $\mathcal{A}$ for some (or, equivalently, for all) $\zeta\in S_{a_j,b_j}$ if $a_j<b_j$, where S_{a_j,b_j} denotes the open sector $\{z\in\mathbb{C}\setminus\{0\}\mid a_j<arg(z)<b_j\}$.

The first step of the construction of the functional calculus consists in obtaining a "normalization" $\mathcal{A}_T$ of the Banach algebra $\mathcal{A}$ with respect to a strongly continuous one-parameter semigroup $(T(t))_{t>0}$ of multipliers on $\mathcal{A}$. The idea behind this normalization process goes back to Feller **[21]**, and we use for this the notion of "QM-homomorphism" between commutative Banach algebras introduced in Section 2, which seems more appropriate than the related notion of "s-homomorphism" introduced by the author in **[15]**. Set $\omega_T=\|T(t)\|$ for $t>0$. A slight improvement of a result proved by P. Koosis and the author in Section 6 of **[14]** shows that the weighted convolution algebra $L^1(\mathbb{R}^+,\omega_T)$ possesses dense principal ideals, which allows to construct in Section 3 a commutative Banach algebra $\mathcal{A}_T\subset\mathcal{QM}_r(\mathcal{A})$ which contains $\mathcal{A}$ as a dense subalgebra and has dense principal ideals such that the injection $\tilde{j}:\mathcal{QM}(\mathcal{A})\to\mathcal{QM}(\mathcal{A}_T)$ associated to the norm-decreasing inclusion map $j:\mathcal{A}\to\mathcal{A}_T$ is onto and such that $\tilde{j}(\mathcal{M}(\mathcal{A}))\subset\mathcal{M}(\mathcal{A}_T)$ for which $\limsup_{t\to 0^+}\|T(t)\|_{\mathcal{M}(\mathcal{A}_T)}<+\infty$.

Set $\Phi_T(f)=\int_0^{+\infty}f(t)T(t)dt$ for $f\in L^1(\mathbb{R}^+,\omega_T)$, where the Bochner integral is computed with respect to the strong operator topology on $\mathcal{M}(\mathcal{A})$, and denote by $\mathcal{I}_T$ the closed subalgebra of $\mathcal{M}(\mathcal{A})$ generated by $\Phi_T(L^1(\mathbb{R}^+,\omega_T))$. In Section 5 we give an interpretation of the generator Δ_T of the semigroup T as a quasimultiplier on $\mathcal{I}_T$, and we define the "Arveson spectrum" $\sigma_{ar}(\Delta_T)$ to be the set $\{\tilde{\chi}(\Delta_T)\}_{\chi\in Spec(\mathcal{I}_T)}$, where $Spec(\mathcal{I}_T)$ denotes the character space if $\mathcal{I}_T$ and where $\tilde{\chi}$ denotes the unique extension to $\mathcal{QM}(\mathcal{I}_T)$ of a character $\chi\in Spec(\mathcal{I}_T)$, with the convention $\sigma_{ar}(\Delta_T)=\emptyset$ if the "Arveson ideal" $\mathcal{I}_T$ is radical.

If $\lambda \in \mathbb{C} \setminus \sigma_{ar}(\Delta_T)$, then the quasimultiplier $\Delta_T - \lambda I$ is invertible in $\mathcal{QM}_r(\mathcal{I}_T)$ and $(\Delta_T - \lambda I)^{-1} \in \mathcal{M}(\mathcal{A}_T) \subset \mathcal{QM}_r(\mathcal{A})$ if $\lambda \in \mathbb{C} \setminus \sigma_{ar}(\Delta_T)$, and we observe in Section 6 that we have, for $\zeta > \limsup_{t \to +\infty} \frac{\log \|T(t)\|}{t}$,

$$(\Delta_T - \zeta I)^{-1} = -\int_0^{+\infty} e^{-\zeta t} T(t) dt \in \mathcal{M}(\mathcal{A}_T) \subset \mathcal{QM}_r(\mathcal{A}),$$

which is the usual "resolvent formula" extended to strongly continuous semigroups not necessarily bounded near the origin.

In Section 4 we construct a more sophisticated normalization of the Banach algebra $\mathcal{A}$ with respect to a semigroup $T = (T(\zeta))_{\zeta \in S_{a,b}}$ which is holomorphic on an open sector $S_{a,b}$, where $a < b \leq a + \pi$. In this case the normalization $\mathcal{A}_T$ of $\mathcal{A}$ with respect to the semigroup T satisfies two more conditions.

- $T(\zeta)u\mathcal{A}_T$ is dense in $\mathcal{A}_T$ for $\zeta \in S_{a,b}$ if $u\mathcal{A}$ is dense in $\mathcal{A}$,
- $\limsup_{\substack{\zeta \to 0 \\ \alpha \leq \arg(\zeta) \leq \beta}} \|T(\zeta)\|_{\mathcal{M}(\mathcal{A}_T)} < +\infty$ for $a < \alpha < \beta < b$.

The generator of the holomorphic semigroup T is interpreted as in **[8]** as a quasimultiplier on the closed subalgebra of $\mathcal{A}$ generated by the semigroup, which is equal to the Arveson ideal $\mathcal{I}_{T_0}$ where T_0 denotes the restriction of T to the half-line $(0, e^{i\frac{a+b}{2}}.\infty)$, and the resolvent $\zeta \to (\Delta_T - \zeta I)^{-1}$, which is defined and holomorphic outside a closed sector of the form $z + \overline{S}_{-ie^{ia}, ie^{ib}}$ is studied in Section 7.

Consider again $(a_1, \ldots, a_k), (b_1, \ldots, b_k)$ satisfying $a_j \leq b_j \leq a_j + \pi$ for $j \leq k$ and a finite family $T = (T_1, \ldots, T_k)$ of commuting semigroups of multipliers on $\mathcal{A}$ satisfying the conditions given above.

By iterating the normalization process of $\mathcal{A}$ with respect to $T_1, \ldots, T_k$ given in Sections 3 and 4, we obtain a "normalization" of $\mathcal{A}$ with respect to the family T, see Definition 8.1, which is a commutative Banach algebra $\mathcal{B}$ contained in $\mathcal{QM}_r(\mathcal{A})$ for which the injection $j : \mathcal{A} \to \mathcal{B}$ is norm-decreasing, has dense range and extends to a norm-decreasing homomorphism from $\mathcal{M}(\mathcal{A})$ into $\mathcal{M}(\mathcal{B})$, for which the natural embedding $\tilde{j} : \mathcal{QM}(\mathcal{A}) \to \mathcal{QM}(\mathcal{B})$ is onto, and for which

$$\begin{cases} \limsup_{t \to 0^+} \|T(te^{ia_j}\|_{\mathcal{M}(\mathcal{B})} < +\infty \ \text{ if } a_j = b_j, \\ \limsup_{\substack{\zeta \to 0 \\ \alpha_j \leq arg(\zeta) \leq \beta_j}} \|T(\zeta)\| < +\infty \text{ for } a_j < \alpha_j \leq \beta_j < b_j \ \text{ if } a_j < b_j \end{cases} .$$

Set $e_z(\zeta) = e^{z_1\zeta_1 + \cdots + z_k\zeta_k}$ for $z = (z_1, \ldots, z_k), \zeta = (\zeta_1, \ldots, \zeta_k) \in \mathbb{C}^k$. Denote by $M_{a,b}$ the set of all pairs $(\alpha, \beta) \in \mathbb{R}^k \times \mathbb{R}^k$ such that $\alpha_j = \beta_j = a_j$ if $a_j = b_j$ and such that $a_j < \alpha_j \leq \beta_j < b_j$ if $a_j < b_j$, and set $\overline{S}_{\alpha,\beta} := \Pi_{j \leq k} \overline{S}_{\alpha_j, \beta_j}$ for $(\alpha, \beta) \in M_{a,b}$.

Let $W_{a,b}$ be the algebra of continuous functions f on $\cup_{(\alpha,\beta) \in M_{a,b}} \overline{S}_{\alpha,\beta}$ such that $e_z(\zeta) f(\zeta) \to 0$ as $|\zeta| \to 0$ in $\Pi_{j \leq k} \overline{S}_{\alpha_j, \beta_j}$ for every $z = (z_1, \ldots, z_k) \in \mathbb{C}^k$ and every $(\alpha, \beta) \in M_{a,b}$, and such that the maps $\zeta \to f(\zeta_1, \ldots, \zeta_{j-1}, \zeta, \zeta_{j+1}, \ldots \zeta_k)$ are holomorphic on S_{a_j, b_j} if $a_j < b_j$. For every element ϕ of the dual space $\mathcal{G}_{a,b} = W'_{a,b}$ there exists $(\alpha, \beta) \in M_{a,b}$, $z \in \mathbb{C}^k$ and a measure ν of bounded variation on $\overline{S}_{\alpha,\beta} := \Pi_{j \leq k} \overline{S}_{\alpha_j, \beta_j}$ such that

$$\langle f, \phi \rangle = \int_{\overline{S}_{\alpha,\beta}} e^{z\zeta} f(\zeta) d\nu(\zeta) \quad (f \in W_{a,b}).$$

This formula allows to extend the action of ϕ to $e_{-z}\mathcal{V}_{\alpha,\beta}(X) \supset e_{-z}\mathcal{U}_{\alpha,\beta}(X)$, where X denotes a separable Banach space and where $\mathcal{U}_{\alpha,\beta}(X)$ (resp. $\mathcal{V}_{\alpha,\beta}(X)$) denotes the algebra of continuous functions $f : \overline{S}_{\alpha,\beta} \to X$ which converge to 0 as

$|\zeta| \to +\infty$ (resp. bounded continuous functions $f : S_{\alpha,\beta} \to X$) such that the maps $\zeta \to f(\zeta_1, \ldots, \zeta_{j-1}, \zeta, \zeta_{j+1}, \ldots, \zeta_k)$ are holomorphic on S_{α_j,β_j} when $\alpha_j < \beta_j$.

Set $\mathcal{U}_{\alpha,\beta} := \mathcal{U}_{\alpha,\beta}(\mathbb{C})$. We describe in Appendix 1 some certainly well-known ways to implement the action of $\mathcal{U}'_{\alpha,\beta}$ on $\mathcal{V}_{\alpha,\beta}(X)$ when $(\alpha, \beta) \in M_{a,b}$ by using Cauchy transforms and Fourier–Borel transforms, and these formulae are extended to the action of elements of $(e_{-z}\mathcal{U}_{\alpha,\beta})'$ to spaces $e_{-z}\mathcal{V}_{\alpha,\beta}(X)$ in Appendix 2.

If $\phi \in (\cap_{z\in\mathbb{C}^k} e_{-z}\mathcal{U}_{\alpha,\beta})'$, define the domain $Dom(\mathcal{FB}(\phi))$ of the Fourier–Borel transform $\mathcal{FB}(\phi)$ of ϕ to be the set of all $z \in \mathbb{C}^k$ such that $\phi \in (e_{-z}\mathcal{U}_{\alpha,\beta})'$, and set $\mathcal{FB}(\phi)(z) = \langle e_{-z}, \phi\rangle$ for $z \in Dom(\mathcal{FB}(\phi))$. One can also define in a natural way the Fourier–Borel transform of $f \in e^{-z}\mathcal{V}_{\alpha,\beta}(X)$.

For $\lambda \in \cup_{(\gamma,\delta)\in M_{a-\alpha,b-\beta}} \overline{S}_{\gamma,\delta}$, set $T_{(\lambda)}(\zeta) = T(\lambda_1\zeta_1, \ldots, \lambda_k\zeta_k)$ $(\zeta \in S_{\alpha,\beta})$. If $\limsup_{\substack{|\zeta|\to+\infty \\ \zeta\in\tilde{\partial}\overline{S}_{\alpha,\beta}}} |e^{-z\zeta}|\|T(\lambda)(\zeta)\| < +\infty$, where $\tilde{\partial}\overline{S}_{\alpha,\beta}$ denotes the "distinguished boundary" of $S_{\alpha,\beta}$, then

$$\sup_{\zeta\in\overline{S}_{\alpha,\beta}} \left\| e^{z\zeta} T_1(\lambda_1\zeta_1) \ldots T_k(\lambda_k\zeta_k) \right\|_{\mathcal{M}(\mathcal{B})} < +\infty,$$

and one can define the action of ϕ on $T_{(\lambda)}$ by using the formula

$$\langle T_{(\lambda)}, \phi\rangle u = \langle T_{(\lambda)} u, \phi\rangle = \int_{\overline{S}_{\alpha,\beta}} e^{z\zeta} T_{(\lambda)}(\zeta) u d\nu(\zeta) \quad (u \in \mathcal{B}),$$

where ν is a representing measure for $\phi e_{-z} : f \to \langle e_{-z} f, \phi\rangle$ $(f \in \mathcal{U}_{\alpha,\beta})$.

Then $\langle T, \phi\rangle \in \mathcal{M}(\mathcal{B}) \subset \mathcal{QM}_r(\mathcal{A})$.

The Fourier–Borel transform of $e_z T_{(\lambda)}$ takes values in $\mathcal{M}(\mathcal{B})$ and extends analytically to $-Res_{ar}(\Delta_{T_{(\lambda)}}) := \Pi_{1\le j\le k}(\mathbb{C} \setminus \sigma_{ar}(-\lambda_j\Delta_{T_j}))$, which gives the formula

$$\mathcal{FB}(e_z T_{(\lambda)})(\zeta) = (-1)^k \prod_{1\le j\le k} (\lambda_j\Delta_j + (z_j + \zeta_j)I)^{-1}.$$

Set $S^*_{\alpha,\beta} = \Pi_{j\le k} S_{-\frac{\pi}{2}-\alpha_j, \frac{\pi}{2}-\beta_j}$, and set

$$W_n(\zeta) = \prod_{1\le j\le k} \frac{n^2}{\left(n + \zeta_j e^{i\frac{\alpha_j+\beta_j}{2}}\right)^2}$$

for $n \ge 1, \zeta = (\zeta_1, \ldots, \zeta_n) \in \overline{S}^*_{\alpha,\beta}$. The results of Section 2 give for $u \in \mathcal{B}$, if $z \in Dom(\phi)$, and if $\limsup_{\substack{|\zeta|\to+\infty \\ \zeta\in\tilde{\partial}\overline{S}_{\alpha,\beta}}} |e^{-z\zeta}|\|T(\lambda)(\zeta)\| < +\infty$, where $\tilde{\partial}\overline{S}_{\alpha,\beta}$ denotes the "distinguished boundary" of $\overline{S}_{\alpha,\beta}$,

$$\langle T_{(\lambda)}, \phi\rangle u$$

$$= \lim_{\substack{\epsilon\to0 \\ \epsilon\in S^*_{\alpha,\beta}}} \left(\lim_{n\to+\infty} \frac{(-1)^k}{(2\pi i)^k} \int_{z+\tilde{\partial}S^*_{\alpha,\beta}} W_n(\sigma - z)\mathcal{FB}(\phi)(\sigma) \prod_{j=1}^{k} ((\sigma_j - \epsilon_j)I + \lambda_j\Delta_{T_j})^{-1} u d\sigma \right),$$

where $\tilde{\partial}S^*_{\alpha,\beta} := \Pi_{1\le j\le k}\partial\overline{S}_{\alpha_j,\beta_j}$ denotes the "distinguished boundary" of $\overline{S}^*_{\alpha,\beta}$, and where $\partial S^*_{\alpha_j,\beta_j}$ is oriented from $-ie^{-i\alpha_j}.\infty$ to $ie^{-i\beta_j}.\infty$.

If, further, $\int_{z+\tilde{\partial}S^*_{\alpha,\beta}} \|\mathcal{FB}(\phi)(\sigma)\||d\sigma| < +\infty$, then we have, for $u \in \mathcal{B}$,

$$\langle T_{\lambda)}, \phi\rangle u = \lim_{\substack{\epsilon \to (0,\ldots,0) \\ \epsilon \in S^*_{\alpha,\beta}}} \langle e_{-\epsilon} T_{(\lambda)}, \phi\rangle u$$

$$= \lim_{\substack{\epsilon \to (0,\ldots,0) \\ \epsilon \in S^*_{\alpha,\beta}}} \frac{(-1)^k}{(2\pi i)^k} \int_{z+\tilde{\partial}S^*_{\alpha,\beta}} \mathcal{FB}(\phi)(\sigma) \prod_{j=1}^{k} ((\sigma_j - \epsilon_j)I + \lambda_j \Delta_{T_j})^{-1} u d\sigma.$$

Finally, if $z \in Dom(\phi)$, if $\int_{z+\tilde{\partial}S^*_{\alpha,\beta}} \|\mathcal{FB}(\phi)(\sigma)\| |d\sigma| < +\infty$, and if

$$\limsup_{\substack{|\zeta| \to +\infty \\ \zeta \in \tilde{\partial}\overline{S}_{\alpha,\beta}}} |e^{-z\zeta}| \|T(\lambda)(\zeta)\| = 0,$$

then we have, for $u \in \mathcal{B}$,

$$\langle T_{(\lambda)}, \phi\rangle u = \frac{(-1)^k}{(2\pi i)^k} \int_{z+\tilde{\partial}S^*_{\alpha,\beta}} \mathcal{FB}(\phi)(\sigma) \prod_{j=1}^{k} (\sigma_j I + \lambda_j \Delta_{T_j})^{-1} u d\sigma.$$

The convolution product of two elements of $(e_{-z}\mathcal{U}_{\alpha,\beta})'$ may be defined in a natural way, and if λ, ϕ_1, ϕ_2 satisfies the conditions above we have

$$\langle T_{(\lambda)}, \phi_1 * \phi_2\rangle = \langle T_{(\lambda)}, \phi_1\rangle \langle T_{(\lambda)}, \phi_2\rangle,$$

but there is no direct extension of this formula to the convolution product of two arbitrary elements of $\mathcal{G}_{a,b}$, see the comments at the end of Section 8.

In Section 9 of the paper we introduce a class $\mathcal{U}$ of "admissible open sets" U, with piecewise $\mathcal{C}^1$-boundary, of the form $(z + S^*_{\alpha,\beta}) \setminus K$, where K is bounded and where $(\alpha, \beta) \in M_{a,b}$.

These open sets U have the property that $U + \epsilon \subset U$ for $\epsilon \in \overline{S}^*_{\alpha,\beta}$ and that $\overline{U} + \epsilon \subset Res_{ar}(-\lambda\Delta_T)$ for some $\epsilon \in \overline{S}^*_{\alpha,\beta}$. Also $\Pi_{j=1}^k(-\lambda\Delta_{T_j} - .I)^1$ is bounded on the distinguished boundary of $U + \epsilon$ for $\epsilon \in S^*_{\alpha,\beta}$ when $|\epsilon|$ is sufficiently small.

Standard properties of the class $H^{(1)}(U)$ of all holomorphic functions F on U such that $\|F\|_{H^{(1)}(U)} := \sup_{\epsilon \in S^*_{\alpha,\beta}} \int_{\epsilon+\tilde{\partial}U} \|F(\sigma\| |d\sigma| < +\infty$ are given in Appendix 3 (when $a_j = b_j$ for $j \le k$, this space is the usual Hardy space H^1 on a product of open half-planes). When an open set $U \subset \mathbb{C}^k$ is admissible with respect to $(\alpha, \beta) \in M_{a,b}$ and satisfies some more suitable admissibility conditions with respect to $T = (T_1, \ldots, T_k)$ and $\lambda \in \cup_{(\gamma,\delta) \in M_{a-\alpha,b-\beta}} \overline{S}_{\gamma,\delta}$, a quasimultiplier $F(-\lambda_1\Delta_{T_1}, \ldots, -\lambda_k\Delta_{T_k}) \in \mathcal{M}(\mathcal{B}) \subset \mathcal{QM}_r(\mathcal{A})$ is defined for $F \in H^{(1)}(U)$ in Section 9 by using the formula

$$F(-\lambda_1\Delta_{T_1}, \ldots, -\lambda_k\Delta_{T_k})$$

$$= \frac{1}{(2\pi i)^k} \int_{\epsilon+\tilde{\partial}U} F(\zeta_1, \ldots, \zeta_k) \prod_{j=1}^{k} (\lambda_j \Delta_{T_j} + \zeta_j I)^{-1} d\zeta_1 \ldots d\zeta_k,$$

where $\tilde{\partial}U$ denotes the distinguished boundary of U, and where $\epsilon \in S^*_{\alpha,\beta}$ is chosen so that $\epsilon + U$ still satisfies the required admissibility conditions with respect to T and λ.

Given T and $\lambda \in \cup_{(\alpha,\beta) \in M_{a,b}} \overline{S}_{a-\alpha,b-\beta}$, denote by $\mathcal{W}_{T,\lambda}$ the family of all open sets $U \subset \mathbb{C}^k$ satisfying these admissibility conditions with respect T and λ. Then

$\mathcal{W}_{T,\lambda}$ is stable under finite intersections, $\cup_{U\in\mathcal{W}_{T,\lambda}} H^{(1)}(U)$ is stable under products and we have

$$(F_1F_2)(-\lambda_1\Delta_{T_1},\ldots,-\lambda_k\Delta_{T_k}) = F_1(-\lambda_1\Delta_{T_1},\ldots,-\lambda_k\Delta_{T_k})F_2(-\lambda_1\Delta_{T_1},\ldots,-\lambda_k\Delta_{T_k}).$$

This homomorphism extends in a natural way to a bounded algebra homomorphism from $\cup_{U\in\mathcal{W}_{T,\lambda}} H^{\infty}(U)$ into $\mathcal{QM}_r(\mathcal{B}) = \mathcal{QM}_r(\mathcal{A})$, and we have, if $\phi \in \mathcal{F}_{\alpha,\beta}$ for some $(\alpha,\beta) \in M_{a,b}$ such that $\lambda \in \overline{S}_{\gamma,\delta}$ for some $(\gamma,\delta) \in M_{a-\alpha,b-\beta}$, and if $\lim_{\substack{|\zeta|\to+\infty \\ \zeta\in\partial\overline{S}_{\alpha,\beta}}} \|e^{-z\zeta}T_1(\lambda_1\zeta_1)\ldots T_k(\lambda_k\zeta_k)\| = 0$ for some $z \in Dom(\mathcal{FB}(\phi))$,

$$\mathcal{FB}(\phi)(-\lambda_1\Delta_{T_1},\ldots,-\lambda_k\Delta_{T_k}) = \langle T_{(\lambda)},\phi\rangle,$$

so that $F(-\lambda_1\Delta_{T_1},\ldots,-\lambda_k\Delta_{T_k}) = T(\nu\lambda_j)$ if $F(\zeta) = e^{-\nu\zeta_j}$, where $\nu\lambda_j$ is in the domain of definition of T_j.

A function $F \in H^{\infty}(U)$ will be said to be strongly outer if there exists a sequence $(F_n)_{n\ge1}$ of invertible elements of $H^{(\infty)}(U)$ such that $|F(\zeta)| \le |F_n(\zeta)|$ and $\lim_{n\to+\infty} F(\zeta)F_n^{-1}(\zeta) = 1$ for $\zeta \in U$.

If U is admissible with respect to some $(\alpha,\beta) \in M_{a,b}$ then there is a conformal map θ from $\mathbb{D}^k$ onto U and the map $F \to F\circ\theta$ is a bijection from the set of strongly outer bounded functions on U onto the set of strongly outer bounded functions on $\mathbb{D}^k$. Every bounded outer function on the open unit disc $\mathbb{D}$ is strongly outer, but the class of strongly outer bounded functions on $\mathbb{D}^k$ is smaller than the usual class of bounded outer functions on $\mathbb{D}^k$ if $k \ge 2$.

We then define the Smirnov class $\mathcal{S}(U)$ to be the class of holomorphic functions F on U such that there exists a strongly outer function $G \in H^{\infty}(U)$ for which $FG \in H^{\infty}(U)$.

The bounded algebra homomorphism $F \to F(-\lambda_1\Delta_{T_1},\ldots,-\lambda_k\Delta_{T_k})$ from $\cup_{U\in\mathcal{W}_{T,\lambda}} H^{\infty}(U)$ into $\mathcal{QM}_r(\mathcal{B}) = \mathcal{QM}_r(\mathcal{A})$ extends to a bounded homomorphism from $\cup_{U\in\mathcal{W}_{T,\lambda}} \mathcal{S}(U)$ into $\mathcal{QM}(\mathcal{B}) = \mathcal{QM}(\mathcal{A})$. If $F : \zeta \to \zeta_j$, is the j-th coordinate projection then of course $F(-\lambda_1\Delta_{T_1},\ldots,-\lambda_k\Delta_{T_k}) = -\lambda_j\Delta_{T_j}$.

The author wishes to thank Isabelle Chalendar and Jonathan Partington for valuable discussions during the preparation of this paper. He also wishes to thank the referees, whose comments and corrections helped to clarify some proofs and to considerably improve the presentation of the paper.

2. Quasimultipliers on weakly cancellative commutative Banach algebras with dense principal ideals

We will say that a Banach algebra $\mathcal{A}$ is weakly cancellative if $u\mathcal{A} \neq \{0\}$ for every $u \in \mathcal{A}\setminus\{0\}$. In the whole paper we will consider weakly cancellative commutative Banach algebras with dense principal ideals, i.e. weakly cancellative commutative Banach algebras such that the set $\Omega(\mathcal{A}) := \{u \in \mathcal{A} \mid [u\mathcal{A}]^- = \mathcal{A}\}$ is not empty.

A quasimultiplier on such an algebra $\mathcal{A} \neq \{0\}$ is a closed operator

$$S = S_{u/v} : \mathcal{D}_S \to \mathcal{A},$$

where $u \in \mathcal{A}$, $v \in \Omega(\mathcal{A})$, where $\mathcal{D}_S := \{x \in \mathcal{A} \mid ux \in v\mathcal{A}\}$, and where Sx is the unique $y \in \mathcal{A}$ such that $vy = ux$ for $x \in \mathcal{D}_S$. Let $\mathcal{QM}(\mathcal{A})$ be the algebra of all quasimultipliers on $\mathcal{A}$. A set $U \subset \mathcal{QM}(\mathcal{A})$ is said to be pseudobounded if $\sup_{S\in U}\|Su\| < +\infty$ for some $u \in \Omega(\mathcal{A}) \cap (\cap_{S\in U}\mathcal{D}(S))$, and a quasimultiplier $S \in \mathcal{QM}(\mathcal{A})$ is said to be regular if the family $\{\lambda^nS^n\}_{n\ge1}$ is pseudobounded for some $\lambda > 0$. The algebra of all regular quasimultipliers on $\mathcal{A}$ will be denoted by $\mathcal{QM}_r(\mathcal{A})$. A multiplier on $\mathcal{A}$ is a bounded linear operator S on $\mathcal{A}$ such that $S(uv) = (Su)v$ for $u \in \mathcal{A}, v \in \mathcal{A}$, and the multiplier algebra $\mathcal{M}(\mathcal{A})$ of all multipliers on $\mathcal{A}$, which is a closed subalgebra of the Banach algebra of bounded operators on $\mathcal{A}$, is also the algebra of all quasimultipliers on $\mathcal{A}$ such that $\mathcal{D}_S = \mathcal{A}$, and $\mathcal{M}(\mathcal{A}) \subset \mathcal{QM}_r(\mathcal{A})$. Also if $S = S_{u/v} \in \mathcal{QM}(\mathcal{A}), w \in \mathcal{D}(S), R \in \mathcal{M}(\mathcal{A})$, then

$$u(Rw) = R(v(Sw)) = v(R(Sw)),$$

so $Rw \in \mathcal{D}(S)$, and we have

$$R(Sw) = S(Rw). \tag{1}$$

If $\mathcal{A}$ is unital then $\Omega(\mathcal{A}) = \mathcal{G}(\mathcal{A})$, where $\mathcal{G}(\mathcal{A})$ denotes the group of invertible elements of $\mathcal{A}$, and $\mathcal{QM}(\mathcal{A}) = \mathcal{M}(\mathcal{A}) = \mathcal{A}$.

We will denote by $Spec(A)$ the space of all characters on a commutative Banach algebra $\mathcal{A}$, equipped with the Gelfand topology. Recall that $\mathcal{A}$ is said to be radical when $Spec(A) = \emptyset$.

DEFINITION 2.1. : *Let $\mathcal{A}$ be a weakly cancellative commutative Banach algebra with dense principal ideals.*

For $\chi \in Spec(\mathcal{A})$, $S = S_{u/v} \in \mathcal{QM}(\mathcal{A})$, set $\tilde{\chi}(S) = \frac{\chi(u)}{\chi(v)}$, and set

$$\sigma_{\mathcal{A}}(S) := \{\tilde{\chi}(S)\}_{\chi \in Spec(\mathcal{A})}, \quad (S \in \mathcal{QM}(\mathcal{A})),$$

with the convention $\sigma_{\mathcal{A}}(S) = \emptyset$ if $\mathcal{A}$ is radical.

Clearly, $\tilde{\chi}$ is a character on $\mathcal{QM}(\mathcal{A})$ for $\chi \in Spec(\mathcal{A})$, and the map $\chi \to \tilde{\chi}(S)$ is continuous on $Spec(\mathcal{A})$ for $S \in \mathcal{QM}(\mathcal{A})$.

The following notion seems slightly more flexible than the notion of s-homomorphism introduced by the author in [**15**].

DEFINITION 2.2. : *Let $\mathcal{A}$ be a weakly cancellative commutative Banach algebra with dense principal ideals, and let $\mathcal{B}$ be a weakly cancellative Banach algebra. A homomorphism $\Phi : \mathcal{A} \to \mathcal{B}$ is said to be a $\mathcal{QM}$-homomorphism if the following conditions are satisfied*

(i) Φ is one-to-one, and $\Phi(\mathcal{A})$ is dense in $\mathcal{B}$.

(ii) $\Phi(u)\mathcal{B} \subset \Phi(\mathcal{A})$ for some $u \in \Omega(\mathcal{A})$.

If the conditions of Definition 2.2 are satisfied, we will say that Φ is a $\mathcal{QM}$-homomorphism with respect to u. Notice that $\Phi(u) \in \Omega(\mathcal{B})$, and so the existence of such an homomorphism implies that $\mathcal{B}$ is a weakly cancellative commutative Banach algebra with dense principal ideals. Notice also that condition (ii) shows that $\mathcal{B}$ may be identified to a subalgebra of $\mathcal{QM}(\Phi(\mathcal{A})) \approx \mathcal{QM}(\mathcal{A})$.

PROPOSITION 2.3. *Let $\Phi : \mathcal{A} \to \mathcal{B}$ be a homomorphism between weakly cancellative commutative Banach algebras with dense principal ideals, and assume that Φ is a $\mathcal{QM}$-homomorphism with respect to some $u_0 \in \Omega(\mathcal{A})$.*

(i) There exists $M > 0$ such that $\|\Phi^{-1}(\Phi(u_0)v)\| \le M\|v\|$ for $v \in \mathcal{B}$.

(ii) $\Phi^{-1}(\Phi(u_0)v) \in \Omega(\mathcal{A})$ for $v \in \Omega(\mathcal{B})$.

(iii) Set $\tilde{\Phi}(S_{u/v}) = S_{\Phi(u)/\Phi(v)}$ for $S_{u/v} \in \mathcal{QM}(\mathcal{A})$.

Then $\tilde{\Phi} : \mathcal{QM}(\mathcal{A}) \to \mathcal{QM}(\mathcal{B})$ is a pseudobounded isomorphism, and

$$\tilde{\Phi}^{-1}(S_{u/v}) = S_{\Phi^{-1}(\Phi(u_0)u)/\Phi^{-1}(\Phi(u_0)v)}$$

for $S_{u/v} \in \mathcal{QM}(\mathcal{B})$.

PROOF. (i) Set $\Psi(v) = \Phi^{-1}(\Phi(u_0)v)$ for $v \in \mathcal{B}$. If $\lim_{n\to+\infty} v_n = v \in \mathcal{B}$, and if $\lim_{n\to+\infty} \Psi(v_n) = w \in \mathcal{A}$, then $\Phi(u_0)v = \Phi(w)$, so that $w = \Psi(v)$ and (i) follows from the closed graph theorem.

(ii) Let $v \in \Omega(\mathcal{B})$, and let $(w_n)_{n\geq 1}$ be a sequence of elements of $\mathcal{A}$ such that $\lim_{n\to+\infty} v\Phi(w_n) = \Phi(u_0)$. Then $\lim_{n\to+\infty} \Phi^{-1}(\Phi(u_0)v)w_n = u_0^2 \in \Omega(\mathcal{A})$, and so $\Phi^{-1}(\Phi(u_0)v) \in \Omega(\mathcal{A})$.

(iii) Let $U \subset \mathcal{QM}(\mathcal{A})$ be a pseudobounded set, and let $w \in \Omega(\mathcal{A}) \cap (\cap_{S\in U}\mathcal{D}(S))$ be such that $\sup_{S\in U} \|Sw\| < +\infty$. Then $\Phi(w) \in \Omega(\mathcal{B})$, and $\Phi(w) \in \cap_{S\in U}\mathcal{D}(\tilde{\phi}(S))$. Since $\sup_{S\in U} \|\tilde{\Phi}(S)\Phi(w)\| \leq \|\Phi\| \sup_{S\in U} \|Sw\| < +\infty$, this shows that the homomorphism $\tilde{\Phi} : \mathcal{QM}(\mathcal{A}) \to \mathcal{QM}(\mathcal{B})$ is pseudobounded.

Now set $\theta(S_{u/v}) = S_{\Phi^{-1}(\Phi(u_0)u)/\Phi^{-1}(\Phi(u_0)v)} = S_{\Psi(u)/\Psi(v)}$ for $S_{u/v} \in \mathcal{QM}(\mathcal{B})$. It follows from (ii) that $\theta : \mathcal{QM}(\mathcal{B}) \to \mathcal{QM}(\mathcal{A})$ is well-defined. Let $U \subset \mathcal{QM}(\mathcal{B})$ be pseudobounded, and let $w \in \Omega(\mathcal{B}) \cap (\cap_{S\in U}\mathcal{D}(S))$ be such that

$$\sup_{S\in U} \|Sw\| < +\infty.$$

We have, for $S = S_{u/v} \in U$,

$$\begin{aligned} v(S_{u/v}w) &= uw, \\ v\Phi(u_0)(S_{u/v}w\Phi(u_0)) &= u\Phi(u_0)w\Phi(u_0), \\ \Phi^{-1}\left(v\Phi(u_0)\right)\left(\Phi^{-1}(S_{u/v}w\Phi(u_0))\right) &= \Phi^{-1}(v\Phi(u_0))\Phi^{-1}(w\Phi(u_0)). \end{aligned}$$

So $\Phi^{-1}(w\Phi(u_0)) \in \mathcal{D}(\theta(S))$, and $\theta(S)\Phi^{-1}(w\Phi(u_0)) = \Phi^{-1}\left(S_{u/v}w\Phi(u_0)\right)$. Since

$$\sup_{S\in U} \|\,(Sw\Phi(u_0))\,\| \leq M \sup_{S\in U} \|Sw\| < +\infty,$$

this shows that $\theta : \mathcal{QM}(\mathcal{B}) \to \mathcal{QM}(\mathcal{A})$ is pseudobounded. We have, for $S = S_{u/v} \in \mathcal{QM}(\mathcal{A})$,

$$(\theta \circ \tilde{\Phi})(S) = S_{\Phi^{-1}(\Phi(u)\Phi(u_0))/\Phi^{-1}(\Phi(v)\Phi(u_0))} = S_{uu_0/vu_0} = S_{u/v} = S.$$

We also have, for $S = S_{u/v} \in \mathcal{QM}(\mathcal{A})$,

$$(\tilde{\Phi} \circ \theta)(S_{u/v}) = S_{\Phi(\Phi^{-1}(u\Phi(u_0))/\Phi(\Phi^{-1}(v\Phi(u_0))} = S_{u\Phi(u_0)/v\Phi(u_0)} = S_{u/v} = S.$$

Hence $\tilde{\Phi} = \mathcal{QM}(\mathcal{A}) \to \mathcal{QM}(\mathcal{B})$ is bijective, and $\tilde{\Phi}^{-1} : \mathcal{QM}(\mathcal{B}) \to \mathcal{QM}(\mathcal{A})$ is pseudobounded, since $\tilde{\Phi}^{-1} = \theta$. □

The following result is a simplified version of Theorem 7.11 of **[15]**, which will be used in the next two sections.

PROPOSITION 2.4. *Let $\mathcal{A}$ be a weakly cancellative commutative Banach algebra with dense principal ideals, and let $U \subset \mathcal{QM}(\mathcal{A})$ be a pesudobounded set stable under products. Set $\mathcal{L} := \{u \in \cap_{S\in U}\mathcal{D}(S) \mid \sup_{S\in U} \|Su\| < +\infty\}$, and set $\|u\|_{\mathcal{L}} := \sup_{S\in U\cup\{I\}} \|Su\|$ for $u \in \mathcal{L}$.*

Then $(\mathcal{L}, \|.\|_{\mathcal{L}})$ is a Banach algebra, $Ru \in \mathcal{L}$ and $\|Ru\|_{\mathcal{L}} \leq \|R\|_{\mathcal{M}(\mathcal{A})}\|u\|_{\mathcal{L}}$ for $R \in \mathcal{M}(\mathcal{A})$, $u \in \mathcal{L}$, and if we denote by $\mathcal{B}$ the closure of $\mathcal{A}$ in $(\mathcal{M}(\mathcal{L}), \|.\|_{\mathcal{M}(\mathcal{L})})$, the following properties hold

(i) $\mathcal{B}$ is a weakly cancellative commutative Banach algebra, and the inclusion map $j : \mathcal{A} \to \mathcal{B}$ is a $\mathcal{QM}$-homomorphism with respect to w for $w \in \Omega(\mathcal{A}) \cap \mathcal{L}$.

(ii) $\tilde{j}(\mathcal{M}(\mathcal{A})) \subset \mathcal{M}(\mathcal{B})$, *and* $\|\tilde{j}(R)\|_{\mathcal{M}(\mathcal{B})} \leq \|R\|_{\mathcal{M}(\mathcal{A})}$ *for* $R \in \mathcal{M}(\mathcal{A})$, *where* $\tilde{j} : \mathcal{QM}(\mathcal{A}) \to \mathcal{QM}(\mathcal{B})$ *is the pseudobounded isomorphism associated to* j *in Proposition* 2.3*(iii)*.

(iii) $S \in \mathcal{M}(\mathcal{B})$, *and* $\|S\|_{\mathcal{M}(\mathcal{B})} \leq \|S\|_{\mathcal{M}(\mathcal{L})} \leq 1$ *for every* $S \in U$.

PROOF. The fact that $(\mathcal{L}, \|.\|_{\mathcal{L}})$ is a Banach space follows from a standard argument given in the proof of Theorem 7.11 of **[15]**. Clearly, $\mathcal{L}$ is an ideal of $\mathcal{A}$, and it follows from the definition of $\|.\|_{\mathcal{L}}$ that $\|u\| \leq \|u\|_{\mathcal{L}}$ for $u \in \mathcal{L}$. We have, for $u \in \mathcal{L}, v \in \mathcal{L}$,

$$\|uv\|_{\mathcal{L}} \leq \|u\|_{\mathcal{L}}\|v\| \leq \|u\|_{\mathcal{L}}\|v\|_{\mathcal{L}},$$

and so $(\mathcal{L}, \|.\|_{\mathcal{L}})$ is a Banach algebra. If $R \in \mathcal{M}(\mathcal{A})$, $u \in \mathcal{L}$, then it follows from the definition of $\mathcal{L}$ that $Ru \in \cap_{S \in U}\mathcal{D}(S)$, and that we have

$$\sup_{S \in U \cup \{I\}} \|S(Ru)\| = \sup_{S \in U \cup \{I\}} \|R(Su)\| \leq \|R\|_{\mathcal{M}(\mathcal{A})}\|u\|_{\mathcal{L}}.$$

Hence $Ru \in \mathcal{L}$, and $\|Ru\|_{\mathcal{L}} \leq \|R\|_{\mathcal{M}(\mathcal{A})}\|u\|_{\mathcal{L}}$.

Now denote by $\mathcal{B}$ the closure of $\mathcal{A}$ in $(\mathcal{M}(\mathcal{L}), \|.\|_{\mathcal{M}(\mathcal{L})})$, and let $w \in \Omega(\mathcal{A}) \cap \mathcal{L}$. Since $\mathcal{L} \subset \mathcal{B}$, $\mathcal{B}$ is weakly cancellative, and $\mathcal{B}$ is commutative and has dense principal ideals since $j(\mathcal{A})$ is dense in $\mathcal{B}$. Since $w\mathcal{B} \subset \mathcal{L} \subset \mathcal{A}$, we see that the inclusion map $j : \mathcal{A} \to \mathcal{B}$ is a $\mathcal{QM}$-homomorphism with respect to w, which proves (i).

Let $R \in \mathcal{M}(\mathcal{A})$, and denote by R_1 the restriction of R to $\mathcal{L}$.

Then $R_1 \in \mathcal{M}(\mathcal{L})$, and $\|R_1\|_{\mathcal{M}(\mathcal{L})} \leq \|R\|_{\mathcal{M}(\mathcal{A})}$. Set $R_2 u = R_1 u$ for $u \in \mathcal{B}$. Then $R_2 u \in \mathcal{B}$, and $\|R_2 u\|_{\mathcal{M}(\mathcal{B})} \leq \|R_1\|_{\mathcal{M}(\mathcal{L})}\|u\|_{\mathcal{M}(\mathcal{L})} \leq \|R\|_{\mathcal{M}(\mathcal{A})}\|\|u\|_{\mathcal{M}(\mathcal{L})}$. Hence $R_2 \in \mathcal{M}(\mathcal{B})$, and $\|R_2\|_{\mathcal{M}(\mathcal{B})} \leq \|R\|_{\mathcal{M}(\mathcal{A})}$.

Now let $S_0 \in U$. we have, for $u \in \mathcal{L}$, since U is stable under products,

$$\|S_0 u\|_{\mathcal{L}} = \sup_{S \in U} \|S_0 S u\| \leq \sup_{S \in U} \|Su\| = \|u\|_{\mathcal{L}},$$

and so $S_0 u \in \mathcal{L}$, and $\|S_0\|_{\mathcal{M}(\mathcal{L})} \leq 1$. This implies that $S_0(\mathcal{B}) \subset \mathcal{B}$, so that $S_0 \in \mathcal{M}(\mathcal{B})$, and $\|S\|_{\mathcal{M}(\mathcal{B})} \leq \|S_0\|_{\mathcal{M}(\mathcal{L})} \leq 1$, which proves (iii). □

We have the following very easy observation.

PROPOSITION 2.5. : *Let* $\mathcal{A}_0, \mathcal{A}_1$ *and* $\mathcal{A}_2$ *be weakly cancellative commutative Banach algebras, and assume that* $\Phi_0 : \mathcal{A}_0 \to \mathcal{A}_1$ *is a* $\mathcal{QM}$*-homomorphism with respect to* $u_0 \in \Omega(\mathcal{A}_0)$ *and that* $\Phi_1 : \mathcal{A}_1 \to \mathcal{A}_2$ *is a* $\mathcal{QM}$*-homomorphism with respect to* $u_1 \in \Omega(\mathcal{A}_1)$. *Then* $\Phi_1 \circ \Phi_0 : \mathcal{A}_0 \to \mathcal{A}_2$ *is a* $\mathcal{QM}$*-homomorphism with respect to* $\Phi_0^{-1}(\Phi_0(u_0)u_1)$.

PROOF. The homomorphism $\Phi_1 \circ \Phi_0$ is one-to-one and has dense range, and it follows from Proposition 2.4 (ii) that $\Phi_0^{-1}(\Phi_0(u_0)u_1) \in \Omega(\mathcal{A}_0)$. Let $u \in \mathcal{A}_2$, let $v \in \mathcal{A}_2$ be such that $\Phi_1(v) = \Phi_1(u_1)u$, and let $w \in \mathcal{A}_0$ be such that $\Phi_0(w) = \Phi_0(u_0)v$. Then

$$\begin{aligned}(\Phi_1 \circ \Phi_0)\left(\Phi_0^{-1}(\Phi_0(u_0)u_1)\right)u &= (\Phi_1 \circ \Phi_0)(u_0)\Phi_1(u_1)u = \Phi_1(\Phi_0(u_0)v) \\ &= (\Phi_1 \circ \Phi_0)(w) \subset (\Phi_1 \circ \Phi_0)(\mathcal{A}_0),\end{aligned}$$

and so $\Phi_1 \circ \Phi_0$ is a $\mathcal{QM}$-homomorphism with respect to $\Phi_0^{-1}(\Phi_0(u_0)u_1)$. □

We will use the following result in the study of the resolvent of semigroups.

PROPOSITION 2.6. *Let $\mathcal{A}$ be a weakly cancellative commutative Banach algebra with dense principal ideals, and let $S \in \mathcal{QM}(\mathcal{A})$. If $\lambda_0 - S$ has an inverse $(\lambda_0 I - S)^{-1}$ in $\mathcal{QM}(\mathcal{A})$ which belongs to $\mathcal{A}$ for some $\lambda_0 \in \mathbb{C}$, where I denotes the unit element of $\mathcal{M}(\mathcal{A})$, then $\sigma_{\mathcal{A}}(S)$ is closed, $\lambda I - S$ has an inverse $(\lambda I - S)^{-1}$ in $\mathcal{QM}(\mathcal{A})$ which belongs to $\mathcal{A}$ for every $\lambda \in \mathbb{C} \setminus \sigma_{\mathcal{A}}(S)$, and the $\mathcal{A}$-valued map $\lambda \to (\lambda I - S)^{-1}$ is holomorphic on $\mathbb{C} \setminus \sigma_{\mathcal{A}}(S)$.*

PROOF. If $\mathcal{A}$ is unital, then $\mathcal{QM}(\mathcal{A}) = \mathcal{A}$, and there is nothing to prove. So assume that $\mathcal{A}$ is not unital, and set $\tilde{\mathcal{A}} := \mathcal{A} \oplus \mathbb{C}I$. Then

$$Spec(\tilde{\mathcal{A}}) = \{\chi_0\} \cup \{\tilde{\chi}_{|\tilde{\mathcal{A}}}\}_{\chi \in Spec(\mathcal{A})},$$

where $\chi_0(a + \lambda I) = \lambda$ for $a \in \mathcal{A}, \lambda \in \mathbb{C}$.

Set $a = (\lambda_0 I - S)^{-1} \in \mathcal{A}$. Then $\chi(a) = \tilde{\chi}(a) = \frac{1}{\lambda_0 - \tilde{\chi}(S)}$, so that $\tilde{\chi}(S) = \lambda_0 - \frac{1}{\chi(a)}$ for $\chi \in Spec(\mathcal{A})$. Since $\sigma_{\mathcal{A}}(a) \cup \{0\} = \sigma_{\tilde{\mathcal{A}}}$ is a compact subset of $\mathbb{C}$, this shows that $\sigma_{\mathcal{A}}(S)$ is closed. We have, for $\lambda \in \mathbb{C}$,

$$spec_{\tilde{\mathcal{A}}}(I + (\lambda - \lambda_0)a) = \{1\} \cup \left\{\frac{\lambda - \tilde{\chi}(S)}{\lambda_0 I - \tilde{\chi}(S)}\right\}_{\chi \in Spec(\mathcal{A})}.$$

So $I + (\lambda - \lambda_0)a$ is invertible in $\tilde{\mathcal{A}}$ for $\lambda \in \mathbb{C} \setminus \sigma_{\mathcal{A}}(S)$, and the $\mathcal{A}$-valued map $\lambda \to a(I + (\lambda - \lambda_0)a)^{-1}$ is holomorphic on $\mathbb{C} \setminus \sigma_{\mathcal{A}}(S)$. We have, for $\lambda \in \mathbb{C} \setminus \sigma_{\mathcal{A}}(S)$,

$$\begin{aligned}(\lambda I - S)a(I + (\lambda - \lambda_0)a)^{-1} &= ((\lambda - \lambda_0)I + (\lambda_0 I - S))a(I + (\lambda - \lambda_0)a)^{-1} \\ &= (I + (\lambda - \lambda_0)a)(I + (\lambda - \lambda_0)a)^{-1} = I.\end{aligned}$$

Hence $\lambda I - S$ has an inverse $(\lambda I - S)^{-1} \in \mathcal{A}$ for $\lambda \in \mathbb{C} \setminus \sigma_{\mathcal{A}}(S)$, and the map $\lambda \to (\lambda I - S)^{-1} = a(I + (\lambda - \lambda_0)a)^{-1}$ is holomorphic on $\mathbb{C} \setminus \sigma_{\mathcal{A}}(S)$. □

3. Normalization of a commutative Banach algebra with respect to a strongly continuous semigroup of multipliers

A semigroup $T = (T(t))_{t>0}$ of multipliers on a commutative Banach algebra $\mathcal{A}$ is said to be strongly continuous if the map $t \to T(t)u$ is continuous on $(0, +\infty)$ for every $u \in \mathcal{A}$. This implies that $\sup_{\alpha \le t \le \beta} \|T(t)\| < +\infty$ for $0 < \alpha \le \beta < +\infty$, and so $\|T(t)\|^{\frac{1}{t}}$ has a limit ρ_T as $t \to +\infty$. In the remainder of the section $T = (T(t))_{t>0}$ will denote a strongly continuous group of multipliers on a weakly cancellative commutative Banach algebra $\mathcal{A}$ such that $\cup_{t>0} T(t)\mathcal{A}$ is dense in $\mathcal{A}$. Hence if $u \in \cap_{t>0} \ker(T(t))$, then $uv = 0$ for every $v \in \cup_{t>0} T(t)\mathcal{A}$, so $u\mathcal{A} = \{0\}$ and $u = 0$.

Notice that in this situation if $\mathcal{A}$ has a unit element 1 then if we set

$$\tilde{T}(t) := (T(t).1)_{t>0},$$

then $\tilde{T}$ is a norm-continuous semigroup of elements of $\mathcal{A}$.

Since $\cup_{t>0} T(t)\mathcal{A}$ is dense in $\mathcal{A}$, $T(t_0).1$ is invertible in $\mathcal{A}$ for some $t_0 > 0$, and so $\lim_{t \to 0^+} \|T(t).1 - 1\| = 0$, which implies that the generator of $\tilde{T}$ is bounded. So there exist $R \in \mathcal{M}(\mathcal{A}) \approx \mathcal{A}$ such that $T(t) = e^{tR}$ for $t > 0$ if $\mathcal{A}$ is unital.

Let ω be a positive measurable weight on $(0, +\infty)$, and denote by $L^1_\omega(\mathbb{R}^+)$ the space of all (classes of) measurable functions on $[0, +\infty)$ satisfying the condition $\|f\|_\omega := \int_0^{+\infty} \|f(t)|\omega(t)dt < +\infty$, so that $(L^1_\omega(\mathbb{R}^+), \|.\|_\omega)$ is a Banach space. Recall that if $\omega(s + t) \le \omega(s)\omega(t)$ for $s > 0, t > 0$, then $L^1_\omega(\mathbb{R}^+)$ is a Banach algebra with respect to convolution.

Set $\omega_T(t) = \|T(t)\|$ for $t > 0$. For $f \in L^1_{\omega_T}(\mathbb{R}^+)$, define $\Phi_T(f) \in \mathcal{M}(\mathcal{A})$ by the formula

$$\Phi_T(f)u = \int_0^{+\infty} f(t)T(t)udt \quad (u \in \mathcal{A}). \tag{2}$$

Denote by $\mathcal{I}_T$ the closure of $\Phi_T(L^1_{\omega_T}(\mathbb{R}^+))$ in $\mathcal{M}(\mathcal{A})$. Let $(f_n)_{n\geq 1}$ be a Dirac sequence, i.e. a sequence (f_n) of nonnegative integrable functions on $\mathbb{R}^+$ such that $\int_0^{+\infty} f_n(t)dt = 1$ and such that $f_n(t) = 0$ a.e. on $(\alpha_n, +\infty)$ with $\lim_{n\to+\infty} \alpha_n = 0$. Since the semigroup T is strongly continuous on $\mathcal{A}$, a standard argument shows that $\lim_{n\to+\infty} \|\Phi_T(f_n * \delta_t)u - T(t)u\| = 0$ for every $t > 0$ and every $u \in \mathcal{A}$. This shows that if $v \in \Omega(\mathcal{I}_T)$, and if $w \in \Omega(\mathcal{A})$, then $vw \in \Omega(\mathcal{A})$.

The following result is then a consequence of Theorem 6.8 of **[14]** and of Proposition 5.4 of **[17]**.

LEMMA 3.1. *There exists $w \in \Omega(\mathcal{A})$ such that* $\limsup_{t\to 0^+} \|T(t)w\| < +\infty$.

PROOF. Let $\lambda > \log(\rho_T)$, and set $\omega_\lambda(t) = e^{\lambda t}\sup_{s\geq t} e^{-\lambda s}\|T(s)\|$ for $t > 0$. An extension to lower semicontinuous weights of Theorem 6.8 of **[14]** given in **[17]** shows that $\Omega(L^1_{\omega_\lambda}(\mathbb{R}^+)) \neq \emptyset$. It follows also from Proposition 5.4 of **[17]** that $\|\Phi_T(g)T(t)\| \leq e^{\lambda t}\|g\|_{L^1_{\omega_\lambda}}$ for every $g \in L^1_{\omega_\lambda}(\mathbb{R}^+)$ and every $t > 0$, and that $\Phi_T(g) \in \Omega(\mathcal{I}_T)$ for every $g \in \Omega\left(L^1_{\omega_\lambda}(\mathbb{R}^+)\right)$. Hence if $g \in \Omega\left(L^1_{\omega_\lambda}(\mathbb{R}^+)\right)$ and if $v \in \Omega(\mathcal{A})$, then $\Phi_T(g)v \in \Omega(\mathcal{A})$, and $\limsup_{t\to 0^+} \|T(t)\Phi_T(g)v\| < +\infty$. □

The following result is a version specific to one-parameter semigroups of Proposition 2.4.

PROPOSITION 3.2. *Let $T := (T(t))_{t>0}$ be a strongly continuous of multipliers on a weakly cancellative commutative Banach algebra $\mathcal{A}$ with dense principal ideals such that $\cup_{t>0}T(t)\mathcal{A}$ is dense in $\mathcal{A}$.*

Let $\mathcal{L}_T := \{u \in \mathcal{A} \mid \limsup_{t\to 0^+} \|T(t)u\| < +\infty\} \supset \cup_{t>0}T(t)\mathcal{A}$, choose $\lambda > \log(\rho_T)$, set $\|u\|_\lambda := sup_{s\geq 0}e^{-\lambda s}\|T(s)u\|$ for $u \in \mathcal{L}_T$, with the convention $T(0) = I$, and set $\|R\|_{\lambda,op} := \sup\{\|Ru\|_\lambda \mid u \in \mathcal{L}_T, \|u\|_\lambda \leq 1\} = \|R\|_{\mathcal{M}(\mathcal{L}_T)}$ for $R \in \mathcal{M}(\mathcal{L}_T)$. Denote by $\mathcal{A}_T$ the closure of $\mathcal{A}$ in $(\mathcal{M}(\mathcal{L}_T), \|.\|_{\lambda,op})$.

Then $(\mathcal{L}_T, \|.\|_\lambda)$ is a Banach algebra, the norm topology on $\mathcal{L}_T$ does not depend on the choice of λ, and the following properties hold

(i) The inclusion map $j : \mathcal{A} \to \mathcal{A}_T$ is a $\mathcal{QM}$-homomorphism with respect to w for every $\omega \in \Omega(\mathcal{A}) \cap \mathcal{L}_T$, the tautological map $\tilde{j} : S_{u/v} \to S_{u/v}$ is a pseudobounded isomorphism from $\mathcal{QM}(\mathcal{A})$ onto $\mathcal{QM}(\mathcal{A}_T)$ and if $w \in \Omega(\mathcal{A}) \cap \mathcal{L}_T$, then $\tilde{j}^{-1}(S) = S_{uw/vw}$ for $S = S_{u/v} \in \mathcal{QM}(\mathcal{A}_T)$.

(ii) $\tilde{j}(\mathcal{M}(\mathcal{A})) \subset \mathcal{M}(\mathcal{A}_T)$, and $\|R\|_{\mathcal{M}(\mathcal{A}_T)} \leq \|R\|_{\mathcal{M}(\mathcal{A})}$ for $R \in \mathcal{M}(\mathcal{A})$.

(iii) $\|T(t)\|_{\mathcal{M}(\mathcal{A}_T)} \leq \|T(t)\|_{\lambda,op} \leq e^{\lambda t}$ for $t > 0$, and we have, for $u \in \mathcal{A}_T$,

$$\limsup_{t\to 0^+} \|T(t)u - u\|_{\lambda,op} = 0.$$

PROOF. It follows from Lemma 3.1 that the family $U = \{e^{-\lambda t}T(t)\}_{t>0}$ is pseudobounded for $\lambda > \log(\rho_T)$. The fact that $(\mathcal{L}_T, \|.\|_\lambda)$ is a Banach algebra, and assertions (i) and (ii) follow from Proposition 2.4 and Proposition 2.5 applied to U, and an elementary argument given in the proof of Theorem 7.1 of **[17]** shows that there exists $k > 0$ and $K > 0$ such that $k\|u\|_\lambda \leq \sup_{0\leq t\leq 1} \|T(t)u\| \leq K\|u\|_\lambda$ for

$u \in \mathcal{L}_T$, which shows that the norm topology on $\mathcal{L}_T$ does not depend on the choice of λ.

It follows also from Proposition 2.4 applied to U that

$$\|T(t)\|_{\mathcal{M}(\mathcal{A}_T)} \leq \|T(t)\|_{\lambda,op} \leq e^{\lambda t}$$

for $t > 0$, and $\lim_{t\to 0^+} \|T(t)u - u\|_{\lambda,op} = 0$ for every $u \in \cup_{t>0} T(t)\mathcal{A}$.

Since $\cup_{t>0} T(t)\mathcal{A}_T$ is dense in $\mathcal{A}_T$, a standard density argument shows that $\lim_{t\to 0^+} \|T(t)u - u\|_{\lambda,op} = 0$ for every $u \in \mathcal{A}_T$. □

This suggests the following definition.

DEFINITION 3.3. *Let $\mathcal{A}$ be a weakly cancellative commutative Banach algebra with dense principal ideals, let $T = ((T(t))_{t>0}$ be a strongly continuous semigroup of multipliers on $\mathcal{A}$ such that $\cup_{t>0} T(t)\mathcal{A}$ is dense in $\mathcal{A}$ for $t > 0$. A normalization $\mathcal{B}$ of $\mathcal{A}$ with respect to T is a subalgebra $\mathcal{B}$ of $\mathcal{QM}(\mathcal{A})$ which is a Banach algebra with respect to a norm $\|.\|_{\mathcal{B}}$ and satisfies the following conditions.*

(i) We have $\|j(u)\|_{\mathcal{B}} \leq \|u\|_{\mathcal{A}}$ for $u \in \mathcal{A}$, and the inclusion map $j : \mathcal{A} \to \mathcal{B}$ is a $\mathcal{QM}$-homomorphism.

(ii) $\tilde{j}(R) \subset \mathcal{M}(\mathcal{B})$, and $\|\tilde{j}(R)\|_{\mathcal{M}(\mathcal{B})} \leq \|R\|_{\mathcal{M}(\mathcal{A})}$ for $R \in \mathcal{M}(\mathcal{A})$, where $\tilde{j} : \mathcal{QM}(\mathcal{A}) \to \mathcal{QM}(\mathcal{B})$ is the pseudobounded isomorphism associated to j introduced in Proposition 2.3 (ii).

(iii) $\limsup_{t\to 0^+} \|\tilde{j}(T(t))\|_{\mathcal{M}(\mathcal{B})} < +\infty$.

For example the algebra $\mathcal{A}_T$ constructed in Proposition 3.2 is a normalization of the given Banach algebra $\mathcal{A}$ with respect to the semigroup T. Notice that if $\mathcal{B}$ is a normalization of $\mathcal{A}$ with respect to T, the same density argument as above shows that $\lim_{t\to 0^+} \|T(t)u - u\|_{\mathcal{B}} = 0$ for every $u \in \mathcal{B}$.

4. Normalization of a commutative Banach algebra with respect to a holomorphic semigroup of multipliers

For $a < b \leq a+\pi$, denote by $S_{a,b}$ the open sector $\{z \in \mathbb{C}\setminus\{0\} \mid a < arg(z) < b\}$, with the convention $\overline{S}_{a,a} = \{re^{ia} \mid r \geq 0\}$. In this section we consider again a weakly cancellative commutative Banach algebra $\mathcal{A}$ with dense principal ideals and we consider a semigroup $T = (T(\zeta))_{\zeta\in S_{a,b}}$ of multipliers on $\mathcal{A}$ such that $\cup_{\zeta\in S_{a,b}} T(\zeta)\mathcal{A}$ is dense in $\mathcal{A}$ which is holomorphic on $S_{a,b}$, which implies that $T(\zeta)\mathcal{A}$ is dense in $\mathcal{A}$ for every $\zeta \in S_{a,b}$. So $T(\zeta)u \in \Omega(\mathcal{A})$ for every $\zeta \in S_{a,b}$ and every $u \in \Omega(\mathcal{A})$. We state as a lemma the following easy observations.

LEMMA 4.1. *Let $\alpha, \beta \in \mathbb{R}$ such that $a < \alpha \leq \beta < b$, set $\gamma := \frac{\beta-\alpha}{2}$, and let $u \in \mathcal{A}$ such that* $\limsup_{\substack{\zeta\to 0\\ \zeta\in\overline{S}_{a,b}}} \|T(\zeta)\| < +\infty$.

(i) If $\lambda > \frac{1}{\cos(\gamma)} \lim_{r\to+\infty} r^{-1}\log(\max(\|T(re^{i\alpha}), \|T(re^{i\beta}\|))$ *then*

$$\sup_{\zeta\in\overline{S}_{\alpha,\beta}} \|e^{-\lambda\zeta e^{-i\frac{\alpha+\beta}{2}}} T(\zeta)u\| < +\infty.$$

$$(ii) \limsup_{\lambda\to+\infty}\left[\sup_{\zeta\in\overline{S}_{\alpha,\beta}} \|e^{-\lambda\zeta e^{-i\frac{\alpha+\beta}{2}}} T(\zeta)u\|\right] = \limsup_{\substack{\zeta\to 0\\ \zeta\in\overline{S}_{\alpha,\beta}}} \|T(\zeta)u\|.$$

PROOF. We have, for $r_1 \geq 0, r_2 \geq 0$,

$$\begin{aligned} &\left\| e^{-\lambda e^{-i\frac{\alpha+\beta}{2}}(r_1 e^{i\alpha}+r_2 e^{i\beta})} T(r_1 e^{i\alpha} + r_2 e^{i\beta})u \right\| \\ \leq\ & \inf \Big[e^{-\lambda\cos(\gamma) r_1} \|T(r_1 e^{i\alpha})u\| e^{-\lambda \cos(\gamma) r_2} \|T(r_2 e^{i\beta})\|, \\ & \qquad e^{-\lambda\cos(\gamma) r_1} \|T(r_1 e^{i\alpha})\| e^{-\lambda \cos(\gamma) r_2} \|T(r_2 e^{i\beta})u\| \Big]. \end{aligned}$$

Set $m := \limsup_{\substack{\zeta\to 0 \\ \zeta \in \overline{S}_{\alpha,\beta}}} \|T(t)u\|$, and let $\epsilon > 0$. There exists $\delta > 0$ such that $\|T(\zeta)u\| \leq m + \epsilon$ for $0 \leq |\zeta| \leq \delta, \zeta \in \overline{S}_{\alpha,\beta}$.

We obtain, considering the cases

$$\sup(r_1, r_2) \leq \delta/2, \inf(r_1, r_2) \leq \delta/2 \text{ and } \sup(r_1, r_2) \geq \frac{\delta}{2},$$

and the case where $\inf(r_1, r_2) \geq \frac{\delta}{2}$,

$$\begin{aligned} &\sup_{\zeta \in \overline{S}_{\alpha,\beta}} \|e^{-\lambda \zeta e^{i\frac{\alpha+\beta}{2}}} T(\zeta)u\| \\ \leq\ & \max \left[(m+\epsilon) \sup_{\substack{|\zeta|\leq\delta \\ \zeta\in\overline{S}_{\alpha,\beta}}} |e^{-\lambda\zeta}| \left(1 + \sup_{r \geq \delta/2} e^{-\lambda r\cos(\gamma)} \max(\|T(re^{i\alpha})\|, \|T(re^{i\beta})\| \right), \right. \\ & \left. \|u\| \left(\sup_{r\geq\delta/2} e^{-\lambda r \cos(\gamma)} \|T(re^{i\alpha})\| \right) \left(\sup_{r\geq\delta/2} e^{-\lambda r\cos(\gamma)} \|T(re^{i\beta})\| \right) \right], \end{aligned}$$

and (i) and (ii) follow from this inequality. □

We now follow a method used in Proposition 3.6 of [8] to construct a $\mathcal{QM}$-homomorphism from $\mathcal{A}$ into a weakly cancellative commutative Banach algebra $\mathcal{B}$ such that $\sup_{\substack{t\in\overline{S}_{\alpha,\beta} \\ |t|\leq 1}} \|T(t)\| < +\infty$ for every α, β satisfying $a < \alpha \leq \beta < b$. The following result is more precise than Proposition 3.6 of [8].

PROPOSITION 4.2. *Let $\mathcal{A}$ be a weakly cancellative commutative Banach algebra with dense principal ideals, and let $T = ((T(\zeta))_{\zeta\in S_{a,b}}$ be a holomorphic semigroup of multipliers on $\mathcal{A}$ such that $T(\zeta)\mathcal{A}$ is dense in $\mathcal{A}$ for $\zeta \in S_{a,b}$. Set $\alpha_n = a + \frac{b-a}{2(n+1)}$, $\beta_n := b - \frac{b-a}{2(n+1)}$ for $n \geq 0$, and let $\mu = (\mu_n)_{n\geq 0}$ be a nondecreasing sequence of positive real numbers satisfying the following conditions*

$$\mu_n > \frac{1}{\cos\left(\frac{\beta_n - \alpha_n}{2}\right)} \lim_{r\to+\infty} \frac{\log(\max(\|T(re^{i\alpha_n})\|, \|(T(re^{i\beta_n})\|))}{r} \quad (n \geq 0), \tag{3}$$

and

$$\sup_{\zeta\in\overline{S}_{\alpha_n,\beta_n}} \|e^{-\mu_n\zeta e^{-i\frac{a+b}{2}}} T(\zeta + 2^{-n} e^{i\frac{a+b}{2}})\|_{\mu_0, op} \leq e^{2^{-n}(\mu_0+1)} \quad (n \geq 1), \tag{4}$$

where $\|.\|_{\mu_0,op}$ is the norm on the normalization $\mathcal{A}_{T_0}$ of $\mathcal{A}$ with respect to the semigroup $T_0 := (T(te^{i\frac{(a+b)}{2}}))_{t>0}$ associated to μ_0 in Theorem 2.2.

For $n \geq 1$ set $V_n := \{e^{-\mu_n \zeta e^{-i\frac{(\alpha_n+\beta_n)}{2}}} T(\zeta)\}_{\zeta\in\overline{S}_{\alpha_n,\beta_n}}$, set $W_n := V_1 \dots V_n$ and set $W = \cup_{n\geq 1} W_n$, so that W is stable under products.

Set

$$\mathcal{L}_\mu := \{u \in \mathcal{A}_{T_0} \mid \sup_{w\in W} \|wu\|_{\mu_0,op} < +\infty, \|u\|_\mu = \sup_{w\in W\cup\{I\}} \|wu\|_{\mu_0,op} (u \in \mathcal{L}_\mu).$$

Denote by $\mathcal{A}_{\mu,T}$ the closure of $\mathcal{A}_{T_0}$ in $\mathcal{M}(\mathcal{L}_\mu)$.

Then $T(\zeta)u \in \mathcal{L}_\mu$ for $\zeta \in S_{a,b}, u \in \mathcal{A}_{T_0}$, $(\mathcal{L}_\mu, \|.\|_\mu)$ is a Banach algebra, the inclusion map $j : \mathcal{A} \to \mathcal{A}_{\mu,T}$ is a $\mathcal{QM}$-homomorphism with respect to $T(\zeta)w^2$ for $w \in \Omega(\mathcal{A})$, and we have the following properties

(i) The tautological map $\tilde{j} : S_{u/v} \to S_{u/v}$ is a pseudobounded isomorphism from $\mathcal{QM}(\mathcal{A})$ onto $\mathcal{QM}(\mathcal{A}_{\mu,T})$ and if $\zeta \in S_{a,b}$ then $\tilde{j}^{-1}(S) = S_{T(\zeta)w^2u/T(\zeta)w^2v}$ for $S = S_{u/v} \in \mathcal{QM}(\mathcal{A}_{\mu,T})$, $w \in \Omega(\mathcal{A})$.

(ii) $\tilde{j}(\mathcal{M}(\mathcal{A})) \subset \mathcal{M}(\mathcal{A}_{\mu,T})$ and $\|R\|_{\mathcal{M}(\mathcal{A}_{\mu,T})} \leq \|R\|_{\mathcal{M}(\mathcal{A})}$ for $R \in \mathcal{M}(\mathcal{A})$.

(iii) $\|T(\zeta)\|_{\mathcal{M}(\mathcal{A}_{\mu,T})} \leq \|T(\zeta)\|_{\mathcal{M}(\mathcal{L}_\mu)} \leq e^{\mu_n Re(\zeta e^{-i\frac{a+b}{2}})}$ for $\zeta \in \overline{S}_{\alpha_n,\beta_n}, n \geq 1$,

(iv) If $a < \alpha < \beta < b$, then we have, for $v \in \mathcal{A}_{\mu,T}$,

$$\limsup_{\substack{\zeta\to 0\\ \zeta\in S_{\alpha,\beta}}} \|T(\zeta)v - v\|_{\mathcal{M}(\mathcal{A}_{\mu,T})} \leq \limsup_{\substack{\zeta\to 0\\ \zeta\in S_{\alpha,\beta}}} \|T(\zeta)v - v\|_{\mathcal{M}(\mathcal{L}_\mu)} = 0.$$

PROOF. Since $\|T_0(2^{-n})\|_{\mu_0,op} \leq e^{2^{-n}\mu_0}$, the existence of a sequence $(\mu_n)_{n\geq 1}$ satisfying the required conditions follows from the lemma.

Let $t > 0$, let $n_0 \geq 2$ be such that $2^{-n_0+1} < t$, and let $v \in W$. Since $1 \in V_n$ for $n \geq 1$, we can assume that $v = v_1 \dots v_n$, where $n \geq n_0$ and where $v_j \in V_j$ for $1 \leq j \leq n$. Then $\|T_0(2^{-j})v_j\|_{\mu_0,op} \leq e^{2^{-j}(\mu_0+1)}$ for $j \geq n_0$, and so $\|T_0(2^{-n_0+1} - 2^{-n})v_{n_0} \dots v_n\|_{\mu_0,op} \leq e^{(2^{-n_0+1}-2^{-n})(\mu_0+1)}$. We obtain

$$\begin{aligned}
&\|T_0(2^{-n_0+1})v_{n_0} \dots v_n\|_{\mu_0,op}\\
\leq\ & \|T_0(2^{-n})\|_{\mu_0,op}\|T((2^{-n_0+1} - 2^{-n}))v_{n_0} \dots v_n\|_{\mu_0,op}\\
\leq\ & e^{2^{-n_0+1}(\mu_0+1)-2^{-n}} \leq e^{2^{-n_0+1}(\mu_0+1)}.
\end{aligned}$$

Set $r = \frac{t-2^{-n_0+1}}{n_0-1}$. It follows from Lemma 4.1(i) that for every $j \leq n_0 - 1$ there exists $k_j > 0$ such that $sup_{v\in V_j}\|T(r)v\|_{\mu_0,op} \leq k_j$. This gives

$$\begin{aligned}
\|T_0(t)v\|_{\mu_0,op} &\leq \|T_0(r)v_1\|_{\mu_0,op} \dots \|T_0(r)v_{n_0-1}\|_{\mu_0,op}\|T_0(2^{-n_0+1})v_{n_0} \dots v_n\|_{\mu_0,op}\\
&\leq k_1 \dots k_{n_0-1}e^{2^{-n_0+1}(\mu_0+1)},
\end{aligned}$$

and so $T_0(t)\mathcal{A}_{T_0} \subset \mathcal{L}_\mu$. Now let $\zeta \in S_{a,b}$. Then $\zeta - te^{i\frac{a+b}{2}} \in S_{a,b}$ for some $t > 0$, and so $T(\zeta)\mathcal{A}_{T_0} = T(\zeta - te^{i\frac{a+b}{2}})T_0(t)\mathcal{A}_{T_0} \subset T(\zeta - te^{i\frac{a+b}{2}})\mathcal{L}_\mu \subset \mathcal{L}_\mu$.

Since W is stable under products, the fact that $\mathcal{L}_\mu$ is a Banach algebra follows from Proposition 2.4, which also implies (ii) and (iii). Let $\zeta_0 \in S_{a,b}$. Since (iv) holds for $u \in T(\zeta_0)\mathcal{A}_{\mu,T}$, and since $T(\zeta_0)\mathcal{A}_{\mu,T}$ is dense in $(\mathcal{A}_{\mu,T}, \|.\|_{\mathcal{L}_\mu})$, (iv) follows from (iii) by a standard density argument.

Let $\zeta \in S_{a,b}$, and let $w \in \Omega(\mathcal{A}) \subset \Omega(\mathcal{A}_{T_0})$. Since $T(\zeta/2)w \in \Omega(\mathcal{A})$, and since $\limsup_{t\to 0^+} \|T_0(t)T(\zeta)w\| < +\infty$, it follows from Proposition 2.5 that the inclusion map $j_0 : \mathcal{A} \to \mathcal{A}_{T_0}$ is a $\mathcal{QM}$-homomorphism with respect to $T(\zeta/2)w$. Since $T(\zeta/2)w \subset \mathcal{M}(\mathcal{L}_\mu) \cap \Omega(\mathcal{A}_{T_0})$, it follows also from Proposition 2.5 that the inclusion map $j_1 : \mathcal{A}_{T_0} \to \mathcal{A}_{\mu,T}$ is a $\mathcal{QM}$-homomorphism with respect to $T(\zeta/2)w$.

It follows from Proposition 2.5 that the inclusion map $j = j_1 \circ j_0 : \mathcal{A} \to \mathcal{A}_{\mu,T}$ is a $\mathcal{QM}$-homomorphism with respect to $T(\zeta/2)wT(\zeta/2)w = T(\zeta)w^2$.

It follows from Proposition 2.3 that the tautological map $\tilde{j} : S_{u/v} \to S_{u/v}$ is a pseudobounded homomorphism from $\mathcal{QM}(\mathcal{A})$ onto $\mathcal{QM}(\mathcal{A}_{\mu,T})$ and that $\tilde{j}^{-1}(S_{u/v})$ $= S_{T(\zeta)uw^2/T(\zeta)vw^2}$ for $S = S_{u/v} \in \mathcal{QM}(\mathcal{A}_{\mu,T})$.

Since $T(\zeta)\mathcal{A}_{\mu,T} \subset T(\zeta/2)\left[T(\zeta/2)\mathcal{A}_{\mu,T}\right] \subset T(\zeta/2)\mathcal{A}_{T_0} \subset \mathcal{A}$, and since $T(\zeta)v\mathcal{A}$ contains $T(\zeta)vw^2\mathcal{A}$ which is dense in $\mathcal{A}$ for $v \in \Omega(\mathcal{A}_{\mu,T})$, we have $T(\zeta)v \in \Omega(\mathcal{A})$ for $\zeta \in S_{a,b}, v \in \Omega(\mathcal{A}_{\mu,T})$, which shows that $\tilde{j}^{-1}(S) = S_{T(\zeta)uw^2/T(\zeta)vw^2}$ for $S = S_{u/v} \in \mathcal{QM}(\mathcal{A}_{\mu,T})$. □

We will use the following notion.

DEFINITION 4.3. *Let $\mathcal{A}$ be a weakly commutative Banach algebra with dense principal ideals, and let $T := (T(\zeta))_{\zeta \in S_{a,b}}$ be an analytic semigroup of multipliers on $\mathcal{A}$ such that $T(\zeta)\mathcal{A}$ is dense in $\mathcal{A}$ for $\zeta \in S_{a,b}$. A normalization of the algebra $\mathcal{A}$ with respect to the semigroup T is a subalgebra $\mathcal{B}$ of $\mathcal{QM}(\mathcal{A})$ which is a Banach algebra with respect to a norm $\|.\|_{\mathcal{B}}$ and satisfies the following conditions.*

(i) There exists $u \in \Omega(\mathcal{A})$ such that the inclusion map $j : \mathcal{A} \to \mathcal{B}$ is a $\mathcal{QM}$-homomorphism with respect to $T(\zeta)u$ for every $\zeta \in S_{a,b}$, and $\|j(u)\|_{\mathcal{B}} \le \|u\|_{\mathcal{A}}$ for $u \in \mathcal{A}$.

(ii) $\tilde{j}(R) \subset \mathcal{M}(\mathcal{B})$, and $\|\tilde{j}(R)\|_{\mathcal{M}(\mathcal{B})} \le \|R\|_{\mathcal{M}(\mathcal{A})}$ for $R \in \mathcal{M}(\mathcal{A})$, where $\tilde{j} : \mathcal{QM}(\mathcal{A}) \to \mathcal{QM}(\mathcal{B})$ is the pseudobounded isomorphism associated to j introduced in Proposition 2.3 (ii).

(iii) $\limsup_{\substack{t\to 0\\ \zeta\in S_{\alpha,\beta}}} \|\tilde{j}(T(t))\|_{\mathcal{M}(\mathcal{B})} < +\infty$ *for $a < \alpha < \beta < b$.*

If $\mathcal{B}$ is a normalization of $\mathcal{A}$ with respect to the holomorphic semigroup $T = (T(\zeta))_{\zeta \in S_{a,b}}$, a standard density argument shows that if $a < \alpha < \beta < b$ then $\limsup_{\substack{\zeta\to 0\\ \zeta in S_{\alpha,\beta}}} \|T(\zeta)u - u\|_{\mathcal{B}} = 0$ for every $u \in \mathcal{B}$

Notice that the algebra $\mathcal{A}_{0,T} = \mathcal{A}_{\mu_0,\mathbf{T}}$ and its norm topology associated to the norm $\|.\|_{\mu_0,op}$ discussed above do not depend on the choice of μ_0. This is no longer the case for the Banach algebra $\mathcal{A}_{\mu,T}$ and its norm topology, which may depend on the choice of the sequence μ.

In order to get a more intrinsic renormalization one could consider the Fréchet algebra $\mathcal{L} := \cap_{n\ge 1}\{u \in \mathcal{A}_T \mid \sup_{\substack{\zeta\in \overline{S}_{\alpha_n,\beta_n}\\ \|u\|_{\mu_0,op}\le 1}} \|T(\zeta)u\|_{\mu_0,op} < +\infty\}$, then consider the closed subalgebra $\mathcal{U}$ of $\mathcal{L}$ generated by the semigroup and introduce an intrinsic normalization of $\mathcal{A}_T$ to be the closure of $\mathcal{U}$ in $\mathcal{M}(\mathcal{U})$ with respect to the Mackey-convergence associated to a suitable notion of boundedness on subsets of $\mathcal{M}(\mathcal{U})$, but it seems more convenient to adopt the point of view used in Proposition 4.2.

5. Generator of a strongly continuous semigroup of multipliers and Arveson spectrum

In this section we consider a weakly cancellative commutative Banach algebra $\mathcal{A}$ with dense principal ideals and a strongly continuous semigroup $T = (T(t))_{t>0}$ of multipliers on $\mathcal{A}$ such that $\cup_{t>0}T(t)\mathcal{A}$ is dense in $\mathcal{A}$. We set again $\omega_T(t) = \|T(t)\|$ for $t > 0$. Denote by $\mathcal{M}_{\omega_T}$ the space of all measures ν on $(0, +\infty)$ such that $\int_0^{+\infty} \omega_T(t)d|\nu|(t) < +\infty$, and for $\nu \in \mathcal{M}_{\omega_T}$ define $\Phi_T : \mathcal{M}_{\omega_T} \to \mathcal{M}(\mathcal{A})$ by the formula

$$\Phi_T(\nu)u := \int_0^{+\infty} T(t)u\, d\nu(t) \ \ (u \in \mathcal{A}). \tag{2'}$$

The Bochner integral is well-defined since the semigroup is strongly continuous, $\mathcal{M}_{\omega_T}$ is a Banach algebra with respect to convolution of measures on the half-line, and we will identify again the space $L^1_{\omega_T}$ of (classes of) measurable functions on $[0,+\infty)$ satisfying $\int_0^{+\infty} |f(t)| \|T(t)\| dt < +\infty$ to the ideal of all $\nu \in \mathcal{M}_{\omega_T}$ which are absolutely continuous with respect to Lebesgue measure. Denote by $\mathcal{B}_T$ the closure of $\phi_T(\mathcal{M}_\omega)$ in $\mathcal{M}(\mathcal{A})$, and denote again by $\mathcal{I}_T$ the closure of $\Phi_T(L^1_{\omega_T})$ in $\mathcal{M}(\mathcal{A})$, so that the "Arveson ideal" $\mathcal{I}_T$ is a closed ideal of $\mathcal{B}_T$.

The idea of considering the generator of a semigroup as a quasimultiplier on some suitable Banach algebra goes back to **[22]** and **[23]** for groups of bounded operators and, more generally, for groups of regular quasimultipliers. An obvious such interpretation was given by I. Chalendar, J. R. Partington and the author in **[8]** for analytic semigroups, and the author interpreted in Section 8 of **[17]** the generator of a semigroup of bounded operators which is weakly continuous in the sense of Arveson **[2]** as a quasimultiplier on the corresponding Arveson ideal $\mathcal{I}_T$. Since in the present context $\Phi_T\left(\Omega(L^1_{\omega_T})\right) \subset \Omega(\mathcal{I}_T)$ and $uv \in \Omega(\mathcal{A})$ for $u \in \Omega(\mathcal{I}_T), v \in \Omega(\mathcal{A})$, the map $j_T : S_{u/v} \to S_{uw/vw}$ is a pseudobounded homomorphism from $\mathcal{QM}(\mathcal{I}_T)$ into $\mathcal{QM}(\mathcal{A})$ for every $w \in \Omega(\mathcal{A})$, and the definition of j_T does not depend on the choice of w.

The generator $\Delta_{T,\mathcal{I}_T}$ of T considered as a strongly continuous semigroup of multipliers on $\mathcal{I}_T$ has been defined in **[17]**, def. 8.1 by the formula

$$\Delta_{T,\mathcal{I}_T} = S_{-\Phi_T(f_0')/\Phi_T(f_0)}, \tag{5}$$

where $f_0 \in \mathcal{C}^1([0,+\infty)) \cap \Omega\left(L^1_{\omega_T}\right)$ satisfies $f_0(0) = 0$, $f_0' \in L^1_{\omega_T}$,

and an easy verification given in **[17]** shows that this definition does not depend on the choice of f_0. This suggests the following definition.

DEFINITION 5.1. *: The infinitesimal generator $\Delta_{T,\mathcal{A}}$ of T is the quasimultiplier on $\mathcal{A}$ defined by the formula*

$$\Delta_{T,\mathcal{A}} = j_T(\Delta_{T,\mathcal{I}_T}) = S_{-\Phi_T(f_0')u_0/\Phi_T(f_0)u_0},$$

where $f_0 \in \mathcal{C}^1([0,+\infty)) \cap \Omega\left(L^1_{\omega_T}\right)$ satisfies $f_0(0) = 0$, $f_0' \in L^1_{\omega_T}$, and where $u_0 \in \Omega(\mathcal{A})$.

Assume that f_1 and u_1 also satisfy the conditions of the definition, and set $f_2 = f_0 * f_1, u_2 = u_0u_1$ Since $\Omega\left(L^1_{\omega_T}\right)$ is stable under convolution, $f_2 \in \Omega\left(L^1_{\omega_T}\right)$, and $f_2' = f_0' * f_1 = f_0 * f_1' \in L^1_{\omega_T}$ is continuous. Also $f_2(0) = 0$, and we have

$$\Phi_T(f_2')u_2\Phi_T(f_0)u_0 = \Phi_T(f_0')\Phi_T(f_1)u_2\Phi_T(f_0)u_0 = \Phi_T(f_0')u_0\Phi_T(f_2)u_2$$

and similarly $\Phi_T(f_2')u_2\Phi_T(f_1)u_1 = \Phi_T(f_1')u_1\Phi_T(f_2)u_2$, which shows that the definition of $\Delta_{\mathbf{T},\mathcal{A}}$ does not depend on the choice of f_0 and u_0.

We now give a link between the quasimultiplier approach and the classical approach based on the study of $\frac{T(t)u-u}{t}$ as $t \to 0^+$. A proof of the following folklore result is given for example in **[17]**, Lemma 8.4.

LEMMA 5.2. *Let ω be a lower semicontinuous submultiplicative weight on $(0,+\infty)$, and let $f \in \mathcal{C}^1([0,+\infty)) \cap L^1_\omega$. If $f(0)=0$, and if $f' \in L^1_\omega$, then the Bochner integral $\int_t^{+\infty}(f' * \delta_s)ds$ is well-defined in L^1_ω for $t \geq 0$, and we have*

$$f * \delta_t - f = -\int_0^t (f' * \delta_s)ds, \quad \text{and} \lim_{t\to 0^+} \left\| \frac{f*\delta_t - f}{t} + f' \right\|_{L^1_\omega} = 0.$$

It follows from the lemma that we have if $f \in \mathcal{C}^1([0,+\infty)) \cap L^1_{\omega_T}$, and if $f(0)=0, f' \in L^1_{\omega_T}$,

$$T(t)\Phi_T(f) - \Phi_T(f) = -\int_0^t T(s)\Phi_T(f')ds \quad (t \geq 0). \tag{6}$$

PROPOSITION 5.3. *(i) Let $u \in \mathcal{A}$. If $\lim_{t\to 0^+} \|\frac{T(t)u-u}{t} - v\| = 0$ for some $v \in \mathcal{A}$, then $u \in \mathcal{D}_{\Delta_{T,\mathcal{A}}}$, and $\Delta_{T,\mathcal{A}}u = v$.*

(ii) Conversely if $u \in \mathcal{D}_{\Delta_{T,\mathcal{A}}}$, and if $\lim_{t\to 0^+} \|T(t)\Delta_{T,\mathcal{A}}u - \Delta_{T,\mathcal{A}}u\| = 0$, then $\limsup_{t\to 0^+} \|\frac{T(t)u-u}{t} - \Delta_{T,\mathcal{A}}u\| = 0$.

PROOF. (i) If $u \in \mathcal{A}$, and if $\lim_{t\to 0^+} \|\frac{T(t)u-u}{t} - v\| = 0$ for some $v \in \mathcal{A}$, let $f_0 \in \mathcal{C}^1([0,+\infty)) \cap \Omega\left(L^1_{\omega_T}\right)$ satisfying $f_0(0)=0$, $f_0' \in L^1_{\omega_T}$, and let $u_0 \in \Omega(\mathcal{A})$. It follows from the lemma that we have, with respect to the norm topology on $\mathcal{A}$,

$$\begin{aligned}
-\Phi_T(f_0')u_0u &= \left[\lim_{t\to 0^+} \frac{T(t)\Phi_T(f_0) - \Phi_T(f_0)}{t}\right] u_0u \\
&= \Phi_T(f_0)u_0 \left[\lim_{t\to 0^+} \frac{T(t)u-u}{t}\right] \\
&= \Phi_T(f_0)u_0v,
\end{aligned}$$

and so $u \in \mathcal{D}_{\Delta_{T,\mathcal{A}}}$, and $\Delta_{T,\mathcal{A}}u = v$.

(ii) Conversely let $u \in \mathcal{D}_{\Delta_{T,\mathcal{A}}}$, let $f_0 \in \mathcal{C}^1([0,+\infty)) \cap \Omega\left(L^1_{\omega_T}\right)$ such that $f_0(0) = 0, f_0' \in L^1_{\omega_T}$, and let $u_0 \in \Omega(\mathcal{A})$. Set $v = \Delta_{T,\mathcal{A}}u$, and assume that $\lim_{t\to 0^+} \|T(t)v - v\| = 0$. It follows from (6) that we have, for $t \geq 0$,

$$\begin{aligned}
\Phi_T(f_0)u_0 \int_0^t T(s)vds &= \int_0^t T(s)\Phi_T(f_0)vu_0ds \\
&= -\int_0^t T(s)\Phi_T(f_0')uu_0ds \\
&= -\left[\int_0^t T(s)\Phi_T(f_0')ds\right] u_0u \\
&= [T(t)\Phi_T(f_0)u_0 - \Phi_T(f_0)u_0]\, u \\
&= \Phi_T(f_0)u_0(T(t)u-u).
\end{aligned}$$

Since $\Phi_T(f_0)u_0 \in \Omega(\mathcal{A})$, this shows that $T(t)u - u = \int_0^t T(s)vds$, and so $\lim_{t\to 0^+} \left\|\frac{T(t)u-u}{t} - v\right\| = 0$. □

Let $T_1 = (T_1(t))_{t>0}$ and $T_2 = (T_2(t))_{t>0}$ be two strongly continuous semigroups of multipliers on $\mathcal{A}$ such that $\cup_{t>0}T_1(t)\mathcal{A}$ and $\cup_{t>0}T_2(t)\mathcal{A}$ are dense in $\mathcal{A}$. We will say that T_1 and T_2 *commute* if $T_1(t)T_2(t) = T_2(t)T_1(t)$ for $t > 0$. Then

$T_1(t)T_2(nt) = T_1(t)T_2(t)^n = T_2(t)^nT_1(t) = T_2(nt)T_1(t)$ for $n \geq 1, t > 0$. This implies that $T_1(t)T_2(\frac{m}{n}t) = T_2(\frac{m}{n}t)T_1(t)$ for $m \geq 1, n \geq 1$. Let $s > 0$, and let $(r_p)_{p\geq 1}$ be a sequence of positive rational numbers such that $\lim_{p\to+\infty} r_p = s$. It follows from the strong continuity of T_2 that we have, for $u \in \mathcal{A}$,

$$T_2(s)T_1(t)u = \lim_{p\to+\infty} T_2(r_p)T_1(t)u = \lim_{p\to+\infty} T_1(t)T_2(r_p)u = T_1(t)T_2(s)u.$$

So $T_2(s)T_1(t) = T_1(t)T_2(s)$ for $s > 0, t > 0$, and $T_1T_2 := (T_1(t)T_2(t))_{t>0}$ is itself a semigroup of multipliers, a folklore observation.

Since $\sup_{x\leq t\leq y} \|T_1(t)\| < +\infty$ for $0 < x < y < +\infty$, a standard argument shows that T_1T_2 is strongly continuous. Let $\epsilon > 0$, and let $u \in \mathcal{A}$. There exists $s > 0$ and $u_1 \in \mathcal{A}$ such that $\|T_1(s)u_1 - u\| < \epsilon/2$, and there exists $t \in (0, s)$ and $u_2 \in \mathcal{A}$ such that $\|T_1(t)T_2(s)u_2 - u\| < \epsilon$. Since $T_1(t)T_2(s)u_2 \in T_1(t)T_2(t)\mathcal{A}$, this shows that $\cup_{t>0}(T_1T_2)(t)\mathcal{A}$ is dense in $\mathcal{A}$. The following folklore result shows that the generator of T_1T_2 is the sum of the generators of T_1 and T_2.

COROLLARY 5.4. *Let T_1 and T_2 be two strongly continuous semigroups of multipliers on $\mathcal{A}$ such that $\cup_{t>0}T_1(t)\mathcal{A}$ and $\cup_{t>0}T_2(t)\mathcal{A}$ are dense in $\mathcal{A}$. If T_1 and T_2 commute, then then $\Delta_{T_1T_2,\mathcal{A}} = \Delta_{T_1,\mathcal{A}} + \Delta_{T_2,\mathcal{A}}$.*

PROOF. Let $u_1 \in \mathcal{A}$, $u_2 \in \mathcal{A}$, and $v_1 \in \Omega(\mathcal{A})$, $v_2 \in \Omega(\mathcal{A})$ be such that $\Delta_{T_1,\mathcal{A}} = S_{u_1/v_1}, \Delta_{T_2,\mathcal{A}} = S_{u_2/v_2}$, let $s > 0$ and set $w_s = T_1(s)T_2(s)v_1v_2$.

Then $w_s \in Dom(\Delta_{T_1,\mathcal{A}})\cap Dom(\Delta_{T_2,\mathcal{A}})$, $\Delta_{T_1,\mathcal{A}}w_s = T_1(s)T_2(s)u_1v_2$ and $\Delta_{T_2,\mathcal{A}}w_s = T_1(s)T_2(s)v_1u_2$, and so $\lim_{t\to 0^+} \|T_j(t)\Delta_{T_j,\mathcal{A}}w_s - \Delta_{T_j,\mathcal{A}}w_s\| = 0$ for $j = 1, 2$.

It follows then from Proposition 5.3 (ii) that we have, for $j = 1, 2$,

$$\lim_{t\to 0+} \left\| \frac{T_j(t)w_s - w_s}{t} - \Delta_{T_j,\mathcal{A}}w_s \right\| = 0.$$

So we have, for $u \in \mathcal{A}$,

$$\lim_{t\to 0^+} \left\| \frac{(T_1T_2)(t)uw_s - uw_s}{t} - \Delta_{T_1,\mathcal{A}}uw_s - \Delta_{T_2,\mathcal{A}}uw_s \right\|$$

$$= \lim_{t\to 0^+} \left\| u\frac{T_1(t)w_s - w_s}{t} - u\Delta_{T_1,\mathcal{A}} + T_1(t)u\frac{T_2(t)w_s - w_s}{t} - u\Delta_{T_2}w_s \right\| = 0,$$

and so it follows from Proposition 5.3 (i) that $uw_s \in \Delta_{T_1T_2,\mathcal{A}}$, and

$$\Delta_{T_1T_2,\mathcal{A}}uw_s = \Delta_{T_1,\mathcal{A}}uw_s + \Delta_{T_2,\mathcal{A}}uw_s = T_1(s)T_2(s)u(v_1u_2 + v_2u_1).$$

Now write $\Delta_{T_1T_2,\mathcal{A}} = S_{u_3/v_3}$, where $u_3 \in \mathcal{A}, v_3 \in \Omega(\mathcal{A})$. We obtain

$$T_1(s)T_2(s)uv_1v_2u_3 = T_1(s)T_2(s)uv_3(v_1u_2 + v_2u_1).$$

Since $\cup_{s>0}T_1(s)T_2(s)(\mathcal{A})$ is dense in $\mathcal{A}$, and since $\mathcal{A}$ is weakly cancellative, we have $v_1v_2u_3 = v_3(v_1u_2 + v_2u_1)$, which gives

$$\Delta_{T_1T_2,\mathcal{A}} = S_{v_1v_2u_3/v_1v_2v_3} = S_{(v_1u_2+v_2u_1)/v_1v_2} = S_{u_1/v_1+u_2/v_2} = \Delta_{T_1,\mathcal{A}} + \Delta_{T_2,\mathcal{A}}$$

□

We now consider a normalization $\mathcal{A}$ with respect to T, see Definition 3.3.

PROPOSITION 5.5. *Let $\mathcal{B}$ be a normalization of $\mathcal{A}$ with respect to T.*
Set $v_\lambda(t) = te^{-\lambda t}$ for $\lambda \in \mathbb{R}, t \geq 0$, and let $\lambda > \log(\rho_T)$.
(i) If $u \in \Omega(\mathcal{B})$, then

$$\Delta_{T,\mathcal{B}} = -S_{\int_0^{+} v'_\lambda(t)T(t)udt / \int_0^{+\infty} v_\lambda(t)T(t)udt}.$$

(ii) Let $\tilde{j} : \mathcal{QM}(\mathcal{A}) \to \mathcal{QM}(\mathcal{B})$ be the pseudobounded isomorphism given in Proposition 3.2 (i). Then $\tilde{j}^{-1}(\Delta_{T,\mathcal{B}}) = \Delta_{T,\mathcal{A}}$. So if $u \in \Omega(\mathcal{A})$, and if $\limsup_{t\to 0^+} \|T(t)u\| < +\infty$, then

$$\Delta_{T,\mathcal{A}} = -S_{\int_0^{+\infty} v'_\lambda(t)T(t)udt / \int_0^{+\infty} v_\lambda(t)T(t)udt}.$$

PROOF. (i) Set $e_\lambda(t) = e^{\lambda t}$ for $\lambda \in \mathbb{R}, t \geq 0$. If $\lambda > \mu > \log(\rho_T)$, then $v_\lambda \in L^1_{e_\mu} \subset L^1_{\tilde{\omega}_T}$, where $\tilde{\omega}_T(t) = \|T(t)\|_{\mathcal{M}(\mathcal{A}_T)}$, and $L^1_{e_\mu}$ is dense in $L^1_{\tilde{\omega}_T}$ since it contains the characteristic function of $[\alpha, \beta]$ for $0 < \alpha < \beta < +\infty$. It follows for example from Nyman's theorem **[28]** about closed ideals of $L^1(\mathbb{R}^+)$ that $v_\lambda e_\mu \in \Omega(L^1(\mathbb{R}^+))$ and so $v_\lambda \in \Omega(L^1_{e_\mu}) \subset \Omega(L^1_{\tilde{\omega}_T})$. So v_λ and u satisfy the conditions of Definition 5.1 with respect to T and $\mathcal{B}$, and (i) holds.

(ii) The map $\tilde{j}$ is the tautological map $S_{u/v} \to S_{u/v}$, where $u \in \mathcal{A} \subset \mathcal{B}$ and $v \in \Omega(\mathcal{A}) \subset \Omega(\mathcal{B})$. Now let $f_0 \in L^1_{\omega_T} \cap \mathcal{C}^1([0, +\infty))$ satisfying Definition 5.1 with respect to T and $\mathcal{A}$ and let $u_0 \in \Omega(\mathcal{A})$. Since $\Omega(L^1_{\omega_T}) \subset \Omega(L^1_{\tilde{\omega}_T})$, and since $\Omega(\mathcal{A}) \subset \Omega(\mathcal{B})$, it follows from Definition 5.1 that $\tilde{j}(\Delta_{T,\mathcal{A}}) = \Delta_{T,\mathcal{B}}$, and so $\tilde{j}^{-1}(\Delta_{T,\mathcal{B}}) = \Delta_{T,\mathcal{A}}$.

Let $u \in \Omega(\mathcal{A}) \subset \Omega(\mathcal{B})$, and assume that $\limsup_{t\to 0^+} \|T(t)u\| < +\infty$. Let $w \in \Omega(\mathcal{A})$ be such that $w\mathcal{B} \subset \mathcal{A}$. Since $v_\lambda \in \Omega(L^1_{\omega_\lambda})$, we see as in the proof of Lemma 3.1 that $\int_0^{+\infty} v_\lambda(t)T(t)udt \in \Omega(\mathcal{B})$, and it follows from Proposition 2.4 that $w\int_0^{+\infty} v_\lambda(t)T(t)udt \in \Omega(\mathcal{A})$. Using the characterization of $\tilde{j}^{-1}$ given in Proposition 2.3, we obtain

$$\Delta_{T,\mathcal{B}} = -S_{\int_0^{+\infty} v'_\lambda(t)T(t)udt / \int_0^{+\infty} v_\lambda(t)T(t)udt},$$

and

$$\begin{aligned}\Delta_{T,\mathcal{A}} &= \tilde{j}^{-1}(\Delta_{T,\mathcal{B}}) \\ &= -S_{w\int_0^{+\infty} v'_\lambda(t)T(t)udt / w\int_0^{+\infty} v_\lambda(t)T(t)udt} \\ &= -S_{\int_0^{+\infty} v'_\lambda(t)T(t)udt / \int_0^{+\infty} v_\lambda(t)T(t)udt}.\end{aligned}$$

□

We will denote by $Spec(\mathcal{I}_T)$ the space of characters of $\mathcal{I}_T$, equipped with the usual Gelfand topology. Notice that if $\chi \in Spec(\mathcal{I}_T)$ then there exists a unique character $\tilde{\chi}$ on $\mathcal{QM}(\mathcal{I}_T)$ such that $\tilde{\chi}_{|\mathcal{I}_T} = \chi$, which is defined by the formula $\tilde{\chi}(S_{u/v}) = \frac{\chi(u)}{\chi(v)}$ for $u \in \mathcal{I}_T, v \in \Omega(\mathcal{I}_T)$.

DEFINITION 5.6. *Assume that $\mathcal{I}_T$ is not radical, and let $S \in \mathcal{QM}(\mathcal{I}_T)$. The Arveson spectrum $\sigma_{ar}(S)$ is defined by the formula*

$$\sigma_{ar}(S) = \{\lambda = \tilde{\chi}(S) \ : \ \chi \in Spec(\mathcal{I}_T)\}.$$

If ν is a measure on $[0, +\infty)$, the Laplace transform of ν is defined by the usual formula $\mathcal{L}(\nu)(z) = \int_0^{+\infty} e^{-zt}d\nu(t)$ when $\int_0^{+\infty} e^{-Re(z)t}d|\nu|(t) < +\infty$.

We have the following easy observation.

PROPOSITION 5.7. *Let $\nu \in \mathcal{M}_{\omega_T}$. Then we have, for $\chi \in Spec(\mathcal{I}_T)$,*

$$\tilde{\chi}\left(\int_0^{+\infty} T(t)d\nu(t)\right) = \mathcal{L}(\nu)(-\tilde{\chi}(\Delta_{T,\mathcal{I}_T})). \tag{7}$$

Similarly we have, for $\nu \in \mathcal{M}_{\tilde{\omega}_T}, \chi \in Spec(\mathcal{I}_T)$,

$$\tilde{\chi}(\Phi_T(\nu)) = \tilde{\chi}\left(\int_0^{+\infty} T(t)d\nu(t)\right) = \mathcal{L}(\nu)(-\tilde{\chi}(\Delta_{T,\mathcal{I}_T})) = \mathcal{L}(\nu)(-\tilde{\chi}(\Delta_{T,\mathcal{I}_{T,\mathcal{B}}})). \tag{8}$$

In particular $\tilde{\chi}(T(t)) = e^{\tilde{\chi}(\Delta_{T,\mathcal{I}_T})t}$ for $t > 0$.

PROOF. If $\chi \in Spec(\mathcal{I}_T)$, then $\tilde{\chi}_{|\mathcal{A}_T}$ is a character on $\mathcal{A}_T$, the map $t \to \tilde{\chi}(T(t))$ is continuous on $(0, +\infty)$ and so there exists $\lambda \in \mathbb{C}$ such that $\tilde{\chi}(T(t)) = e^{-\lambda t}$ for $t > 0$, and $|e^{-\lambda t}| \le \|T(t)\|$, which shows that $Re(\lambda) \ge -\log(\rho_T)$.

Let $u \in \Omega(\mathcal{I}_T)$, and let $\nu \in \mathcal{M}_{\omega_T}$. We have

$$\begin{aligned}
\chi(u)\tilde{\chi}\left(\int_0^{+\infty} T(t)d\nu(t)\right) &= \chi\left(u\int_0^{+\infty} T(t)d\nu(t)\right)\\
&= \chi\left(\int_0^{+\infty} T(t)ud\nu(t)\right)\\
&= \int_0^{+\infty} \chi(T(t)u)d\nu(t)\\
&= \chi(u)\int_0^{+\infty} e^{-\lambda t}d\nu(t)\\
&= \chi(u)\mathcal{L}(\nu)(\lambda),
\end{aligned}$$

and so $\tilde{\chi}(\Phi_T(\nu)) = \mathcal{L}(\nu)(\lambda)$.

Let $f_0 \in \mathcal{C}^1((0,+\infty)) \cap \Omega(\mathcal{I}_T)$ such that $f_0(0) = 0, f_0' \in L^1_{\omega_T}$. We have

$$\begin{aligned}
\lambda\mathcal{L}(f_0)(\lambda) &= \mathcal{L}(f_0')(\lambda)\\
&= \chi(\Phi_T(f_0'))\\
&= -\tilde{\chi}\left(\Delta_{T,\mathcal{I}_T}\Phi_T(f_0)\right)\\
&= -\tilde{\chi}(\Delta_{T,\mathcal{I}_T})\chi(\phi_T(f_0))\\
&= -\tilde{\chi}(\Delta_{T,\mathcal{I}_T})\mathcal{L}(f_0)(\lambda),
\end{aligned}$$

and so $\lambda = -\tilde{\chi}(\Delta_{T,\mathcal{I}_T})$, which proves (7), and formula (8) follows from a similar argument. In particular $\chi(T(t)) = \mathcal{L}(\delta_t)(-\tilde{\chi}(\Delta_{T,\mathcal{I}_T})) = e^{\tilde{\chi}(\Delta_{T,\mathcal{I}_T})t}$ for $t > 0$. □

The following consequence of Proposition 5.7 pertains to folklore.

COROLLARY 5.8. *Assume that $\mathcal{I}_T$ is not radical. Then the map $\chi \to \tilde{\chi}(\Delta_{T,\mathcal{I}_T})$ is a homeomorphism from $Spec(\mathcal{I}_T)$ onto $\sigma_{ar}(\Delta_{T,\mathcal{I}_T})$, and if we set set*

$$\Lambda_t := \{\lambda \in \sigma_{ar}(\Delta_{T,\mathcal{I}_T}) \mid Re(\lambda) \le t\},$$

then Λ_t is compact for every $t \in \mathbb{R}$, so that $\sigma_{ar}(\Delta_{T,\mathcal{I}_T})$ is closed.

PROOF. Let $f_0 \in \mathcal{C}^1((0,+\infty)) \cap \Omega(\mathcal{I}_T)$ such that $f_0(0) = 0, f_0' \in L^1_{\omega_T}$. We have $\chi(\Phi(f_0)) \neq 0$ and $\tilde{\chi}(\Delta_{T,\mathcal{I}_T}) = -\frac{\chi(\Phi_T(f_0'))}{\chi(\Phi_T(f_0))}$ for $\chi \in Spec(\mathcal{I}_T)$, and so the map $\chi \to \tilde{\chi}(\Delta_T)$ is continuous with respect to the Gelfand topology on $Spec(\mathcal{I}_T)$.

Conversely let $f \in L^1_{\omega_T}$. It follows from Proposition 5.6 that we have, for $\chi \in Spec(\mathcal{I}_T)$,

$$\chi(\Phi_T(f)) = \mathcal{L}(f)(-\tilde{\chi}(\Delta_{T,\mathcal{I}_T})).$$

Since the set $\{u = \Phi_T(f) : f \in L^1_{\omega_T}\}$ is dense in $\mathcal{I}_T$, this shows that the map $\chi \to \tilde{\chi}(\Delta_{T,\mathcal{I}_T})$ is one-to-one on $Spec(\mathcal{I}_T)$, and that the inverse map $\sigma_{ar}(\Delta_{T,\mathcal{I}_T}) \to Spec(\mathcal{I}_T)$ is continuous with respect to the Gelfand topology.

Now let $t \in \mathbb{R}$, and set $U_t := \{\chi \in Spec(\mathcal{I}_T) : Re(\chi(\Delta_{T,\mathcal{I}_T})) \le t\}$. Then $|\tilde{\chi}(T(1))| \ge e^{-t}$ for $\chi \in U_t$, and so 0 does not belong to the closure of U_t with respect to the weak* topology on the unit ball of the dual of $\mathcal{I}_T$. Since $Spec(\mathcal{I}_T) \cup \{0\}$ is compact with respect to this topology, U_t is a compact subset of $Spec(\mathcal{I}_T)$, and so the set Λ_t is compact, which implies that $\sigma_{ar}(\Delta_{T,\mathcal{I}_T}) = \cup_{n\ge 1}\Lambda_n$ is closed. □

6. The resolvent

We now wish to discuss the resolvent of the generator of a strongly continuous semigroup $T = (T(t))_{t>0}$ of multipliers on $\mathcal{A}$, where $\mathcal{A}$ is a weakly cancellative commutative Banach algebra with dense principal ideals, and where $\cup_{t>0}T(t)\mathcal{A}$ is dense in $\mathcal{A}$. From now on we will write $\Delta_T = \Delta_{T,\mathcal{A}}$ and we will denote by $\mathcal{D}_{\Delta_T,\mathcal{A}}$ the domain of Δ_T considered as a quasimultiplier on $\mathcal{A}$. The Arveson ideal $\mathcal{I}_T$ is as above the closed subalgebra of $\mathcal{M}(\mathcal{A})$ generated by $\Phi_T(L^1_{\omega_T})$.

The Arveson resolvent set is defined by the formula

$$Res_{ar}(\Delta_{T,\mathcal{I}_T}) = \mathbb{C} \setminus \sigma_{ar}(\Delta_{T,\mathcal{I}_T}),$$

with the convention $\sigma_{ar}(\Delta_{T,\mathcal{I}_T}) = \emptyset$ if $\mathcal{I}_T$ is radical.

The usual "resolvent formula," interpreted in terms of quasimultipliers, shows that $\lambda I - \Delta_{T,\mathcal{I}_T} \in \mathcal{QM}(\mathcal{I}_T)$ is invertible in $\mathcal{QM}(\mathcal{I}_T)$ and that its inverse $(\lambda I - \Delta_{T,\mathcal{I}_T})^{-1}$ belongs to the Banach algebra $\mathcal{I}_{T,\mathcal{B}} \subset \mathcal{QM}_r(\mathcal{I}_T)$ obtained by applying Theorem 2.2 to $\mathcal{I}_T$ with respect to the semigroup T, and that we have, for $Re(\lambda) > \log(\rho_T)$,

$$(\lambda I - \Delta_{T,\mathcal{I}_T})^{-1} = \int_0^{+\infty} e^{-\lambda s}T(s)ds \in \mathcal{I}_{T,\mathcal{B}},$$

where the Bochner integral is computed with respect to the strong operator topology on $\mathcal{M}(\mathcal{I}_{T,\mathcal{B}})$. Also the $\mathcal{I}_{T,\mathcal{B}}$-valued map $\lambda \to (\lambda I - \Delta_{T,\mathcal{I}_T})^{-1}$ is holomorphic on $Res_{ar}(T, \mathcal{I}_T)$. The details of the adaptation to the context of quasimultipliers of this classical part of semigroup theory are given in **[17]**, Proposition 10.2.

We now give a slightly more general version of this result, which applies in particular to the case where $\mathcal{B}$ is the normalization $\mathcal{A}_T$ of $\mathcal{A}$ with respect to the semigroup T introduced in Proposition 3.2.

In the following we will identify the algebras $\mathcal{QM}(\mathcal{A})$ and $\mathcal{QM}(\mathcal{B})$ using the isomorphism $\tilde{j}$ intoduced in Proposition 2.3 (iii) if $\mathcal{B}$ is a normalization of $\mathcal{A}$ with respect to T. We set $\Phi_{T,\mathcal{B}}(\nu)u = \int_0^{+\infty} T(t)ud\nu(t)$ for $u \in \mathcal{B}, \nu \in \mathcal{M}_{\omega_{T,\mathcal{B}}}$, where $\omega_{T,\mathcal{B}}(t) = \|T(t)\|_{\mathcal{M}(\mathcal{B})}$ for $t > 0$, and we denote by $\mathcal{I}_{T,\mathcal{B}}$ the closure of $\Phi_{T,\mathcal{B}}(L^1_{\omega_{T,\mathcal{B}}})$ in $\mathcal{M}(\mathcal{B})$.

PROPOSITION 6.1. *Let $\mathcal{A}$ be a weakly cancellative commutative Banach algebra with dense principal ideals, let $T = ((T(t))_{t>0}$ be a strongly continuous semigroup of multipliers on $\mathcal{A}$ such that $T(t)\mathcal{A}$ is dense in $\mathcal{A}$ for $t > 0$, and let $\mathcal{B}$ be a normalization of $\mathcal{A}$ with respect to T.*

Set $Res_{ar}(\Delta_T) = Res_{ar}(\Delta_{T,\mathcal{I}_T}) = \mathbb{C} \setminus \sigma_{ar}(\Delta_{T,\mathcal{I}_T})$. *Then the quasimultiplier* $\lambda I - \Delta_T \in \mathcal{QM}(\mathcal{A})$ *admits an inverse* $(\lambda I - \Delta_T)^{-1} \in \mathcal{I}_{T,\mathcal{B}} \subset \mathcal{M}(\mathcal{B}) \subset \mathcal{QM}_r(\mathcal{A})$ *for* $\lambda \in Res_{ar}(\Delta_T)$, *and the map* $\lambda \to (\lambda I - \Delta_T)^{-1}$ *is an holomorphic map from* $Res_{ar}(\Delta_T)$ *into* $\mathcal{I}_{T,\mathcal{B}}$. *Moreover we have, for* $Re(\lambda) > \log(\rho_T)$,

$$(\lambda I - \Delta_T)^{-1} = \int_0^{+\infty} e^{-\lambda s} T(s) ds \in \mathcal{I}_{T,\mathcal{B}},$$

where the Bochner integral is computed with respect to the strong operator topology on $\mathcal{M}(\mathcal{B})$, *and* $\|(\lambda I - \Delta_T)^{-1}\|_{\mathcal{M}(\mathcal{B})} \le \int_0^{+\infty} e^{-Re(\lambda)t} \|T(t)\|_{\mathcal{M}(\mathcal{B})} dt$.

PROOF. We could deduce this version of the resolvent formula from Proposition 10.2 of [**17**], but we give a proof for the sake of completeness. Set again $e_\lambda(t) = e^{\lambda t}$ for $t \ge 0, \lambda \in \mathbb{C}$. Assume that

$$Re(\lambda) > \log(\rho_T) \ge \lim_{t \to +\infty} \frac{\log \|T(t)\|_{\mathcal{M}(\mathcal{B})}}{t},$$

let $v \in \mathcal{I}_{T,\mathcal{B}}$, and set $a = \Phi_{T,\mathcal{B}}(e_{-\lambda})$. We have

$$\begin{aligned} av = \int_0^{+\infty} & e^{-\lambda s} T(s) v ds, T(t)av - av \\ &= \int_0^{+\infty} e^{-\lambda s} T(s+t) v ds - \int_0^{+\infty} e^{-\lambda s} T(s) v ds \\ &= e^{\lambda t} \int_t^{+\infty} e^{-\lambda s} T(s) v ds - \int_0^{+\infty} e^{-\lambda s} T(s) v ds \\ &= (e^{\lambda t} - 1) av - e^{\lambda t} \int_0^t e^{-\lambda s} T(s) v ds. \end{aligned}$$

Since $\lim_{t \to 0^+} \|T(t)v - v\|_{\mathcal{I}_{T,\mathcal{B}}} = 0$, we obtain

$$\lim_{t \to 0^+} \left\| \frac{T(t)av - av}{t} - \lambda av + v \right\|_{\mathcal{I}_{T,\mathcal{B}}} = 0,$$

and so $av \in \mathcal{D}_{\Delta_T, \mathcal{I}_{T,\mathcal{B}}}$, and $\Delta_{T,\mathcal{I}_{T,\mathcal{B}}}(av) = \lambda av - v$.

This shows that $a\mathcal{I}_{T,\mathcal{B}} \subset \mathcal{D}_{\Delta_{T,\mathcal{I}_{T,\mathcal{B}}}}$, and that $(\lambda I - \Delta_{T,\mathcal{I}_{T,\mathcal{B}}})av = v$ for every $v \in \mathcal{I}_{T,\mathcal{B}}$. We have $\lambda I - \Delta_{T,\mathcal{I}_{T,\mathcal{B}}} = S_{u/v}$, where $u \in \mathcal{I}_{T,\mathcal{B}}, v \in \Omega(\mathcal{I}_{T,\mathcal{B}})$, and we see that $ua = v$. Hence $u \in \Omega(\mathcal{I}_{T,\mathcal{B}}), \lambda I - \Delta_{T,\mathcal{I}_{T,\mathcal{B}}}$ is invertible in $\mathcal{QM}(\mathcal{I}_{T,\mathcal{B}})$, and $(\lambda I - \Delta_{T,\mathcal{B}})^{-1} = a = \Phi_{T,\mathcal{B}}(e_{-\lambda}) = \int_0^{+\infty} e^{-\lambda t} T(t) dt \in \mathcal{I}_{T,\mathcal{B}}$, where the Bochner integral is computed with respect to the strong operator topology on $\mathcal{M}(\mathcal{I}_{T,\mathcal{B}})$.

Let $\chi \in Spec(\mathcal{I}_{T,\mathcal{B}})$. Then $\chi \circ \tilde{j} \in Spec(\mathcal{I}_T)$, and $\sigma_{ar}(\Delta_{T,\mathcal{I}_{T,\mathcal{B}}}) \subset \sigma_{ar}(\Delta_{T,\mathcal{I}_T})$. It follows then from Proposition 2.6 that $\lambda I - \Delta_{T,\mathcal{B}}$ has in $\mathcal{QM}_r(\mathcal{I}_{T,\mathcal{B}})$ an inverse $(\lambda I - \Delta_{T,\mathcal{B}})^{-1} \in \mathcal{I}_{T,\mathcal{B}}$ for $\lambda \in Res_{ar}(\Delta_{T,\mathcal{I}_T})$ and that the $\mathcal{I}_{T,\mathcal{B}}$-valued map $\lambda \to (\lambda I - \Delta_{T,\mathcal{B}})^{-1}$ is holomorphic on $Res_{ar}(\Delta_{T,\mathcal{B}})$.

Fix $u_0 \in \Omega(\mathcal{A}) \subset \Omega(\mathcal{B})$, and set $j_T(S) = S_{uu_0/vu_0}$ for $S = S_{u/v} \in \mathcal{QM}(\mathcal{I}_{T,\mathcal{B}})$. Then $j_T : \mathcal{QM}(\mathcal{I}_{T,\mathcal{B}}) \to \mathcal{QM}(\mathcal{B})$ is a pseudobounded homomorphism, and $j_T(\Delta_{T,\mathcal{I}_{T,\mathcal{B}}}) = \Delta_{T,\mathcal{B}}$.

Identifying $\mathcal{I}_{T,\mathcal{B}}$ to a subset of $\mathcal{QM}(\mathcal{I}_{T,\mathcal{B}})$ as above in the obvious way, we see that the restriction of j_T to $\mathcal{I}_{T,\mathcal{B}}$ is the identity map, and so $\lambda I - \Delta_T$ is invertible in $\mathcal{QM}(\mathcal{B})$ for $\lambda \in Res_{ar}(\Delta_T)$, we have $(\lambda I - \Delta_T)^{-1} = (\lambda I - \Delta_{T,\mathcal{I}_{T,\mathcal{B}}})^{-1} \in \mathcal{I}_{T,\mathcal{B}}$, and the $\mathcal{I}_{T,\mathcal{B}}$-valued map $\lambda \to (\lambda I - \Delta_{T,\mathcal{B}})^{-1}$ is holomorphic on $Res_{ar}(\Delta_T)$.

If $Re(\lambda) > \log(\rho_T) \geq \lim_{t\to+\infty} \frac{\log \|T(t)\|_{\mathcal{M}(\mathcal{B})}}{t}$, then if $u \in \mathcal{I}_{T,\mathcal{B}}, v \in \mathcal{B}$, we have

$$(\lambda I - \Delta_T)^{-1}uv = ((\lambda I - \Delta_{T,\mathcal{B}})^{-1}u)v = \int_0^{+\infty} e^{-\lambda t}T(t)uvdt.$$

Since $uv \in \Omega(\mathcal{B})$ for $u \in \Omega(\mathcal{I}_{T,\mathcal{B}}), v \in \Omega(\mathcal{B})$, $u\mathcal{B}$ is dense in $\mathcal{B}$ for $u \in \Omega(\mathcal{I}_{T,\mathcal{B}})$, and we obtain $(\lambda I - \Delta_T)^{-1} = \int_0^{+\infty} e^{-\lambda t}T(t)dt \in \mathcal{I}_{T,\mathcal{B}}$, where the Bochner integral is computed with respect to the strong operator topology on $\mathcal{M}(\mathcal{B})$, so that

$$\|(\lambda I - \Delta_T)^{-1}\|_{\mathcal{M}(\mathcal{B})} \leq \int_0^{+\infty} e^{-Re(\lambda)t}\|T(t)\|_{\mathcal{M}(\mathcal{B})}dt. \qquad \square$$

If we consider Δ_T as a quasimultiplier on $\mathcal{B}$, the fact that the multiplier $(\lambda I - \Delta_T)^{-1} \in \mathcal{M}(\mathcal{B})$ is the inverse of $\lambda I - \Delta_T$ for $\lambda \in Res(\Delta_T)$ means that $(\lambda I - \Delta_T)^{-1}v \in \mathcal{D}_{\Delta_{T,\mathcal{B}}}$ and that $(\lambda I - \Delta_T)\left((\lambda I - \Delta_T)^{-1}v\right) = v$ for every $v \in \mathcal{B}$, and that if $w \in \mathcal{D}_{\Delta_{T,\mathcal{B}}}$, then $(\lambda I - \Delta_T)^{-1}\left((\lambda I - \Delta_T)w\right) = w$. The situation is slightly more complicated if we consider Δ_T as a quasimultiplier on $\mathcal{A}$ when $\limsup_{t\to0^+}\|T(t)\| = +\infty$. In this case the domain $\mathcal{D}_{(\lambda I - \Delta_T)^{-1},\mathcal{A}}$ of $(\lambda I - \Delta_T)^{-1} \in \mathcal{QM}(\mathcal{A})$ is a proper subspace of $\mathcal{A}$ containing $\mathcal{L}_T \supset \cup_{t>0}T(t)\mathcal{A}$, and we have $(\lambda I - \Delta_T)^{-1}v \in \mathcal{D}_{\Delta_{T,\mathcal{A}}}$ and $(\lambda I - \Delta_T)\left((\lambda I - \Delta_T)^{-1}v\right) = v$ for every $v \in \mathcal{D}_{(\lambda I - \Delta_T)^{-1},\mathcal{A}}$. Also if $w \in \mathcal{D}_{\Delta_{T,\mathcal{A}}}$, then $(\lambda I - \Delta_T)w \in \mathcal{D}_{(\lambda I - \Delta_T)^{-1},\mathcal{A}}$, and we have $(\lambda I - \Delta_T)^{-1}\left((\lambda I - \Delta_T)w\right) = w$.

In order to interpret $(\lambda I - \Delta_T)^{-1}$ as a partially defined operator on $\mathcal{A}$ for $Re(\lambda) > \log(\rho_T)$, we can use the formula

$$(\lambda I - \Delta_T)^{-1}v = \int_0^{+\infty} e^{-\lambda t}T(t)vdt \quad (v \in \mathcal{L}_T), \tag{9}$$

which defines a quasimultiplier on $\mathcal{A}$ if we apply it to some $v \in \Omega(\mathcal{A})$ such that $\limsup_{t\to0^+}\|T(t)v\| < +\infty$. The fact that this quasimultiplier is regular is not completely obvious but follows from the previous discussion since $(\lambda I - \Delta_T)^{-1} \in \mathcal{M}(\mathcal{B}) \subset \mathcal{QM}_r(\mathcal{A})$. Notice that since $\cup_{t>0}T(t)\mathcal{A}$ is dense in $(\mathcal{U}_T, \|.\|_{\mathcal{U}_T})$, $(\lambda I - \Delta_T)^{-1}$ is characterized by the simpler formula

$$(\lambda I - \Delta_T)^{-1}T(s)v = e^{\lambda s}\int_s^{+\infty} e^{-\lambda t}T(t)vdt \quad (s > 0, v \in \mathcal{A}). \tag{10}$$

7. The generator of a holomorphic semigroup and its resolvent

Assume that $a < b \leq a + \pi$. In this section we consider a holomorphic semigroup $T = (T(\zeta))_{\zeta\in S_{a,b}}$ of multipliers on a weakly cancellative commutative Banach algebra $\mathcal{A}$ having dense principal ideals such that $T(\zeta)\mathcal{A}$ is dense in $\mathcal{A}$ for some, or, equivalently, for every $\zeta \in S_{a,b}$.

Denote by $\mathcal{I}_T$ the closed span of $\{T(\zeta)\}_{\zeta\in S_{a,b}}$ in $\mathcal{M}(\mathcal{A})$, which is equal to the closed span of $\{T(t\zeta)\}_{t>0}$ for $\zeta \in S_{a,b}$. For $\zeta \in S_{a,b}$, set $T_\zeta = (T(t\zeta))_{t>0}$, let $\Phi_{T_\zeta} : \mathcal{M}_{\omega_{T_\zeta}} \to \mathcal{M}(\mathcal{A})$ be the homomorphism defined by (2').

Set $\omega_T(\zeta) = \|T(\zeta)\|$ for $\zeta \in S_{a,b}$, denote by $\mathcal{M}_{\omega_T}(S_{a,b})$ the space of all measures μ on $S_{a,b}$ such that $\|\mu\|_{\omega_T} := \int_{S_{a,b}} \omega_T(\zeta)d|\mu|(\zeta) < +\infty$, which is a Banach algebra with respect to convolution on the additive semigroup $S_{a,b}$, see **[27]** for convolution of measures on semigroups. The convolution algebra $L^1_{\omega_T}(S_{a,b})$ is defined in a similar way and will be identified to the closed ideal of $\mathcal{M}_{\omega_T}$ consisting of measures which are absolutely continuous with respect to Lebesgue measure in $S_{a,b}$.

Define the Banach algebra homomorphism $\Phi_T : \mathcal{M}_{\omega_T} \to \mathcal{I}_T \subset \mathcal{M}(\mathcal{A})$ by the formula

$$\Phi_T(\mu) = \int_{S_{a,b}} T(\zeta)d\mu(\zeta),$$

which is well-defined since the map $\zeta \to T(\zeta)$ is continuous with respect to the norm topology on $\mathcal{M}(\mathcal{A})$ and since $\mathcal{I}_T$ is separable.

Let $\zeta \in S_{a,b}$. Since the semigroup T_ζ is continuous with respect to the norm topology on $\mathcal{M}(\mathcal{A})$, a standard argument shows that we have, for every Dirac sequence $(f_n)_{n\geq 1}$,

$$\limsup_{n\to+\infty} \left\| \int_0^{+\infty} (f_n * \delta_s)(t)T(t\zeta)dt - T(s\zeta) \right\| = 0,$$

and so $T(s\zeta) \in \mathcal{I}_{T_\zeta}$ for every $s > 0$, which implies that $\mathcal{I}_{T_\zeta} = \mathcal{I}_T$.

A similar argument shows that $\mathcal{I}_T$ is the closure in $(\mathcal{M}(\mathcal{A}), \|.\|_{\mathcal{M}(\mathcal{A})})$ of $\Phi_T(\mathcal{M}_{\omega_T}(S_{a,b}))$, as well as the closure of $\Phi_T(L^1_{\omega_T}(S_{a,b}))$ and the closure of $\Phi_{T_\zeta}(L^1_{\omega_{T_\zeta}})$ in $(\mathcal{M}(\mathcal{A}), \|.\|_{\mathcal{M}(\mathcal{A})})$, and the notation $\mathcal{I}_T$ is consistent with the notation used to denote the Arveson ideal associated to a strongly continuous semigroup of multipliers on the half-line.

The following interpretation of the generator of a holomorphic semigroup as a quasimultiplier follows the interpretation given in **[8]** in the case where $\mathcal{A} = \mathcal{I}_T$.

Proposition 7.1. *Set*

$$\Delta_{T,\mathcal{A}} := S_{T'(\zeta_0)u_0/T(\zeta_0)u_0} \in \mathcal{QM}(\mathcal{A}),$$

where $\zeta_0 \in S_{a,b}$, $u_0 \in \Omega(\mathcal{A})$.

Then this definition does not depend on the choice of ζ_0 *and* u_0, *and we have, for* $\zeta \in S_{a,b}$,

$$\Delta_{T_\zeta,\mathcal{A}} = \zeta\Delta_{T,\mathcal{A}}, \tag{11}$$

where the generator $\Delta_{T_\zeta,\mathcal{A}}$ *of the semigroup* T_ζ *is the quasimultiplier on* $\mathcal{A}$ *introduced in Definition 4.1.*

Moreover if $T_1 = (T_1(\zeta))_{\zeta\in S_{a,b}}$ *and* $T_2 = (T_2(\zeta))_{\zeta\in S_{a,b}}$ *are two holomorphic semigroups of multipliers on* $\mathcal{A}$ *such that* $T_1(\zeta)\mathcal{A}$ *and* $T_2(\zeta)\mathcal{A}$ *are dense in* $\mathcal{A}$ *and such that* $T_1(\zeta)T_2(\zeta) = T_2(\zeta)T_1(\zeta)$ *for* $\zeta \in S_{\alpha,\beta}$, *then* $T_1T_2 := (T_1(\zeta)T_2(\zeta)_{\zeta\in S_{a,b}}$ *is a semigroup holomorphic on* $S_{a,b}$, $(T_1T_2)(\zeta)\mathcal{A}$ *is dense in* $\mathcal{A}$ *for* $\zeta \in S_{a,b}$, *and we have*

$$\Delta_{T_1T_2,\mathcal{A}} = \Delta_{T_1,\mathcal{A}} + \Delta_{T_2,\mathcal{A}}.$$

Proof. We have, for $\zeta \in S_{a,b}$,

$$T'(\zeta_0)T(\zeta) = T'(\zeta_0 + \zeta) = T'(\zeta)T(\zeta_0),$$

and so the definition of $\Delta_{T,\mathcal{A}}$ does not depend on the choice of ζ_0, and an easy argument given in the comments following Definition 4.1 shows that this definition does not depend on the choice of $u_0 \in \Omega(\mathcal{A})$ either.

Now let $\zeta_0 \in S_{a,b}$, and let $f \in \mathcal{C}^1([0,+\infty)) \cap \Omega(L^1_{\omega_{T_{\zeta_0}}})$ such that $f(0) = 0$ and $f' \in L^1_{\omega_{T_{\zeta_0}}}$. We have, integrating by parts, since $\lim_{p\to+\infty} |f(n_p)|\|T(n_p\zeta_0)\| = 0$ for some strictly increasing sequence $(n_p)_{p\geq 1}$ of integers,

$$\begin{aligned} & T(\zeta_0)\int_0^{+\infty} f'(t)T(t\zeta_0)dt \\ = & \lim_{p\to+\infty} \int_0^{n_p} f'(t)T(\zeta_0+t\zeta_0)dt \\ = & \lim_{p\to+\infty} \left([f(t)T(\zeta_0+t\zeta_0)]_0^{n_p} - \zeta_0 \int_0^{n_p} f(t)T'(\zeta_0+t\zeta_0)dt \right) \\ = & -\zeta_0 T'(\zeta_0) \int_0^{+\infty} f(t)T(t\zeta_0)dt, \end{aligned}$$

and formula (11) follows since $\left(\int_0^{+\infty} f(t)T(t\zeta_0)dt\right) u = \phi_{T_{\zeta_0}}(f)u \in \Omega(\mathcal{A})$ for $u \in \Omega(\mathcal{A})$.

Now let $T_1 = (T_1(\zeta))_{\zeta\in S_{a,b}}$ and $T_2 = (T_2(\zeta))_{\zeta\in S_{a,b}}$ be two holomorphic semigroups of multipliers on $\mathcal{A}$ such that $T_1(\zeta)\mathcal{A}$ and $T_2(\zeta)\mathcal{A}$ are dense in $\mathcal{A}$ and such that $T_1(\zeta)T_2(\zeta) = T_2(\zeta)T_1(\zeta)$ for $\zeta \in S_{\alpha,\beta}$. As in section 5 we see that if $\zeta \in S_{a,b}$ then $T_1(\zeta)T_2(t\zeta) = T_2(t\zeta)T_1(\zeta)$ for $t > 0$, and it follows from the analyticity of T_2 that $T_1(\zeta)T_2(\zeta') = T_2(\zeta')T_1(\zeta)$ for $\zeta', \in S_{a,b}$. This shows that T_1T_2 is a semigroup on $S_{a,b}$, which is obviously holomorphic. Since $T_1(\zeta)\mathcal{A}$ and $T_2(\mathcal{A})$ are dense in $\mathcal{A}$, $(T_1T_2)(\zeta)\mathcal{A} = T_1(\zeta)T_2(\zeta)\mathcal{A}$ is dense in $\mathcal{A}$ for $\zeta \in S_{a,b}$.

The last assertion of the corollary follows then immediately from the Leibniz rule. □

The following corollary follows then from Proposition 5.3.

COROLLARY 7.2. *(i) Let $u \in \mathcal{A}$, and let $\zeta \in S_{a,b}$. If* $\lim_{t\to0^+} \left\| \frac{T(t\zeta)u-u}{t} - v \right\| = 0$ *for some $v \in \mathcal{A}$, then $u \in \mathcal{D}_{\Delta_{T,\mathcal{A}}}$, and $\zeta\Delta_{T,\mathcal{A}}u = v$.*

(ii) Conversely if $\zeta \in S_{a,b}$, then $\lim_{t\to0^+} \left\| \frac{T(t)u-u}{t} - \zeta\Delta_{T,\mathcal{A}}u \right\| = 0$ *for every $u \in \mathcal{D}_{\Delta_{T,\mathcal{A}}}$ satisfying the condition* $\lim_{t\to0^+} \|T(t\zeta)u - u\| = 0$.

In the remainder of the section we will denote by $\mathcal{B}$ a normalization of $\mathcal{A}$ with respect to the semigroup T, see Definition 4.3. Since $\mathcal{QM}(\mathcal{A})$ is isomorphic to $\mathcal{QM}(\mathcal{B})$, we can consider the generator $\Delta_{T,\mathcal{A}}$ as a quasimultiplier on $\mathcal{B}$, and it follows immediately from Definition 7.1 that this quasimultiplier on $\mathcal{B}$ is the generator of the semigroup T considered as a semigroup of multipliers on $\mathcal{B}$. From now on we will thus set $\Delta_T = \Delta_{T,\mathcal{A}} = \Delta_{T,\mathcal{B}}$. Applying Corollary 7.2 to T and $\mathcal{B}$, we obtain the following result.

COROLLARY 7.3. *(i) Let $u \in \mathcal{B}$ Then the following conditions imply each other*

(i) There exists $\zeta_0 \in S_{a,b}$ and $v \in \mathcal{B}$ such that $\lim_{t\to0^+} \left\| \frac{T(\zeta_0 t)u-u}{t} - v \right\|_{\mathcal{B}} = 0$,

(ii) $u \in \mathcal{D}_{\Delta_T,\mathcal{B}}$, and in this situation $\lim_{t\to0^+} \left\| \frac{T(\zeta t)u-u}{t} - \zeta\Delta_T u \right\|_{\mathcal{B}} = 0$ *for every $\zeta \in S_{a,b}$.*

Denote by $Spec(\mathcal{I}_T)$ the space of characters on $\mathcal{I}_T$, equipped with the usual Gelfand topology. If $\chi \in Spec(\mathcal{I}_T)$, the map $\zeta \to \chi(T(\zeta))$ is holomorphic on $S_{a,b}$, and so there exists a unique complex number c_χ such that $\chi(T(\zeta)) = e^{\zeta c_\chi}$ for $\zeta \in S_{a,b}$. We see as in Section 5 that there exists a unique character $\tilde{\chi}$ on $\mathcal{QM}(\mathcal{I}_T)$ such that $\tilde{\chi}_{|\mathcal{I}_T} = \chi$, and since $\Delta_{T_\zeta,\mathcal{I}_T} = \zeta\Delta_{T,\mathcal{I}_T}$ it follows from Proposition 5.7 and Proposition 7.1 that $\chi(T(t\zeta)) = e^{t\tilde{\chi}(\Delta_{T_\zeta,\mathcal{I}_T})} = e^{t\zeta\tilde{\chi}(\Delta_{T,\mathcal{I}_T})}$ for $\zeta \in S_{a,b}, t > 0$, and so $c_\chi = \tilde{\chi}(\Delta_{T,\mathcal{I}_T})$.

Since $\Delta_{T_\zeta,\mathcal{I}_T} = \zeta\Delta_{T,\mathcal{I}_T}$ for $\zeta \in S_{a,b}$, we deduce from Corollary 5.8 and Proposition 6.1 the following result.

PROPOSITION 7.4. *Let $T = (T(\zeta)_{\zeta\in S_{a,b}} \subset \mathcal{M}(\mathcal{A})$ be a holomorphic semigroup. Set $\sigma_{ar}(\Delta_{T,\mathcal{I}_T}) = \{\tilde{\chi}(\Delta_{T,\mathcal{I}_T})\}_{\chi\in Spec(\mathcal{I}_T)}$, with the convention $\sigma_{ar}(\Delta_{T,\mathcal{I}_T}) = \emptyset$ if the semigroup is quasinilpotent.*

Set $Res_{ar}(\Delta_T) = Res_{ar}(\Delta_{T,\mathcal{I}_T}) = \mathbb{C}\setminus\sigma_{ar}(\Delta_{T,\mathcal{I}_T})$. Let $\mathcal{B}$ be a normalization of $\mathcal{A}$ with respect to the holomorphic semigroup T, and let $\mathcal{I}_{T,\mathcal{B}}$ be the closed subalgebra of $\mathcal{M}(\mathcal{B})$ generated by the semigroup.

(i) The set $\Lambda_{t,\zeta} := \{\lambda \in \sigma_{ar}(\Delta_T, \mathcal{I}_T) \mid Re(\lambda\zeta) \le t\}$ is compact for $\zeta \in S_{a,b}, t \in \mathbb{R}$.

(ii) If $\lambda \in Res_{ar}(\Delta_T)$, then the quasimultiplier $\lambda I - \Delta_T$ has an inverse in $\mathcal{QM}(\mathcal{A})$, $(\lambda I - \Delta_T)^{-1} \in \mathcal{I}_{T,\mathcal{B}} \subset \mathcal{M}(\mathcal{B}) \subset \mathcal{QM}_r(\mathcal{A})$, and the $\mathcal{I}_{T,\mathcal{B}}$-valued map $\lambda \to (\lambda I - \Delta_T)^{-1}$ is holomorphic on $Res_{ar}(\Delta_T)$.

(iii) If $\zeta \in S_{a,b}$, then $\lambda \in Res_{ar}(\Delta_T)$ for $Re(\lambda\zeta) > \lim_{t\to+\infty} \frac{\log(\|T(t\zeta)\|)}{t}$, and we have

$$(\lambda I - \Delta_T)^{-1} = \int_0^{\zeta.\infty} e^{-s\lambda} T(s)ds, \tag{12}$$

so that

$$\|(\lambda I - \Delta_T)^{-1}\|_{\mathcal{M}(\mathcal{B})} \le |\zeta| \int_0^{+\infty} e^{-tRe(\lambda\zeta)} \|T(t\zeta)\|_{\mathcal{M}(\mathcal{B})} dt. \tag{13}$$

PROOF. (i) Let $\zeta \in S_{a,b}$, $t > 0$, and set

$$V = \{\lambda \in \zeta\sigma_{ar}(\Delta_{T,\mathcal{B}_T}) \mid Re(\lambda) \le t\} = \{\lambda \in \sigma_{ar}(\Delta_{T_\zeta,\mathcal{I}_T}) \mid Re(\lambda) \le t\}.$$

It follows from Corollary 4.7 that V is compact, and so $\Lambda_{t,\zeta} = \zeta^{-1}V$ is compact.

(ii) Fix $\zeta_0 \in S_{a,b}$. We have $\lambda I - \Delta_T = \lambda I - \zeta_0^{-1}\Delta_{T_{\zeta_0}} = \zeta_0^{-1}\left(\lambda\zeta_0 I - \Delta_{T_{\zeta_0}}\right)$. If $\lambda \in Res_{ar}(\Delta_T)$, then $\lambda I - \Delta_T$ is invertible in $\mathcal{QM}(\mathcal{A})$, and we have

$$(\lambda I - T)^{-1} = \zeta_0^{-1}(\lambda\zeta_0 I - \Delta_{T_{\zeta_0}})^{-1} \in \mathcal{I}_{T,\mathcal{B}} \subset \mathcal{M}(\mathcal{B}) \subset \mathcal{QM}_r(\mathcal{A}),$$

since in this situation $\lambda\zeta_0 \in Res(\Delta_{T_{\zeta_0}})$, and it follows also from Proposition 6.1 that the $\mathcal{I}_{T,\mathcal{B}}$-valued map $\lambda \to (\lambda I - T)^{-1} = \zeta_0^{-1}(\lambda\zeta_0 I - \Delta_{T_{\zeta_0}})^{-1}$ is holomorphic on $Res_{ar}(\Delta_T)$.

(iii) This follows from Proposition 6.1 applied to $\lambda\zeta$ and T_ζ. □

8. Multivariable functional calculus associated to linear functionals

In the following definition, we write by convention $T_j(0) = I$ for $1 \le j \le k$. Set $\sigma\zeta = \sigma_1\zeta_1 + \cdots + \sigma_k\zeta_k$ for $\sigma = (\sigma_1, \ldots, \sigma_k), \zeta = (\zeta_1, \ldots, \zeta_k) \in \mathbb{C}^k$.

Let $a = (a_1, \dots, a_k) \in \mathbb{R}^k, b = (b_1, \dots, b_k) \in \mathbb{R}^k$ such that $a_j \le b_j \le a_j + \pi$ for $1 \le j \le k$. As in Appendix 2 we set

$$M_{a,b} = \{(\alpha, \beta) \in \mathbb{R}^k \times \mathbb{R}^k \mid a_j < \alpha_j \le \beta_j < b_j \text{ if } a_j < b_j, \alpha_j = \beta_j = a_j \text{ if } a_j = b_j\}.$$

We will say that a family $T = (T_1, \dots, T_k)$ of semigroups satisfying the conditions below is *commuting* if $T_j(\zeta)T_j'(\zeta') = T_j'(\zeta')T_j(\zeta)$ for $j' \neq j$ whenever ζ is in the domain of definition of T_j and ζ' is in the domain of definition of T_j'.

DEFINITION 8.1. : *Let $a = (a_1, \dots, a_k) \in \mathbb{R}^k, b = (b_1, \dots, b_k) \in \mathbb{R}^k$ such that $a_j \le b_j \le a_j + \pi$ for $j \le k$, let $\mathcal{A}$ be a weakly cancellative commutative Banach algebra with dense principal ideals, and let $T = (T_1, \dots, T_k)$ be a commuting family of semigroups of multipliers on $\mathcal{A}$ which possesses the following properties*

$$\begin{cases} T_j = (T_j(\zeta))_{\zeta \in (0, e^{a_j}.\infty)} \text{ is strongly continuous on } (0, e^{ia_j}.\infty), \text{ and} \\ \cup_{t>0} T(te^{ia_j})\mathcal{A} \text{ is dense in } \mathcal{A} \text{ if } a_j = b_j, \\ T_j = (T(\zeta))_{\zeta \in S_{a_j,b_j}} \text{ is holomorphic on } S_{a_j,b_j}, \text{ and } T(\zeta)\mathcal{A} \text{ is dense in } \mathcal{A} \\ \text{for every } \zeta \in S_{a_j,b_j} \text{ if } a_j < b_j. \end{cases}$$

For $\zeta = (\zeta_1, \dots, \zeta_k) \in \cup_{(\alpha,\beta) \in M_{a,b}} \overline{S}_{\alpha,\beta}$ set

$$T(\zeta) = T_1(\zeta_1) \dots T_k(\zeta_k).$$

A subalgebra $\mathcal{B}$ of $\mathcal{QM}(\mathcal{A})$ is said to be a normalization of $\mathcal{A}$ with respect to T if the following conditions are satisfied

(a) $(\mathcal{B}, \|.\|_{\mathcal{B}})$ is a Banach algebra with respect to a norm $\|.\|_{\mathcal{B}}$ satisfying $\|u\|_{\mathcal{B}} \le \|u\|_{\mathcal{A}}$ for $u \in \mathcal{A}$, and there exists a family $(w_1, \dots, w_k)$ of elements of $\Omega(\mathcal{A})$ such that the inclusion map $j : \mathcal{A} \to \mathcal{B}$ is a $\mathcal{QM}$-homomorphism with respect to $T_1(\zeta_1) \dots T_k(\zeta_k)w_1 \dots w_k$ for every family $(\zeta_1, \dots, \zeta_k)$ of complex numbers such that $\zeta_j \in S_{a_j,b_j}$ if $a_j < b_j$ and such that $\zeta_j = 0$, if $a_j = b_j$.

(b) $\tilde{j}(\mathcal{M}(\mathcal{A})) \subset \mathcal{M}(\mathcal{B}))$, and $\|\tilde{j}(R)\|_{\mathcal{M}(\mathcal{B})} \le \|R\|_{\mathcal{M}(\mathcal{A})}$ for every $R \in \mathcal{M}(\mathcal{A})$, where $\tilde{j} : \mathcal{QM}(\mathcal{A}) \to \mathcal{QM}(\mathcal{B})$ is the pseudobounded isomorphism associated to j in Proposition 2.3 (ii).

(c) $\limsup_{\substack{\zeta \to 0 \\ \zeta \in S_{\gamma,\delta}}} \|T(\zeta)\|_{\mathcal{M}(\mathcal{B})} < +\infty$ for $a_j < \gamma < \delta < b_j$ if $a_j < b_j$, and $\limsup_{t \to 0^+} \|T(te^{ia_j})\|_{\mathcal{M}(\mathcal{B})} < +\infty$ if $a_j = b_j$.

It follows from Proposition 3.2 and Proposition 4.2 that there exists a normalization $\mathcal{B}_1$ of $\mathcal{A}$ with respect to T_1. Also if $\mathcal{B}_m$ is a normalization of $\mathcal{A}$ with respect to $(T_1, \dots, T_m)$ and if $\mathcal{B}_{m+1}$ is a normalization of $\mathcal{B}_m$ with respect to T_{m+1}, it follows from Proposition 2.5 and Definitions 3.3 and 4.3 that $\mathcal{B}_{m+1}$ is a normalization of $\mathcal{A}$ with respect to $(T_1, \dots, T_{m+1})$. It is thus immediate to construct a normalization of $\mathcal{A}$ with respect to T by a finite induction. Notice that if $\mathcal{B}$ is a normalization of $\mathcal{A}$ with respect to T, then $\mathcal{B}$ is a normalization of $\mathcal{A}$ with respect to $T_\sigma := (T(t\sigma))_{t>0}$ for every $\sigma \in \cup_{(\alpha,\beta) \in M_{a,b}} \overline{S}_{\alpha,\beta}$.

Since $\cup_{t>0} T(te^{ia_j})\mathcal{B}$ is dense in $\mathcal{B}$, when $a_j = b_j$, and since $T(\zeta)\mathcal{A}$ is dense in $\mathcal{A}$ for $\zeta \in S_{a_j,b_j}$ if $a_j < b_j$, it follows from Definition 10.1 that the map $\zeta \to T(\zeta)u_1 \dots u_k$ is continuous on $\overline{S}_{\alpha,\beta}$ for $(\alpha, \beta) \in M_{a,b}$, $u_1, \dots, u_k \in \mathcal{B}$. Since $u_1 \dots u_k \in \Omega(\mathcal{B})$ for $u_1, \dots, u_k \in \Omega(\mathcal{B})$, it follows again from Definition 10.1 that the map $\zeta \to T(\zeta)u$ is continuous on $\overline{S}_{\alpha,\beta}$ for $(\alpha, \beta) \in M_{a,b}$ for every $u \in \mathcal{B}$. Let $(\alpha, \beta) \in M_{a,b}$ and assume that $a_j < b_j$. Since the semigroup T_j is holomorphic on S_{a_j,b_j} the map $\eta \to T(\zeta_1, \zeta_2, \dots, \zeta_{j-1}, \eta, \zeta_{j+1}, \zeta_k)u$ is holomorphic on S_{α_j,β_j} for every $(\zeta_1, \dots, \zeta_{j-1}, \zeta_{j+1}, \zeta_k) \in \Pi_{\substack{1\le s\le k\\ s\neq j}} \overline{S}_{\alpha_s,\beta_s}$.

Notice that if $u \in \mathcal{B}$, where $\mathcal{B}$ is a normalization of $\mathcal{A}$ with respect to T, then the closed subspace $\mathcal{B}_{T,u}$ spanned by the set $\{T(\zeta)u \mid \zeta \in S_{a,b}\}$ is separable, and so the function $\zeta \to T(\zeta)u$ takes its values in a closed separable subspace of $\mathcal{B}$.

With the convention $T_j(0) = I$ for $1 \le j \le k$, we see that if $(\alpha, \beta) \in M_{a,b}$ and if $\lambda \in \cup_{(\gamma,\delta)\in M_{a-\alpha,b-\beta}} \overline{S}_{\gamma,\delta}$ then $T_{(\lambda)} : \zeta \to T(\lambda\zeta) = (T_1(\lambda_1\zeta_1), \dots, T_k(\lambda_k\zeta_k))$ is well-defined for $\zeta \in \overline{S}_{\alpha,\beta}$.

PROPOSITION 8.2. *Let $(\alpha, \beta) \in M_{a,b}$. For $\lambda \in \cup_{(\gamma,\delta)\in M_{a-\alpha,b-\beta}} \overline{S}_{\gamma,\delta}$, denote by $N(T, \lambda, \alpha, \beta)$ the set of all $z \in \mathbb{C}^k$ such that*

$$\limsup_{t\to+\infty} |e^{tz_je^{i\omega}}| \|T_j(t\lambda_je^{i\omega})\| < +\infty \quad \text{for } \alpha_j \le \omega \le \beta_j, 1 \le j \le k$$

, and denote by $N_0(T, \lambda, \alpha, \beta)$ the set of all $z \in \mathbb{C}^k$ such that

$$\lim_{t\to+\infty} |e^{tz_je^{i\omega}}| \|T_j(t\lambda_je^{i\omega})\| = 0 \quad \text{for } \alpha_j \le \omega \le \beta_j, 1 \le j \le k.$$

Then $z \in N(T, \lambda, \alpha, \beta)$ if, and only if, we have, for $1 \le j \le k$,

$$\limsup_{t\to+\infty} |e^{tz_je^{i\alpha_j}}| \|T_j(t\lambda_je^{i\alpha_j})\| < +\infty$$

and

$$\limsup_{t\to+\infty} |e^{tz_je^{i\beta_j}}| \|T_j(t\lambda_je^{i\beta_j})\| < +\infty.$$

Also $z \in N_0(T, \lambda, \alpha, \beta)$ if and only if $Re(z_je^{i\alpha_j}) < -\lim_{t\to+\infty} \frac{\log\|T(t\lambda_je^{i\alpha_j})\|}{t}$ and $Re(z_je^{i\beta_j}) < -\lim_{t\to+\infty} \frac{\log\|T(t\lambda_je^{i\beta_j})\|}{t}$ for $1 \le j \le k$.

PROOF. Let $j \le k$ such that $\alpha_j < \beta_j$. If $\alpha_j \le \omega \le \beta_j$, there exists $r_0 > 0$ and $s_0 > 0$ such that $e^{i\omega} = r_0e^{i\alpha_j} + s_0e^{i\beta_j}$, and we have, for $z_j \in \mathbb{C}$,

$$|e^{tz_je^{i\omega}}| \|T_j(t\lambda_je^{i\omega}\| \le |e^{r_0tz_je^{i\alpha_j}}| \|T_j(r_0t\lambda_je^{i\alpha_j}\| |e^{s_0tz_je^{i\beta_j}}| \|T_j(s_0t\lambda_je^{i\beta_j}\|, \tag{14}$$

and we see that $z \in N(T, \lambda, \alpha, \beta)$ if and only if we have, for $1 \le j \le k$,

$$\limsup_{t\to+\infty} |e^{tz_je^{i\alpha_j}}| \|T_j(t\lambda_je^{i\alpha_j})\| < +\infty \text{ and } \limsup_{t\to+\infty} |e^{tz_je^{i\beta_j}}| \|T_j(t\lambda_je^{i\beta_j})\| < +\infty,$$

which implies that we have

$$Re(z_je^{i\alpha_j}) \le -\lim_{t\to+\infty} \frac{\log\|T(t\lambda_je^{i\alpha_j}\|}{t} \text{ and } Re(z_je^{i\beta_j}) \le -\lim_{t\to+\infty} \frac{\log\|T(t\lambda_je^{i\beta_j}\|}{t}.$$

A similar argument shows that $z \in N_0(T, \lambda, \alpha, \beta)$ if and only if

$$Re(z_je^{i\alpha_j}) < -\lim_{t\to+\infty} \frac{\log\|T(t\lambda_je^{i\alpha_j}\|}{t} \text{ and } Re(z_je^{i\beta_j}) < -\lim_{t\to+\infty} \frac{\log\|T(t\lambda_je^{i\beta_j}\|}{t},$$

so that $Re(z_je^{i\omega}) < -\lim_{t\to+\infty} \frac{\log\|T(t\lambda_je^{i\omega}\|}{t}$ for $\alpha_j \le \omega \le \beta_j$, $1 \le j \le k$. □

Notice that it follows from equations (16) and (17) in Section 10 that we have the inclusions

$$N(T, \lambda, \alpha, \beta) - \overline{S}^*_{\alpha,\beta} \subset N(T, \lambda, \alpha, \beta) \text{ and } N(T, \lambda, \alpha, \beta) - S^*_{\alpha,\beta} \subset N_0(T, \lambda, \alpha, \beta).$$

Set again $e_z(\zeta) = e^{z\zeta}$ for $z \in \mathbb{C}^k, \zeta \in \mathbb{C}^k$. If $\mathcal{B}$ is a normalization of $\mathcal{A}$ with respect to T, then $\sup_{\substack{|\zeta|\le 1\\ \zeta\in\overline{S}_{\alpha,\beta}}} \|T(\zeta)\|_{\mathcal{M}(\mathcal{B})} < +\infty$ for $(\alpha,\beta) \in M_{a,b}$.

It follows then from (14) that if $\lambda \in \cup_{(\gamma,\delta)\in M_{a-\alpha,b-\delta}} \overline{S}_{\gamma,\delta}$, we have

$$\sup_{\zeta\in S_{\alpha,\beta}} |e_z(\zeta)|\|T(\lambda\zeta)\|_{\mathcal{M}(\mathcal{B})} < +\infty \text{ for } z \in N(T,\alpha,\beta,\lambda),$$

$$\lim_{\substack{|\zeta|\to+\infty\\ \zeta\in\overline{S}_{\alpha,\beta}}} |e_z(\zeta)|\|T(\lambda\zeta)\|_{\mathcal{M}(\mathcal{B})} = 0 \text{ for } z \in N_0(T,\lambda,\alpha,\beta).$$

With the notations of Sections 10 and 11, we obtain the following result, which involves the Fourier–Borel transform introduced in Section 10.

PROPOSITION 8.3. *Let $(\alpha,\beta) \in M_{a,b}$, and let $\lambda \in \cup_{(\gamma,\delta)\in M_{a-\alpha,b-\beta}} \overline{S}_{\gamma,\delta}$.*

(i) If $z \in N(T,\lambda,\alpha,\beta)$, then $e_zT(\lambda.)u_{|\overline{S}_{\alpha,\beta}} \in \mathcal{V}_{\alpha,\beta}(\mathcal{B})$, $\zeta_j - z_j \in Res_{ar}(\lambda_j\Delta_{T_j})$ for $\zeta \in S^*_{a,b}$, $u \in \mathcal{B}$, $1 \le j \le k$, and we have

$$\mathcal{FB}(e_zT(\lambda.)u_{|\overline{S}_{\alpha,\beta}})(\zeta) = (-1)^k \prod_{1\le j\le k} ((z_j - \zeta_j)I + \lambda_j\Delta_{T_j})^{-1}u.$$

(ii)If $z \in N_0(T,\lambda,\alpha,\beta)$ then $e_zT(\lambda.)u_{|\overline{S}_{\alpha,\beta}} \in \mathcal{U}_{\alpha,\beta}(\mathcal{B})$, $\mathcal{FB}(e_zT(\lambda.)u_{|\overline{S}_{\alpha,\beta}}$ has a continuous extension to $\overline{S}^*_{\alpha,\beta}$, $-z_j + \zeta_j \in Res_{ar}(\lambda_j\Delta_j)$ for $1 \le j \le k$, and we have, for $\zeta \in \overline{S}^*_{a,b}$, $u \in \mathcal{B}$,

$$\mathcal{FB}(e_zT(\lambda.)u_{|\overline{S}_{\alpha,\beta}})(\zeta) = (-1)^k \prod_{1\le j\le k} ((z_j - \zeta_j)I + \lambda_j\Delta_{T_j})^{-1}u.$$

PROOF. It follows from the discussion above that $e_zT(\lambda.)u_{|\overline{S}_{\alpha,\beta}} \in \mathcal{V}_{\alpha,\beta}(\mathcal{B})$ if $z \in N(T,\lambda,\alpha,\beta)$, and that $e_zT(\lambda.)u_{|\overline{S}_{\alpha,\beta}} \in \mathcal{U}_{\alpha,\beta}(\mathcal{B})$ if $z \in N_0(T,\lambda,\alpha,\beta)$. Let $z \in N(T,\lambda,\alpha,\beta)$, and let $u \in \mathcal{B}$. It follows from Definition 10.3 (iii) that we have, for $\zeta \in S^*_{a,b}$,

$$\mathcal{FB}(e_zT(\lambda.)u_{|\overline{S}_{\alpha,\beta}})(\zeta)$$
$$= \int_0^{e^{i\omega_1}.\infty} \cdots \int_0^{e^{i\omega_k}.\infty} \prod_{1\le j\le k} e^{(z_j-\zeta_j)\sigma_j}T_j(\lambda_j\sigma_j)ud\sigma_1\ldots d\sigma_k,$$

where $\alpha_j \le \omega_j \le \beta_j$ and where $Re(\zeta_je^{i\omega_j}) > 0$ for $1 \le j \le k$.

Since

$$Re((\zeta_j - z_j)e^{i\omega_j}) > \lim_{t\to+\infty} \frac{\log(\|T(t\lambda_j\omega_j\|}{t}$$

by Proposition 8.2, it follows from Proposition 6.1 and Proposition 7.4 that $\zeta_j - z_j \in Res_{ar}(\lambda_j\Delta_{T_j})$ for $j \le k$, and that we have, for $v \in \mathcal{B}$,

$$\int_0^{e^{i\omega_j}.\infty} e^{(z_j-\zeta_j)\sigma_j}T(\lambda_j\sigma_j)vd\sigma_j = -((z_j - \zeta_j)I + \lambda_j\Delta_{T_j})^{-1}v.$$

Using Fubini's theorem, we obtain

$$\begin{aligned}&\mathcal{FB}(e_zT(\lambda.)u_{|_{\overline{S}_{\alpha,\beta}}})(\zeta)\\ &= (-1)^k((z_1-\zeta_1)I+\lambda_1\Delta_{T_1})^{-1}\dots((z_k-\zeta_k)I+\lambda_k\Delta_{T_k})^{-1}u.\end{aligned}$$

Since $N_0(T,\lambda,\alpha,\beta) \subset N(T,\lambda,\alpha,\beta) - S^*_{\alpha,\beta}$, (ii) follows then from (i). □

We now consider the space $\mathcal{F}_{\alpha,\beta} = (\cap_{z\in\mathbb{C}^k} e_{-z}\mathcal{U}_{\alpha,\beta})' = \cup_{z\in\mathbb{C}^k}(e_{-z}\mathcal{U}_{\alpha,\beta})'$ introduced in Section 11, Definition 11.3. If $\phi \in \mathcal{F}_{\alpha,\beta}$, then $Dom(\mathcal{FB}(\phi))$ is the set of all $z \in \mathbb{C}^k$ such that $\phi \in (e_{-z}\mathcal{U}_{\alpha,\beta})'$, and the z-Cauchy transform $\mathcal{C}_z(\phi)$ is defined according to Definition 11.3 (iv) for $z \in Dom(\mathcal{FB}(\phi))$. In the following definition the action of $\phi \in \mathcal{F}_{\alpha,\beta}$ on an element f of $e_{-z}\mathcal{V}_{\alpha,\beta}(X)$ taking values in a closed separable subspace of $\mathcal{B}$, where $z \in Dom(\mathcal{FB}(\phi))$, is defined according to Definition 11.3. by the formula

$$\langle f, \phi\rangle = \langle e_z f, \phi e_{-z}\rangle,$$

where $\langle g, \phi e_{-z}\rangle = \langle e_{-z}g, \phi\rangle$ for $g \in \mathcal{U}_{\alpha,\beta}$. It follows from the remarks following Definition 11.3 that the above definition does not depend on the choice of z.

DEFINITION 8.4. *Let $(\alpha,\beta) \in M_{a,b}$, let $\lambda \in \cup_{(\gamma,\delta)\in M_{a-\alpha,b-\beta}}\overline{S}_{\gamma,\delta}$, let $\phi \in \mathcal{F}_{\alpha,\beta}$, and let $\mathcal{B}$ be a normalization of $\mathcal{A}$ with respect to T.*

For $\zeta \in \overline{S}_{\alpha,\beta}$, set $T_{(\lambda)}(\zeta) = T(\lambda_1\zeta_1,\dots,\lambda_k\zeta_k) = T_1(\lambda_1\zeta_1)\dots T_k(\lambda_k\zeta_k)\}$, with the convention $T_j(0) = I$.

If $N(T,\lambda,\alpha,\beta) \cap Dom(\mathcal{FB}(\phi)) \neq \emptyset$, set, for $u \in \mathcal{B}$,

$$\langle T_{(\lambda)}, \phi\rangle u = \langle T(\lambda.)u_{|_{\overline{S}_{\alpha,\beta}}}, \phi\rangle.$$

For $(\alpha,\beta) \in M_{a,b}$, $z^{(1)} \in \mathbb{C}^k$, $z^{(2)} \in \mathbb{C}^k$, we set as in Definition 11.1

$$\sup(z^{(1)}, z^{(2)}) := \{z \in \mathbb{C}^k \mid z + \overline{S}^*_{\alpha,\beta} = (z^{(1)} + \overline{S}^*_{\alpha,\beta}) \cap (z^{(2)} + \overline{S}^*_{\alpha,\beta}),$$

so that $\sup(z^{(1)}, z^{(2)})$ is a singleton if $a_j < b_j$ for $j \leq k$.

LEMMA 8.5. *Let $\phi_1, \phi_2 \in \mathcal{F}_{\alpha,\beta}$, and assume that we have*

$$N(T,\lambda,\alpha,\beta) \cap Dom(\mathcal{FB}(\phi_1)) \neq \emptyset, N(T,\lambda,\alpha,\beta) \cap Dom(\mathcal{FB}(\phi_2)) \neq \emptyset.$$

Then if $z^{(1)} \in N(T,\lambda,\alpha,\beta) \cap Dom(\mathcal{FB}(\phi_1))$, $z^{(2)} \in N(T,\lambda,\alpha,\beta) \cap Dom(\mathcal{FB}(\phi_2))$, we have

$$\begin{aligned}\sup(z^{(1)}, (z^{(2)}) &\subset N(T,\lambda,\alpha,\beta) \cap Dom(\mathcal{FB}(\phi_1)) \cap Dom(\mathcal{FB}(\phi_2))\\ &\subset N(T,\lambda,\alpha,\beta) \cap Dom(\mathcal{FB}(\phi_1 * \phi_2)),\end{aligned}$$

and the same property holds for $N_0(T,\lambda,\alpha,\beta)$.

PROOF. Assume that $z^{(j)} \in N(T,\lambda,\alpha,\beta) \cap Dom(\mathcal{FB}(\phi_j))$, for $j = 1,2$, set $z \in \sup(z^{(1)}, z^{(2)})$, and let $j \leq k$. There exists $s_1 \in \{1,2\}$ and $s_2 \in \{1,2\}$ such that $z_j \in \left(z_j^{(s_1)} + [0, e^{(-\frac{\pi}{2}-\alpha_j)i}.\infty)\right) \cap \left([z_j^{(s_2)} + [0, e^{(\frac{\pi}{2}-\beta_j)i}.\infty)\right)$, and it follows from (17) and from Proposition 8.2 that $z \in N(T,\lambda,\alpha,\beta)$. The fact that $z \in Dom(\phi_1 * \phi_2)$ follows from Proposition 11.6. A similar argument shows that the same property holds for $N_0(T,\lambda,\alpha,\beta)$. □

The following theorem involves the notion of z-representative measures and the notion of z-Cauchy transform, which are introduced in Section 11.

THEOREM 8.6. *Let $\mathcal{A}$ be a weakly cancellative commutative Banach algebra with dense principal ideals, let $a, b \in \mathbb{R}^k$, let $T = (T_1, \dots, T_k)$ be a family of semigroups of multipliers on $\mathcal{A}$ satisfying the conditions of Definition 8.1, let $\mathcal{B}$ be a normalization of $\mathcal{A}$ with respect to T. Also let $(\alpha, \beta) \in M_{a,b}$ and let $\lambda \in \cup_{(\gamma,\delta)\in M_{a-\alpha,b-\beta}} \overline{S}_{\gamma,\delta}$.*

If $N(T, \lambda, \alpha, \beta) \cap Dom(\mathcal{FB}(\phi)) \neq \emptyset$ for some $\phi \in \mathcal{F}_{\alpha,\beta}$, then the following properties hold

(i) $\langle T_{(\lambda)}, \phi\rangle \in \mathcal{M}(\mathcal{B}) \subset \mathcal{QM}_r(\mathcal{A})$, and we have, if ν is a z-representative measure for ϕ,

$$\langle T_{(\lambda)}, \phi\rangle = \int_{\overline{S}_{\alpha,\beta}} e^{z\zeta} T(\lambda\zeta) d\nu(\zeta) \ (z \in N(T, \lambda, \alpha, \beta) \cap Dom(\mathcal{FB}(\phi))),$$

where the Bochner integral is computed with respect to the strong operator topology on $\mathcal{M}(\mathcal{B})$. Also, if χ is a character on $\mathcal{A}$, then we have

$$\tilde{\chi}\left(\langle T_{(\lambda)}, \phi\rangle\right) = \mathcal{FB}(\phi)(-\lambda_1\tilde{\chi}(\Delta_{T_1}), \dots, -\lambda_k\tilde{\chi}(\Delta_{T_k})),$$

where $\tilde{\chi}$ denotes the unique character on $\mathcal{QM}(\mathcal{A})$ such that $\tilde{\chi}_{|\mathcal{A}} = \chi$.

(ii)

$$\lim_{\substack{\eta\to(0,\dots,0),\eta\in\overline{S}_{\alpha,\beta} \\ \epsilon\to(0,\dots,0),\epsilon\in\overline{S}^*_{\alpha,\beta}}} \|\langle e_{-\epsilon}T_{(\lambda)}, \phi * \delta_\eta\rangle u \ - \langle T_{(\lambda)}, \phi\rangle u\|_{\mathcal{B}} = 0 \text{ for } u \in \mathcal{B}.$$

(iii) If $\alpha_j < \beta_j < \alpha_j + \pi$ for $1 \le j \le k$, then we have, for $\eta \in S_{\alpha,\beta}, \epsilon \in S^*_{\alpha,\beta}$, $z \in N(T, \lambda, \alpha, \beta) \cap Dom(\mathcal{FB}(\phi))$,

$$\langle e_{-\epsilon}T_{(\lambda)}, \phi * \delta_\eta\rangle = e^{-z\eta} \int_{\partial\overline{S}_{\alpha,\beta}} e^{(z-\epsilon)\sigma} \mathcal{C}_z(\phi)(\sigma - \eta) T(\lambda\sigma) d\sigma$$

where the Bochner integral is computed with respect to the strong operator topology on $\mathcal{M}(\mathcal{B})$.

(iv) In the general case, set

$$W_n(\zeta) = \prod_{1\le j\le k} \frac{n^2}{\left(n + \zeta_j e^{i\frac{\alpha_j+\beta_j}{2}}\right)^2} \ (n \ge 1, \zeta \in \overline{S}^*_{\alpha\beta}).$$

Then we have, for $z \in N(T, \lambda, \alpha, \beta) \cap Dom(\mathcal{FB}(\phi))$,

$$\begin{aligned} &\langle T_{(\lambda)}, \phi\rangle \\ = &\lim_{\substack{\epsilon\to 0\\ \epsilon\in S^*_{\alpha,\beta}}} \left(\lim_{n\to+\infty} \frac{(-1)^k}{(2\pi i)^k} \int_{z+\tilde{\partial} S^*_{\alpha,\beta}} W_n(\sigma - z)\mathcal{FB}(\phi)(\sigma) \prod_{1\le j\le k} ((\sigma_j - \epsilon_j)I + \lambda_j\Delta_{T_j})^{-1} d\sigma \right) \end{aligned}$$

where the Bochner integral is computed with respect to the norm topology on $\mathcal{M}(\mathcal{B})$.

(v) Assume, further, that $\int_{z+\tilde{\partial}S^*_{\alpha,\beta}} |\mathcal{FB}(\phi)(\sigma)||d\sigma)| < +\infty$. Then we have, for $z \in N(T, \lambda, \alpha, \beta) \cap Dom(\mathcal{FB}(\phi))$,

$$\begin{aligned} &\langle T_{(\lambda)}, \phi\rangle \\ = &\lim_{\substack{\epsilon\to 0\\ \epsilon\in S^*_{\alpha,\beta}}} \langle e_{-\epsilon}T_{(\lambda)}, \phi\rangle \\ = &\lim_{\substack{\epsilon\to 0\\ \epsilon\in S^*_{\alpha,\beta}}} \frac{(-1)^k}{(2\pi i)^k} \int_{z+\tilde{\partial}S^*_{\alpha,\beta}} \mathcal{FB}(\phi)(\sigma) \prod_{1\le j\le k} ((\sigma_j - \epsilon_j)I + \lambda_j\Delta_{T_j})^{-1} d\sigma, \end{aligned}$$

where the Bochner integral is computed with respect to the norm topology on $\mathcal{M}(\mathcal{B})$.

(vi) If the condition of (v) is satisfied by $z \in N_0(T, \lambda, \alpha, \beta) \cap Dom(\mathcal{FB}(\phi))$, *then we have*

$$\langle T_{(\lambda)}, \phi\rangle = \frac{(-1)^k}{(2\pi i)^k} \int_{z+\tilde{\delta}S^*_{\alpha,\beta}} \mathcal{FB}(\phi)(\sigma) \prod_{1\le j\le k} ((\sigma_j I + \lambda_j \Delta_{T_j})^{-1} d\sigma.$$

(vii) If $\phi_1 \in \mathcal{F}_{\alpha,\beta}, \phi_2 \in \mathcal{F}_{\alpha,\beta}$, *and if* $N(T, \lambda, \alpha, \beta) \cap Dom(\mathcal{FB}(\phi_1)) \neq \emptyset$ *and* $N(T, \lambda, \alpha, \beta) \cap Dom(\mathcal{FB}(\phi_2)) \neq \emptyset$, *then* $N(T, \lambda, \alpha, \beta) \cap Dom(\mathcal{FB}(\phi_1 * \phi_2)) \neq \emptyset$, *and*

$$\langle T_{(\lambda)}, \phi_1 * \phi_2\rangle = \langle T_{(\lambda)}, \phi_1\rangle\langle T_{(\lambda)}, \phi_2\rangle.$$

PROOF. (i) Let $z \in N(T, \lambda, \alpha, \beta) \cap Dom(\mathcal{FB}(\phi))$, and set

$$m := \sup_{\zeta\in S_{\alpha,\beta}} |e^{z\zeta}| \|T_\lambda(\zeta)\|_{\mathcal{M}(\mathcal{B})} < +\infty.$$

We have

$$\|\langle T_\lambda, \phi\rangle u\|_{\mathcal{B}} \le m\|\phi e_z\|_{\mathcal{U}'_{\alpha,\beta}} \|u\|_{\mathcal{B}},$$

and so $\langle T_\lambda, \phi\rangle \in \mathcal{M}(\mathcal{B})$. The integral formula in (i) follows then immediately from the definition given in Proposition 10.2 and from Definition 11.3.

Assume that $\mathcal{A}$ is not radical, let χ be a character on $\mathcal{A}$, and let $\tilde{\chi}$ be the unique character on $\mathcal{QM}(\mathcal{A})$ such that $\tilde{\chi}(u) = \chi(u)$ for every $u \in \mathcal{A}$. Set $f_n(t) = 0$ if $0 \le t < \frac{1}{n+1}$ or if $t > \frac{1}{n}$, and $f_n(t) = n(n+1)$ if $\frac{1}{n+1} \le t \le \frac{1}{n}$, and let ζ be an element of the domain of definition of T_j. Set $T_{j,\zeta} := (T_j(t\zeta))_{t>0}$. Then $(f_n)_{n\ge1} \subset L^1_{\omega_{T_{j,\zeta}}}(\mathbb{R}^+)$ is a Dirac sequence, and since the map $t \to T_j(t\zeta)u$ is continuous on $(0, +\infty)$, a standard argument shows that we have, for $s > 0$, $u \in \mathcal{A}$,

$$\begin{aligned} &\limsup_{n\to+\infty} \left\|\Phi_{T_{j,\zeta}}(f_n)T_j(s\zeta)u - T_j(s\zeta)u\right\| \\ = &\limsup_{n\to+\infty} \left\|\int_0^{+\infty} (f_n * \delta_s)(t)T_j(t\zeta)udt - T_j(s\zeta)u\right\| = 0. \end{aligned}$$

Since $\cup_{t>0}T_j(t\zeta)(\mathcal{A})$ is dense in $\mathcal{A}$, t $\tilde{\chi}(\Phi_{T_{j,\zeta}}(f_n)) \neq 0$, for some $n \ge 1$, and the restriction of $\tilde{\chi}$ to the Arveson ideal $\mathcal{I}_{T_{j,\zeta}}$ is a character on $\mathcal{I}_{T_{j,\zeta}}$. It follows then from Propositions 5.7 and 7.1 that $\tilde{\chi}(T_j(t\zeta)) = e^{t\tilde{\chi}(\Delta_{T_{j,\zeta}})} = e^{t\zeta\tilde{\chi}(\Delta_{T_j})}$ for $t > 0$, and so $\tilde{\chi}(T_j(\zeta)) = e^{\zeta\tilde{\chi}(\Delta_{T_j})}$ for every ζ in the domain of definition of T_j. Let $u \in \mathcal{M}(\mathcal{B})$. By continuity, we see that $\tilde{\chi}(T_j(\zeta)u) = e^{\zeta\tilde{\chi}(\Delta_{T_j})}\tilde{\chi}(u)$ for every $\zeta \in \overline{S}_{\alpha_j,\beta_j}$. Set $\lambda\zeta\tilde{\chi}(\Delta_T) = \lambda_1\zeta_1\tilde{\chi}(\Delta_{T_1}) + \cdots + \lambda_k\zeta_k\tilde{\chi}(\Delta_{T_k})$. Consider again $z \in N(T, \lambda, \alpha, \beta) \cap Dom(\mathcal{FB}(\phi))$. Since

$$\mathcal{FB}(\phi e_{-z})(\zeta) = \langle e_{-\zeta}, \phi e_{-z}\rangle = \langle e_{-z-\zeta}, \phi\rangle = \mathcal{FB}(\phi)(\zeta + z)$$

for $\zeta \in \overline{S}^*_{\alpha,\beta}$, we obtain

$$\begin{aligned}
\tilde{\chi}(\langle T_{(\lambda)}, \phi\rangle)\tilde{\chi}(u) &= \tilde{\chi}\left(\int_{\overline{S}_{\alpha,\beta}} e^{z\zeta}T(\lambda\zeta)ud\nu(\zeta)\right)\\
&= \int_{\overline{S}_{\alpha,\beta}} e^{z\zeta}\tilde{\chi}(T(\lambda\zeta))\tilde{\chi}(u)d\nu(\zeta)\\
&= \left(\int_{\overline{S}_{\alpha,\beta}} e^{z\zeta}e^{\lambda\zeta\tilde{\chi}(\Delta_T)}d\nu(\zeta)\right)\tilde{\chi}(u)\\
&= \mathcal{FB}(\phi e_{-z})(-\lambda\chi(\Delta_T) - z)\tilde{\chi}(u)\\
&= \mathcal{FB}(\phi)(-\lambda\chi(\Delta_T))\tilde{\chi}(u),
\end{aligned}$$

which concludes the proof of (i), since $\tilde{\chi}(u) \neq 0$.

(ii) Let $u \in \mathcal{B}$, and set $f(\zeta) = T(\lambda\zeta)u$ for $\zeta \in \overline{S}_{\alpha,\beta}$. Using Definition 11.3, we see that (ii) follows from (34) applied to f.

(iii) Define f as above. We have, for $\epsilon \in S^*_{\alpha,\beta}, \eta \in S_{\alpha,\beta}$, $u \in \mathcal{B}$,

$$\langle e_{-\epsilon}T_{(\lambda)}, \phi * \delta_\eta\rangle u = \langle e_{-\epsilon}f, \phi * \delta_\eta\rangle = \langle (e_{-\epsilon})_\eta f_\eta, \phi\rangle = e^{-\epsilon\eta}\langle e_{-\epsilon}f_\eta, \phi\rangle,$$

so (iii) follows from (37).

(iv) The result follows from Proposition 11.9 (i) applied to $T_{(\lambda)}u_{|\overline{S}_{\alpha,\beta}}$, z and ϕ for $u \in \mathcal{B}$.

(v) The result follows from Proposition 11.9 (ii) applied to $T_{(\lambda)}u_{|\overline{S}_{\alpha,\beta}}$, z and ϕ for $u \in \mathcal{B}$.

(vi) Now assume that $z \in N_0(T, \lambda, \alpha, \beta) \cap Dom(\mathcal{FB}(\phi))$ satisfies the condition of (v).

There exists $\epsilon \in S^*_{\alpha,\beta}$ such that $z + \epsilon \in N_0(T, \lambda, \alpha, \beta) \cap Dom(\mathcal{FB}(\phi e_\epsilon))$. We have

$$\begin{aligned}
&\int_{\partial\overline{S}_{\alpha,\beta}} e^{Re(z\sigma)}\|T_{(\lambda)}(\sigma)\|_{\mathcal{M}(\mathcal{B})}|d\sigma|\\
&\leq \left(\sup_{\zeta\in\overline{S}_{\alpha,\beta}} e^{Re((z+\epsilon)\zeta)}\|T_{(\lambda)}(\zeta)\|_{\mathcal{M}(\mathcal{B})}\right)\int_{\partial\overline{S}_{\alpha,\beta}} e^{-Re(\epsilon\sigma)}|d\sigma| < +\infty,
\end{aligned}$$

and (vi) follows from Proposition 11.9 (iii) applied to $T_{(\lambda)}u_{|\overline{S}_{\alpha,\beta}}$ z and ϕ for $u \in \mathcal{B}$.

(vii) Now assume that $\phi_1 \in \mathcal{F}_{\alpha,\beta}, \phi_2 \in \mathcal{F}_{\alpha,\beta}$ satisfy the hypothesis of (vi) with respect to T and λ, and let $z^{(1)} \in N(T, \lambda, \alpha, \beta) \cap Dom(\mathcal{FB}(\phi_1))$ and $z^{(2)} \in N(T, \lambda, \alpha, \beta) \cap Dom(\mathcal{FB}(\phi_2))$. Set $z = \sup(z^{(1)}, z^{(2)})$.

It follows from Lemma 8.5 that

$$z \in N(T, \lambda, \alpha, \beta) \cap Dom(\mathcal{FB}(\phi_1)) \cap Dom(\mathcal{FB}(\phi_2)) \subset N(T, \lambda, \alpha, \beta) \cap Dom(\mathcal{FB}(\phi_1 * \phi_2)).$$

Let ν_1 be a z-representative measure for ϕ_1 and let ν_2 be a z-representative measure for ϕ_2. Then $\nu_1 * \nu_2$ is a z-representative measure for $\phi_1 * \phi_2$, and we have, for $u \in \mathcal{B}$,

$$
\begin{aligned}
&\langle T_{(\lambda)}, \phi_1 * \phi_2 \rangle u \\
= & \int_{\overline{S}_{\alpha,\beta}} e^{z\zeta} T(\lambda\zeta) u d(\nu_1 * \nu_2)(\zeta) \\
= & \int_{\overline{S}_{\alpha,\beta}\times\overline{S}_{\alpha,\beta}} e^{z(\zeta_1+\zeta_2)} T(z(\zeta_1+\zeta_2)) u d\nu_1(\zeta_1) d\nu_2(\zeta_2) \\
& \left[\int_{\overline{S}_{\alpha,\beta}} e^{z\zeta_1} T(\lambda\zeta_1) d\nu_1(\zeta_1)\right]\left[\int_{\overline{S}_{\alpha,\beta}} e^{z\zeta_2} T(\lambda\zeta_2) u d\nu_2(\zeta_2)\right] \\
= & \langle T_{(\lambda)}, \phi_1\rangle \left(\langle T_{(\lambda)}, \phi_2\rangle u\right),
\end{aligned}
$$

which proves (vii). □

Let $\mathcal{G}_{a,b} = \cup_{(\alpha,\beta)\in M_{a,b}} \mathcal{F}_{\alpha,\beta}$ be the dual space introduced in Definition 11.2, which is an algebra with respect to convolution according to Proposition 11.13. If $\phi_1 \in \mathcal{F}_{\alpha^{(1)},\beta^{(1)}}$, and if $\phi_1 \in \mathcal{F}_{\alpha^{(2)},\beta^{(2)}}$, where $(\alpha^{(j)},\beta^{(j)}) \in M_{a,b}$ for $j = 1, 2$, then $\phi_1 * \phi_2$ is well-defined.

But in general the fact that $N(T,\lambda,\alpha^{(j)},\beta^{(j)}) \cap Dom(\phi_j) \neq \emptyset$ for $j = 1, 2$ does not seem to imply that $N(T,\lambda,\inf(\alpha^{(1)},\alpha^{(2)}),\sup(\beta^{(1)},\beta^{(2)})) \cap Dom(\phi_1 * \phi_2)$ is not empty, which prevents from obtaining a direct extension of (vi) to the case where $\phi_1 \in \mathcal{G}_{a,b}, \phi_2 \in \mathcal{G}_{a,b}$. This difficulty will be circumvented in the next section by using Fourier–Borel transforms.

9. Multivariable functional calculus associated to holomorphic functions of several complex variables

In the following definition, the generator Δ_{T_j} of the strongly continuous semigroup T_j and its Arveson spectrum $\sigma_{ar}(\Delta_{T_j})$ are defined according to Section 5 if $a_j = b_j$, and the generator Δ_{T_j} of the holomorphic semigroup T_j and its Arveson spectrum $\sigma_{ar}(\Delta_{T_j})$ are defined according to Section 7 if $a_j < b_j$.

DEFINITION 9.1. *Let $a = (a_1,\dots,a_k) \in \mathbb{R}^k, b = (b_1,\dots,b_k) \in \mathbb{R}^k$ such that $a_j \le b_j \le a_j + \pi$ for $j \le k$, let $\mathcal{A}$ be a weakly cancellative commutative Banach algebra having dense principal ideals, and let $T = (T_1,\dots,T_k)$ be a family of semigroups of multipliers on $\mathcal{A}$ satisfying the conditions of Definition 8.1. Let $(\alpha,\beta) \in M_{a,b}$ and let $\lambda \in \cup_{(\gamma,\delta)\in M_{a-\alpha,b-\beta}} \overline{S}_{\gamma,\delta}$.*

An open set $U \subset \mathbb{C}^k$ is said to be admissible with respect to (T,λ,α,β) if $U = \Pi_{1\le j\le k} U_j$ where the open sets $U_j \subset \mathbb{C}$ satisfy the following conditions for some $z = (z_1,\dots,z_k) \in N_0(T,\alpha,\beta,\lambda)$

(i) $U_j + \overline{S}^*_{\alpha_j,\beta_j} \subset U_j$

(ii) $U_j \subset z_j + S^*_{\alpha_j,\beta_j}$, and

$$
\begin{aligned}
& \partial U_j \\
= & (z_j + e^{(-\frac{\pi}{2}-\alpha_j)i}.\infty, z_j + e^{(-\alpha_j-\frac{\pi}{2})i}s_{0,j}) \cup \gamma([0,1]) \\
& \cup (z_j + e^{(\frac{\pi}{2}-\beta_j)i}s_{1,j}, z_j + e^{(\frac{\pi}{2}-\beta_j)i}.\infty),
\end{aligned}
$$

where $s_{0,j} \ge 0, s_{1,j} \ge 0$, and where

$$\gamma : [0,1] \to z_j + \overline{S}^*_{\alpha_j,\beta_j} \setminus \left(e^{(-\frac{\pi}{2}-\alpha_j)i}.\infty, e^{(-\alpha_j-\frac{\pi}{2})i}s_{0,j}) \cup (e^{(\frac{\pi}{2}-\beta_j)i}s_{1,j}, e^{(\frac{\pi}{2}-\beta_j)i}.\infty)\right)$$

is a one-to-one piecewise-$\mathcal{C}^1$ *curve such that*

$$\gamma(0) = e^{(-\alpha_j - \frac{\pi}{2})i} s_{0,j} \quad and\{\gamma(1) = e^{(\frac{\pi}{2} - \beta_j)i} s_{1,j}.$$

(iii) $\lambda_j \sigma_{ar}(-\Delta_{T_j}) = \sigma_{ar}(-\Delta_{T_j(\lambda_j .)}) \subset U_j$.

Conditions (i) and (ii) mean that U is admissible with respect to (α, β) in the sense of Definition 12.1 and that some, hence all elements $z \in \mathbb{C}^k$ with respect to which U satisfies condition (ii) of Definition 12.1 belong to $N_0(T, \alpha, \beta, \lambda)$. Hence U_j is a open half-plane if $\alpha_j = \beta_j$, and the geometric considerations about ∂U_j made in Section 12 when $\alpha_j < \beta_j$ apply.

For $\alpha = (\alpha_1, \dots, \alpha_k) \in \mathbb{R}^k, \beta = (\beta_1, \dots, \beta_k) \in \mathbb{R}^k$, we will use as in Appendix 3 the obvious conventions

$$\inf(\alpha, \beta) = (\inf(\alpha_1, \beta_1), \dots, \inf(\alpha_k, \beta_k)),\ \sup(\alpha, \beta) = (\sup(\alpha_1, \beta_1), \dots, \sup(\alpha_k, \beta_k)).$$

PROPOSITION 9.2. *If* $U^{(1)}$ *is admissible with respect to* $(T, \lambda, \alpha^{(1)}, \beta^{(1)})$ *and if* $U^{(2)}$ *is admissible with respect to* $(T, \lambda, \alpha^{(2)}, \beta^{(2)})$*, then* $U^{(1)} \cap U^{(2)}$ *is admissible with respect to* $(T, \lambda, \inf(\alpha^{(1)}, \alpha^{(2)}), \sup(\beta^{(1)}, \beta^{(2)}))$.

PROOF. Set

$$\alpha^{(3)} = \inf(\alpha^{(1)}, \alpha^{(2)}), \beta^{(3)} = \sup(\beta^{(1)}, \beta^{(2)}), U^{(3)} = U^{(1)} \cap U^{(2)}.$$

Then

$$\left[\cup_{(\gamma,\delta)\in M_{a-\alpha^{(1)},b-\beta^{(1)}}} \overline{S}_{\gamma,\delta}\right] \cap \left[\cup_{(\gamma,\delta)\in M_{a-\alpha^{(2)},b-\beta^{(2)}}} \overline{S}_{\gamma,\delta}\right] \subset \left[\cup_{(\gamma,\delta)\in M_{a-\alpha^{(3)},b-\beta^{(3)}}} \overline{S}_{\gamma,\delta}\right],$$

so it makes sense to check whether $U^{(1)} \cap U^{(2)}$ is admissible with respect to $(T, \lambda, \alpha^{(3)}, \beta^{(3)})$. The fact that $U^{(3)}$ satisfies (i) and (ii) follows from Proposition 12.2, and the fact that $U^{(3)}$ satisfies (iii) is obvious. $\square$

If an open set $U \subset \mathbb{C}^k$ is admissible with respect to $(T, \lambda, \alpha, \beta)$, we denote as in Section 12 by $H^{(1)}(U)$ the set of all holomorphic functions F on U satisfying the condition

$$\|F\|_{H^{(1)}(U)} := \sup_{\epsilon \in S^*_{\alpha,\beta}} \int_{\sigma \in \tilde{\partial} U + \epsilon} |F(\sigma)||d\sigma| < +\infty.$$

Notice that $\cup_{(\alpha,\beta)\in M_{a,b}} \left(\cup_{(\gamma,\delta)\in M_{a-\alpha,b-\beta}} \overline{S}_{\beta,\gamma}\right) = \cup_{(\alpha,\beta)\in M_{a,b}} \overline{S}_{a-\alpha,b-\beta}$. The inclusion $\cup_{(\gamma,\delta)\in M_{a-\alpha,b-\beta}} \overline{S}_{\beta,\gamma} \subset \overline{S}_{a-\alpha,b-\beta}$ for $(\alpha, \beta) \in M_{a,b}$ is obvious. Conversely assume that $\lambda \in \overline{S}_{a-\alpha,b-\beta}$ for some $(\alpha, \beta) \in M_{a,b}$. If $a_j = b_j$ then $a_j = \alpha_j = \beta_j = b_j$, and so λ_j is a nonnegative real number. In this situation set $\alpha'_j = \beta'_j = a_j, \gamma_j = \delta_j = 0$. If $a_j < b_j$, then $a_j < \alpha_j \le \beta_j < b_j$, and $a_j - \alpha_j \le arg(\lambda_j) \le b_j - \beta_j$ if $\lambda_j \neq 0$.

In this situation set $\alpha'_j = \frac{a_j + \alpha_j}{2}, \gamma_j = a_j - \alpha_j, \delta_j = b_j - \beta_j$ and $\beta'_j = \frac{b_j + \beta_j}{2}$. Then $(\alpha', \beta') \in M_{a,b}$, $(\gamma, \delta) \in M_{a-\alpha',b-\beta'}$, and $\lambda \in \overline{S}_{\gamma,\delta}$, which concludes the proof of the reverse inclusion.

DEFINITION 9.3. *For* $\lambda \in \cup_{(\alpha,\beta)\in M_{a,b}} \overline{S}_{a-\alpha,b-\beta} = \cup_{\alpha,\beta\in M_{a,b}} \left(\cup_{(\gamma,\delta)\in M_{a-\alpha,b-\beta}} \overline{S}_{\gamma,\delta}\right)$, *denote by* $\mathcal{N}_\lambda$ *the set of all* $(\alpha, \beta) \in M_{a,b}$ *such that* $\lambda \in \cup_{(\gamma,\delta)\in M_{a-\alpha,b-\beta}} \overline{S}_{\gamma,\delta}$*, and denote by* $\mathcal{W}_{T,\lambda}$ *the set of all open sets* $U \subset \mathbb{C}^k$ *which are admissible with respect to* $(T, \lambda, \alpha, \beta)$ *for some* $(\alpha, \beta) \in \mathcal{N}_\lambda$.

COROLLARY 9.4. *Let*

$$\lambda \in \cup_{(\alpha,\beta)\in M_{a,b}} \overline{S}_{a-\alpha,b-\beta} = \cup_{\alpha,\beta\in M_{a,b}} \left(\cup_{(\gamma,\delta)\in M_{a-\alpha,b-\beta}} \overline{S}_{\gamma,\delta}\right).$$

The family $\mathcal{W}_{T,\lambda}$ is stable under finite intersections, and $\cup_{U\in\mathcal{W}_{T,\lambda}} H^{(1)}(U)$ is stable under products.

PROOF. The first assertion follows from the proposition and the second assertion follows from the fact that the restriction of $F \in H^{(1)}(U)$ is bounded on $U+\epsilon$ if U is admissible with respect to $(\alpha,\beta) \in M_{a,b}$ and if $\epsilon \in S^*_{\alpha,\beta}$, see Corollary 12.4. □

A set $\mathcal{E} \subset \cup_{U\in\mathcal{W}_{T,\lambda}} H^{(1)}(U)$ will be said to be bounded if there exists $U \in \mathcal{W}_{T,\lambda}$ such that $\mathcal{E} \subset H^{(1)}(U)$ and such that $\sup_{F\in\mathcal{E}} \|F\|_{H^{(1)}(U)} < +\infty$, and bounded subsets of $\cup_{U\in\mathcal{W}_{T,\lambda}} H^\infty(U)$ are defined in a similar way. A homomorphism $\phi : \cup_{U\in\mathcal{W}_{T,\lambda}} H^{(1)}(U) \to \mathcal{M}(\mathcal{B})$ will be said to be bounded if $\phi(\mathcal{E})$ is bounded for every bounded subset $\mathcal{E}$ of $\cup_{U\in\mathcal{W}_{T,\lambda}} H^{(1)}(U)$. Also a homomorphism $\phi : \cup_{U\in\mathcal{W}_{T,\lambda}} H^\infty(U) \to \mathcal{QM}_r(\mathcal{B}) = \mathcal{QM}_r(\mathcal{A})$ will be said to be bounded if $\phi(\mathcal{E})$ is pseudobounded for every bounded subset $\mathcal{E}$ of $\cup_{U\in\mathcal{W}_{T,\lambda}} H^\infty(U)$.

Similarly let $\mathcal{S}(U)$ be the Smirnov class on $U \in \mathcal{W}_{T,\lambda}$ introduced in Definition 12.6.

A set $\mathcal{E} \subset \cup_{U\in\mathcal{W}_{T,\lambda}} \mathcal{S}(U)$ will be said to be bounded if there exists $U \in \mathcal{W}_{T,\lambda}$ such that $\mathcal{E} \subset \mathcal{S}(U)$ and such that $\sup_{F\in\mathcal{E}} \|FG\|_{H^\infty(U)} < +\infty$ for some strongly outer function $G \in H^\infty(U)$.

Also a homomorphism $\phi : \cup_{U\in\mathcal{W}_{T,\lambda}} \mathcal{S}(U) \to \mathcal{QM}(\mathcal{B}) = \mathcal{QM}(\mathcal{A})$ will be said to be bounded if $\phi(\mathcal{E})$ is pseudobounded for every bounded subset $\mathcal{E}$ of $\cup_{U\in\mathcal{W}_{T,\lambda}} \mathcal{S}(U)$.

Let $U = \Pi_{j\le k} U_j \in \mathcal{W}_{T,\lambda}$, and let $(\alpha,\beta) \in \mathcal{N}_\lambda$ such that U is admissible with respect to (T,λ,α,β). Let ∂U_j be oriented from $e^{-\frac{\pi}{2}-\alpha_j}.\infty$ to $e^{\frac{\pi}{2}-\beta_j}.\infty$. This gives an orientation on the distinguished boundary $\tilde{\partial} U = \Pi_{j\le k} \partial U_j$ of U, to be used in the following theorem.

THEOREM 9.5. *Let $a = (a_1,\dots,a_k) \in \mathbb{R}^k$, let $b = (b_1,\dots,b_k) \in \mathbb{R}^k$ such that $a_j \le b_j \le a_j + \pi$ for $j \le k$, let $\mathcal{A}$ be a weakly cancellative commutative Banach algebra with dense principal ideals, let $T = (T_1,\dots,T_k)$ be a family of semigroups of multipliers on $\mathcal{A}$ satisfying the conditions of Definition 8.1 with respect to (a,b) and $\mathcal{A}$ and let $\mathcal{B}$ be a normalization of $\mathcal{A}$ with respect to T.*

(i) For $\lambda \in \cup_{(\alpha,\beta)\in M_{a,b}} \overline{S}_{\alpha,\beta}$, $U \in \mathcal{W}_{T,\lambda}$, $F \in H^{(1)}(U)$, set

$$\begin{aligned} &F(-\lambda_1\Delta_{T_1},\dots,-\lambda_k\Delta_{T_k}) \\ &= \frac{1}{(2\pi i)^k}\int_{\tilde{\partial} U+\epsilon} F(\zeta_1,\dots,\zeta_k)(\lambda_1\Delta_{T_1}+\zeta_1 I)^{-1}\dots(\lambda_1\Delta_{T_k}+\zeta_k I)^{-1} d\zeta_1\dots d\zeta_k, \end{aligned}$$

where U is admissible to respect to (T,λ,α,β), with $(\alpha,\beta) \in \mathcal{N}_\lambda$, and where $\epsilon \in S^*_{\alpha,\beta}$ is such that $U+\epsilon \in \mathcal{W}_{T,\lambda}$. Then this definition does not depend on the choice of U and ϵ, and the map $F \to F(-\lambda_1\Delta_{T_1},\dots,-\lambda_k\Delta_{T_k})$ is a bounded algebra homomorphism from $\cup_{U\in\mathcal{W}_{T,\lambda}} H^{(1)}(U)$ into $\mathcal{M}(\mathcal{B}) \subset \mathcal{QM}_r(\mathcal{A})$.

(ii) For every $U \in \mathcal{W}_{T,\lambda}$ there exists $G \in H^{(1)}(U) \cap H^\infty(U)$ such that

$$G(-\lambda_1\Delta_{T_1},\dots,-\lambda_k\Delta_{T_k})(\mathcal{B})$$

is dense in $\mathcal{B}$, and for every $F \in H^\infty(U)$ there exists a unique $R_F \in \mathcal{QM}_r(\mathcal{B}) = \mathcal{QM}_r(\mathcal{A})$ satisfying

$$R_F G(-\lambda_1 T_1,\dots,-\lambda_k T_k) = (FG)(-\lambda_1 T_1,\dots,-\lambda_k T_k) \quad (G \in H^{(1)}(U)).$$

The definition of R_F does not depend on the choice of U, and if we set

$$F(-\lambda_1 T_1, \dots, -\lambda_k T_k) = R_F,$$

the definition of $F(-\lambda_1\Delta_{T_1}, \dots, -\lambda_k\Delta_{T_k})$ agrees with the definition given in (i) if $F \in \cup_{U\in\mathcal{W}_{T,\lambda}} H^{(1)}(U)$, and the map $F \to F(-\lambda_1 T_1, \dots, -\lambda_k T_k)$ is a bounded homomorphism from $\cup_{U\in\mathcal{W}_{T,\lambda}} H^\infty(U)$ into $\mathcal{QM}_r(\mathcal{B}) = \mathcal{QM}_r(\mathcal{A})$.

(iii) If $(\alpha,\beta) \in \mathcal{N}_\lambda$, if $\phi \in \mathcal{F}_{\alpha,\beta}$, and if $N_0(T,\lambda,\alpha,\beta) \cap Dom(\mathcal{FB}(\phi)) \neq \emptyset$, then

$$\mathcal{FB}(\phi)(-\lambda_1\Delta_{T_1}, \dots, -\lambda_k\Delta_{T_k}) = \langle T_{(\lambda)}, \phi\rangle.$$

In particular if $F(\zeta) = e^{-\nu\zeta_j}$, where $\nu \in \mathbb{C}$ satisfies $\nu\lambda_j \in \cup_{(\gamma_j,\delta_j)\in M_{a_j,b_j}} \overline{S}_{\gamma_j,\delta_j}$ then $F(-\lambda_1\Delta_{T_1}, \dots, -\lambda_k\Delta_{T_k}) = T_j(\nu\lambda_j)$.

(iv) If $(\alpha,\beta) \in \mathcal{N}_\lambda$, if $\phi \in \mathcal{F}_{\alpha,\beta}$, and if $N(T,\lambda,\alpha,\beta) \cap Dom(\mathcal{FB}(\phi)) \neq \emptyset$, then

$$\langle T_{(\lambda)}, \phi\rangle u = \lim_{\substack{\epsilon\to(0,\dots,0)\\ \epsilon\in S^*_{\alpha,\beta}}} \mathcal{FB}(\phi)(-\lambda_1\Delta_{T_1} + \epsilon_1 I, \dots, -\lambda_k\Delta_{T_k} + \epsilon_k I)u \quad (u \in \mathcal{B}).$$

(v) If $U \in \mathcal{W}_{T,\lambda}$, and if $F \in H^\infty(U)$ is strongly outer on U, then there exists $u \in \Omega(\mathcal{B}) \cap Dom(F(-\lambda_1\Delta_{T_1}, \dots, -\lambda_k\Delta_{T_k}))$ such that

$$F(-\lambda_1\Delta_{T_1}, \dots, -\lambda_k\Delta_{T_k})u \in \Omega(\mathcal{B}).$$

(vi) For every $U \in \mathcal{W}_{T,\lambda}$ and every $F \in \mathcal{S}(U)$ there exists a unique quasimultiplier $R_F \in \mathcal{QM}(\mathcal{B}) = \mathcal{QM}(\mathcal{A})$ satisfying

$$R_F G(-\lambda_1 T_1, \dots, -\lambda_k T_k) = (FG)(-\lambda_1 T_1, \dots, -\lambda_k T_k)$$

for every $G \in H^\infty(U)$ such that $FG \in H^\infty(U)$. The definition of R_F does not depend on the choice of U, and if we set

$$F(-\lambda_1 T_1, \dots, -\lambda_k T_k) = R_F,$$

the definition of $F(-\lambda_1\Delta_{T_1}, \dots, -\lambda_k\Delta_{T_k})$ agrees with the definition given in (ii) if $F \in \cup_{U\in\mathcal{W}_{T,\lambda}} H^\infty(U)$, the map $F \to F(-\lambda_1 T_1, \dots, -\lambda_k T_k)$ is a bounded homomorphism from $\cup_{U\in\mathcal{W}_{T,\lambda}} \mathcal{S}(U)$ into $\mathcal{QM}(\mathcal{B}) = \mathcal{QM}(\mathcal{A})$, and we have, for $\chi \in Spec(\mathcal{A})$,

$$\tilde{\chi}(F(-\lambda_1 T_1, \dots, -\lambda_k T_k)) = F(-\lambda_1\tilde{\chi}(\Delta_{T_1}), \dots, -\lambda_k\tilde{\chi}(\Delta_{T_k})) \quad (F \in \cup_{U\in\mathcal{W}_{T,\lambda}} \mathcal{S}(U)),$$

where $\tilde{\chi}$ is the character on $\mathcal{QM}(\mathcal{A})$ such that $\tilde{\chi}_{|\mathcal{A}} = \chi$.

(vii) If $F(\zeta_1, \dots, \zeta_k) = -\zeta_j$ then $F(-\lambda_1\Delta_1, \dots, -\lambda_k\Delta_k) = \lambda_j\Delta_{T_j}$.

PROOF. In the following we will use the notations

$$d\zeta = d\zeta_1 \dots d\zeta_k, \lambda\Delta_T = (\lambda_1\Delta_{T_1}, \dots, \lambda_k\Delta_{T_k}),$$

$$R(-\lambda\Delta_T, \zeta) = (-1)^k(\lambda_1\Delta_{T_1} + \zeta_1 I)^{-1} \dots (\lambda_k\Delta_{T_k} + \zeta_k I)^{-1})$$

for $\zeta = (\zeta_1, \dots, \zeta_k) \in -Res_{ar}(\lambda\Delta_T) := -\Pi_{j=1}^k Res_{ar}(\Delta_{T_j(\lambda_j\cdot)})$ With these notations, the formula given in (i) takes the form

$$F(-\lambda\Delta_T) = \frac{(-1)^k}{(2\pi i)^k} \int_{\tilde{\partial} U+\epsilon} F(\zeta) R(-\lambda\Delta_T, \zeta) d\zeta.$$

Clearly, $F(-\lambda\Delta_T) \in \mathcal{M}(\mathcal{B}) \subset \mathcal{QM}_r(\mathcal{A})$. Let $U, U' \in \mathcal{W}_{T,\lambda}$, let (α,β) and (α',β') be the elements of $M_{a,b}$ associated to U and U' and let $\epsilon \in S^*_{\alpha,\beta}$ and $\epsilon' \in S^*_{\alpha',\beta'}$ such that $U + \epsilon \in \mathcal{W}_{T,\lambda}$ and $U' + \epsilon' \in \mathcal{W}_{T,\lambda}$. Set

$$V = U + \epsilon, V' = U' + \epsilon', V'' = V \cap V'.$$

Then the function $G : \zeta \to F(\zeta)R(-\lambda\Delta_T, \zeta)$ is holomorphic on a neighborhood of $\overline{V} \setminus V''$, and it follows from (41) that there exists $M > 0$ such that $|G(\zeta)| \leq M$ for $\zeta \in \overline{V} \setminus V''$. The open sets $V = \Pi_{j\leq k} V_j$ and $V'' = \Pi_{j\leq k} V_j''$ have the form $(z + S^*_{\alpha,\beta}) \setminus K$ and $(z'' + S^*_{\alpha'',\beta''}) \setminus K''$ where K and K'' are compact subsets of $\mathbb{C}^k$, and where $\alpha'' = \inf(\alpha, \alpha')$ and $\beta'' = sup(\beta, \beta')$. Choose $\epsilon'' \in S^*_{\alpha'',\beta''}$, and denote by $V_{L,k}$ the intersection of $V_k \setminus \overline{V}_k''$ with the strip having for boundaries the lines $D^1_L = Le^{i(-\frac{\pi}{2}-\alpha_k)} + \mathbb{R}\epsilon_k''$ and $D^2_L = Le^{i(\frac{\pi}{2}+\beta_k)} + \mathbb{R}\epsilon_k''$. Set

$$W_{n,j}(\zeta_j) = \frac{n^2}{\left(n + 1 + (\zeta_j - z_j)e^{i\frac{\alpha_j+\beta_j}{2}}\right)^2},$$

and set $W_n(\zeta) = W_{n,1}(\zeta_1) \ldots W_{n,k}(\zeta_k)$. Then $|W_n(\zeta)| < 1$ and $\lim_{n\to+\infty} W_n(\zeta) = 1$ for $\zeta \in \overline{V}$.

It follows from Cauchy's theorem that we have, when L is sufficiently large

$$0 = \int_{\Pi_{j\leq k-1}\partial V_j} \left[\int_{\partial V_{L,k}} W_n(\zeta)G(\zeta)d\zeta_k\right] d\zeta_1 \ldots d\zeta_{k-1}.$$

We have, for $s = 1, 2$,

$$\begin{aligned} &\left\| \int_{(\Pi_{j\leq k-1}\partial V_j)\times(\partial V_k \cap D^s_L)} W_n(\zeta)G(\zeta)d\zeta \right\| \\ &\leq M \left[\Pi_{j\leq k-1} \int_{\partial V_j} |W_{n,j}(\zeta_j)||d\zeta_j|\right] \int_{\partial V_k \cap D^s_L} |W_{n,k}||d\zeta_k|, \end{aligned}$$

and so $\lim_{L\to+\infty} \int_{(\Pi_{j\leq k-1}\partial V_j)\times(\partial V_k\cap D^s_L)} W_n(\zeta)G(\zeta)d\zeta = 0$. We obtain

$$\int_{\tilde{\partial}V} W_n(\zeta)G(\zeta)d\zeta = \int_{\Pi_{j\leq k-1}\partial V_j \times \partial V_k''} W_n(\zeta)G(\zeta)d\zeta.$$

It follows then from the Lebesgue dominated convergence theorem that we have

$$\int_{\tilde{\partial}V} G(\zeta)d\zeta = \int_{\Pi_{j\leq k-1}\partial V_j \times \partial V_k''} G(\zeta)d\zeta.$$

Using the same argument and a finite induction, we obtain

$$\int_{\tilde{\partial}V} G(\zeta)d\zeta = \int_{\tilde{\partial}V''} G(\zeta)d\zeta.$$

Similarly $\int_{\tilde{\partial}V'} G(\zeta)d\zeta = \int_{\tilde{\partial}V''} G(\zeta)d\zeta$, which shows that the definition of

$$F(-\lambda_1\Delta_{T_1}, \ldots, -\lambda_k\Delta_{T_k})$$

does not depend on the choice of U and ϵ.

Now let $F \in \cup_{U\in\mathcal{W}_{T,\lambda}} H^{(1)}(U)$, let $G \in \cup_{U\in\mathcal{W}_{T,\lambda}} H^{(1)}(U)$. There exists an open set $U \in \mathcal{W}_{T,\lambda}$ such that $F_{|_U} \in H^{(1)}(U)$ and $G_{|_U} \in H^{(1)}(U)$.

Choose $\epsilon \in S^*_{\alpha,\beta}$, where (α, β) is the element of $M_{a,b}$ associated to U, such that $U + \epsilon \in \mathcal{W}_{T,\lambda}$, and set $V = U + \frac{\epsilon}{2}$, $V' = U + \epsilon$. For $M \subset \{1, \ldots, k\}$, denote by $|M|$ the cardinal of M, and set $M^* := \{1, \ldots, k\} \setminus M$. Then $|M^*| = k - |M|$. Since

$$\begin{aligned}&(\lambda_j\Delta_{T_j}+\zeta_jI)^{-1}(\lambda_j\Delta_{T_j}+\sigma_jI)^{-1}\\=\ &\frac{1}{\sigma_j-\zeta_j}\left((\lambda_j\Delta_{T_j}+\zeta_jI)^{-1}-(\lambda_j\Delta_{T_j}+\sigma_jI)^{-1}\right),\end{aligned}$$

we have

$$\begin{aligned}F(-\lambda\Delta_T)G(-\lambda\Delta_T)&=\frac{1}{(2\pi i)^{2k}}\int_{\tilde{\partial}V\times\tilde{\partial}V'}F(\zeta)G(\sigma)R(-\lambda\Delta_T,\zeta)R(-\lambda\Delta_T,\sigma)d\zeta d\sigma\\&=\frac{1}{(2\pi i)^{2k}}\sum_{M\subset\{1,\dots,k\}}L_M,\end{aligned}$$

where

$$(-1)^{|M|}L_M:=$$
$$\int_{\tilde{\partial}V\times\tilde{\partial}V'}\prod_{1\le j\le k}\frac{1}{\sigma_j-\zeta_j}F(\zeta)G(\sigma)\prod_{j\in M}(\lambda_j\Delta_{T_j}+\zeta_jI)^{-1}\prod_{j'\in C_M}(\lambda_{j'}\Delta_{T_{j'}}+\sigma_{j'}I)^{-1}d\zeta d\sigma.$$

Assume that $M\neq\emptyset$, and for $\sigma_M=(\sigma_j)_{j\in M}$, set

$$W_{n,M}(\sigma_M)=\prod_{j\in M}\frac{n^2}{\left(n+1+(\sigma_j-z_j)e^{i\frac{\alpha_j+\beta_j}{2}}\right)^2},$$

where $z\in\mathbb{C}^k$ is choosen so that $z+S^*_{\alpha,\beta}\supset U$. It follows from Corollary 12.4 that G is bounded on V, and so the function

$$\sigma_M\to W_{n,M}(\sigma_M)\prod_{j\in M}\frac{1}{\sigma_j-\zeta_j}G(\sigma)$$

belongs to $H^{(1)}(\prod_{j\in M}V_j+\frac14(\epsilon_j)_{j\in M})$ for $\sigma_{M^*}=(\sigma_j)_{j\in M^*}\in\prod_{j\in M^*}\partial V'_j$ and $\zeta\in\tilde{\partial}V$. Here we associate to $\sigma_M=(\sigma_j)_{j\in M}$ and $\sigma_{M^*}=(\sigma_j)_{j\in M^*}$ the k-uplet $\sigma=(\sigma_j)_{1\le j\le k}$.

The open set $\Pi_{j\in M}V_j$ is admissible with respect to the family $\{(\alpha_j,\beta_j)\}_{j\in M}$, and it follows from Theorem 12.5 that we have, for every $\sigma_{M^*}\in\Pi_{j\in M^*}\partial V'_j$ and every $\zeta\in\tilde{\partial}V$

$$\int_{\Pi_{j\in M}\partial V'_j}W_{n,M}(\sigma_M)\prod_{j\in M}\frac{1}{\sigma_j-\zeta_j}G(\sigma)d\sigma_M=0,$$

where $d\sigma_M:=\Pi_{j\in M}d\sigma_j$. We then deduce from the Lebesgue dominated convergence theorem that we have

$$P_M(\zeta,\sigma_{M^*}))=0,\quad\left(\sigma_{M^*}\in\prod_{j\in M^*}\partial V'_j,\zeta\in\tilde{\partial}V\right),$$

where

$$P(\zeta,\sigma_{M^*}):=\int_{\Pi_{j\in M}\partial V_j}\prod_{j\in M}\frac{1}{\sigma_j-\zeta_j}G(\sigma)d\sigma_M,$$

Define $\zeta_M, d\zeta_M$ and $d\sigma_{M^*}$ as above and set

$$Q_M(\zeta_M,\sigma_{M*}):=\prod_{j\in M}(\lambda_j\Delta_{T_j}+\zeta_jI)^{-1}\prod_{j'\in M^*}(\lambda_{j'}\Delta_{T_{j'}}+\sigma_{j'}I)^{-1}.$$

We obtain

$$
\begin{aligned}
&(-1)^{|M|}L_M \\
= &\int_{\tilde{\partial}V\times(\prod_{j'\in M^*}\partial V'_{j'})}\prod_{j'\in M^*}\frac{1}{\sigma_{j'}-\zeta_{j'}}F(\zeta)P_M(\zeta,\sigma_{M^*})Q_M(\zeta_M,\sigma_{M^*})d\zeta_M d\sigma_{M^*} \\
= &\ 0.
\end{aligned}
$$

We obtain

$$
\begin{aligned}
&F(-\lambda\Delta_T)G(-\lambda\Delta_T) \\
= &\frac{1}{(2\pi i)^{2k}}L_\emptyset \\
= &\frac{1}{(2\pi i)^{2k}}\int_{\tilde{\partial}V'}\left[\int_{\tilde{\partial}V}\frac{F(\zeta)}{(\sigma_1-\zeta_1)\dots(\sigma_k-\zeta_k)}d\zeta\right]G(\sigma)(\lambda_1\Delta_1+\sigma_1 I)^{-1}(\lambda_k\Delta_k+\sigma_k I)^{-1}d\sigma \\
= &\frac{1}{(2\pi i)^{k}}\int_{\tilde{\partial}V'}F(\sigma)G(\sigma)(\lambda_1\Delta_1+\sigma_1 I)^{-1}(\lambda_k\Delta_k+\sigma_k I)^{-1}d\sigma \\
= &(FG)(-\lambda\Delta_T).
\end{aligned}
$$

So the map $F\to F(-\lambda\Delta_T)$ is an algebra homomorphism from $\cup_{U\in\mathcal{W}_{T,\lambda}}H^{(1)}(U)$ into $\mathcal{M}(\mathcal{B})$.

Let $\mathcal{E}$ be a bounded subset of $\cup_{U\in\mathcal{W}_{T,\lambda}}H^{(1)}(U)$, let $U\in W_{T,\lambda}$ such that $\mathcal{E}$ is a bounded subset of $H^{(1)}(U)$, let (α,β) be the element of $M_{a,b}$ associated to U, and let $\epsilon\in S^*_{\alpha,\beta}$ be such that $U+\epsilon\in\mathcal{W}_{T,\lambda}$. Set $K=\sup_{\zeta\in\tilde{\partial}U+\epsilon}\|R(-\lambda\Delta_T,\zeta)\|_{\mathcal{M}(\mathcal{B})}$. We have

$$
\sup_{F\in\mathcal{E}}\|F(-\lambda\Delta_T)\|_{\mathcal{M}(\mathcal{B})}\le\frac{K}{(2\pi)^k}\sup_{F\in\mathcal{E}}\|F\|_{H^{(1)}(U)}<+\infty,
$$

which shows that the map $F\to F(-\lambda\Delta_T)$ is a bounded homomorphism from $\cup_{U\in\mathcal{W}_{T,\lambda}}H^{(1)}(U)$ into $\mathcal{M}(\mathcal{B})\subset\mathcal{QM}_r(\mathcal{A})$.

(ii) Let $U\in\mathcal{W}_{T,\lambda}$, and let $(\alpha,\beta)\in M_{a,b}$ and $z\in\mathbb{C}^k$ be such that $U\subset z+S^*_{\alpha,\beta}$ and $(z+S^*_{\alpha,\beta})\setminus U$ is bounded. For $j\le k$, set

$$
s_j=1+\sup\left(\lim_{t\to+\infty}\frac{\log\left(\left\|T\left(t\lambda_j e^{i\frac{\alpha_j+\beta_j}{2}}\right)\right\|\right)}{t},-Re\left(z_j e^{i\frac{\alpha_j+\beta_j}{2}}\right)\right).
$$

Set $\tilde{T}_j(t)=T(t\lambda_j e^{i\frac{\alpha_j+\beta_j}{2}})$ for $t>0$, with the convention $\tilde{T}_j(0)=I$, and set, for $f\in\cap_{\zeta\in\mathbb{C}^k}e_{-\zeta}\mathcal{U}_{\alpha,\beta}$,

$$
\langle f,\phi\rangle=\int_{(\mathbb{R}^+)^k}f(t_1e^{i\frac{\alpha_1+\beta_1}{2}},\dots,t_ke^{i\frac{\alpha_k+\beta_k}{2}})e^{-s_1t_1-\dots-s_kt_k}dt_1\dots dt_k.
$$

Then $z\in Dom(\mathcal{FB}(\phi))$, and we have, for $\zeta\in Dom(\mathcal{FB}(\phi))$,

$$
\begin{aligned}
\mathcal{FB}(\phi)(\zeta)&=\int_{(R^+)^k}e^{-t_1\zeta_1e^{i\frac{\alpha_1+\beta_1}{2}}\dots-t_k\zeta_ke^{i\frac{\alpha_k+\beta_k}{2}}}e^{-s_1t_1+\dots-s_kt_k}dt_1\dots dt_k \\
=&\ \frac{1}{(\zeta_1e^{i\frac{\alpha_1+\beta_1}{2}}+s_1)\dots(\zeta_ke^{i\frac{\alpha_k+\beta_k}{2}}+s_k)},
\end{aligned}
$$

$$\begin{aligned}
&\langle T_{(\lambda)}, \phi\rangle \\
&= \int_{(\mathbb{R}^+)^k} T_1(t_1\lambda_1 e^{i\frac{\alpha_1+\beta_1}{2}})\dots T_k(t_1\lambda_k e^{i\frac{\alpha_k+\beta_k}{2}})e^{-s_1t_1-\dots-s_kt_k}dt_1\dots dt_k \\
&= \left[\int_0^{+\infty}\tilde{T}_1(t_1)e^{-s_1t_1}dt_1\right]\dots\left[\int_0^{+\infty}\tilde{T}_k(t_1)e^{-s_kt_k}dt_k\right],
\end{aligned}$$

where the Bochner integrals are computed with respect to the strong operator topology on $\mathcal{M}(\mathcal{B})$.

It follows from the observations in Section 5 that $\left(\int_0^{+\infty}\tilde{T}_j(t)e^{-s_jt}dt\right)(\mathcal{B})$ is dense in $\mathcal{B}$ for $1\le j\le k$, and so $< T_{(\lambda)}, \phi > (\mathcal{B})$ is dense in $\mathcal{B}$. Now set $\phi_1 = \phi * \phi$. It follows from Theorem 8.6 (vii) that we have

$$\langle T_{(\lambda)}, \phi_1\rangle = \langle T_\lambda, \phi\rangle^2,$$

and so $\langle T_{(\lambda)}, \phi_1\rangle(\mathcal{B})$ is dense in $\mathcal{B}$.

Set $F = \mathcal{FB}(\phi_1) = \mathcal{FB}(\phi)^2$. Then $F \in H^{(1)}(U-\epsilon)\cap H^\infty(U-\epsilon)$ for some $\epsilon \in S^*_{\alpha,\beta}$, and we have, using assertion (vi) of Theorem 8.6

$$\begin{aligned}
F(-\lambda\Delta_T) &= \frac{(-1)^k}{(2\pi i)^k}\int_{z+\tilde{\partial}S^*_{\alpha,\beta}} F(\zeta)(\lambda_1\Delta_{T_1}+\sigma_1)^{-1}\dots(\lambda_k\Delta_{T_k}+\sigma_k)^{-1}d\sigma_1\dots d\sigma_k \\
&= \langle T_{(\lambda)}, \phi_1\rangle,
\end{aligned}$$

which shows that $F(-\lambda\Delta_T)(\mathcal{B})$ is dense in $\mathcal{B}$.

Now consider again $U \in \mathcal{W}_{T,\lambda}$, and let $F \in H^\infty(U)$. Let $G_0 \in H^{(1)}(U)$ be such that $G_0(-\lambda\Delta_T)(\mathcal{B})$ is dense in $\mathcal{B}$, and let $u \in \Omega(\mathcal{B})$.

Then $G_0(-\lambda\Delta_T)u \in \Omega(\mathcal{B})$, $FG_0 \in H^{(1)}(U)$, and so there exists a unique $R_F \in \mathcal{QM}_r(\mathcal{B}) = \mathcal{QM}_r(\mathcal{A})$ such that

$$R_FG_0(-\lambda\Delta_T)u = (FG_0)(-\lambda\Delta_T)u \in \mathcal{M}(\mathcal{B}),$$

and $R_F = F(-\lambda\Delta_T)$ if $F \in H^{(1)}(U)$.

Let $U' \in \mathcal{W}_{T,\lambda}$, and let $G \in H^{(1)}(U')$. We have

$$\begin{aligned}
R_FG(-\lambda\Delta_T)G_0(-\lambda\Delta_T) &= R_FG_0(-\lambda\Delta_T)G(-\lambda\Delta_T) \\
&= (FG_0)(-\lambda\Delta_T)G(-\lambda\Delta_T) \\
&= (FG_0G)(-\lambda\Delta_T) \\
&= (FG)(-\lambda\Delta_T)G_0(-\lambda\Delta_T),
\end{aligned}$$

and so $R_FG(-\lambda\Delta_T) = (FG)(-\lambda\Delta_T)$, which shows that the definition of R_F does not depend on the choice of U. The map $F \to R_F$ is clearly linear. Now let $F_1 \in \cup_{U\in\mathcal{W}_{T,\lambda}}H^\infty(U)$, let $F_2 \in \cup_{U\in\mathcal{W}_{T,\lambda}}H^\infty(U)$, and let $G \in \cup_{U\in\mathcal{W}_{T,\lambda}}H^{(1)}(U)$ such that $G(-\lambda\Delta_T)(\mathcal{B})$ is dense in $\mathcal{B}$. We have

$$\begin{aligned}
R_{F_1F_2}G^2(-\lambda\Delta_T) &= (F_1F_2G^2)(-\lambda\Delta_T) \\
&= (F_1G)(-\lambda\Delta_T)(F_2G)(-\lambda\Delta_T) \\
&= R_{F_1}R_{F_2}G^2(-\lambda\Delta_T),
\end{aligned}$$

and so $R_{F_1F_2} = R_{F_1}R_{F_2}$ since $G^2(-\lambda\Delta_T)\mathcal{B}$ is dense in $\mathcal{B}$.

Now let $\mathcal{E}$ be a bounded family of elements of $\cup_{U\in\mathcal{W}_{T,\lambda}}H^\infty(U)$. There exists $U\in\mathcal{W}_{T,\lambda}$ and $M>0$ such that $F\in H^{(\infty)}(U)$ and $\|F\|_{H^\infty(U)}\le M$ for every $F\in\mathcal{E}$. Let $G\in H^{(1)}(U)$ such that $G(-\lambda\Delta_T)(\mathcal{B})$ is dense in $\mathcal{B}$. Then the family $\{FG\}_{F\in\mathcal{E}}$ is bounded in $H^{(1)}(U)$, and it follows from (i) that there exists $u\in\Omega(\mathcal{B})$ such that $\sup_{F\in\mathcal{E}}\|(FG)(-\lambda\Delta_T)u\|_{\mathcal{B}}<+\infty$.

We obtain

$$\sup_{F\in\mathcal{E}}\|R_FG(-\lambda\Delta_T)u\|_{\mathcal{B}}=\sup_{F\in\mathcal{E}}\|(FG)(-\lambda\Delta_T)u\|_{\mathcal{B}}<+\infty,$$

and so the family $\{R_F\}_{F\in\mathcal{E}}$ is pseudobounded in $\mathcal{QM}(\mathcal{B})=\mathcal{QM}(\mathcal{A})$ since $G(-\lambda\Delta_T)u\in\Omega(\mathcal{B})$. Since the family $\{\lambda^{-n}F^n\}_{n\ge1}$ is bounded in $H^\infty(U)$ for $F\in H^\infty(U)$, $\lambda>(1+\|F\|_{H^\infty(U)})^{-1}$, this shows that $R_F\in\mathcal{QM}_r(\mathcal{B})=\mathcal{QM}_r(\mathcal{A})$ for $F\in\cup_{U\in\mathcal{F}}H^\infty(U)$, and that the map $F\to R_F$ is a bounded algebra homomorphism from $\cup_{U\in\mathcal{F}}H^\infty(U)$ into $\mathcal{QM}_r(\mathcal{B})=\mathcal{QM}_r(\mathcal{A})$, which concludes the proof of (ii).

(iii) Let $(\alpha,\beta)\in\mathcal{N}_\lambda$, let $\phi\in\mathcal{F}_{\alpha,\beta}$, with $N_0(T,\lambda,\alpha,\beta)\cap Dom(\mathcal{FB}(\phi))\ne\emptyset$, and let $z\in N_0(T,\lambda,\alpha,\beta)\cap Dom(\mathcal{FB}(\phi))$. Then $z+S^*_{\alpha,\beta}$ is admissible with respect to (T,λ,α,β). As in the proof of (ii) we can construct $\phi_1\in\mathcal{F}_{\alpha,\beta}$ having the following properties

- $z\in N_0(T,\lambda,\alpha,\beta)\cap Dom(\mathcal{FB}(\phi_1))$,
- $G:=\mathcal{FB}(\phi_1)\in H^{(1)}(z+S^*_{\alpha,\beta})\cap H^\infty(z+S^*_{\alpha,\beta})$,
- $\langle T_{(\lambda)},\phi_1\rangle=G(-\lambda\Delta_T)$, and $G(-\lambda\Delta_T)(\mathcal{B})$ is dense in $\mathcal{B}$.

Let $\epsilon\in S^*_{\alpha,\beta}$ be such that $z+\epsilon+S^*_{\alpha,\beta}$ is admissible with respect to (T,λ,α,β). It follows from assertions (v) and (vi) of Theorem 8.6 and from (i) and (ii) that we have

$$\begin{aligned}
&\langle T_{(\lambda)},\phi\rangle\mathcal{FB}(\phi_1)(-\lambda\Delta_T)\\
=\;&\langle T_{(\lambda)},\phi\rangle\langle T_{(\lambda)},\phi_1\rangle\\
=\;&\langle T_{(\lambda)},\phi*\phi_1\rangle\\
=\;&\frac{1}{(2\pi i)^k}\int_{z+\epsilon+S^*_{\alpha,\beta}}\mathcal{FB}(\phi)(\sigma)\mathcal{FB}(\phi_1)(\sigma)\prod_{1\le j\le k}(\lambda_j\Delta_{T_j}+\sigma_jI)^{-1}d\sigma_1\dots d\sigma_k\\
=\;&(\mathcal{FB}(\phi)\mathcal{FB}(\phi_1))(-\lambda\Delta_T)=\mathcal{FB}(\phi)(-\lambda\Delta_T)\mathcal{FB}(\phi_1)(-\lambda\Delta_T),
\end{aligned}$$

and so $\langle T_{(\lambda)},\phi\rangle=\mathcal{FB}(\phi)(-\lambda\Delta_T)$ since $\mathcal{FB}(\phi_1)(-\lambda\Delta_T)(\mathcal{B})$ is dense in $\mathcal{B}$.

Now let $\nu\in\mathbb{C}$ such that $\nu\lambda_j\in\cup_{(\gamma_j,\delta_j)\in M_{a_j,b_j}}\overline{S}_{\gamma_j,\delta_j}$; let $\nu_j=(\nu_{j,1},\dots,\nu_{j,k})$ be the k-tuple defined by the conditions $\nu_{j,s}=0$ if $s\ne j$, $\nu_{j,j}=\nu$. There exists $(\gamma_j,\delta_j)\in\mathcal{N}_{\lambda_j}$ such that $\nu\in\overline{S}_{\gamma_j,\delta_j}$, and there exists $(\alpha,\beta)\in\mathcal{N}_\lambda$ such that $\alpha_j=\gamma_j$ and $\beta_j=\delta_j$. Set $F(\zeta)=e^{-\nu\zeta_j}$ for $\zeta\in\mathbb{C}^k$, and set $\langle f,\phi\rangle=f(\nu_j)$ for $f\in\cap_{z\in\mathbb{C}^k}e_{-z}\mathcal{U}_{\alpha,\beta}$. Then $Dom(\mathcal{FB}(\phi))=\mathbb{C}^k$, and we have, for $\zeta\in\mathbb{C}^k$,

$$\mathcal{FB}(\phi)(\zeta)=\langle e_{-\zeta},\phi\rangle=e^{-\nu_j\zeta}=e^{-\nu\zeta_j},$$

and so $F=\mathcal{FB}(\phi)$. Let $z\in N_0(T,\lambda,\alpha,\beta)=N_0(T,\lambda,\alpha,\beta)\cap Dom(\mathcal{FB}(\phi))$. Let δ_{ν_j} be the Dirac measure at ν_j. Since $e_{-z}\delta_{\nu_j}$ is a representing measure for ϕe_{-z} we have

$$F(-\lambda\Delta_T)=\langle T_{(\lambda)},\phi\rangle=T_j(\nu\lambda_j),$$

which concludes the proof of (iii).

(iv) Let $(\alpha,\beta)\in\mathcal{N}_\lambda$, let $\phi\in\mathcal{F}_{\alpha,\beta}$, with $N(T,\lambda,\alpha,\beta)\cap Dom(\mathcal{FB}(\phi))\neq\emptyset$. Set $e_{-\epsilon}T=(e_{-\epsilon_1}T_1,\dots,e_{-\epsilon_k}T_k)$. Then $N(T,\lambda,\alpha,\beta)\subset N_0(e_{-\epsilon}T,\lambda,\alpha,\beta)$ for $\epsilon\in S^*_{\alpha,\beta}$, and it follows from Theorem 8.6 (ii) and from (iii) that we have, for $u\in\mathcal{B}$,

$$\begin{aligned}\langle T_{(\lambda)},\phi\rangle u &= \lim_{\substack{\epsilon\to(0,\dots,0)\\ \epsilon\in S^*_{\alpha,\beta}}}\langle e_{-\epsilon}T_\lambda,\phi\rangle u\\ &= \lim_{\substack{\epsilon\to(0,\dots,0)\\ \epsilon\in S^*_{\alpha,\beta}}}\mathcal{FB}(\phi)(e_{-\epsilon}T_{(\lambda)})u\\ &= \lim_{\substack{\epsilon\to(0,\dots,0)\\ \epsilon\in S^*_{\alpha,\beta}}}\mathcal{FB}(\phi)(-\lambda_1T_1+\epsilon_1I,\dots,-\lambda_kT_k+\epsilon_kI)u,\end{aligned}$$

which concludes the proof of (iv).

(v) Let $U\in\mathcal{W}_{T,\lambda}$, let $F\in H^\infty(U)$ be strongly outer, and let $(F_n)_{n\geq1}$ be a sequence of invertible elements of $H^\infty(U)$ satisfying the conditions of Definition 12.6 with respect to F. It follows from (ii) that there exists $G\in H^{(1)}(U)\cap H^\infty(U)$ such that $G(-\lambda\Delta_T)(\mathcal{B})$ is dense in $\mathcal{B}$. Let $(\alpha,\beta)\in M_{a,b}$ and $z\in N_0(T,\lambda,\alpha,\beta)$ such that $U\subset z+S^*_{\alpha,\beta}$ and such that $(z+S^*_{\alpha,\beta})\setminus U$ is bounded. There exists $\epsilon\in\mathbb{C}^k$ such that $U+\epsilon\subset U$ is admissible with respect to (T,λ,α,β) and we have

$$\begin{aligned}&F(-\lambda\Delta_T)F_n^{-1}(-\lambda\Delta_T)G^2(-\lambda\Delta_T)\\ =\ &\frac{(-1)^k}{(2\pi i)^k}\int_{\epsilon+\tilde\partial U}F(\sigma)F_n^{-1}(\sigma)G^2(\sigma)\prod_{1\leq j\leq k}(\lambda_j\Delta_{T_j}+\sigma_jI)^{-1}d\sigma_1\dots d\sigma_k,\end{aligned}$$

and it follows from the Lebesgue dominated convergence theorem that

$$\lim_{n\to+\infty}\|F(-\lambda\Delta_T)F_n^{-1}(-\lambda\Delta_T)G^2(-\lambda\Delta_T)-G^2(-\lambda\Delta_T)\|_{\mathcal{M}(\mathcal{B})}=0.$$

Let $u\in\Omega(\mathcal{B})$. Then $G(-\lambda\Delta_T)u\in Dom(F(-\lambda\Delta_T))\cap\Omega(\mathcal{B})$. Set

$$u_n=F_n^{-1}(-\lambda\Delta_T)G(-\lambda\Delta_T)u\in\mathcal{B}.$$

We have

$$G(-\lambda\Delta_T)^2u^2=\lim_{n\to+\infty}F(-\lambda\Delta_T)G(-\lambda\Delta_T)uu_n.$$

Since $G(-\lambda\Delta)^2u^2\in\Omega(\mathcal{B})$, we have $F(-\lambda\Delta_T)G(-\lambda\Delta_T)u\in\Omega(\mathcal{B})$, which proves (v).

(vi) Let $U\in\mathcal{W}_{T,\lambda}$, let $F\in\mathcal{S}(U)$, let $G_0\in H^\infty(U)$ be a strongly outer function such that $FG_0\in H^\infty(U)$, and let $u\in Dom(G_0(-\lambda\Delta_T))$ such that $G_0(-\lambda\Delta_T)u\in\Omega(\mathcal{B})$. Let $v\in\Omega(\mathcal{B})\cap Dom(FG_0(-\lambda\Delta_T))$. There exists a unique $R_F\in\mathcal{QM}(\mathcal{B})=\mathcal{QM}(\mathcal{A})$ satisfying the equation

$$(FG_0)(-\lambda\Delta_T)uv=R_FG_0(-\lambda\Delta_T)uv,$$

and we have

$$(FG_0)(-\lambda\Delta_T)=R_FG_0(-\lambda\Delta_T),$$

so that $R_F=F(-\lambda\Delta_T)$ if $F\in H^\infty(U)$.

Let $G\in\cup_{V\in\mathcal{W}_{T,\lambda}}H^\infty(V)$ such that $FG\in H^\infty(W)$ for some $W\in\mathcal{W}_{T,\lambda}$, and let $w\in\Omega(\mathcal{B})\cap Dom(G(-\lambda\Delta_T))$. We have

$$\begin{aligned}((FG)(-\lambda\Delta_T)vw)G_0(-\lambda\Delta_T)u &= (FG_0)(-\lambda\Delta_T)G(-\lambda\Delta_T)uvw\\ &= R_F G_0(-\lambda\Delta_T)G(-\lambda\Delta_T)uvw\\ &= (R_F G(-\lambda\Delta_T)vw)G_0(-\lambda\Delta_T)u.\end{aligned}$$

Since $vw\,(G_0(-\lambda\Delta_T)u) \in \Omega(\mathcal{B})$, we have $(FG)(-\lambda\Delta_T) = R_F G(-\lambda\Delta_T)$. So if we set $F(-\lambda\Delta_T) = R_F$, we obtain $F(-\lambda\Delta_T)G(-\lambda\Delta_T) = (FG)(-\lambda\Delta_T)$ for every $F \in \cup_{U\in\mathcal{W}_{T,\lambda}}\mathcal{S}(U)$ and for every $G \in \cup_{U\in\mathcal{W}_{T,\lambda}}H^\infty(U)$ such that $FG \in \cup_{U\in\mathcal{W}_{T,\lambda}}H^\infty(U)$. The map $F \to F(-\lambda\Delta_T)$ is clearly linear.

Now let $F_1 \in \cup_{U\in\mathcal{W}_{T,\lambda}}\mathcal{S}(U), F_2 \in \cup_{U\in\mathcal{W}_{T,\lambda}}\mathcal{S}(U)$. There exist strongly outer functions $G_1 \in \cup_{U\in\mathcal{W}_{T,\lambda}}H^\infty(U)$ and $G_2 \in \cup_{U\in\mathcal{W}_{T,\lambda}}H^\infty(U)$ for which we have $F_1G_1 \in \cup_{U\in\mathcal{W}_{T,\lambda}}H^\infty(U)$ and $F_2G_2 \in \cup_{U\in\mathcal{W}_{T,\lambda}}H^\infty(U)$. Then

$$\begin{aligned}&(F_1F_2)(-\lambda\Delta_T)G_1(-\lambda\Delta_T)G_2(-\lambda\Delta_T)\\ =\;& (F_1F_2G_1G_2)(-\lambda\Delta_T)\\ =\;& (F_1G_1)(-\lambda\Delta_T)(F_2G_2)(-\lambda\Delta_T)\\ =\;& F_1(-\lambda\Delta_T)F_2(-\lambda\Delta_T)G_1(-\lambda\Delta_T)G_2(-\lambda\Delta_T).\end{aligned}$$

Since $Dom(G_1(-\lambda\Delta_T)) \cap \Omega(\mathcal{B}) \neq \emptyset$ and $Dom(G_2(-\lambda\Delta_T)) \cap \Omega(\mathcal{B}) \neq \emptyset$, this shows that $(F_1F_2)(-\lambda\Delta_T) = F_1(-\lambda\Delta_T)F_2(-\lambda\Delta_T)$.

So the map $F \to F(-\lambda\Delta_T)$ is an algebra homomorphism from $\cup_{U\in\mathcal{W}_{T,\lambda}}\mathcal{S}(U)$ into $\mathcal{QM}(\mathcal{B}) = \mathcal{QM}(\mathcal{A})$.

Now let $\mathcal{E}$ be a bounded family of elements of $\cup_{U\in\mathcal{W}_{T,\lambda}}\mathcal{S}(U)$. There exists $U \in \mathcal{W}_{T,\lambda}$ and a strongly outer function $G \in H^\infty(U)$ such that $FG \in H^\infty(U)$ for every $F \in \mathcal{E}$ and such that $\sup_{F\in\mathcal{E}}\|FG\|_{H^\infty(U)} < +\infty$.

So the family $\{(FG)(-\lambda\Delta_T)\}_{F\in\mathcal{E}}$ is a pseudobounded family of elements of $\mathcal{QM}_r(\mathcal{B}) = \mathcal{QM}_r(\mathcal{A})$, and there exists $u \in \Omega(\mathcal{B}) \cap (\cap_{F\in\mathcal{E}}Dom((FG)(-\lambda\Delta_T)))$ such that $\sup_{F\in\mathcal{E}}\|(FG)(-\lambda\Delta_T)u\|_{\mathcal{B}} < +\infty$. Let $v \in Dom(G(-\lambda\Delta_T)) \cap \Omega(\mathcal{B})$, and set $w = G(-\lambda\Delta_T)uv$. Then $w \in \Omega(\mathcal{B}) \cap (\cap_{F\in\mathcal{E}}Dom(F(-\lambda\Delta_T)))$ and

$$\begin{aligned}\sup_{F\in\mathcal{E}}\|F(-\lambda\Delta_T)w\|_{\mathcal{B}} &= \sup_{F\in\mathcal{E}}\|(F(-\lambda\Delta_T)G(-\lambda\Delta_T)uv\|_{\mathcal{B}}\\ &\leq \sup_{F\in\mathcal{E}}\|(FG)(-\lambda\Delta_T)u\|_{\mathcal{B}}\|v\|_{\mathcal{B}} < +\infty,\end{aligned}$$

and so the family $\{F(-\lambda\Delta_T\}_{F\in\mathcal{E}}$ is pseudobounded in $\mathcal{QM}(\mathcal{B}) = \mathcal{QM}(\mathcal{A})$, and the map $F \to F(-\lambda\Delta_T)$ is a bounded algebra homomorphism from $\cup_{U\in\mathcal{W}_{T,\lambda}}\mathcal{S}(U)$ into $\mathcal{QMB}) = \mathcal{QM}(\mathcal{A})$, which concludes the proof of (vi).

Now assume that $\mathcal{A}$ is not radical, let $\chi \in Spec(\mathcal{A})$, and let $\tilde{\chi}$ be the unique character on $\mathcal{QM}(\mathcal{A})$ such that $\tilde{\chi}(u) = \chi(u)$ for every $u \in \mathcal{A}$.

Let $F \in H^{(1)}(U)$, where $U \in \mathcal{W}_{T,\lambda}$, let (α,β) be the element of $M_{a,b}$ associated to U, and let $\epsilon \in S^*_{\alpha,\beta}$ be such that $U+\epsilon$ is admissible with respect to (T,λ,α,β). Since Bochner integrals commute with linear functionals, we have

$$\begin{aligned}&\tilde{\chi}\,(F(-\lambda_1\Delta_{T_1},\dots,-\lambda_k\Delta_{T_k}))\\ =\;& \frac{(-1)^k}{(2\pi i)^k}\int_{\tilde{\partial}U+\epsilon}F(\zeta_1,\dots,\zeta_k)\prod_{1\leq j\leq k}(\lambda_j\tilde{\chi}(\Delta_{T_j})+\zeta_jI)^{-1}d\zeta_1\dots d\zeta_k.\end{aligned}$$

Since $U+\epsilon$ is admissible with respect to (T,λ,α,β), we have

$$(-\lambda_1\tilde{\chi}(\Delta_{T_1}),\dots,-\lambda_k\tilde{\chi}(\Delta_{T_k})) \in U+\epsilon,$$

and it follows from Theorem 12.5 that we have

$$\tilde{\chi}(F(-\lambda_1\Delta_{T_1},\dots,-\lambda_k\Delta_{T_k})) = F(-\lambda_1\tilde{\chi}(\Delta_{T_1}),\dots,-\lambda_k\tilde{\chi}(\Delta_{T_k})).$$

Now let $F \in H^\infty(U)$, where $U \in \mathcal{W}_{T,\lambda}$, and let $G \in H^{(1)}(U)$ such that $G(-\lambda\Delta_T)(\mathcal{B})$ is dense in $\mathcal{B}$. Then $\tilde{\chi}(G(-\lambda\Delta_T)) \neq 0$, and we have

$$\tilde{\chi}(F(-\lambda\Delta_T)) = \frac{\tilde{\chi}((FG)(-\lambda\Delta_T))}{\tilde{\chi}(G(-\lambda\Delta_T))} = \frac{(FG)(-\lambda\tilde{\chi}(\Delta_T))}{G(-\lambda\tilde{\chi}(\Delta_T))} = F(-\lambda\tilde{\chi}(\Delta_T)).$$

Finally let $F \in \mathcal{S}(U)$, where $U \in \mathcal{W}_{T,\lambda}$, and let $G \in H^\infty(U)$ be a strongly outer function such that $FG \in H^\infty(U)$. It follows from (v) that there exists some $u \in \mathcal{B}$ such that $G(-\lambda\Delta_T)u \in \Omega(\mathcal{B})$, and so $\tilde{\chi}(G(-\lambda\Delta_T)) \neq 0$. The same argument as above shows then that $\tilde{\chi}(F(-\lambda\Delta_T)) = F(-\lambda\tilde{\chi}(\Delta_T))$, which concludes the proof of (vi).

(vii) Set $F(\zeta_1,\dots,\zeta_k) = -\zeta_j$, choose $\nu_0 > \nu_1 > \lim_{t\to+\infty}\frac{\log\|T_j(t\lambda_j)\|}{t}$, and set again $v_{\nu_0}(t) = te^{-\nu_0 t}$. It follows from Proposition 12.8(ii) that $F \in \mathcal{S}(U)$ for every $U \in \mathcal{W}_{T,\lambda}$, and it follows from Proposition 5.5(i) that we have

$$\lambda_j\Delta_{T_j}\int_0^{+\infty} v_{\nu_0}(t)T_j(t\lambda_j)dt = -\int_0^{+\infty} v'_{\nu_0}(t)T_j(t\lambda_j)dt,$$

where the Bochner integrals are computed with respect to the strong operator topology on $\mathcal{M}(\mathcal{B})$.

Now choose $(\alpha,\beta) \in \mathcal{N}_\lambda$, and set, for $f \in \cap_{z\in\mathbb{C}^k} e_{-z}\mathcal{U}_{\alpha,\beta}$,

$$\langle f,\phi_0\rangle = \int_{[0,+\infty)^k} f(0,\dots,0,t_j,0,\dots,0)v_{\nu_0}(t_j)dt_j,$$

$$\langle f,\phi_1\rangle = \int_{[0,+\infty)^k} f(0,\dots,0,t_j,0,\dots,0)v'_{\nu_0}(t_j)dt_j.$$

Then $\phi_0 \in \mathcal{F}_{\alpha,\beta}, \phi_1 \in \mathcal{F}_{\alpha,\beta}$.

Also $-\nu_1\lambda_j + S^*_{\alpha,\beta} \in N_0(T,\lambda,\alpha,\beta) \cap Dom(\mathcal{FB}(\phi_0)) \cap Dom(\mathcal{FB}(\phi_1))$, and it follows from (iii) that we have

$$\int_0^{+\infty} v_{\nu_0}(t)T_j(t\lambda_j)dt = \langle T_{(\lambda)},\phi_0\rangle = \mathcal{FB}(\phi_0)(-\lambda\Delta_T),$$

$$\int_0^{+\infty} v'_{\nu_0}(t)T_j(t\lambda_j)dt = \langle T_{(\lambda)},\phi_1\rangle = \mathcal{FB}(\phi_1)(-\lambda\Delta_T).$$

But $\mathcal{FB}(\phi_0)(\zeta) = \frac{1}{(\nu_0+\zeta_j)^2}, \mathcal{FB}(\phi_1)(\zeta) = \frac{\zeta_j}{(\nu_0+\zeta_j)^2} = -F(\zeta)\mathcal{FB}(\phi_0)(\zeta)$, which gives

$$\lambda_j\Delta_{T_j}\int_0^{+\infty} v_{\nu_0}(t)T_j(t\lambda_j)dt = F(-\lambda\Delta_T)\int_0^{+\infty} v_{\nu_0}(t)T_j(t\lambda_j)dt,$$

and so $F(-\lambda\Delta_T) = \lambda_j\Delta_{T_j}$, since $(\int_{[0,\infty)^n} v_{\nu_0}(t)T_j(t\lambda_j)dt)(\mathcal{B})$ is dense in $\mathcal{B}$, as observed in Section 5. $\square$

10. Appendix 1: Fourier–Borel and Cauchy transforms

In this section we present some certainly well-known results about Fourier–Borel and Cauchy transforms of linear functionals on some spaces of holomorphic functions on sectors. The notion of Fourier–Borel transform is classical for elements θ of the dual of $\mathcal{H}(\mathbb{C}^k)$, **[3]**, **[25]**, and the Cauchy transform of $\theta \in \mathcal{H}(\mathbb{C}^k)'$ can be interpreted as the "indicatrice" of its Fourier–Borel transform $\mathcal{FB}(\theta)$, see **[25]**, Lemme 3, p.85.

For $\alpha < \beta \le \alpha + \pi$ denote as usual by $\overline{S}_{\alpha,\beta}$ the closure of the open sector $S_{\alpha,\beta}$, and set by convention $\overline{S}_{\alpha,\alpha} := \{te^{i\alpha}\}_{t\ge 0}$.

We set

$$S^*_{\alpha,\beta} = S_{-\pi/2-\alpha,\pi/2-\beta}, \overline{S}^*_{\alpha,\beta} = \overline{S}_{-\pi/2-\alpha,\pi/2-\beta}. \tag{15}$$

Notice that $S^*_{\alpha,\alpha+\pi} = \emptyset$, while $\overline{S}^*_{\alpha,\alpha+\pi} = \overline{S}_{-\pi/2-\alpha,-\pi/2-\alpha} = \{-tie^{-i\alpha}\}_{t\ge 0}$.

Now asssume that $\alpha \le \beta < \alpha + \pi$. Let $\lambda = |\lambda|e^{i\omega} \in \overline{S}^*_{\alpha,\beta}$ and let $\zeta = |\zeta|e^{i\theta} \in \overline{S}_{\alpha,\beta}$, with $-\frac{\pi}{2} - \alpha \le \omega \le \frac{\pi}{2} - \beta$, $\alpha \le \theta \le \beta$. We have $-\frac{\pi}{2} \le \omega + \theta \le \frac{\pi}{2}$, $|e^{-\lambda\zeta}| = e^{-|\lambda||\zeta|\cos(\omega+\theta)}$, and we obtain

$$|e^{-\lambda\zeta}| < 1 \ \ (\lambda \in S^*_{\alpha,\beta}, \zeta \in \overline{S}_{\alpha,\beta} \setminus \{0\}). \tag{16}$$

$$|e^{-\lambda\zeta}| \le 1 \ \ (\lambda \in \overline{S}^*_{\alpha,\beta}, \zeta \in \overline{S}_{\alpha,\beta}). \tag{17}$$

DEFINITION 10.1. *Let $\alpha, \beta \in \mathbb{R}^k$ such that $\alpha_j \le \beta_j < \alpha_j + \pi$ for $1 \le j \le k$. Set $\overline{S}_{\alpha,\beta} := \prod_{j=1}^k \overline{S}_{\alpha_j,\beta_j}$, $S^*_{\alpha,\beta} := \prod_{j=1}^k S^*_{\alpha_j,\beta_j}$, $\overline{S}^*_{\alpha,\beta} := \prod_{j=1}^k \overline{S}^*_{\alpha_j,\beta_j}$. If, further, $\alpha_j < \beta_j$ for $1 \le j \le k$, set $S_{\alpha,\beta} := \prod_{j=1}^k S_{\alpha_j,\beta_j}$.*

Let X be a Banach space. We denote by $\mathcal{U}_{\alpha,\beta}(X)$ the set of all continuous X-valued functions f on $\overline{S}_{\alpha,\beta}$ satisfying $\lim_{\substack{|z|\to+\infty \\ z\in\overline{S}_{\alpha,\beta}}} \|f(z)\|_X = 0$ such that the map $\zeta \to f(\zeta_1, \zeta_2, \dots, \zeta_{j-1}, \zeta, \zeta_{j+1}, \dots, \zeta_k)$ is holomorphic on S_{α_j,β_j} for every $(\zeta_1, \dots, \zeta_{j-1}, \zeta_{j+1}, \dots, \zeta_k) \in \prod_{\substack{1\le s\le k \\ s\ne j}} \overline{S}_{\alpha_s,\beta_s}$ when $\alpha_j < \beta_j$.

Similarly we denote by $\mathcal{V}_{\alpha,\beta}(X)$ the set of all continuous bounded X-valued functions f on $\overline{S_{\alpha,\beta}}$ such that the map $\zeta \to f(\zeta_1, \zeta_2, \dots, \zeta_{j-1}, \zeta, \zeta_{j+1}, \dots, \zeta_k)$ is holomorphic on S_{α_j,β_j} for every $(\zeta_1, \dots, \zeta_{j-1}, \zeta_{j+1}, \dots, \zeta_k) \in \prod_{\substack{1\le s\le k \\ s\ne j}} \overline{S}_{\alpha_s,\beta_s}$ when $\alpha_j < \beta_j$.

The spaces $\mathcal{U}_{\alpha,\beta}(X)$ and $\mathcal{V}_{\alpha,\beta}(X)$ are equipped with the norm $\|f\|_\infty = \sup_{z\in\overline{S}_{\alpha,\beta}} \|f(z)\|_X$, and we will write $\mathcal{U}_{\alpha,\beta} := \mathcal{U}_{\alpha,\beta}(\mathbb{C})$, $\mathcal{V}_{\alpha,\beta} := \mathcal{V}_{\alpha,\beta}(\mathbb{C})$.

A representing measure for $\phi \in \mathcal{U}'_{\alpha,\beta}$ is a measure of bounded variation ν on $\overline{S}_{\alpha,\beta}$ satisfying

$$\langle f, \phi \rangle = \int_{\overline{S}_{\alpha,\beta}} f(\zeta) d\nu(\zeta) \ \ (f \in \mathcal{U}_{\alpha,\beta}). \tag{18}$$

.

Set $I := \{j \le k \mid \alpha_j = \beta_j\}$, $J := \{j \le k \mid \alpha_j < \beta_j\}$. Since separate holomorphy with respect to each of the variables $z_j, j \in J$ implies holomorphy with respect to $z_J = (z_j)_{j \in J}$, the map $z_J \to f(z_I, z_J)$ is holomorphic on $\Pi_{j \in J} S_{\alpha_j,\beta_j}$ for every $z_I \in \Pi_{j \in I} \overline{S}_{\alpha_j,\alpha_j}$.

For $z = (z_1, \dots, z_k), \zeta = (\zeta_1, \dots, \zeta_k) \in \mathbb{C}^k$, set again $e_z(\zeta) = e^{z_1\zeta_1 \cdots + z_k\zeta_k}$. Also set, if X is a separable Banach space, and if $\alpha, \beta \in \mathbb{R}^k$ satisfy the conditions above

$$\mathcal{U}^*_{\alpha,\beta}(X) = \mathcal{U}_{(-\pi/2-\alpha_1,\dots,-\pi/2-\alpha_k),(\pi/2-\beta_1,\dots,\pi/2-\beta_k)}(X), \tag{19}$$

$$\mathcal{V}^*_{\alpha,\beta}(X) = \mathcal{V}_{(-\pi/2-\alpha_1,\dots,-\pi/2-\alpha_k),(\pi/2-\beta_1,\dots,\pi/2-\beta_k)}(X), \tag{20}$$

with the conventions $\mathcal{U}^*_{\alpha,\beta} = \mathcal{U}^*_{\alpha,\beta}(\mathbb{C})$, $\mathcal{V}^*_{\alpha,\beta} = \mathcal{V}^*_{\alpha,\beta}(\mathbb{C})$.

PROPOSITION 10.2. *Let $\phi \in \mathcal{U}'_{\alpha,\beta}$, and let X be a separable Banach space. Set, for $f \in \mathcal{V}_{\alpha,\beta}(X)$,*

$$\langle f, \phi \rangle = \int_{\overline{S}_{\alpha,\beta}} f(\zeta) d\nu(\zeta),$$

where ν is a representing measure for ϕ. Then this definition does not depend on the choice of ν, and we have

$$\langle f, \phi \rangle = \lim_{\substack{\epsilon \to 0 \\ \epsilon \in \overline{S}^*_{\alpha,\beta}}} \langle e_{-\epsilon} f, \phi \rangle. \tag{21}$$

PROOF. It follows from (16) and (17) that if $f \in \mathcal{V}_{\alpha,\beta}(X), \epsilon \in S^*_{\alpha,\beta}$, then $e_{-\epsilon} f \in \mathcal{U}_{\alpha,\beta}(X)$. If $f \in \mathcal{U}_{\alpha,\beta}(X)$, then we have, for $l \in \mathcal{U}_{\alpha,\beta}(X)'$,

$$\left\langle \int_{\overline{S}_{\alpha,\beta}} f(\zeta) d\nu(\zeta), l \right\rangle = \int_{\overline{S}_{\alpha,\beta}} \langle f(\zeta), l \rangle d\nu(\zeta) = \langle \langle f(\zeta), l \rangle, \phi \rangle,$$

which shows that the definition of $\langle f, \phi \rangle$ does not depend on the choice of ν. Now if $f \in \mathcal{V}_{\alpha,\beta}(X)$, it follows from the Lebesgue dominated convergence theorem that we have

$$\int_{\overline{S}_{\alpha,\beta}} f(\zeta) d\nu(\zeta) = \lim_{\substack{\epsilon \to 0 \\ \epsilon \in \overline{S}^*_{\alpha,\beta}}} \int_{\overline{S}_{\alpha,\beta}} e_{-\epsilon}(\zeta) f(\zeta) d\nu(\zeta) = \lim_{\substack{\epsilon \to 0 \\ \epsilon \in \overline{S}^*_{\alpha,\beta}}} \langle e_{-\epsilon} f, \phi \rangle,$$

and we see again that the definition of $\langle f, \phi \rangle$ does not depend on the choice of the measure ν. □

We now introduce the classical notions of Cauchy transforms and Fourier–Borel transforms.

DEFINITION 10.3. *Let $\phi \in \mathcal{U}'_{\alpha,\beta}$, and let $f \in \mathcal{V}_{\alpha,\beta}(X)$.*

(i) The Fourier–Borel transform of ϕ is defined on $\overline{S}^*_{\alpha,\beta}$ by the formula

$$\mathcal{FB}(\phi)(\lambda) = \langle e_{-\lambda}, \phi \rangle \quad (\lambda \in \overline{S}^*_{\alpha,\beta}).$$

(ii) The Cauchy transform of ϕ is defined on $\Pi_{1 \le j \le k}(\mathbb{C} \setminus \overline{S}_{\alpha_j,\beta_j})$ by the formula

$$\mathcal{C}(\phi)(\lambda) = \frac{1}{(2\pi i)^k}\langle \frac{1}{(\zeta-\lambda)}, \phi_\zeta\rangle$$

$$:= \frac{1}{(2\pi i)^k}\langle \frac{1}{\zeta_1-\lambda_1}\cdots\frac{1}{\zeta_k-\lambda_k}, \phi_{\zeta_1,\dots,\zeta_k}\rangle$$

$$(\lambda = (\lambda_1,\dots,\lambda_k) \in \prod_{1\le j\le k}(\mathbb{C}\setminus \overline{S}_{\alpha_j,\beta_j})).$$

(iii) The Fourier–Borel transform of f is defined on $\Pi_{1\le j\le k}(\mathbb{C}\setminus -\overline{S}^*_{\alpha_j,\beta_j})$ by the formula

$$\mathcal{FB}(f)(\lambda) = \int_0^{e^{i\omega}.\infty} e^{-\lambda\zeta}f(\zeta)d\zeta$$

$$:= \int_0^{e^{i\omega_1}.\infty}\cdots\int_0^{e^{i\omega_k}.\infty} e^{-\lambda_1\zeta_1-\cdots-\lambda_k\zeta_k}f(\zeta_1,\dots,\zeta_k)d\zeta_1\dots d\zeta_k$$

$$(\lambda = (\lambda_1,\dots,\lambda_k) \in \prod_{1\le j\le k}(\mathbb{C}\setminus -\overline{S}^*_{\alpha_j,\beta_j})),$$

where $\alpha_j \le \omega_j \le \beta_j$ and where $Re(\lambda_j e^{i\omega_j}) > 0$ for $1\le j\le k$.

It follows from these definitions that $\mathcal{C}(\phi)$ is holomorphic on $\Pi_{1\le j\le k}(\mathbb{C}\setminus \overline{S}_{\alpha_j,\beta_j})$ for $\phi \in \mathcal{U}'_{\alpha,\beta}$, and that $\mathcal{FB}(f)$ is holomorphic on $\Pi_{1\le j\le k}(\mathbb{C}\setminus -\overline{S}^*_{\alpha_j,\beta_j})$ for $f\in\mathcal{V}_{\alpha,\beta}(X)$. Also using Proposition 10.2 we see that $\mathcal{FB}(\phi)\in\mathcal{V}^*_{\alpha,\beta} := \mathcal{V}_{-\frac{\pi}{2}-\alpha,\frac{\pi}{2}-\beta}$ for $\phi\in\mathcal{U}'_{\alpha,\beta}$.

PROPOSITION 10.4. *Let $\phi\in\mathcal{U}'_{\alpha,\beta}$. For $1\le j\le k$, set $I_{\eta,j} = (\frac{\pi}{2}-\eta, \frac{\pi}{2}-\beta_j]$ for $\eta\in(\beta_j,\alpha_j+\pi]$, $I_{\eta,j} = (-\frac{\pi}{2}-\alpha_j, \frac{\pi}{2}-\beta_j)$ for $\eta\in(\alpha_j+\pi,\beta_j+\pi]$, and set $I_{\eta,j} := (-\frac{\pi}{2}-\alpha_j, \frac{3\pi}{2}-\eta)$ for $\eta\in(\beta_j+\pi,\alpha_j+2\pi)$. Then $I_{\eta,j}\subset[-\frac{\pi}{2}-\alpha_j,\frac{\pi}{2}-\beta_j]$, $\cos(\eta+s)<0$ for $s\in I_{\eta,j}$, and if $\lambda=(\lambda_1,\dots,\lambda_k)\in\prod_{1\le j\le k}(\mathbb{C}\setminus\overline{S}_{\alpha_j,\beta_j})$, we have for $\omega=(\omega_1,\dots,\omega_k)\in\prod_{1\le j\le k}I_{arg(\lambda_j),j}$,*

$$\mathcal{C}(\phi)(\lambda) = \frac{1}{(2\pi i)^k}\int_0^{e^{i\omega}.\infty} e^{\lambda\sigma}\mathcal{FB}(\phi)(\sigma)d\sigma$$

$$:= \frac{1}{(2\pi i)^k}\int_0^{e^{i\omega_1}.\infty}\cdots\int_0^{e^{i\omega_k}.\infty} e^{\lambda_1\sigma_1+\cdots+\lambda_k\sigma_k}\mathcal{FB}(\phi)(\sigma_1,\dots,\sigma_k)d\sigma_1\dots d\sigma_k.$$

PROOF. It follows from the definition of $I_{\eta,j}$ that $I_{\eta,j}\subset[-\frac{\pi}{2}-\alpha_j,\frac{\pi}{2}-\beta_j]$. In the second case we have obviously $\frac{\pi}{2}<\eta+s<\frac{3\pi}{2}$ for $s\in I_{\eta_j}$. In the first case we have $\frac{\pi}{2}<\eta+s<\frac{\pi}{2}+\eta-\beta_j\le\frac{3\pi}{2}+\alpha_j-\beta_j<\frac{3\pi}{2}$ for $\omega\in I_{\eta,j}$ and in the third case we have $\frac{3\pi}{2}>\eta+s>\pi+\beta_j-\frac{\pi}{2}-\alpha_j>\frac{\pi}{2}$ for $\omega\in I_{\eta,j}$. We thus see that $\cos(\eta+s)<0$ for $\eta\in(\beta_j,2\pi+\alpha_j)$, $\omega\in I_{\eta,j}$.

Now assume that $\lambda\in\Pi_{1\le j\le k}(\mathbb{C}\setminus\overline{S}_{\alpha_j,\beta_j})$, let $\eta_j\in(\beta_j,2\pi+\alpha_j)$ be a determination of $arg(\lambda_j)$, let ν be a representing measure for ϕ and let $\omega\in\Pi_{1\le j\le k}I_{\eta_j}$. Then $\mathcal{FB}(\phi)$ is bounded on $\overline{S}_{-\frac{\pi}{2}-\alpha,\frac{\pi}{2}-\beta}$, and since $\cos(\eta_j+\omega_j)<0$ for $j\le k$, we have

$$
\begin{aligned}
\frac{1}{(2\pi i)^k}\int_0^{e^{i\omega}.\infty} e^{\lambda\sigma}\mathcal{FB}(\phi)(\sigma)d\sigma &= \frac{1}{(2\pi i)^k}\int_0^{e^{i\omega}.\infty} e^{\lambda\sigma}\left[\int_{\overline{S}_{\alpha,\beta}} e^{-\sigma\zeta}d\nu(\zeta)\right]d\sigma \\
&= \frac{1}{(2\pi i)^k}\int_{\overline{S}_{\alpha,\beta}}\left[\int_0^{e^{i\omega}.\infty} e^{\sigma(\lambda-\zeta)}d\sigma\right]d\nu(\zeta) \\
&= \int_{\overline{S}_{\alpha,\beta}} \frac{1}{\zeta-\lambda}d\nu(\zeta) \\
&= \mathcal{C}(\phi)(\lambda).
\end{aligned}
$$

□

Now identify the space $\mathcal{M}(\overline{S}_{\alpha,\beta})$ of all measures of bounded variation on $\overline{S}_{\alpha,\beta}$ to the dual space of the space $\mathcal{C}_0(\overline{S}_{\alpha,\beta})$ of continuous functions on $\overline{S}_{\alpha,\beta}$ vanishing at infinity via the Riesz representation theorem.The convolution product of two elements of $\mathcal{M}(\overline{S}_{\alpha,\beta})$ is defined by the usual formula

$$\int_{\overline{S}_{\alpha,\beta}} f(\zeta)d(\nu_1*\nu_2)(\zeta) := \int_{\overline{S}_{\alpha,\beta}\times\overline{S}_{\alpha,\beta}} f(\zeta+\zeta')d\nu_1(\zeta)d\nu_2(\zeta') \quad (f\in\mathcal{C}_0(\overline{S}_{\alpha,\beta})).$$

PROPOSITION 10.5. *Let X be a separable Banach space.*

(i) For $f\in\mathcal{V}_{\alpha,\beta}(X), \lambda\in\overline{S}_{\alpha,\beta}$, set $f_\lambda(\zeta)=f(\zeta+\lambda)$. Then $f_\lambda\in\mathcal{V}_{\alpha,\beta}(X)$ for $f\in\mathcal{V}_{\alpha,\beta}(X)$, $f_\lambda\in\mathcal{U}_{\alpha,\beta}(X)$ and the map $\lambda\to f_\lambda$ belongs to $\mathcal{U}_{\alpha,\beta}(\mathcal{U}_{\alpha,\beta}(X))$ for $f\in\mathcal{U}_{\alpha,\beta}(X)$. Moreover if we set, for $\phi\in\mathcal{U}'_{\alpha,\beta}$,

$$f_\phi(\lambda)=\langle f_\lambda,\phi\rangle,$$

then $f_\phi\in\mathcal{V}_{\alpha,\beta}(X)$ for $f\in\mathcal{V}_{\alpha,\beta}(X)$, and $f_\phi\in\mathcal{U}_{\alpha,\beta}(X)$ for $f\in\mathcal{U}_{\alpha,\beta}(X)$.

(ii) For $\phi_1\in\mathcal{U}'_{\alpha,\beta}, \phi_2\in\mathcal{U}'_{\alpha,\beta}$, set

$$\langle f,\phi_1*\phi_2\rangle=\langle f_{\phi_1},\phi_2\rangle \quad (f\in\mathcal{U}_{\alpha,\beta}).$$

Then $\phi_1*\phi_2\in\mathcal{U}'_{\alpha,\beta}$, $\nu_1*\nu_2$ is a representing measure for $\phi_1*\phi_2$ if ν_1 is a representing measure for ϕ_1 and if ν_2 is a representing measure for ϕ_2, and we have

$$
\begin{aligned}
\langle f,\phi_1*\phi_2\rangle &= \langle f_{\phi_1},\phi_2\rangle \quad (f\in\mathcal{V}_{\alpha,\beta}(X)),\\
\mathcal{FB}(\phi_1*\phi_2) &= \mathcal{FB}(\phi_1)\mathcal{FB}(\phi_2).
\end{aligned}
$$

PROOF. These results follow from standard easy verifications which are left to the reader. We will just prove the last formula. Let $\phi_1\in\mathcal{U}'_{\alpha,\beta}$, $\phi_2\in\mathcal{U}'_{\alpha,\beta}$. We have, for $z\in\overline{S}^*_{\alpha,\beta}, \lambda\in\overline{S}_{\alpha,\beta}, \zeta\in\overline{S}_{\alpha,\beta}$,

$$(e_{-z})_\lambda(\zeta)=e^{-z_1(\lambda_1+\zeta_1)\cdots-z_k(\lambda_k+\zeta_k)}=e_{-z}(\lambda)e_{-z}(\zeta),$$

so $(e_{-z})_\lambda=e_{-z}(\lambda)e_{-z}$, $(e_{-z})_{\phi_1}(\lambda)=\langle(e_{-z})_\lambda,\phi_1\rangle=e_{-z}(\lambda)\mathcal{FB}(\phi_1)(z)$, $(e_{-z})_{\phi_1}=\mathcal{FB}(\phi_1)(z)e_{-z}$, and

$$\mathcal{FB}(\phi_1 * \phi_2)(z) = \langle (e_{-z})_{\phi_1}, \phi_2\rangle = \mathcal{FB}(\phi_1)(z)\langle e_{-z}, \phi_2\rangle = \mathcal{FB}(\phi_1)(z)\mathcal{FB}(\phi_2)(z).$$

□

For $\eta \in \overline{S}_{\alpha,\beta}$, denote by δ_η the Dirac measure at η. We identify δ_η to the linear functional $f \to f(\eta)$ on $\mathcal{U}_{\alpha,\beta}$. With the above notations, we have, for $f \in \mathcal{V}_{\alpha,\beta}(X)$, $\phi \in \mathcal{U}'_{\alpha,\beta}$,

$$f_{\delta_\eta} = f_\eta, \langle f, \phi * \delta_\eta\rangle = \langle f_\eta, \phi\rangle.$$

If $f \in \mathcal{U}_{\alpha,\beta}(X)$, we have $\lim_{\substack{\epsilon\to 0\\ \epsilon\in\overline{S}^*_{\alpha,\beta}}} \|e_{-\epsilon}f - f\|_\infty = 0 = \lim_{\substack{\eta\to 0\\ \eta\in\overline{S}_{\alpha,\beta}}} \|f_\eta - f\|_\infty$. We obtain, since $\|e_{-\epsilon}f\|_\infty \le \|f\|_\infty$ for $f \in \mathcal{U}_{\alpha,\beta}, \epsilon \in \overline{S}^*_{\alpha,\beta}$,

$$\lim_{\substack{\epsilon\to 0, \epsilon\in\overline{S}^*_{\alpha,\beta}\\ \eta\to 0, \eta\in\overline{S}_{\alpha,\beta}}} \|e_{-\epsilon}f_\eta - f\|_\infty = 0 \quad (f \in \mathcal{U}_{\alpha,\beta}(X). \tag{22}$$

Now let $f \in \mathcal{V}_{\alpha,\beta}(X)$, let $\phi \in \mathcal{U}'_{\alpha,\beta}$, and let ν be a representative measure for ϕ. Since

$$\langle e_{-\epsilon}, f_\eta\rangle = \int_{\overline{S}_{\alpha,\beta}} e^{-\epsilon\zeta} f(\zeta + \eta)d\nu(\zeta),$$

and since $\langle e_{-\epsilon}f, \phi * \delta_\eta\rangle = e^{-\epsilon\eta}\langle e_{-\epsilon}f_\eta, \phi\rangle$, it follows from the Lebesgue dominated convergence theorem that we have

$$\langle f, \phi\rangle = \lim_{\substack{\epsilon\to 0, \epsilon\in\overline{S}^*_{\alpha,\beta}\\ \eta\to 0, \eta\in\overline{S}_{\alpha,\beta}}} \langle e_{-\epsilon}f_\eta, \phi\rangle = \lim_{\substack{\epsilon\to 0, \epsilon\in\overline{S}^*_{\alpha,\beta}\\ \eta\to 0, \eta\in\overline{S}_{\alpha,\beta}}} \langle e_{-\epsilon}f, \phi * \delta_\eta\rangle \quad (f \in \mathcal{V}_{\alpha,\beta}(X), \phi \in \mathcal{U}'_{\alpha,\beta}). \tag{23}$$

In the following we will denote by $\tilde{\partial}\overline{S}_{\alpha,\beta} = \Pi_{1\le j\le k}\partial\overline{S}_{\alpha_j,\beta_j}$ the distinguished boundary of $\overline{S}_{\alpha,\beta}$, where $\partial\overline{S}_{\alpha_j,\beta_j} = (e^{i\alpha_j}.\infty, 0] \cup [0, e^{i\beta_j}.\infty)$ is oriented from $e^{i\alpha_j}.\infty$ towards $e^{i\beta_j}.\infty$.

The following standard computations allow to compute in some cases $\langle f, \phi\rangle$ by using the Cauchy transform when $\alpha_j < \beta_j$ for $j \le k$.

PROPOSITION 10.6. *Assume that $\alpha_j < \beta_j < \alpha_j + \pi$ for $1 \le j \le k$, and let $\phi \in \mathcal{U}'_{\alpha,\beta}$. If $f \in \mathcal{V}_{\alpha,\beta}(X)$, and if*

$$\int_{\tilde{\partial}\overline{S}_{\alpha,\beta}} \|f(\sigma)\|_X |d\sigma| < +\infty,$$

then we have, for $\eta \in S_{\alpha,\beta}$,

$$\langle f_\eta, \phi\rangle = \langle f, \phi * \delta_\eta\rangle = \int_{\tilde{\partial}\overline{S}_{\alpha,\beta}} \mathcal{C}(\phi)(\sigma - \eta)f(\sigma)d\sigma. \tag{24}$$

In particular we have, for $f \in \mathcal{V}_{\alpha,\beta}(X), \epsilon \in S^*_{\alpha,\beta}, \eta \in S_{\alpha,\beta}$,

$$e^{-\epsilon\eta}\langle e_{-\epsilon}f_\eta, \phi\rangle = \langle e_{-\epsilon}f, \phi * \delta_\eta\rangle = \int_{\tilde{\partial}\overline{S}_{\alpha,\beta}} e^{-\epsilon\sigma}\mathcal{C}(\phi)(\sigma - \eta)f(\sigma)d\sigma \tag{25}$$

PROOF. Let $f \in \mathcal{V}_{\alpha,\beta}(X)$ such that $\sup_{\sigma\in\overline{S}_{\alpha,\beta}}(1+|\sigma|)^{2k}\|f(\sigma)\| < +\infty$, and let $\nu \in \mathcal{M}(\overline{S}_{\alpha,\beta})$ be a representing measure for ϕ. For $R > 0, j \le k$, we denote by $\Gamma_{R,j}$ the Jordan curve $\{Re^{i\omega}\}_{\alpha_j\le\omega\le\beta_j} \cup [Re^{i\beta_j}, 0] \cup [0, Re^{i\alpha_j}]$, oriented counterclockwise. We have, for $\eta \in S_{\alpha,\beta}$, $\sigma \in \Pi_{1\le j\le k}\partial S_{\alpha_j,\beta_j}$,

$$|\mathcal{C}(\phi)(\sigma-\eta)| \le \frac{1}{(2\pi)^k}\|\phi\|_{\mathcal{U}'_{\alpha,\beta}} \prod_{1\le j\le k} dist(\partial S_{\alpha_j,\beta_j}-\eta_j, \partial S_{\alpha_j,\beta_j})^{-1}.$$

It follows then from Fubini's theorem and Cauchy's formula that we have

$$\begin{aligned}
&\int_{\tilde{\partial}\overline{S}_{\alpha,\beta}} \mathcal{C}(\phi)(\sigma-\eta)f(\sigma)d\sigma\\
=\ & \int_{\overline{S}_{\alpha,\beta}} \left[\frac{1}{2\pi i)^k}\int_{\tilde{\partial}S_{\alpha,\beta}} \frac{f(\sigma)}{\zeta-\sigma+\eta}d\sigma\right] d\nu(\zeta)\\
=\ & \int_{\overline{S}_{\alpha,\beta}} \lim_{R\to+\infty}\frac{1}{(2\pi i)^k}\left[\int_{\Gamma_{R,1}}\cdots\int_{\Gamma_{R,k}} \frac{f(\sigma)}{(\sigma_1-\zeta_1-\eta_1)\dots(\sigma_k-\zeta_k-\eta_k)}d\sigma\right] d\nu(\zeta)\\
=\ & \int_{\overline{S}_{\alpha,\beta}} f(\zeta+\eta)d\nu(\zeta) = \langle f, \phi * \delta_\eta\rangle.
\end{aligned}$$

Formula (26) follows from this equality applied to $e_{-\epsilon}f$. Taking the limit as $\epsilon \to 0, \epsilon \in S^*_{\alpha,\beta}$ in formula (26), we deduce formula (25) from the Lebesgue dominated convergence theorem. □

The following result is indeed standard, but we give a proof for the convenience of the reader.

PROPOSITION 10.7. *Set* $E_{\alpha,\beta} := \{f = e_{-\sigma} : \sigma \in \Pi_{j\le k}(0, e^{-i\frac{\alpha_j+\beta_j}{2}}.\infty)\}$. *Then the linear span of* $E_{\alpha,\beta}$ *is dense in* $\mathcal{U}_{\alpha,\beta}$, *and the Fourier–Borel transform is one-to-one on* $\mathcal{U}'_{\alpha,\beta}$.

PROOF. Set $J_1 = \{j \in \{1,\dots,k\} \mid \alpha_j = \beta_j\}$, $J_2 := \{j \in \{1,\dots,k\} \mid \alpha_j < \beta_j\}$, denote by $\mathcal{U}_1$ the space of continuous functions on $\overline{S}_1 = \Pi_{j\in J_1}\overline{S}_{\alpha_j,\beta_j}$ vanishing at infinity, set $S_2 := \Pi_{j\in J_2}S_{\alpha_j,\beta_j}$, and denote by $\mathcal{U}_2$ the space of continuous functions on $\overline{S}_2$ vanishing at infinity which satisfy the same analyticity condition as in Definition 10.1 with respect to $\overline{S}_2$. Also set

$$E_1 := \left\{ f = e_{-\sigma} : \sigma \in \prod_{j\in J_1}(0, e^{-i\frac{\alpha_j+\beta_j}{2}}.\infty)\right\},$$

$$E_2 := \left\{ f = e_{-\sigma} : \sigma \in \prod_{j\in J_2}(0, e^{-i\frac{\alpha_j+\beta_j}{2}}.\infty)\right\}.$$

Assume that $J_1 \ne \emptyset$. Then the complex algebra $span(E_1)$ is self-adjoint and separates the point on $\mathcal{U}_1$, and it follows from the Stone–Weierstrass theorem applied to the one-point compactification of S_1 that $span(E_1) \oplus \mathbb{C}.1$ is dense in $\mathcal{U}_1 \oplus \mathbb{C}.1$, which implies that $span(E_1)$ is dense in $\mathcal{U}_1$ since $\mathcal{U}_1$ is the kernel of a character on $\mathcal{U}_1 \oplus \mathbb{C}.1$.

Now assume that $J_2 \ne \emptyset$, set $S_2^* = \Pi_{j\in J_2}S_{-\frac{\pi}{2}-\alpha_j,\frac{\pi}{2}-\beta_j}$, let $\phi \in U'_2$, and define the Cauchy transform and the Fourier–Borel transform of ϕ as in Definition 10.3.

Assume that $\langle f, \phi\rangle = 0$ for $\phi \in E_2$. If $j \in J_2$, then $g = 0$ for every holomorphic function g on $S^*_{\alpha_j,\beta_j}$ which vanishes on $(0, e^{-i\frac{\alpha_j+\beta_j}{2}}.\infty)$. An immediate finite induction shows then that $\mathcal{FB}(\phi) = 0$ since $\mathcal{FB}(\phi)$ is holomorphic on S^*_2. It follows then from Proposition 10.4 that $\mathcal{C}(\phi) = 0$, and it follows from (23) and (26) that $\langle f, \phi\rangle = 0$ for every $f \in \mathcal{U}_2$. Hence $\phi = 0$, which shows that $span(E_2)$ is dense in $\mathcal{U}_2$. This shows that $span(E_{\alpha,\beta})$ is dense in $\mathcal{U}_{\alpha,\beta}$ if $J_1 = \emptyset$ or if $J_2 = \emptyset$.

Now assume that $J_1 \neq \emptyset$ and $J_2 \neq \emptyset$, and denote by $E \subset \mathcal{U}_{\alpha,\beta}$ the set of products $f = gh$, where $g \in \mathcal{U}_1$ and $h \in \mathcal{U}_2$. The space $\mathcal{U}_1 = \mathcal{C}_0(\overline{S}_1)$ is a closed subspace of codimension one of $\mathcal{C}(\overline{S}_1 \cup \{\infty\})$. Since the space $\mathcal{C}(K)$ has a Schauder basis for every compact space K, **[4]**,**[33]**, the space $\mathcal{U}_1$ has a Schauder basis. Identifying the dual space of $\mathcal{U}_1$ to the space of measures of bounded variation on $\overline{S}_1$, this means that there exists a sequence $(g_n)_{n\geq 1}$ of elements of $\mathcal{U}_1$ and a sequence $(\nu_n)_{n\geq 1}$ of measures of bounded variation on $\overline{S}^1$ such that we have

$$g = \sum_{n=1}^{+\infty}\left(\int_{\overline{S}_1} g(\eta)d\nu_n(\eta)\right) g_n \quad (g \in \mathcal{U}_1),$$

where the series is convergent in $(\mathcal{U}_1, \|.\|_\infty)$.

Set $P_m(g) = \sum\limits_{n=1}^{m}\left(\int_{\overline{S}_1} g(\eta)d\nu_n(\eta)\right) g_n$ for $g \in \mathcal{U}_1, m \geq 1$. Then $P_m : \mathcal{U}_1 \to \mathcal{U}_1$ is a bounded linear operator, and $\limsup_{m\to+\infty} \|P_m(g)\| \leq \|g\| < +\infty$ for every $g \in \mathcal{U}_1$. It follows then from the Banach–Steinhaus theorem that there exists $M > 0$ such that $\|P_m\|_{\mathcal{B}(\mathcal{U}_1)} \leq M$ for $m \geq 1$, a standard property of Schauder bases in Banach spaces.

Now let $\phi \in \mathcal{U}'_{\alpha,\beta}$ such that $\langle f, \phi\rangle = 0$ for $f \in E$, let ν be a representing measure for ϕ, and let $f \in \mathcal{U}_{\alpha,\beta}$. The function $f_\zeta = \eta \to f(\eta,\zeta)$ belongs to U_1 for $\zeta \in \overline{S}_2$, and a routine verification shows that the function $h_n : \zeta \to \int_{\overline{S}_1} f_\zeta(\sigma)d\nu_n(\sigma) = \int_{\overline{S}_1} f(\zeta,\eta)d\nu_n(\eta)$ belongs to U_2 for $n \geq 1$. Since the evaluation map $g \to g(\eta)$ is continuous on $\mathcal{U}_1$ for $\eta \in \overline{S}_1$, we obtain, for $\eta \in \overline{S}_1, \zeta \in \overline{S}_2$,

$$f(\eta,\zeta) = \lim_{m\to+\infty}\sum_{n=1}^{m} g_n(\eta)h_n(\zeta).$$

We have, for $m \geq 1$, $\eta \in \overline{S}_1, \zeta \in \overline{S}_2$,

$$\left|\sum_{n=1}^{m} g_n(\eta)h_n(\zeta)\right| \leq \|P_m(f_\zeta)\|_\infty \leq M\|f_\zeta\|_\infty \leq M\|f\|_\infty.$$

It follows then from the Lebesgue dominated convergence theorem that

$$\int_{\overline{S}_{\alpha,\beta}} f(\eta,\zeta)d\nu(\eta,\zeta) = \lim_{m\to+\infty}\sum_{n=1}^{m}\int_{\overline{S}_{\alpha,\beta}} g_n(\eta)h_n(\zeta)d\nu(\eta,\zeta) = 0.$$

This shows that $span(E)$ is dense in $\mathcal{U}_{\alpha,\beta}$. Since $span(E_1)$ is dense in $\mathcal{U}_1$ and $span(E_2)$ is dense in $\mathcal{U}_2$, $span(E_{\alpha,\beta})$ is dense in $span(E)$, and so $span(E_{\alpha,\beta})$ is dense in $\mathcal{U}_{\alpha,\beta}$.

Now let $\phi \in \mathcal{U}'_{\alpha,\beta}$. If $\mathcal{FB}(\phi) = 0$, then $\langle f, \phi\rangle = 0$ for every $f \in E_{\alpha,\beta}$, and so $\phi = 0$ since $span(E_{\alpha,\beta})$ is dense in $\mathcal{U}_{\alpha,\beta}$, which shows that the Fourier–Borel transform is one-to-one on $\mathcal{U}_{\alpha,\beta}$. $\square$

We will now give a way to compute $\langle f, \phi\rangle$ for $f \in \mathcal{V}_{\alpha,\beta}(X), \phi \in \mathcal{U}'_{\alpha,\beta}$ by using Fourier–Borel transforms. For $\sigma \in \Pi_{1\le j\le k}\left(\mathbb{C}\setminus \overline{S}^*_{\alpha_j,\beta_j}\right)$, define $e^*_\sigma \in \mathcal{U}'_{\alpha,\beta}$ by using the formula

$$\langle f, e^*_\sigma\rangle = \mathcal{FB}(f)(-\sigma). \tag{26}$$

Also for $\phi \in \mathcal{U}'_{\alpha,\beta}$, $g \in \mathcal{U}_{\alpha,\beta}$, define $\phi g \in \mathcal{U}'_{\alpha,\beta}$ by using the formula

$$\langle f, \phi g\rangle = \langle fg, \phi\rangle \quad (f \in \mathcal{U}_{\alpha,\beta}).$$

It follows from Definition 10.3 that if $\sigma = (\sigma_1, \dots, \sigma_k) \in \Pi_{1\le j\le k}\left(\mathbb{C}\setminus \overline{S}^*_{\alpha_j,\beta_j}\right)$, we have, for $f \in \mathcal{U}_{\alpha,\beta}$,

$$\langle f, e^*_\sigma\rangle = \int_0^{e^{i\omega}.\infty} e^{\sigma\zeta} f(\zeta)d\zeta,$$

where $\omega = (\omega_1, \dots, \omega_k)$ satisfies $\alpha_j \le \omega_j \le \beta_j, Re(\sigma_j e^{i\omega_j}) < 0$ for $1 \le j \le k$, which gives

$$\|e^*_\sigma\|_\infty \le \prod_{1\le j\le k} \int_0^\infty e^{tRe(\sigma_j e^{i\omega_j})}dt = \frac{1}{\prod\limits_{1\le j\le k} (-Re(\sigma_j \omega_j))}.$$

The same formula as above holds with the same ω to compute $\langle f, e^*_{\sigma'}\rangle$ for $\sigma' \in \Pi_{1\le j\le k}\left(\mathbb{C}\setminus \overline{S}^*_{\alpha_j,\beta_j}\right)$ when $|\sigma - \sigma'|$ is sufficiently small, and so the map $\sigma \to e^*_\sigma \in \mathcal{U}'_{\alpha,\beta}$ is holomorphic on $\Pi_{1\le j\le k}\left(\mathbb{C}\setminus \overline{S}^*_{\alpha_j,\beta_j}\right)$ since the $L^1(\mathbb{R}^+)$-valued map $\lambda \to e_{-\lambda}$ is holomorphic on the open half-plane $P^+ := \{\lambda \in \mathbb{C} \mid Re(\lambda) > 0\}$.

Now let $\epsilon \in S^*_{\alpha,\beta}$ and let $\omega \in \Pi_{1\le j\le k}[\alpha_j, \beta_j]$ such that $Re(\epsilon_j e^{i\omega_j}) > 0$ for $1 \le j \le k$. Then $\sigma - \epsilon \in \Pi_{1\le j\le k}\left(\mathbb{C}\setminus \overline{S}^*_{\alpha_j,\beta_j}\right)$ for $\sigma \in \tilde{\partial} S^*_{\alpha,\beta}$, and we have $Re((\sigma_j - \epsilon_j)e^{i\omega_j}) \le -Re(\epsilon_j e^{i\omega_j}) < 0$ for $1 \le j \le k$. We obtain

$$\|e^*_{\sigma-\epsilon}\| \le \frac{1}{\prod\limits_{1\le j\le k} Re(\epsilon_j e^{i\omega_j})},$$

and so $\sup_{\sigma\in\tilde{\partial} S^*_{\alpha,\beta}} \|e^*_{\sigma-\epsilon}\|_\infty < +\infty \quad (\epsilon \in S^*_{\alpha,\beta})$.

We now give the following certainly well-known natural result.

PROPOSITION 10.8. *Let $\phi \in \mathcal{U}'_{\alpha,\beta}$. Assume that*

$$\int_{\tilde{\partial} S^*_{\alpha,\beta}} |\mathcal{FB}(\phi)(\sigma)||d\sigma| < +\infty.$$

Then we have, for $\epsilon \in S^*_{\alpha,\beta}$,

$$\phi e_{-\epsilon} = \frac{1}{(2\pi i)^k} \int_{\tilde{\partial} S^*_{\alpha,\beta}} \mathcal{FB}(\phi)(\sigma) e^*_{\sigma-\epsilon} d\sigma,$$

where the Bochner integral is computed in $(\mathcal{U}'_{\alpha,\beta}, \|.\|_\infty)$, which gives

$$(27)\qquad \langle fe_{-\epsilon},\phi\rangle=\frac{1}{(2\pi i)^k}\int_{\tilde\partial S^*_{\alpha,\beta}}\mathcal{FB}(\phi)(\sigma)\mathcal{FB}(f)(-\sigma+\epsilon)d\sigma\quad (f\in\mathcal{V}_{\alpha,\beta}(X)).$$

PROOF. Since the map $\sigma\to e^*_{\sigma-\epsilon}\in\mathcal{U}'_{\alpha,\beta}$ is continuous on $\tilde\partial S^*_{\alpha,\beta}$, and since $\sup_{\sigma\in\tilde\partial S^*_{\alpha,\beta}}\|e^*_{\sigma-\epsilon}\|_\infty<+\infty$, the Bochner integral $\int_{\tilde\partial S^*_{\alpha,\beta}}\mathcal{FB}(\phi)(\sigma)e^*_{\sigma-\epsilon}d\sigma$ is well-defined in $(\mathcal{U}'_{\alpha,\beta},\|.\|_\infty)$. Set $\phi_\epsilon:=\frac{1}{(2\pi i)^k}\int_{\tilde\partial S^*_{\alpha,\beta}}\mathcal{FB}(\phi)(\sigma)e^*_{\sigma-\epsilon}d\sigma\in\mathcal{U}'_{\alpha,\beta}$. Since the map $\phi\to\mathcal{FB}(\phi)(\zeta)$ is continuous on $\mathcal{U}'_{\alpha,\beta}$, we have, for $\zeta\in\overline{S}_{\alpha,\beta}$,

$$\begin{aligned}\mathcal{FB}(\phi_\epsilon)(\zeta)&=\frac{1}{(2\pi i)^k}\int_{\tilde\partial S^*_{\alpha,\beta}}\mathcal{FB}(\phi)(\sigma)\mathcal{FB}(e^*_{\sigma-\epsilon})(\zeta)d\sigma\\&=\frac{1}{(2\pi i)^k}\int_{\tilde\partial S^*_{\alpha,\beta}}\mathcal{FB}(\phi)(\sigma)\langle e_{-\zeta},e^*_{\sigma-\epsilon}\rangle d\sigma.\end{aligned}$$

It follows from (27) that $\langle e_{-\zeta},e^*_{\sigma-\epsilon}\rangle=\mathcal{FB}(e_{-\zeta})(\epsilon-\sigma)$. Let $\omega\in\Pi_{1\le j\le k}[\alpha_j,\beta_j]$ such that $Re(\epsilon_je^{i\omega_j})>0$ for $j\le k$. Since $Re((\sigma_j-\epsilon_j)e^{i\omega_j})\le -Re(\epsilon_je^{i\omega_j})<0$ for $1\le j\le k$, we have, for $\sigma\in\tilde\partial S^*_{\alpha,\beta}$,

$$\begin{aligned}\mathcal{FB}(e_{-\zeta})(\epsilon-\sigma)&=\int_0^{e^{i\omega}.\infty}e^{(\sigma-\epsilon)\eta}e_{-\zeta}(\eta)d\eta\\&=\int_0^{e^{i\omega}.\infty}e^{(\sigma-\epsilon-\zeta)\eta}d\eta\\&=\frac{1}{\prod\limits_{1\le j\le k}(\zeta_j+\epsilon_j-\sigma_j)}.\end{aligned}$$

Using the notation $\frac{1}{\zeta+\epsilon-\sigma}:=\frac{1}{\Pi_{1\le j\le k}(\zeta_j+\epsilon_j-\sigma_j)}$, this gives

$$\mathcal{FB}(\phi_\epsilon)(\zeta)=\frac{1}{(2\pi i)^k}\int_{\tilde\partial S^*_{\alpha,\beta}}\frac{\mathcal{FB}(\phi)(\sigma)}{\zeta+\epsilon-\sigma}d\sigma.$$

As in Appendix 3, set $W_{j,n}(\zeta_j)=\frac{n^2}{\left(n+e^{\frac{\alpha_j+\beta_j}{2}i}\zeta_j\right)^2}$ for $n\ge1,\zeta_j\in\overline{S}^*_{\alpha_j,\beta_j}$, and set $W_n(\zeta)=\Pi_{j\le k}W_{n,j}(\zeta_j)$ for $\zeta\in\overline{S}^*_{\alpha,\beta}$.

Then $|W_{n,j}(\zeta_j)|\le1$ for $\zeta_j\in\overline{S}^*_{\alpha_j,\beta_j}$, $W_n(\zeta)\to1$ as $n\to\infty$ uniformly on compact sets of $\overline{S}^*_{\alpha,\beta}$, and $\lim_{\substack{|\zeta|\to\infty\\ \zeta\in S^*_{\alpha,\beta}}}W_n(\zeta)=0$. The open set $S^*_{\alpha,\beta}$ is admissible with respect to (α,β) in the sense of Definition 12.1 and, since $\mathcal{FB}(\phi)$ is bounded on $S^*_{\alpha,\beta}$, $\mathcal{FB}(\phi)W_n\in H^{(1)}(S^*_{\alpha,\beta})$ for $n\ge1$. It follows then from Theorem 12.5 that we have, for $t\in(0,1),\zeta\in\overline{S}^*_{\alpha,\beta}$,

$$\begin{aligned}&\frac{1}{(2\pi i)^k}\int_{\tilde\partial S^*_{\alpha,\beta}}\frac{\mathcal{FB}(\phi)(\sigma+t\epsilon)W_n(\sigma+t\epsilon)}{\zeta+(1-t)\epsilon-\sigma}d\sigma\\=&\frac{1}{(2\pi i)^k}\int_{\tilde\partial S^*_{\alpha,\beta}+t\epsilon}\frac{\mathcal{FB}(\phi)(\sigma)W_n(\sigma)}{\zeta+\epsilon-\sigma}d\sigma=\mathcal{FB}(\phi)(\zeta+\epsilon)W_n(\zeta+\epsilon),\end{aligned}$$

and it follows from the Lebesgue dominated convergence theorem that we have

$$\frac{1}{(2\pi i)^k}\int_{\tilde{\partial} S^*_{\alpha,\beta}} \frac{\mathcal{FB}(\phi)(\sigma)W_n(\sigma)}{\zeta+\epsilon-\sigma}d\sigma = \mathcal{FB}(\phi)(\zeta+\epsilon)W_n(\zeta+\epsilon).$$

Taking the limit as $n \to +\infty$, and using again the Lebesgue dominated convergence theorem, we obtain, for $\zeta \in \overline{S}^*_{\alpha,\beta}$,

$$\begin{aligned}\mathcal{FB}(\phi_\epsilon)(\zeta) &= \frac{1}{(2\pi i)^k}\int_{\tilde{\partial} S^*_{\alpha,\beta}} \frac{\mathcal{FB}(\phi)(\sigma)}{\zeta+\epsilon-\sigma}d\sigma = \mathcal{FB}(\phi)(\zeta+\epsilon)\\ &= \langle e_{-\zeta}e_{-\epsilon}, \phi\rangle\\ &= \langle e_{-\zeta}, \phi e_{-\epsilon}\rangle = \mathcal{FB}(\phi e_{-\epsilon})(\zeta),\end{aligned}$$

and it follows from the injectivity of the Fourier–Borel transform on $\mathcal{U}'_{\alpha,\beta}$ that $\phi_\epsilon = \phi e_{-\epsilon}$.

This gives, for $f \in \mathcal{V}_{\alpha,\beta}(X)$, since $\langle f, e^*_{\sigma-\epsilon}\rangle = \mathcal{FB}(f)(-\sigma+\epsilon)$ for $\sigma \in \tilde{\partial} S^*_{\alpha,\beta}$,

$$\begin{aligned}\langle f e_{-\epsilon}, \phi\rangle = \langle f, \phi e_{-\epsilon}\rangle &= \frac{1}{(2\pi i)^k}\int_{\tilde{\partial} S^*_{\alpha,\beta}} \mathcal{FB}(\phi)(\sigma)\langle f, e^*_{\sigma-\epsilon}\rangle d\sigma\\ &= \frac{1}{(2\pi i)^k}\int_{\tilde{\partial} S^*_{\alpha,\beta}} \mathcal{FB}(\phi)(\sigma)\mathcal{FB}(f)(-\sigma+\epsilon)d\sigma.\end{aligned}$$

□

For $J \subset \{1,\ldots,k\}$, set

- $P_{J,j} = \mathbb{C}\setminus -S_{-\frac{\pi}{2}-\alpha_j,\frac{\pi}{2}-\alpha_j}$, $\omega_{J,j} = \alpha_j$ for $j \in J$,
- $P_{J,j} = \mathbb{C}\setminus -S_{-\frac{\pi}{2}-\beta_j,\frac{\pi}{2}-\beta_j}$, $\omega_{J,j} = \beta_j$ for $j \in \{1,\ldots,k\}\setminus J$,
- $P_J = \Pi_{1\le j\le k}P_{J,j}$, $\omega_J = (\omega_{J,1},\ldots,\omega_{J,k})$.

If $f \in \mathcal{V}_{\alpha,\beta}(X)$, and if $\int_{\tilde{\partial}\overline{S}_{\alpha,\beta}} \|f(\zeta)\|_X|d\zeta| < +\infty$, then the formula

$$\mathcal{FB}(f)(\sigma) = \int_0^{e^{i\omega_J}.\infty} e^{-\zeta\sigma}f(\zeta)d\sigma$$

defines a continuous bounded extension of $\mathcal{FB}(f)$ to P_J, and so $\mathcal{FB}(f)$ has a continuous bounded extension to $\cup_{J\subset\{1,\ldots,k\}}P_J = \Pi_{1\le j\le k}\left(\mathbb{C}\setminus -S^*_{\alpha_j,\beta_j}\right)$. Applying formula (28) to the sequence $(\epsilon_n) = (\frac{\epsilon}{n})$ for some $\epsilon \in S^*_{\alpha,\beta}$, we deduce from the Lebesgue dominated convergence theorem and from formula (23) the following result.

Corollary 10.9. *Let $f \in \mathcal{V}_{\alpha,\beta}(X)$, and let $\phi \in \mathcal{U}'_{\alpha,\beta}$. Assume that the following conditions are satisfied*

$$(i) \int_{\tilde{\partial}\overline{S}_{\alpha,\beta}} \|f(\zeta)\|_X|d\zeta| < +\infty.$$

$$(ii) \int_{\tilde{\partial} S^*_{\alpha,\beta}} |\mathcal{FB}(\phi)(\sigma)||d\sigma| < +\infty.$$

Then

$$\langle f, \phi\rangle = \frac{1}{(2\pi i)^k}\int_{\tilde{\partial} S^*_{\alpha,\beta}} \mathcal{FB}(\phi)(\sigma)\mathcal{FB}(f)(-\sigma)d\sigma. \tag{28}$$

In the following we will denote by $\tilde{\nu}$ the functional $f \to \int_{S_{\alpha,\beta}} f(\zeta)d\nu(\zeta)$ for $\nu \in \mathcal{M}(\overline{S}_{\alpha,\beta})$. In order to give a way to compute $\langle f, \phi\rangle$ for $\phi \in \mathcal{U}'_{\alpha,\beta}, f \in \mathcal{V}_{\alpha,\beta}(X)$, we will use the following easy observation.

PROPOSITION 10.10. *Let ν be a probability measure on $\overline{S}_{\alpha,\beta}$, let $R > 0$, and let X be a separable Banach space. Set $\nu_R(A) = \nu(RA)$ for every Borel set $A \subset \overline{S}_{\alpha,\beta}$. Then $\lim_{R\to+\infty} \|f_{\tilde{\nu}_R} - f\|_\infty = 0$ for every $f \in \mathcal{U}_{\alpha,\beta}(X)$.*

PROOF. Let $f \in \mathcal{U}_{\alpha,\beta}(X)$. Then f is uniformly continuous on $\overline{S}_{\alpha,\beta}$, and so for every $\delta > 0$ there exists $r > 0$ such that $\|f(\zeta+\eta) - f(\zeta)\|_X < \delta$ for every $\zeta \in \overline{S}_{\alpha,\beta}$ and for every $\eta \in \overline{S}_{\alpha,\beta} \cap \overline{B}(0,r)$. It follows from the Lebesgue dominated convergence theorem that

$$\lim_{R\to+\infty} \nu_R(B(0,r)) = \lim_{R\to+\infty} \nu(B(0,rR)) = \nu(\overline{S}_{\alpha,\beta}) = 1.$$

This gives

$$\begin{aligned}\limsup_{R\to+\infty} \|f_{\tilde{\nu}_R} - f\|_\infty &= \limsup_{R\to+\infty}\left(\sup_{\zeta\in S_{\alpha,\beta}} \left\|\int_{\overline{S}_{\alpha,\beta}} (f(\zeta+\eta) - f(\zeta))d\nu_R(\eta)\right\|_X\right)\\ &\le \limsup_{R\to+\infty}\left(\sup_{\zeta\in\overline{S}_{\alpha,\beta}} \int_{\overline{S}_{\alpha,\beta}\cap B(0,r)} \|f(\zeta+\eta) - f(\zeta)\|_X d\nu_R(\eta)\right)\\ &\quad + 2\|f\|_\infty \limsup_{R\to+\infty} \int_{\overline{S}_{\alpha,\beta}\setminus(\overline{S}_{\alpha,\beta}\cap B(0,r))} d\nu_R(\eta) \le \delta.\end{aligned}$$

Hence $\lim_{R\to+\infty} \|f_{\tilde{\nu}_R} - f\|_\infty = 0$. □

It follows from the definition of ν_R that $\langle f, \tilde{\nu}_R\rangle = \langle f_{\frac{1}{R}}, \tilde{\nu}\rangle$ for $f \in \mathcal{V}_{\alpha,\beta}(X)$, where $f_{\frac{1}{R}}(\zeta) = f(R^{-1}\zeta)$ $(\zeta \in S_{\alpha,\beta})$. In particular $\mathcal{FB}(\tilde{\nu}_R) = \mathcal{FB}(\nu)_{\frac{1}{R}}$, and $(\tilde{\nu}_1)_R * (\tilde{\nu}_2)_R = (\tilde{\nu}_1 * \tilde{\nu}_2)_R = (\widetilde{\nu_1 * \nu_2})_R$ for $R > 0$ if ν_1 and ν_2 are two probability measures on $S_{\alpha,\beta}$.

We deduce from Proposition 10.8 and Proposition 10.9 the following corollary, in which the sequence $(W_n)_{n\ge1}$ of functions on $\overline{S}^*_{\alpha,\beta}$ introduced in Appendix 3 and used in the proof of Proposition 10.8 allows to compute $\langle f, \phi\rangle$ for $\phi \in \mathcal{U}'_{\alpha,\beta}$, $f \in \mathcal{V}_{\alpha,\beta}(X)$ in the general case.

COROLLARY 10.11. *Set $W_n(\zeta) = \Pi_{1\le j\le k} \frac{n^2}{\left(n+\zeta_j e^{i\frac{\alpha_j+\beta_j}{2}}\right)^2}$ for $n \ge 1$, $\zeta \in \overline{S}^*_{\alpha,\beta}$. Then we have, for $\phi \in \mathcal{U}'_{\alpha,\beta}$, $f \in \mathcal{V}_{\alpha,\beta}(X)$,*

$$\langle f, \phi\rangle = \lim_{\substack{\epsilon\to0\\ \epsilon\in S_{\alpha,\beta}}} \left(\lim_{n\to+\infty} \frac{1}{(2\pi i)^k} \int_{\tilde{\partial} S^*_{\alpha,\beta}} W_n(\sigma)\mathcal{FB}(\phi)(\sigma)\mathcal{FB}(f)(\epsilon-\sigma)d\sigma\right). \tag{29}$$

PROOF. Define a measure ν_0 on $\overline{S}_{\alpha,\beta}$ by using the formula

$$\langle f, \nu_0\rangle = \int_{[0,+\infty)^k} e^{-t_1 \cdots - t_k} f(t_1 e^{i\frac{\alpha_1+\beta_1}{2}}, \dots, t_k e^{i\frac{\alpha_k+\beta_k}{2}})dt_1 \dots dt_k \quad (f \in \mathcal{C}_0(S_{\alpha,\beta})).$$

Then ν_0 and $\nu = \nu_0 * \nu_0$ are probability measures on $\overline{S}_{\alpha,\beta}$, and we have, for $\zeta \in \overline{S}^*_{\alpha,\beta}$,

$$\begin{aligned}\mathcal{FB}(\tilde{\nu}_0)(\zeta) &= \int_{[0,+\infty)^k} e^{-t_1\cdots -t_k} e^{-t_1\zeta_1 e^{i\frac{\alpha_1+\beta_1}{2}}-\cdots -t_k\zeta_k e^{i\frac{\alpha_k+\beta_k}{2}}} dt_1\dots dt_k \\ &= \prod_{1\le j\le k} \frac{1}{1+\zeta_j e^{i\frac{\alpha_j+\beta_j}{2}}}.\end{aligned}$$

Hence $\mathcal{FB}(\tilde{\nu}) = \mathcal{FB}(\tilde{\nu}_0)^2 = W_1$, and $\mathcal{FB}(\tilde{\nu}_n) = (W_1)_{\frac{1}{n}} = W_n$. It follows from (29) that we have, for $\epsilon \in \overline{S}^*_{\alpha,\beta}$,

$$\begin{aligned}\langle fe_{-\epsilon}, \phi\rangle &= \lim_{n\to+\infty} \langle (fe_{-\epsilon})_{\tilde{\nu}_n}, \phi\rangle \\ &= \lim_{n\to+\infty} \langle fe_{-\epsilon}, \phi * \tilde{\nu}_n\rangle \\ &= \lim_{n\to+\infty} \frac{1}{(2\pi i)^k}\int_{\tilde{\partial} S^*_{\alpha,\beta}} \mathcal{FB}(\phi*\tilde{\nu}_n)(\sigma)\mathcal{FB}(f)(\epsilon-\sigma)d\sigma \\ &= \lim_{n\to+\infty} \frac{1}{(2\pi i)^k}\int_{\tilde{\partial} S^*_{\alpha,\beta}} W_n(\sigma)\mathcal{FB}(\phi)(\sigma)\mathcal{FB}(f)(\epsilon-\sigma)d\sigma,\end{aligned}$$

and the result follows from the fact that $\langle f,\phi\rangle = \lim_{\substack{\epsilon\to 0\\ \epsilon\in S_{\alpha,\beta}}} \langle fe_{-\epsilon},\phi\rangle$. □

11. Appendix 2: An algebra of fast-decreasing holomorphic functions on products of sectors and half-lines and its dual

In this section we will use the notations introduced in Definition 4.1 for $\alpha, \beta \in \mathbb{R}^k$ satisfying $\alpha_j \le \beta_j < \alpha_j + \pi$ for $1 \le j \le k$. Notice that is $x \in \mathbb{C}, y \in \mathbb{C}$, there exists $z \in \mathbb{C}$ such that $\left(x + \overline{S}^*_{\alpha_j,\beta_j}\right) \cap \left(y + \overline{S}^*_{\alpha_j,\beta_j}\right) = z + \overline{S}^*_{\alpha_j,\beta_j}$. Such a complex number z is unique if $\alpha_j < \beta_j$. If $\alpha_j = \beta_j$, then $\overline{S}^*_{\alpha_j,\beta_j} = \overline{S}_{\alpha_j-\pi/2,\alpha_j+\pi/2}$ is a closed half-plane, the family $\left\{x + \overline{S}^*_{\alpha_j,\beta_j}\right\}_{x\in\mathbb{C}}$ is linearly ordered with respect to inclusion and the condition $\left(x + \overline{S}^*_{\alpha_j,\beta_j}\right) \cap \left(y + \overline{S}^*_{\alpha_j,\beta_j}\right) = z + \overline{S}^*_{\alpha_j,\beta_j}$ defines a real line of the form $z_0 + e^{i\alpha_j}\mathbb{R}$, where $z_0 \in \{x, y\}$.

The following partial preorder on $\mathbb{C}^k$ is the partial order associated to the cone $\overline{S}^*_{\alpha,\beta}$ if $\alpha_j < \beta_j$ for $1 \le j \le k$.

DEFINITION 11.1. *(i) For $z = (z_1, \dots, z_k) \in \mathbb{C}^k$ and $z' = (z'_1, \dots, z'_k) \in \mathbb{C}^k$, set $z \preceq z'$ if $z' \in z + \overline{S}^*_{\alpha,\beta}$.*

(ii) If $(z^{(j)})_{1\le j\le m}$ is a finite family of elements of $\mathbb{C}^k$ denote by $\sup_{1\le j\le m} z^{(j)}$ the set of all $z \in \mathbb{C}^k$ such that $\cap_{1\le j\le k}\left(z^{(j)} + \overline{S}^*_{\alpha,\beta}\right) = z + \overline{S}^*_{\alpha,\beta}$.

For $z = (z_1, \dots, z_k) \in \mathbb{C}^k$, set $e^z = (e^{z_1}, \dots, e^{z_k})$, and denote again by $e_z : \mathbb{C}^k \to \mathbb{C}$ the map $(\zeta_1, \dots, \zeta_k) \to e^{z\zeta} = e^{z_1\zeta_1+\cdots+z_k\zeta_k}$.

It follows from (17) that $e_{-z'}\mathcal{U}_{\alpha,\beta} \subseteq e_{-z}\mathcal{U}_{\alpha,\beta}$ if $z \preceq z'$.

For $f \in e_{-z}\mathcal{V}_{\alpha,\beta}$, set $\|f\|_{e_{-z}\mathcal{V}_{\alpha,\beta}} = \|e_z f\|_\infty$, which defines a Banach space norm on $e_{-z}\mathcal{U}_{\alpha,\beta}$ and $e_{-z}\mathcal{V}_{\alpha,\beta}$.

PROPOSITION 11.2. *(i)Set $\gamma_n = ne^{-i\frac{\alpha+\beta}{2}}$ for $n \ge 1$. Then the sequence $(\gamma_n)_{n\ge1}$ is cofinal in $(\mathbb{C}^k, \preceq)$.*

(ii) If $z \preceq z'$, then $e_{-z'}\mathcal{U}_{\alpha,\beta}$ is a dense subset of $(e_{-z}\mathcal{U}_{\alpha,\beta}, \|.\|_{e_{-z}\mathcal{U}_{\alpha,\beta}})$.

(iii) The set $\cap_{z\in\mathbb{C}^k} e_{-z}\mathcal{U}_{\alpha,\beta}$ *is a dense ideal of* $\mathcal{U}_{\alpha,\beta}$*, which if a Fréchet algebra with respect to the family* $(\|.\|_{e_{-\gamma_n}\mathcal{U}_{\alpha,\beta}})_{n\geq 1}$.

(iv) If X is a separable Banach space, and if $z \in sup_{1\leq j\leq m} z^{(j)}$*, then*

$$e_{-z}\mathcal{U}_{\alpha,\beta}(X) \subset \cap_{1\leq j\leq m} e_{-z^{(j)}}\mathcal{U}_{\alpha,\beta}(X), e_{-z}\mathcal{V}_{\alpha,\beta}(X) \subset \cap_{1\leq j\leq m} e_{-z^{(j)}}\mathcal{V}_{\alpha,\beta}(X),$$

and $\max_{1\leq j\leq m} \|f\|_{e_{-z^{(j)}}\mathcal{V}_{\alpha,\beta}(X)} \leq \|f\|_{e_{-z}\mathcal{V}_{\alpha,\beta}(X)}$ *for* $f \in e_{-z}\mathcal{V}_{\alpha,\beta}(X)$.

If, further, $k = 1$*, then*

$$e_{-z}\mathcal{U}_{\alpha,\beta}(X) = \cap_{1\leq j\leq m} e_{-z^{(j)}}\mathcal{U}_{\alpha,\beta}(X), e_{-z}\mathcal{V}_{\alpha,\beta}(X) = \cap_{1\leq j\leq m} e_{-z^{(j)}}\mathcal{V}_{\alpha,\beta}(X),$$

and $\max_{1\leq j\leq m} \|f\|_{e_{-z^{(j)}}\mathcal{V}_{\alpha,\beta}(X)} = \|f\|_{e_{-z}\mathcal{V}_{\alpha,\beta}(X)}$ *for* $f \in e_{-z}\mathcal{V}_{\alpha,\beta}(X)$.

PROOF. (i) Let $z = (z_1, \dots, z_k) \in \mathbb{C}^k$, and let $j \leq k$. Since

$$\left(\frac{\pi}{2} - \beta_j\right) + \left(-\frac{\pi}{2} - \alpha_j\right) = -(\alpha_j + \beta_j),$$

$t_{0,j}e^{-i\frac{(\alpha_j+\beta_j)}{2}} \in \partial(z_j + \overline{S}^*_{\alpha_j,\beta_j})$ for some $t_{0,j} \in \mathbb{R}$, so $te^{-i\frac{(\alpha_j+\beta_j)}{2}} \in z_j + \overline{S}^*_{\alpha_j,\beta_j}$ for every $t \geq t_{0,j}$, and (i) follows.

(ii) Assume that $z \preceq z'$. The fact that $e_{-z'}\mathcal{U}_{\alpha,\beta} \subset e_{-z}\mathcal{U}_{\alpha,\beta}$ follows from (16). Let $z'' \in z' + S^*_{\alpha,\beta} \subset z + S^*_{\alpha,\beta}$. We have $z'' = z + re^{i\eta}$ where $r > 0$, and where $\eta = (\eta_1, \dots, \eta_k)$ satisfies $-\frac{\pi}{2} - \alpha_j < \eta_j < \frac{\pi}{2} - \beta_j$ for $j \leq k$.

The semigroup $(e_{-te^{i\eta}})_{t>0}$ is analytic and bounded in the Banach algebra $\mathcal{U}_{\alpha,\beta}$, and $\lim_{t\to 0^+} \|f - fe_{-te^{i\eta}}\|_\infty = 0$ for every $f \in \mathcal{U}_{\alpha,\beta}$.

It follows then from the analyticity of this semigroup that

$$\left[e_{-re^{i\eta}}\mathcal{U}_{\alpha,\beta}\right]^- = \left[\cup_{t>0} e_{-te^{i\eta}}\mathcal{U}_{\alpha,\beta}\right]^- = \mathcal{U}_{\alpha,\beta}.$$

Hence $e_{-z''}\mathcal{U}_{\alpha,\beta}$ is dense in $e_{-z}\mathcal{U}_{\alpha,\beta}$, which proves (ii) since $e_{-z''}\mathcal{U}_{\alpha,\beta} \subset e_{-z'}\mathcal{U}_{\alpha,\beta}$.

(iii) Denote by $i_{z,z'} : f \to f$ the inclusion map from $e_{-z'}\mathcal{U}_{\alpha,\beta}$ into $e_{-z}\mathcal{U}_{\alpha,\beta}$ for $z \preceq z'$. Equipped with these maps, the family $(e_{-z}\mathcal{U}_{\alpha,\beta})_{z\in\mathbb{C}^k}$ is a projective system of Banach spaces, and we can identify $\left(\cap_{z\in\mathbb{C}^k} e_{-z}\mathcal{U}_{\alpha,\beta}, (\|.\|_{e_z\mathcal{U}_{\alpha,\beta}})_{z\in\mathbb{C}^k}\right)$ to the inverse limit of this system, which defines a structure of complete locally convex topological space on $\left(\cap_{z\in\mathbb{C}^k} e_{-z}\mathcal{U}_{\alpha,\beta}, (\|.\|_{e_{-z}\mathcal{U}_{\alpha,\beta}})_{z\in\mathbb{C}^k}\right)$. It follows from (i) that the sequence $(\|.\|_{e_{-\gamma_n}\mathcal{U}_{\alpha,\beta}})_{n\geq 1}$ of norms defines the same topology as the family $\left(\|.\|_{e_{-z}\mathcal{U}_{\alpha,\beta}})_{z\in\mathbb{C}^k}\right)$ on $\cap_{z\in\mathbb{C}^k} e_{-z}\mathcal{U}_{\alpha,\beta} = \cap_{n\geq 1} e_{-\gamma_n}\mathcal{U}_{\alpha,\beta}$, which defines a Fréchet algebra structure on $\cap_{z\in\mathbb{C}^k} e_{-z}\mathcal{U}_{\alpha,\beta}$.

It follows from (ii) that $e_{-\gamma_{n+1}}\mathcal{U}_{\alpha,\beta}$ is dense in $e_{-\gamma_n}\mathcal{U}_{\alpha,\beta}$ for $n \geq 0$, and a standard application of the Mittag-Leffler theorem of projective limits of complete metric spaces, see for example Theorem 2.14 of **[14]**, shows that $\cap_{z\in\mathbb{C}^k} e_{-z}\mathcal{U}_{\alpha,\beta} = \cap_{n\geq 1} e_{-\gamma_n}\mathcal{U}_{\alpha,\beta}$ is dense in $e_{-\gamma_0}\mathcal{U}_{\alpha,\beta} = \mathcal{U}_{\alpha,\beta}$.

(iv) Let $z, z' \in \mathbb{C}^k$, and let $z" \in sup(z, z')$. Then

$$e_{-z"+z} \in \mathcal{V}_{\alpha,\beta}, e_{-z"+z'} \in \mathcal{V}_{\alpha,\beta}, \|e_{-z"+z}\|_\infty \leq 1, \|e_{-z"+z}\|_\infty \leq 1,$$

and so

$$e_{-z"}\mathcal{U}_{\alpha,\beta}(X) \subset e_{-z}\mathcal{U}_{\alpha,\beta}(X) \cap e_{-z'}\mathcal{U}_{\alpha,\beta}(X), e_{-z"}\mathcal{V}_{\alpha,\beta}(X) \subset e_{-z}\mathcal{V}_{\alpha,\beta}(X) \cap e_{-z'}\mathcal{V}_{\alpha,\beta}(X),$$

and $\max(\|f\|_{e_{-z}\mathcal{V}_{\alpha,\beta}(X)}, \|f\|_{e_{-z'}\mathcal{V}_{\alpha,\beta}(X)}) \leq \|f\|_{e_{-z"}\mathcal{V}_{\alpha,\beta}(X)}$ for $f \in e_{-z"}\mathcal{V}_{\alpha,\beta}(X)$.

Now assume that $k = 1$.

We claim that $|e^{z"\zeta}| = \max(|e^{z\zeta}|, |e^{z'\zeta}|)$ for $\zeta \in \partial\overline{S}_{\alpha,\beta}$. If $z \in z' + \overline{S}^*_{\alpha,\beta}$, or if $z' \in z + \overline{S}^*_{\alpha,\beta}$, this is obviously true. Otherwise we have $\alpha < \beta$ and, say, $z" =$

$z + re^{i(-\alpha-\frac{\pi}{2})} = z' + r'e^{i(-\beta+\frac{\pi}{2})}$, with $r > 0, r' > 0$. Let $\zeta = \rho e^{i\theta} \in \overline{S}_{\alpha,\beta}$, where $\rho \geq 0, \theta \in \{\alpha, \beta\}$. We have

$$Re((z" - z)\zeta) = r\rho\cos(\theta - \alpha - \frac{\pi}{2}) \geq 0, \text{ and } Re((z" - z')\zeta) = r'\rho\cos(\theta - \beta + \frac{\pi}{2}) \geq 0.$$

So $|e^{z'\zeta}| \leq |e^{z\zeta}| = |e^{z"\zeta}|$ if $\theta = \alpha$, and $|e^{z\zeta}| \leq |e^{z'\zeta}| = |e^{z''\zeta}|$ if $\theta = \beta$, which proves the claim.

We now use the Phragmén–Lindelöf principle. Let $s \in (1, \frac{\pi}{\beta-\alpha})$ and let ζ^s be a continuous determination of the s-power of ζ on $\overline{S}_{\frac{\beta-\alpha}{2},\frac{\alpha-\beta}{2}}$ which is holomorphic on $S_{\frac{\beta-\alpha}{2},\frac{\alpha-\beta}{2}}$ if $\alpha < \beta$. Let $f \in e_{-z}\mathcal{V}_{\alpha,\beta}(X) \cap e_{-z'}\mathcal{V}_{\alpha,\beta}(X)$, and let $\epsilon > 0$. Set

$$g_\epsilon(\zeta) = e^{-\epsilon\left(\zeta e^{-i\frac{\alpha+\beta}{2}}\right)^s} e^{z"\zeta} f(\zeta)$$

for $\zeta \in \overline{S_{\alpha,\beta}}$. Then $g_\epsilon \in \mathcal{U}_{\alpha,\beta}(X)$, and it follows from the maximum modulus principle that there exists $\zeta_0 \in \partial S_{\alpha,\beta}$ such that

$$\|g_\epsilon\|_{\mathcal{U}_{\alpha,\beta}(X)} = \|g_\epsilon(\zeta_0)\| \leq |e^{z"\zeta_0}| \|f(\zeta_0)| = \max(|e^{z\zeta_0}| \|f(\zeta_0)|, |e^{z'\zeta_0}| \|f(\zeta_0)|).$$

Since $\lim_{\epsilon\to 0} e^{-\epsilon\zeta^s} = 1$ for every $\zeta \in \overline{S_{\alpha,\beta}}$, this shows that $f \in e^{-z"}\mathcal{V}_{\alpha,\beta}(X)$, and $\|f\|_{e_{-z"}\mathcal{V}_{\alpha,\beta}(X)} = \max(\|f\|_{e_{-z}\mathcal{V}_{\alpha,\beta}(X)}, \|f\|_{e_{-z'}\mathcal{V}_{\alpha,\beta}(X)})$.

Now let $f \in e_{-z}\mathcal{U}_{\alpha,\beta}(X) \cap e_{-z'}\mathcal{U}_{\alpha,\beta}(X)$. Then $f \in e_{-z"}\mathcal{V}_{\alpha,\beta}(X)$, and

$$\lim_{\substack{|\zeta|\to 0 \\ \zeta\in\partial\overline{S}_{\alpha,\beta}}} \|e^{z"\zeta} f(\zeta)\| = 0.$$

The Banach algebra $\mathcal{U}_{\alpha,\beta}$ possesses a sequential bounded approximate identity $(g_n)_{n\geq 1}$, one can take for example $g_n(\zeta) = \frac{n\zeta}{n\zeta + e^{i\frac{\alpha+\beta}{2}}}$. We have

$$\lim_{n\to+\infty} \|e_{z"} f g_n - e_{z"} f\|_\infty = \lim_{n\to+\infty} \max_{\zeta\in\partial\overline{S}_{\alpha,\beta}} \|e_{z"}(\zeta) f(\zeta) g_n(\zeta) - e_{z"}(\zeta) f(\zeta)\| = 0,$$

and so $e_{z"} f \in \mathcal{U}_{\alpha,\beta}(X)$ since $\mathcal{U}_{\alpha,\beta}(X)$ is a closed subspace of $\mathcal{V}_{\alpha,\beta}(X)$. This concludes the proof of (iv) when $m = 2$. The general case follows by an immediate induction, since $\sup(\zeta, z^{(l)}) = \sup_{1\leq j\leq l} z^{(j)})$ for every $\zeta \in \sup_{1\leq j\leq l-1} z^{(j)}$ if $(z^{1)}, \ldots, z^{(l)})$ is a finite family of elements of $\mathbb{C}^k$. □

Notice that assertions (ii) and (iii) of the proposition do not extend to the case where $\beta_j = \alpha_j + \pi$ for some $j \leq k$. It suffices to consider the case where $\alpha_j = -\frac{\pi}{2}, \beta_j = \frac{\pi}{2}$. Set $\lambda_j(t) = (\lambda_{s,t})_{1\leq s\leq k}$, where $\lambda_{s,t} = 0$ for $s \neq j$ and $\lambda_{j,t} = t$. Then the map $f \to e_{-\lambda_j(t)} f$ is an isometry on $\mathcal{U}_{\alpha,\beta}$ for every $t \geq 0$ and $\cap_{t>0} u_{\lambda_j(t)}\mathcal{U}_{\alpha,\beta} = \{0\}$ since the zero function is the only bounded holomorphic function f on the right-hand open half-plane satisfying $\lim_{r\to+\infty} |e^{tr} f(r)| = 0$ for every $t > 0$.

Let $i_\zeta : f \to f$ be the inclusion map from $\cap_{z\in\mathbb{C}^k} e_{-z}\mathcal{U}_{\alpha,\beta}$ into $e_{-\zeta}\mathcal{U}_{\alpha,\beta}$. Since i_ζ has dense range, the map $i^*_\zeta : \phi \to \phi_{|\cap_{z\in\mathbb{C}^k} e_{-z}\mathcal{U}_{\alpha,\beta}}$ is a one-to-one map from $(e_{-\zeta}\mathcal{U}_{\alpha,\beta})'$ into $\cap_{z\in\mathbb{C}^k} e_{-z}\mathcal{U}_{\alpha,\beta})'$, which allows to identify $(e_{-\zeta}\mathcal{U}_{\alpha,\beta})'$ to a subset of $(\cap_{z\in\mathbb{C}^k} e_{-z}\mathcal{U}_{\alpha,\beta})'$, so that we have

$$(\cap_{z\in\mathbb{C}^k} e_{-z}\mathcal{U}_{\alpha,\beta})' = \cup_{z\in\mathbb{C}^k} (e_{-z}\mathcal{U}_{\alpha,\beta})' = \cup_{n\geq 1} (e_{-ne^{-i\frac{\alpha+\beta}{2}}} \mathcal{U}_{\alpha,\beta})'. \tag{30}$$

DEFINITION 11.3. *Set* $\mathcal{F}_{\alpha,\beta} := (\cap_{z\in\mathbb{C}^k} e_{-z}\mathcal{U}_{\alpha,\beta})'$. *Let* $\phi \in \mathcal{F}_{\alpha,\beta}$, *and let* X *be a separable Banach space.*

(i) The domain of the Fourier–Borel transform of ϕ *is defined by the formula*

$$Dom(\mathcal{FB}(\phi)) := \{z \in \mathbb{C}^k \mid \phi \in (e_{-z}\mathcal{U}_{\alpha,\beta})'\}.$$

(ii) For $z \in Dom(\mathcal{FB}(\phi))$ the functional $\phi e_{-z} \in \mathcal{U}'_{\alpha,\beta}$ is defined by the formula

$$\langle f, \phi e_{-z}\rangle = \langle e_{-z}f, \phi\rangle \quad (f \in \mathcal{U}_{\alpha,\beta}),$$

and $\langle g, \phi\rangle$ is defined for $g \in e_{-z}\mathcal{V}_{\alpha,\beta}(X)$ by the formula

$$\langle g, \phi\rangle = \langle e_z g, \phi e_{-z}\rangle.$$

(iii) The Fourier–Borel transform of ϕ is defined for $z \in Dom(\mathcal{FB}(\phi))$ by the formula

$$\mathcal{FB}(\phi)(z) = \langle e_{-z}, \phi\rangle.$$

(iv) The z-Cauchy transform of ϕ is defined on $\mathbb{C}^k \setminus -\overline{S}^*_{\alpha,\beta}$ for $z \in Dom(\mathcal{FB}(\phi))$ by the formula

$$\mathcal{C}_z(\phi) = \mathcal{C}(\phi e_{-z}).$$

(v) If $z \in Dom(\mathcal{FB}(\phi))$ a measure ν of bounded variation on $\overline{S}_{\alpha,\beta}$ is said to be a z-representing measure for ϕ if ν is a representing measure for ϕe_{-z}.

Since the map $\zeta \to e_{-\zeta}$ is holomorphic on $S^*_{\alpha,\beta}$, the map $z \to e_{-z}$ is a holomorphic map from $\lambda + S^*_{\alpha,\beta}$ into $e_{-\lambda}\mathcal{U}_{\alpha,\beta}$ for every $\lambda \in Dom(\mathcal{FB}(\phi))$, and so $\mathcal{FB}(\phi)$ is holomorphic on the interior of $Dom(\mathcal{FB}(\phi))$ for $\phi \in \mathcal{F}_{\alpha,\beta}$. Also the z-Cauchy transform $\mathcal{C}_z(\phi)$ is holomorphic on $\mathbb{C} \setminus \overline{S}_{\alpha,\beta}$ if $z \in Dom(\mathcal{FB}(\phi))$. Notice also that if $\phi \in \mathcal{U}'_{\alpha,\beta}$, then $\overline{S}^*_{\alpha,\beta} \subset Dom(\mathcal{FB}(\phi))$ and so the function $\mathcal{FB}(\phi)$ defined above is an extension to $Dom(\mathcal{FB}(\phi))$ of the Fourier–Borel transform already introduced in Definition 10.3 on $\overline{S}^*_{\alpha,\beta}$.

Assume that $g \in e_{-z}\mathcal{V}_{\alpha,\beta}(X) \cap e_{-z'}\mathcal{V}_{\alpha,\beta}(X)$, where $z, z' \in Dom(\mathcal{FB}(\phi))$. Let $z'' \in \sup(z, z') \subset Dom(\mathcal{FB}(\phi))$. Then $g \in e_{-z''}\mathcal{V}_{\alpha,\beta}(X)$. Let ν be a z-representative measure for ϕ.

We have, for $h \in \cap_{\lambda\in\mathbb{C}^k} e_{-\lambda}\mathcal{U}_{\alpha,\beta}$, since $e_{-z} = e_{-z''}e_{z''-z}$,

$$\langle h, \phi\rangle = \int_{\overline{S}_{\alpha,\beta}} e_z(\zeta)h(\zeta)d\nu(\zeta) = \int_{\overline{S}_{\alpha,\beta}} e_{z''}(\zeta)h(\zeta)e_{z-z''}(\zeta)d\nu(\zeta).$$

Since $e_{z-z''}\nu$ is a measure of bounded variation on $S_{\alpha,\beta}$, $e_{z-z''}\nu$ is a z''-representative measure for ϕ. Similarly if ν' is a z'-representative measure for ϕ then $e_{z'-z''}\nu'$ is a z''-representative measure for ϕ, and we have

$$\begin{aligned}\int_{\overline{S}_{\alpha,\beta}} e_z(\zeta)g(\zeta)d\nu(\zeta) &= \int_{\overline{S}_{\alpha,\beta}} e_{z''}(\zeta)g(\zeta)e_{z-z''}(\zeta)d\nu(\zeta)\\ &= \int_{\overline{S}_{\alpha,\beta}} e_{z''}(\zeta)g(\zeta)e_{z'-z''}(\zeta)d\nu'(\zeta)\\ &= \int_{\overline{S}_{\alpha,\beta}} e_{z'}(\zeta)g(\zeta)d\nu'(\zeta),\end{aligned}$$

which shows that the definition of $\langle g,\phi\rangle$ does not depend on the choice of $z \in Dom(\mathcal{FB}(\phi))$ such that $g \in e_{-z}\mathcal{V}_{\alpha,\beta}(X)$.

PROPOSITION 11.4. *Let $\phi \in \mathcal{F}_{\alpha,\beta}$.*
(i) The set $Dom(\mathcal{FB}(\phi))$ is connected.
(ii) $z+\overline{S}^*_{\alpha,\beta} \subset Dom(\mathcal{FB}(\phi))$, and $\mathcal{FB}(\phi)$ is continuous on $z+\overline{S}^*_{\alpha,\beta}$ and holomorphic on $z+S^*_{\alpha,\beta}$ for every $z \in Dom(\mathcal{FB}(\phi))$.

PROOF. (i) The fact that $Dom(\mathcal{FB}(\phi))$ is connected follows from the fact that the arcwise connected set $(z_1+\overline{S}^*_{\alpha,\beta})\cup(z_2+\overline{S}^*_{\alpha,\beta})$ is contained in $Dom(\mathcal{FB}(\phi))$ for $z_1 \in Dom(\mathcal{FB}(\phi)), z_2 \in Dom(\mathcal{FB}(\phi))$.

(ii) Let $z \in Dom(\mathcal{FB}(\phi))$. It follows from (17) that $z+\overline{S}^*_{\alpha,\beta} \subset Dom(\mathcal{FB}(\phi))$ and so $\mathcal{FB}(\phi)$ is holomorphic on the open set $z+S^*_{\alpha,\beta} \subset Dom(\mathcal{FB}(\phi))$. Let ν be a measure of bounded variation on $\overline{S}_{\alpha,\beta}$ which is z-representing measure for ϕ. We have, for $\eta \in \overline{S}_{\alpha,\beta}$,

$$\mathcal{FB}(\phi)(z+\eta) = \langle e_{-z-\eta},\phi\rangle = \langle e_{-\eta},\phi e_{-z}\rangle = \int_{\overline{S}_{\alpha,\beta}} e^{-\eta\zeta}d\nu(\zeta),$$

and the continuity of $\mathcal{FB}(\phi)$ on $z+\overline{S}^*_{\alpha,\beta}$ follows from the Lebesgue dominated convergence theorem. □

Notice that $Dom(\mathcal{FB}(\phi))$ is not closed in general: for example if we set

$$\langle f,\phi\rangle = \int_{\overline{S}_{-\frac{\pi}{4},\frac{\pi}{4}}} \zeta f(\zeta)dm(\zeta)$$

for $f \in \cap_{z\in\mathbb{C}^k} e_{-z}\mathcal{U}_{-\frac{\pi}{4},\frac{\pi}{4}}$, where m denotes the Lebesgue measure on $\mathbb{C}$, then $t \in Dom(\mathcal{FB}(\phi))$ for every $t>0$, but $0 \notin Dom(\mathcal{FB}(\phi))$. Notice also that if ν is a measure supported by a compact subset of $S_{\alpha,\beta}$, and if we set

$$\langle f,\phi\rangle := \int_{S_{\alpha,\beta}} f(\zeta)d\nu(\zeta)$$

for $f \in \cap_{z\in\mathbb{C}^k} e_{-z}\mathcal{U}_{\alpha,\beta}$, then $\phi \in \cap_{z\in\mathbb{C}^k}(e_{-z}\mathcal{U}_{\alpha,\beta})'$, so that $Dom(\mathcal{FB}(\phi)) = \mathbb{C}^k$, and $\mathcal{FB}(\phi)$ is the entire function defined on $\mathbb{C}^k$ by the formula

$$\mathcal{FB}(\phi)(z) = \int_{S_{\alpha,\beta}} e^{-z\zeta}d\nu(\zeta).$$

We now introduce the convolution product of elements of $\mathcal{F}_{\alpha,\beta}$.

If $\phi \in \mathcal{F}_{\alpha,\beta}, f \in \cap_{z\in\mathbb{C}^k} e_{-z}\mathcal{U}_{\alpha,\beta}$, $\lambda \in \overline{S}_{\alpha,\beta}$, set again $f_\lambda(\zeta) = f(\zeta+\lambda)$ for $\zeta \in \overline{S}_{\alpha,\beta}$. Then $f_\lambda \in \cap_{z\in\mathbb{C}^k} e_{-z}\mathcal{U}_{\alpha,\beta}$, and we can compute $\langle f_\lambda,\phi\rangle$. The map $\lambda \to f_\lambda$ is a continuous map from $\overline{S}_{\alpha,\beta}$ into the Fréchet algebra $\cap_{z\in\mathbb{C}^k} e_{-z}\mathcal{U}_{\alpha,\beta}$ which is holomorphic on $S_{\alpha,\beta}$. We obtain

LEMMA 11.5. *Let* $\phi \in \mathcal{F}_{\alpha,\beta}$. *Then the function* $f_\phi : \lambda \to \langle f_\lambda, \phi\rangle$ *belongs to* $\cap_{z\in\mathbb{C}^k} e_{-z}\mathcal{U}_{\alpha,\beta}$ *for every* f *in* $\cap_{z\in\mathbb{C}^k} e_{-z}\mathcal{U}_{\alpha,\beta}$, *and the linear map* $f \to f_\phi$ *is continuous on* $\cap_{z\in\mathbb{C}^k} e_{-z}\mathcal{U}_{\alpha,\beta}$.

PROOF. Let $f \in \cap_{z\in\mathbb{C}^k} e_{-z}\mathcal{U}_{\alpha,\beta}$, let $z_0 \in Dom(\mathcal{FB}(\phi))$, let ν be a z_0-representing measure for ϕ on $\overline{S}_{\alpha,\beta}$, and let $z \in \mathbb{C}^k$.

Let $z_1 \in \sup(z_0, z)$, so that $(z_0 + \overline{S}^*_{\alpha,\beta}) \cap (z + \overline{S}^*_{\alpha,\beta}) = z_1 + \overline{S}^*_{\alpha,\beta}$, and set $\eta_0 = z_1 - z_0, \eta = z_1 - z$. We have, for $\lambda \in \overline{S}_{\alpha,\beta}$,

$$e^{z\lambda}\langle f_\lambda, \phi\rangle = \int_{\overline{S}_{\alpha,\beta}} e^{z\lambda + z_0\zeta} f(\zeta+\lambda)d\nu(\zeta) = \int_{\overline{S}_{\alpha,\beta}} e^{-\eta\lambda - \eta_0\zeta} e^{z_1(\zeta+\lambda)} f(\zeta+\lambda)d\nu(\zeta).$$

Since $|e^{-\eta\lambda-\eta_0\zeta}e^{z_1(\zeta+\lambda)}f(\zeta+\lambda)| \le \|e_{z_1}f\|_\infty$, it follows from Lebesgue's dominated convergence theorem that $\lim_{\substack{|\lambda|\to+\infty \\ \lambda\in\overline{S}_{\alpha,\beta}}} |e^{z\lambda}\langle f_\lambda, \phi\rangle| = 0$, and so $f_\phi \in \cap_{z\in\mathbb{C}^k} e_{-z}\mathcal{U}_{\alpha,\beta}$. Also $\|e_z f_\phi\|_\infty \le \|e_{z_1}f\|_\infty \int_{\overline{S}_{\alpha,\beta}} d|\nu|(\zeta)$, which shows that the map $f \to f_\phi$ is continuous on $\cap_{z^k\in\mathbb{C}} e_{-z}\mathcal{U}_{\alpha,\beta}$. □

Notice that it follows from the Hahn-Banach theorem that if $\phi \in (e_{-z_0}\mathcal{U}_{\alpha,\beta})'$ there exists a z_0-representing measure ν for ϕ such that

$$\int_{\overline{S}_{\alpha,\beta}} d|\nu|(\zeta) = \|\phi\|_{(e_{-z_0}\mathcal{U}_{\alpha,\beta})'}.$$

The calculation above shows then that we have, for $z \in \mathbb{C}^k$, $f \in \cap_{z\in\mathbb{C}^k} e_{-z}\mathcal{U}_{\alpha,\beta}$, $\phi \in \mathcal{F}_{\alpha,\beta}$, $z_0 \in Dom(\mathcal{FB}(\phi))$, $z_1 \in \sup(z_0, z)$,

$$\|e_z f_\phi\|_\infty \le \|e_{z_1}f\|_\infty \|\phi\|_{(e_{-z_0}\mathcal{U}_{\alpha,\beta})'}. \tag{31}$$

PROPOSITION 11.6. *For* $\phi_1 \in \mathcal{F}_{\alpha,\beta}, \phi_2 \in \mathcal{F}_{\alpha,\beta}$, *define the convolution product* $\phi_1 * \phi_2 \in \mathcal{F}_{\alpha,\beta}$ *by the formula*

$$\langle f, \phi_1 * \phi_2\rangle = \langle f_{\phi_1}, \phi_2\rangle \quad (f \in \cap_{z\in\mathbb{C}^k} e_{-z}\mathcal{U}_{\alpha,\beta}).$$

Then

$$sup(z_1, z_2) \subset Dom(\mathcal{FB}(\phi_1 * \phi_2)) \quad (z_1 \in Dom(\mathcal{FB}(\phi_1)), z_2 Dom(\mathcal{FB}(\phi_2)),$$

and we have, for $z \in \sup(z_1, z_2)$,

$$\|\phi_1 * \phi_2\|_{(e_{-z}\mathcal{U}_{\alpha,\beta})'} \le \|\phi_1\|_{(e_{-z_1}\mathcal{U}_{\alpha,\beta})'}\|\phi_2\|_{(e_{-z_2}\mathcal{U}_{\alpha,\beta})'}.$$

More generally $Dom(\mathcal{FB}(\phi_1)) \cap Dom(\mathcal{FB}(\phi_2)) \subset Dom(\mathcal{FB}(\phi_1 * \phi_2))$, *and if* $z \in Dom(\mathcal{FB}(\phi_1)) \cap Dom(\mathcal{FB}(\phi_2))$ *then* $(\phi_1 * \phi_2)e_{-z} = (\phi_1 e_{-z}) * (\phi_2 e_{-z})$, *so that* $\nu_1 * \nu_2$ *is a* z*-representative measure for* $\phi_1 * \phi_2$ *if* ν_1 *is a* z*-representing measure for* ϕ_1 *and if* ν_2 *is a* z*-representing measure for* ν_2, *and we have*

$$\mathcal{FB}(\phi_1 * \phi_2)(z) = \mathcal{FB}(\phi_1)(z)\mathcal{FB}(\phi_2)(z) \quad (z \in Dom(\mathcal{FB}(\phi_1)) \cap Dom(\mathcal{FB}(\phi_2))).$$

PROOF. Let $z_1 \in Dom(\mathcal{FB}(\phi_1))$, let $z_2 \in Dom(\mathcal{FB}(\phi_2))$, and let $z \in \sup(z_1, z_2)$. It follows from (32) that we have, for $f \in \cap_{z\in\mathbb{C}^k} e_z\mathcal{U}_{\alpha,\beta}$,

$$\begin{aligned}|\langle f, \phi_1 * \phi_2\rangle| &= |\langle f_{\phi_1}, \phi_2\rangle| \\ &\le \|e_{z_2} f_{\phi_1}\|_\infty \|\phi_2\|_{(e_{-z_2}\mathcal{U}_{\alpha,\beta})'} \\ &\le \|e_z f\|_\infty \|\phi_1\|_{(e_{-z_1}\mathcal{U}_{\alpha,\beta})'} \|\phi_2\|_{(e_{-z_2}\mathcal{U}_{\alpha,\beta})'}.\end{aligned}$$

Hence $\phi_1 * \phi_2 \in \mathcal{F}_{\alpha,\beta}$, $sup(z_1, z_2) \subset Dom(\mathcal{FB}(\phi_1 * \phi_2))$, and $\|\phi_1 * \phi_2\|_{(e_{-z}\mathcal{U}_{\alpha,\beta})'} \le \|\phi_1\|_{(e_{-z_1}\mathcal{U}_{\alpha,\beta})'}\|\phi_2\|_{(e_{-z_2}\mathcal{U}_{\alpha,\beta})'}$ for $z \in \sup(z_1, z_2)$.

Let $z \in Dom(\mathcal{FB}(\phi_1)) \cap \mathcal{FB}(\phi_2))$. Then $z \in \sup(z, z) \subset Dom(\mathcal{FB}(\phi_1 * \phi_2))$.

Let ν_1 be a z-representing measure for ϕ_1 and let ν_2 be a z-representing measure for ϕ_2. We have, for $f \in \cap_{s\in\mathbb{C}^k} e_{-s}\mathcal{U}_{\alpha,\beta}$,

$$\begin{aligned}\langle f, \phi_1 * \phi_2\rangle &= \langle f_{\phi_1}, \phi_2\rangle \\ &= \int_{\overline{S}_{\alpha,\beta}} e^{z\lambda} f_{\phi_1}(\lambda) d\nu_2(\lambda) \\ &= \int_{\overline{S}_{\alpha,\beta}} \left[\int_{\overline{S}_{\alpha,\beta}} e^{z\zeta} f(\zeta+\lambda) d\nu_1(\lambda)\right] e^{z\lambda} d\nu_2(\lambda) \\ &= \int\int_{\overline{S}_{\alpha,\beta}\times\overline{S}_{\alpha,\beta}} e^{z(\zeta+\lambda)} f(\zeta+\lambda) d\nu_1(\zeta) d\nu_2(\lambda) \\ &= \int_{\overline{S}_{\alpha,\beta}} e^{zs} d(\nu_1 * \nu_2)(s),\end{aligned}$$

and so $\nu_1 * \nu_2$ is a representing measure for $(\phi_1 * \phi_2)e_{-z}$, which means that $\nu_1 * \nu_2$ is a z-representative measure for $\phi_1 * \phi_2$. Since ν_1 is a representative measure for $\phi_1 e_{-z}$, and since ϕ_2 is a representative measure for $\phi_2 e_{-z}$, it follows from Proposition 10.5 (ii) that $(\phi_1 * \phi_2)e_{-z} = (\phi_1 e_{-z}) * (\phi_2 e_{-z})$.

It follows also from Proposition 10.5(ii) that

$$\begin{aligned}\mathcal{FB}(\phi_1 * \phi_2)(z) &= \mathcal{FB}((\phi_1 * \phi_2)e_{-z})(1) \\ &= \mathcal{FB}((\phi_1 e_{-z}) * (\phi_2 e_{-z}))(1) \\ &= \mathcal{FB}(\phi_1 e_{-z})(1)\mathcal{FB}(\phi_2 e_{-z})(1) \\ &= \mathcal{FB}(\phi_1)(z)\mathcal{FB}(\phi_2)(z).\end{aligned}$$

□

Using Proposition 10.4, we obtain the following link between z-Cauchy transforms and Fourier–Borel transforms of elements of $\mathcal{F}_{\alpha,\beta}$.

PROPOSITION 11.7. *Let $\phi \in \mathcal{F}_{\alpha,\beta}$. For $j \le k$, set $I_{\eta,j} = (\frac{\pi}{2} - \eta, \frac{\pi}{2} - \beta_j]$ for $\eta \in (\beta_j, \alpha_j + \pi]$, $I_{\eta,j} = (-\frac{\pi}{2} - \alpha_j, \frac{\pi}{2} - \beta_j)$ for $\eta \in (\alpha_j + \pi, \beta_j + \pi]$, and set $I_{\eta,j} = (-\frac{\pi}{2} - \alpha_j, \frac{3\pi}{2} - \eta)$ for $\eta \in (\beta_j + \pi, \alpha_j + 2\pi)$. Then $I_{\eta,j} \subset [-\frac{\pi}{2} - \alpha_j, \frac{\pi}{2} - \beta_j]$, $\cos(\eta + s) < 0$ for $s \in I_{\eta,j}$, and if $\lambda = (\lambda_1, \dots, \lambda_k) \in \mathbb{C}^k \setminus \overline{S}_{\alpha,\beta}$, we have for $\omega = (\omega_1, \dots, \omega_k) \in \Pi_{1\le j\le k} I_{arg(\lambda_j),j}$, $z \in Dom(\mathcal{FB}(\phi))$,*

$$\begin{aligned}\mathcal{C}_z(\phi)(\lambda) &= \frac{1}{(2\pi i)^k}\int_0^{e^{i\omega}.\infty} e^{\lambda\sigma}\mathcal{FB}(\phi)(\sigma+z)d\sigma \\ &:= \frac{1}{(2\pi i)^k}\int_0^{e^{i\omega_1}.\infty}\cdots\int_0^{e^{i\omega_k}.\infty} e^{\lambda\sigma}\mathcal{FB}(\phi)(\sigma+z)d\sigma.\end{aligned}$$

PROOF. We have $\mathcal{C}_z(\phi) = \mathcal{C}(\phi e_{-z})$, and, for $\sigma \in \overline{S}^*_{\alpha,\beta}$,

$$\mathcal{FB}(\phi)(\sigma + z) = \langle e_{-\sigma-z}, \phi\rangle = \langle e_{-\sigma}e_{-z}, \phi\rangle = \langle e_{-\sigma}, \phi e_{-z}\rangle = \mathcal{FB}(\phi e_{-z})(\sigma).$$

The result follows then from formula (22) applied to ϕe_{-z}. □

Let X be a separable Banach space. For $\eta \in \overline{S}_{\alpha,\beta}$, $z \in \mathbb{C}^k$, $f \in e_{-z}\mathcal{V}_{\alpha,\beta}(X)$, set $f_\eta(\zeta) = f(\zeta + \eta)$ $(\zeta \in \overline{S}_{\alpha,\beta})$. If $\phi \in \mathcal{F}_{\alpha,\beta}$, and if $z \in Dom(\mathcal{FB}(\phi))$, we have

$$\begin{aligned}\langle f, \phi * \delta_\eta\rangle &= \langle e_z f, (\phi * \delta_\eta)e_{-z}\rangle\\ &= \langle e_z f, (\phi e_{-z}) * (\delta_\eta e_{-z})\rangle\\ &= e^{-z\eta}\langle e_z f, (\phi e_{-z}) * \delta_\eta\rangle\\ &= e^{-z\eta}\langle (e_z f)_\eta, (\phi e_{-z})\rangle\\ &= \langle e_z f_\eta, \phi e\_z\rangle\\ &= \langle f_\eta, \phi\rangle.\end{aligned}$$

We also have, for $f \in e_{-z}\mathcal{U}_{\alpha,\beta}(X)$,

$$\begin{aligned}&\lim_{\substack{\eta\to 0\\ \eta\in\overline{S}_{\alpha,\beta}}} \|f_\eta - f\|_{e_{-z}\mathcal{U}_{\alpha,\beta}(X)}\\ =\;& \lim_{\substack{\eta\to 0\\ \eta\in\overline{S}_{\alpha,\beta}}} \sup_{\zeta\in\overline{S}_{\alpha,\beta}} \|e^{z\zeta}f(\zeta+\eta) - e^{z\zeta}f(\zeta)\|_\infty\\ \le\;& \lim_{\substack{\eta\to 0\\ \eta\in\overline{S}_{\alpha,\beta}}} \left(\|(e_z f)_\eta - e_z f\|_\infty + |1 - e^{-z\eta}|\|(e_z f)_\eta)\|_\infty\right)\\ =\;& 0,\end{aligned}$$

and so, since $(e_{-\epsilon}f)_\eta = e^{-\epsilon\eta}e_{-\epsilon}f_\eta$,

$$\begin{aligned}&\lim_{\substack{\eta\to 0,\eta\in\overline{S}_{\alpha,\beta}\\ \epsilon\to 0,\epsilon\in\overline{S}^*_{\alpha,\beta}}} \|(e_{-\epsilon}f)_\eta - f\|_{e_{-z}\mathcal{U}_{\alpha,\beta}(X)}\\ =\;& \lim_{\substack{\eta\to 0,\eta\in\overline{S}_{\alpha,\beta}\\ \epsilon\to 0,\epsilon\in\overline{S}^*_{\alpha,\beta}}} \|e_{-\epsilon}f_\eta - f\|_{e_{-z}\mathcal{U}_{\alpha,\beta}(X)} = 0 \quad (f \in e_{-z}\mathcal{U}_{\alpha,\beta}(X), z \in \mathbb{C}^k).\end{aligned}$$

Now let $f \in e_{-z}\mathcal{V}_{\alpha,\beta}(X)$, and let $\phi \in (e_{-z}\mathcal{U}_{\alpha,\beta})'$. If ν is a z-representative measure for ϕ, we have, for $\eta \in \overline{S}_{\alpha,\beta}, \epsilon \in \overline{S}^*_{\alpha,\beta}$,

$$\langle (e_{-\epsilon}f)_\eta, \phi\rangle = e^{-\epsilon\eta}\langle e_{-\epsilon}f_\eta, \phi\rangle = e^{-(\epsilon+z)\eta}\int_{\overline{S}_{\alpha,\beta}} e^{-\epsilon\zeta}e^{z(\zeta+\eta)}f(\zeta+\eta)d\nu(\zeta),$$

and it follows from the Lebesgue dominated convergence theorem that we have

$$\begin{aligned}&\lim_{\substack{\eta\to 0,\eta\in\overline{S}_{\alpha,\beta}\\ \epsilon\to 0,\epsilon\in\overline{S}^*_{\alpha,\beta}}} \|\langle (e_{-\epsilon}f)_\eta, \phi\rangle - \langle f, \phi\rangle\|_X\\ =\;& \lim_{\substack{\eta\to 0,\eta\in\overline{S}_{\alpha,\beta}\\ \epsilon\to 0,\epsilon\in\overline{S}^*_{\alpha,\beta}}} \|\langle e_{-\epsilon}f_\eta, \phi\rangle - \langle f, \phi\rangle\|_X = 0 \quad (f \in e_{-z}\mathcal{V}_{\alpha,\beta}(X), \phi \in (e_{-z}\mathcal{U}_{\alpha,\beta})', z \in \mathbb{C}^k).\end{aligned}$$

The following consequences of Proposition 10.6 allow to compute in some cases $\langle f, \phi\rangle$ for $\phi \in \mathcal{U}'_{\alpha,\beta}, f \in \mathcal{V}_{\alpha,\beta}(X), z \in Dom(\mathcal{FB}(\phi))$ by using the z-Cauchy transform.

PROPOSITION 11.8. *Assume that* $\alpha_j < \beta_j < \alpha_j + \pi$ *for* $1 \le j \le k$, *let* $\phi \in \mathcal{F}_{\alpha,\beta}$, *let* $z \in Dom(\mathcal{FB}(\phi))$, *and let* X *be a separable Banach space.*

If $f \in e_{-z}\mathcal{V}_{\alpha,\beta}(X)$, *and if*

$$\int_{\tilde{\partial} S_{\alpha,\beta}} e^{Re(z\sigma)} \|f(\sigma)\|_X |d\sigma| < +\infty,$$

then we have, for $\eta \in S_{\alpha,\beta}$,

$$\langle f_\eta, \phi\rangle = \langle f, \phi * \delta_\eta\rangle = \int_{\tilde{\partial}\overline{S}_{\alpha,\beta}} e^{z(\sigma-\eta)} C_z(\phi)(\sigma-\eta) f(\sigma) d\sigma. \tag{32}$$

In particular we have, for $f \in \mathcal{V}_{\alpha,\beta}(X), \epsilon \in S^*_{\alpha,\beta}, \eta \in S_{\alpha,\beta}$,

$$e^{-\epsilon\eta}\langle e_{-\epsilon} f_\eta, \phi\rangle = \langle e_{-\epsilon} f, \phi * \delta_\eta\rangle = \int_{\tilde{\partial}\overline{S}_{\alpha,\beta}} e^{(z-\epsilon)(\sigma-\eta)} \mathcal{C}_z(\phi)(\sigma-\eta) f(\sigma) d\sigma. \tag{33}$$

PROOF. Assume that $f \in e_{-z}\mathcal{V}_{\alpha,\beta}(X)$ satisfies the condition

$$\int_{\tilde{\partial} S_{\alpha,\beta}} \|f(\sigma)\|_X |d\sigma| < +\infty.$$

We have, for $\eta \in S_{\alpha,\beta}$, $\epsilon \in S^*_{\alpha,\beta}$,

$$\langle f_\eta, \phi\rangle = \langle e_z f_\eta, \phi e_{-z}\rangle = e^{-z\eta}\langle (e_z f)_\eta, \phi e_{-z}\rangle, e_{-\epsilon} f_\eta = e^{\epsilon\eta}(e_{-\epsilon} f)_\eta$$

so (32) follows from (25) applied to $e_z f$ and ϕe_{-z}, and (33) follows from (32) applied to $e_{-\epsilon} f$. □

For $z \in \mathbb{C}^k, f \in e_{-z}\mathcal{V}_{\alpha,\beta}(X)$, recall the Fourier–Borel transform of f is defined for $\zeta = (\zeta_1 \dots, \zeta_k) \in \Pi_{1\le j\le k}\left(\mathbb{C} \setminus (-z_j - \overline{S}^*_{\alpha_j,\beta_j})\right)$ by the formula

$$\mathcal{FB}(f)(\zeta) = \mathcal{FB}(e_z f)(z+\zeta) = \int_0^{e^{i\omega}.\infty} e^{-\zeta\sigma} f(\sigma) d\sigma$$

$$:= \int_0^{e^{i\omega_1}.\infty} \cdots \int_0^{e^{i\omega_k}.\infty} e^{-\zeta_1\sigma_1 \cdots - \zeta_k\sigma_k} f(\sigma_1, \dots, \sigma_k) d\sigma_1 \dots d\sigma_k,$$

where $\alpha_j \le \omega_j \le \beta_j$ and where $Re((z_j + \zeta_j)e^{i\omega_j}) > 0$ for $1 \le j \le k$.

The following consequences of Proposition 10.8, Corollary 10.9 and Corollary 10.11 allow to interpret the action of $\phi \in \mathcal{F}_{\alpha,\beta}$ on $e_{-z}\mathcal{U}_{\alpha,\beta}$ for $z \in Dom(\mathcal{FB}(\phi))$ in terms of Fourier–Borel transforms.

PROPOSITION 11.9. *Let* $\phi \in \mathcal{F}_{\alpha,\beta}$, *let* $z = (z_1, \dots, z_k) \in Dom(\mathcal{FB}(\phi))$, *and let* $f \in e_{-z}\mathcal{V}_{\alpha,\beta}(X)$. *Set again, for* $\zeta = (\zeta_1, \dots, \zeta_k) \in S^*_{\alpha,\beta}, n \ge 1$,

$$W_n(\zeta) = \Pi_{1\le j\le k} \frac{n^2}{\left(n + \zeta_j e^{i\frac{\alpha_j+\beta_j}{2}}\right)^2}.$$

Then

(i) $\langle f, \phi\rangle = \lim_{\substack{\epsilon\to 0\\ \epsilon\in S_{\alpha,\beta}}} \left(\lim_{n\to+\infty} \frac{1}{(2\pi i)^k} \int_{z+\tilde{\partial}\overline{S}^*_{\alpha,\beta}} W_n(\sigma - z)\mathcal{FB}(\phi)(\sigma)\mathcal{FB}(f)(-\sigma+\epsilon) d\sigma\right).$

(ii) If, further, $\int_{\tilde{\partial} S^*_{\alpha,\beta}} |\mathcal{FB}(\phi)(\sigma))||d\sigma| < +\infty$, *then we have, for* $\epsilon \in S^*_{\alpha,\beta}$,

$$\langle e_{-\epsilon} f, \phi \rangle = \frac{1}{(2\pi i)^k} \int_{z+\tilde{\partial}\overline{S}^*_{\alpha,\beta}} \mathcal{FB}(\phi)(\sigma)\mathcal{FB}(f)(-\sigma+\epsilon)d\sigma,$$

and so

$$\langle f, \phi \rangle = \lim_{\substack{\epsilon \to 0 \\ \epsilon \in S_{\alpha,\beta}}} \frac{1}{(2\pi i)^k} \int_{z+\tilde{\partial}\overline{S}^*_{\alpha,\beta}} \mathcal{FB}(\phi)(\sigma)\mathcal{FB}(f)(-\sigma+\epsilon)d\sigma.$$

If, further, $\int_{\tilde{\partial}\overline{S_{\alpha,\beta}}} e^{Re(z\sigma)}\|f(\sigma)\||d\sigma| < +\infty$, *then*
(iii)

$$\langle f, \phi \rangle = \frac{1}{(2\pi i)^k} \int_{z+\tilde{\partial}\overline{S}^*_{\alpha,\beta}} \mathcal{FB}(\phi)(\sigma)\mathcal{FB}(f)(-\sigma)d\sigma.$$

PROOF. We have $\langle f, \phi \rangle = \langle e_z f, \phi e_{-z} \rangle$. Since $\langle fb(f)(-\zeta - z) = \mathcal{FB}(e_z f)(-\zeta)$ for $\zeta \in \Pi_{1\le j\le k}\left(\mathbb{C} \setminus \overline{S}^*_{\alpha,\beta}\right)$, and since $\mathcal{FB}(\phi e_{-z})(\zeta) = \langle e_{-\zeta-z}, \phi \rangle = \mathcal{FB}(\phi)(\zeta+z)$ for $\zeta \in \overline{S}^*_{\alpha,\beta}$, it follows from Corollary 10.11 that we have

$$\begin{aligned}
\langle f, \phi \rangle &= \langle e_z f, \phi e_{-z} \rangle \\
&= \lim_{\substack{\epsilon \to 0 \\ \epsilon \in S^*_{\alpha,\beta}}} \left(\frac{1}{(2\pi i)^k} \int_{\tilde{\partial} S^*_{\alpha,\beta}} W_n(\zeta)\mathcal{FB}(\phi e_{-z})(\zeta)\mathcal{FB}(e_z f)(\epsilon - \zeta)d\zeta \right) \\
&= \lim_{\substack{\epsilon \to 0 \\ \epsilon \in S^*_{\alpha,\beta}}} \left(\frac{1}{(2\pi i)^k} \int_{\tilde{\partial} S^*_{\alpha,\beta}} W_n(\zeta)\mathcal{FB}(\phi)(z+\zeta)\mathcal{FB}(f)(-z+\epsilon-\zeta)d\zeta \right),
\end{aligned}$$

and we obtain (i) by using the change of variables $\sigma = z + \zeta$ for $\zeta \in \tilde{\partial} S^*_{\alpha,\beta}$. Using the same change of variables we deduce (ii) from Proposition 10.8 and (iii) from Corollary 10.9. □

LEMMA 11.10. *Let* $\alpha, \alpha', \beta, \beta' \in \mathbb{R}^k$ *such that* $\alpha'_j \le \alpha_j \le \beta_j \le \beta'_j < \alpha'_j + \pi$ *for* $j \le k$. *Then* $\cap_{z\in\mathbb{C}^k} e_{-z}\mathcal{U}_{\alpha',\beta'}$ *is dense in* $\cap_{z\in\mathbb{C}^k} e_{-z}\mathcal{U}_{\alpha,\beta}$.

PROOF. Let $\phi \in \mathcal{F}_{\alpha,\beta}$, and assume that $\langle f, \phi \rangle = 0$ for every $f \in \cap_{z\in\mathbb{C}^k} e_{-z}\mathcal{U}_{\alpha',\beta'}$. Let $z \in Dom(\mathcal{FB}(\phi))$. Then $\mathcal{FB}(\phi)(z+\zeta) = 0$ for every $\zeta \in \overline{S}^*_{\alpha',\beta'}$. Since $Dom(\mathcal{FB}(\phi))$ is connected, we have $\mathcal{FB}(\phi) = 0$. Hence $\phi = 0$, since the Fourier–Borel transform is one-to-one on $\mathcal{F}_{\alpha,\beta}$. □

So we can identify $\mathcal{F}_{\alpha,\beta}$ to a subset of $\mathcal{F}_{\alpha',\beta'}$ if $\alpha'_j \le \alpha_j \le \beta_j \le \beta'_j < \alpha'_j + \pi$ for $1 \le j \le k$.

A standard application of the Mittag-Leffler theorem of projective limits of complete metric spaces, see for example **[14]**, Theorem 2.14, shows that we have the following result, where as before

$$M_{a,b} = \{(\alpha,\beta) \in \mathbb{R}^k \times \mathbb{R}^k \mid a_j < \alpha_j \le \beta_j < b_j \text{ if } a_j < b_j, \alpha_j = \beta_j = a_j \text{ if } a_j = b_j\}.$$

PROPOSITION 11.11. *Let* $a = (a_1, \dots, a_k) \in \mathbb{R}^k, b = (b_1, \dots, b_k) \in \mathbb{R}^k$ *such that* $a_j \le b_j \le a_j + \pi$ *for* $j \le k$. *Then* $\cap_{(\alpha',\beta')\in M_{a,b},\lambda\in\mathbb{C}^k} e_{-\lambda}\mathcal{U}_{\alpha',\beta'}$ *is dense in* $e_{-z}\mathcal{U}_{\alpha,\beta}$ *for every* $z \in \mathbb{C}^k$ *and every* $(\alpha,\beta) \in M_{a,b}$.

Let $(a,b) \in \mathbb{R}^k \times \mathbb{R}^k$ be as above, and denote by $\Delta_{a,b}$ the set of all triples (α,β,z) where $(\alpha,\beta) \in M_{a,b}$ and $z \in \mathbb{C}^k$. Denote by $\preceq$ the product partial order on $\mathbb{R}^k$ associated to the usual order on $\mathbb{R}$. If $(\alpha,\beta,z) \in \Delta_{a,b}, (\alpha',\beta',z') \in \Delta_{a,b}$, set $(\alpha,\beta,z) \preceq (\alpha',\beta',z')$ if $\alpha' \preceq \alpha, \beta \preceq \beta'$ and $z' \in z + \overline{S}^*_{\alpha',\beta'}$. For every finite family $F = \{(\alpha^{(l)},\beta^{(l)},z^{(l)})\}_{1\le l\le m}$ of elements of $\Delta_{a,b}$, set

$$\sup(F) = \{\inf_{1\le l\le m} \alpha^{(l)}\} \times \{\sup_{1\le l\le m} \beta^{(l)}\} \times \sup_{1\le l\le m} z^{(l)},$$

where $\sup_{1\le l\le m} z^{(l)}$ denotes the set of all $z \in \mathbb{C}^k$ satisfying the condition

$$z + \overline{S}^*_{\inf_{1\le l\le m}\alpha^{(l)},\sup_{1\le l\le m}\beta^{(l)}} = \cap_{1\le l\le m}\left(z^{(l)} + \overline{S}^*_{\inf_{1\le l\le k}\alpha^{(l)},\sup_{1\le l\le k}\beta^{(l)}}\right),$$

so that $\sup_{1\le l\le m} z^{(l)}$ is the set introduced in Definition 9.1(ii) when $\alpha = \inf_{1\le l\le m}\alpha^{(l)}$ and $\beta = \sup_{1\le l\le m}\beta^{(l)}$. Notice that $\sup_{1\le l\le m} z^{(j)}$, is a singleton if $(\inf_{1\le l\le m}\alpha^{(l)})_j < (\sup_{1\le l\le m}\beta^{(l)})_j$ for $1 \le j \le k$.

It follows from the proposition that we can identify the dual of the projective limit $\cap_{(\alpha,\beta,z)\in\Delta_{a,b}} e_{-z}\mathcal{U}_{\alpha,\beta}$ to the inductive limit $\cup_{(\alpha,\beta,z)\in\Delta_{a,b}}(e_{-z}\mathcal{U}_{\alpha,\beta})'$. This suggests the following definition.

DEFINITION 11.12. *Let $a = (a_1,\dots,a_k) \in \mathbb{R}^k, b = (b_1,\dots,b_k) \in \mathbb{R}^k$ such that $a_j \le b_j \le a_j + \pi$ for $j \le k$. Set*

$$\mathcal{G}_{a,b} = (\cap_{(\alpha,\beta,z)\in\Delta_{a,b}} e_{-z}\mathcal{U}_{\alpha,\beta})' = \cup_{(\alpha,\beta,z)\in\Delta_{a,b}}(e_{-z}\mathcal{U}_{\alpha,\beta})'.$$

For $\phi \in \mathcal{G}_{a,b}$, set $dom(\phi) = \{(\alpha,\beta,z) \in \Delta_{\alpha,\beta} \mid \phi \in (e_{-z}\mathcal{U}_{\alpha,\beta})'\}$.

We thus see that the inductive limit $\mathcal{G}_{a,b} = \cup_{(\alpha,\beta)\in M_{a,b}}\mathcal{F}_{\alpha,\beta}$ is an associative unital pseudo-Banach algebra with respect to the convolution product introduced above on the spaces $\mathcal{F}_{\alpha,\beta}$. A subset V of $\mathcal{G}_{a,b}$ is bounded if and only if there exists $(\alpha,\beta) \in M_{a,b}$ and $z \in \mathbb{C}^k$ such that V is a bounded subset of $(e_{-z}\mathcal{U}_{\alpha,\beta})'$.

The proof of the following proposition is left to the reader.

PROPOSITION 11.13. *Let $\phi \in \mathcal{G}_{a,b}$, and let $(\alpha,\beta,z) \in dom(\phi)$.*
Then $(\alpha',\beta',z') \in dom(\phi)$ if $(\alpha,\beta,z) \preceq (\alpha',\beta',z')$.
In particular if $(\phi_j)_{1\le j\le m}$ is a finite family of elements of $\mathcal{G}_{a,b}$, and if $(\alpha^{(j)},\beta^{(j)},z^{(j)}) \in dom(\phi_j)$ for $1 \le j \le m$, then

$$sup_{1\le j\le m}(\alpha^{(j)},\beta^{(j)},z^{(j)}) \subset \cap_{1\le j\le m} dom(\phi_j) \subset dom(\phi_1 * \cdots * \phi_m).$$

12. Appendix 3: Holomorphic functions on admissible open sets

DEFINITION 12.1. *Let $a = (a_1,\dots,a_k) \in \mathbb{R}^k, b = (b_1,\dots,b_k) \in \mathbb{R}^k$ such that $a_j \le b_j \le a_j + \pi$ for $j \le k$.*

An open set $U \subset \mathbb{C}^k$ is said to be admissible with respect to $(\alpha,\beta) \in M_{a,b}$ if $U = \Pi_{1\le j\le k}U_j$, where the open sets $U_j \subset \mathbb{C}$ satisfy the following conditions for some $z = (z_1,\dots,z_k) \in \mathbb{C}^k$,

(i) $U_j + \overline{S}^*_{\alpha_j,\beta_j} \subset U_j$

(ii) $U_j \subset z_j + S^*_{\alpha_j,\beta_j}$, and

$$\partial U_j - z_j = (e^{(-\frac{\pi}{2}-\alpha_j)i}.\infty, e^{(-\alpha_j-\frac{\pi}{2})i}s_{0,j}) \cup \theta_j([0,1]) \cup (e^{(\frac{\pi}{2}-\beta_j)i}s_{1,j}, e^{(\frac{\pi}{2}-\beta_j)i}.\infty),$$

where $s_{0,j} \geq 0, s_{1,j} \geq 0$, *and where*

$$\theta_j : [0,1] \to \overline{S}^*_{\alpha_j,\beta_j} \setminus \left(e^{(-\frac{\pi}{2}-\alpha_j)i}.\infty, e^{(-\alpha_j-\frac{\pi}{2})i}s_{j,0}) \cup (e^{(\frac{\pi}{2}-\beta_j)i}s_{j,1}, e^{(\frac{\pi}{2}-\beta_j)i}.\infty)\right)$$

is a one-to-one piecewise-$\mathcal{C}^1$ *curve such that*

$$\theta_j(0) = e^{(-\alpha_j-\frac{\pi}{2})i}s_{j,0}, \text{ and } \theta_j(1) = e^{(\frac{\pi}{2}-\beta_j)i}s_{j,1}.$$

If U *is an admissible open set with respect to some* $(\alpha,\beta) \in M_{a,b}$, $H^{(1)}(U)$ *denotes the space of all functions* F *holomorphic on* U *such that*

$$\|F\|_{H^{(1)}(U)} := \sup_{\epsilon \in S^*_{\alpha,\beta}} \int_{\tilde{\partial}U+\epsilon} |F(\sigma)||d\sigma| < +\infty.$$

For example if $\alpha_j = \beta_j$ then conditions (i) and (ii) are satisfied if an only if U_j is a half-plane of the form $\{z_j \in \mathbb{C} \mid Re(z_j e^{i\alpha_j}) > \lambda\}$ for some $\lambda \in \mathbb{R}$.

If $\alpha_j < \beta_j$, define $\tilde{x}_j = \tilde{x}_j(\zeta_j)$ and $\tilde{y}_j = \tilde{y}_j(\zeta_j)$ for $\zeta_j \in \mathbb{C}$ by the formula

$$\zeta_j = z_j + \tilde{x}_j e^{(-\frac{\pi}{2}-\alpha_j)i} + \tilde{y}_j e^{(\frac{\pi}{2}-\beta_j)i}. \tag{34}$$

Notice that $\zeta'_j \in z_j + \overline{S}^*_{\alpha_j,\beta_j} \subset U_j$ if $\zeta_j \in U_j$, and if $\tilde{x}_j(\zeta'_j) \geq \tilde{x}_j(\zeta_j)$ and $\tilde{y}_j(\zeta'_j) \geq \tilde{y}_j(\zeta_j)$. This shows that there exists $t_{j,0} \in [0, s_{j,0}]$ and $t_{j,1} \in [0, s_{j,1}]$ and continuous piecewise $\mathcal{C}^1$-functions f_j and g_j defined respectively on $[0, t_{j,0}]$ and $[0, t_{j,1}]$ such that

$$\begin{aligned} & U_j - z_j \\ = \;& \{\zeta_j \in S^*_{\alpha_j,\beta_j} \mid \tilde{x}(\zeta_j) \in (0, t_{j,0}], \tilde{y}(\zeta_j) > f_j(\tilde{x}(\zeta_j))\} \cup \{\zeta_j \in S^*_{\alpha_j,\beta_j} \mid \tilde{x}(\zeta_j) > t_{j,0}\} \\ = \;& \{\zeta_j \in S^*_{\alpha_j,\beta_j} \mid \tilde{y}(\zeta_j) \in (0, t_{j,1}], \tilde{x}(\zeta_j) > g_j(\tilde{y}(\zeta_j))\} \cup \{\zeta_j \in S^*_{\alpha_j,\beta_j} \mid \tilde{y}(\zeta_j) > t_{j,1}\}. \end{aligned}$$

We have $f_j(0) = t_{j,1}, f_j(t_{j,0}) = 0, g_j(0) = t_{j,1}, g_j(t_{j,1}) = 0$, f_j and g_j are strictly decreasing and $f_j = g_j^{-1}$ if $t_{j,s} > 0$ for some, hence for all $s \in \{1,2\}$.

For $\alpha = (\alpha_1, \ldots, \alpha_k) \in \mathbb{R}^k, \beta = (\beta_1, \ldots, \beta_k) \in \mathbb{R}^k$, we will use the obvious conventions

$$\inf(\alpha,\beta) = (\inf(\alpha_1,\beta_1), \ldots, \inf(\alpha_k,\beta_k)), \ \sup(\alpha,\beta) = (\sup(\alpha_1,\beta_1), \ldots, \sup(\alpha_k,\beta_k)).$$

Clearly, if $(\alpha^{(1)},\beta^{(1)}) \in M_{a,b}$ and $(\alpha^{(2)},\beta^{(2)}) \in M_{a,b}$, then

$$(\inf(\alpha^{(1)},\alpha^{(2)}), \sup(\beta^{(1)},\beta^{(2)})) \in M_{a,b}.$$

PROPOSITION 12.2. *If* $U^{(1)}$ *is admissible with respect to* $(\alpha^{(1)},\beta^{(1)}) \in M_{a,b}$ *and if* $U^{(2)}$ *is admissible with respect to* $(\alpha^{(2)},\beta^{(2)}) \in M_{a,b}$, *then* $U^{(1)} \cap U^{(2)}$ *is admissible with respect to* $(\inf(\alpha^{(1)},\alpha^{(2)}), \sup(\beta^{(1)},\beta^{(2)}))$.

PROOF. Set $\alpha^{(3)} = \inf(\alpha^{(1)},\alpha^{(2)}), \beta^{(3)} = \sup(\beta^{(1)},\beta^{(2)})$, and $U^{(3)} = U^{(1)} \cap U^{(2)}$. The fact that $U^{(3)}$ satisfies (i) follows from the fact that

$$\overline{S}^*_{\alpha_j^{(3)},\beta_j^{(3)}} = \overline{S}^*_{\alpha_j^{(1)},\beta_j^{(1)}} \cap \overline{S}^*_{\alpha_j^{(2)},\beta_j^{(2)}}.$$

The fact that $U^{(3)}$ satisfies (ii) follows easily from the fact that

$$\left[z^{(1)} + S^*_{\alpha^{(1)},\beta^{(1)}}\right] \cap \left[z^{(2)} + S^*_{\alpha^{(2)},\beta^{(2)}}\right]$$

is itself admissible with respect to $(\alpha^{(3)},\beta^{(3)})$ if $U^{(1)}$ satisfies Definition 12.1 with respect to $z^{(1)}$ and if $U^{(2)}$ satisfies Definition 12.1 with respect to $z^{(2)}$. □

LEMMA 12.3. *Let U be an admissible open set with respect to $(\alpha,\beta)\in M_{a,b}$, and let $F\in H^{(1)}(U)$.*

(i) We have, for $\epsilon=(\epsilon_1,\dots,\epsilon_k)\in S^*_{\alpha,\beta}$,

$$\int_{\Pi_{j\le k}\left(U_j\setminus(\overline{U}_j+\epsilon_j)\right)}|F(\zeta)|dm(\zeta)\le|\epsilon_1|\dots|\epsilon_k|\|F\|_{H^{(1)}(U)}.$$

where m denotes the Lebesgue measure on $\mathbb{C}^k\approx\mathbb{R}^{2k}$.

(ii) We have, for $\zeta\in U$,

$$|F(\zeta)|\le\frac{2^k}{\pi^k\cos\left(\frac{\beta_1-\alpha_1}{2}\right)\dots\cos\left(\frac{\beta_k-\alpha_k}{2}\right)}\frac{\Pi_{1\le j\le k}dist(\zeta_j,\partial S^*_{\alpha_j,\beta_j})}{[\Pi_{1\le j\le k}dist(\zeta_j,\partial U_j)]^2}\|F\|_{H^{(1)}(U)}.$$

PROOF. (i) Let $F\in H^{(1)}(U)$, let $\epsilon=(\epsilon_1,\dots,\epsilon_k)\in S^*_{\alpha,\beta}$, for $j\le k$ let $\gamma_j\in(-\frac{\pi}{2}-\alpha_j,\frac{\pi}{2}-\beta_j)$ be a determination of $arg(\epsilon_j)$, and set $r_j=|\epsilon_j|>0$.

Set $U_{j,1}=z_j+t_{j,0}e^{(-\frac{\pi}{2}-\alpha_j)i}+S_{-\frac{\pi}{2}-\alpha_j,\gamma_j}$, $U_{j,2}=z_j+t_{j,1}e^{(\frac{\pi}{2}-\beta_j)i}+S_{\gamma_j,\frac{\pi}{2}-\beta_j}$, and $U_{j,3}=z_j+\cup_{\rho>0}\left(\rho e^{\gamma_j i}+\left(\partial U_j\cap S^*_{\alpha_j,\beta_j}\right)\right)$, with the convention $U_{j,3}=\emptyset$ if $t_{j,0}=t_{j,1}=0$. Also for $\zeta_j\in\mathbb{C}$ set $x_i=Re(\zeta_j),y_j=Im(\zeta_j)$.

For $t_j<0$, $0<\rho_j<r_j$, set

$$\zeta_j=\zeta_j(\rho_j,t_j)=\rho_je^{i\gamma_j}+(t_{j,0}-t_j)e^{-i(\frac{\pi}{2}-\alpha_j)}.$$

This gives a parametrization of $U_{j,1}\setminus(U_{j,1}+\epsilon_j)$, and we have

$$dx_jdy_j=\begin{vmatrix}\cos(\gamma_j)&\sin(\alpha_j)\\\sin(\gamma_j)&\cos(\alpha_j)\end{vmatrix}d\rho_jdt_j=\cos(\alpha_j+\gamma_j)d\rho_jdt_j.$$

Similarly for $t_j>t_{j,1}$, $0<\rho_j<r_j$, set

$$\zeta_j=\zeta_j(\rho_j,t_j)=\rho_je^{i\gamma_j}+t_je^{i(\frac{\pi}{2}-\beta_j)}.$$

This gives a parametrization of $U_{j,2}\setminus(U_{j,2}+\epsilon_j)$, and we have

$$dx_jdy_j=\begin{vmatrix}\cos(\gamma_j)&\sin(\beta_j)\\\sin(\gamma_j)&\cos(\beta_j)\end{vmatrix}dt_jd\rho_j=\cos(\beta_j+\gamma_j)d\rho_jdt_j.$$

Now assume that $U_{j,3}\neq\emptyset$, so that $t_{j,0}>0$ and $t_{j,1}>0$. For $0<t_j<t_{j,1}$, $0<\rho_j<r_j$ set

$$\zeta_j=\zeta_j(\rho_j,t_j)=\rho_je^{i\gamma_j}+g_j(t_j)e^{(-\frac{\pi}{2}-\alpha_j)i}+t_je^{(\frac{\pi}{2}-\beta_j)i}.$$

This gives a parametrization of $U_{j,3}\setminus(U_{j,3}+\epsilon_j)$, and we have

$$\begin{aligned}dx_jdy_j&=\begin{vmatrix}\cos(\gamma_j)&-g_j'(t)\sin(\alpha_j)+\sin(\beta_j)\\\sin(\gamma_j)&-g_j'(t)\cos(\alpha_j)+\cos(\beta_j)\end{vmatrix}d\rho_jdt_j\\&=(\cos(\beta_j+\gamma_j)-g_j'(t)\cos(\alpha_j+\gamma_j))d\rho_jdt_j.\end{aligned}$$

We have $0<\cos(\alpha_j+\gamma_j)<1$, $0<\cos(\alpha_j+\gamma_j)<1$, $g_j'(t_j)<0$, and using the Cauchy-Schwartz inequality, we obtain

$$\begin{aligned}0 &< \cos(\beta_j+\gamma_j) - g_j'(t)\cos(\alpha_j+\gamma_j)\\&= \cos(\gamma_j)(\cos(\beta_j) - g_j'(t)\cos(\alpha_j)) - \sin(\gamma_j)(\sin(\beta_j) - g_j'(t)\sin(\alpha_j))\\&\le \sqrt{(\cos(\beta_j) - g_j'(t)\cos(\alpha_j))^2 + (\sin(\beta_j) - g_j'(t)\sin(\alpha_j))^2}\\&= \sqrt{1 - 2g_j'(t)\cos(\beta_j-\alpha_j) + g_j'(t)^2}.\end{aligned}$$

On the other hand we have

$$\begin{aligned}\left|\frac{\partial\zeta_j}{\partial t_j}(\rho_j,t_j)\right|^2 &= \left(g_j'(t)e^{(-\frac{\pi}{2}-\alpha_j)i} + e^{(\frac{\pi}{2}-\beta_j)i}\right)\left(g_j'(t)e^{(\frac{\pi}{2}+\alpha_j)i} + e^{(-\frac{\pi}{2}+\beta_j)i}\right)\\&= 1 - 2g_j'(t)\cos(\beta_j-\alpha_j) + g_j'(t)^2.\end{aligned}$$

The boundary $\partial U_j + \rho_j e^{i\gamma_j}$ being oriented from $e^{(-\frac{\pi}{2}-\alpha_j)i}.\infty$ to $e^{(\frac{\pi}{2}-\beta_j)}.\infty$, we obtain

$$\begin{aligned}&\int_{\Pi_{j\le k}(U_j\setminus(\overline{U}_j+\epsilon_j))} |F(\zeta)|dm(\zeta)\\&\le \int_{(0,r_1)\times\cdots\times(0,r_k)}\left[\int_{\Pi_{j\le k}(\partial U_j+\rho_j e^{i\gamma_j})} |F(\sigma_j)||d\sigma_1|\ldots|d\sigma_k|\right]d\rho_1\ldots d\rho_k\\&\le r_1\ldots r_k\|F\|_{H^{(1)}(U)},\end{aligned}$$

which proves (i).

(ii) Let $F\in H^{(1)}(U)$, let $\zeta\in U$, set $r_j = dist(\zeta_j,\partial U_j)$, set $r=(r_1,\ldots,r_k)$, and set $B(\zeta,r) = \Pi_{j\le k}B(\zeta_j,r_j)$. Using Cauchy's formula and polar coordinates, we obtain the standard formula

$$F(\zeta) = \frac{1}{|B(\zeta,r)|}\int_{B(\zeta,r)} F(\eta)dm(\eta). \tag{35}$$

where $|B(\zeta,r)| = \pi^k r_1^2\ldots r_k^2$ denotes the Lebesgue measure of $B(\zeta,r)$.

Denote by u_j the orthogonal projection of ζ_j on the real line $z_j + \mathbb{R}e^{i(-\frac{\pi}{2}-\alpha_j)}$, denote by v_j the orthogonal projection of ζ_j on the real line $z_j+\mathbb{R}e^{(\frac{\pi}{2}-\beta_j)i}$, and let $w_j\in\{u_j,v_j\}$ be such that $|\zeta_j-w_j| = \min(|\zeta_j-u_j|,|\zeta_j-v_j|)$. An easy topological argument shows that $w_j\in\partial S^*_{\alpha_j,\beta_j}$, so that

$$|\zeta_j - w_j| = dist(\zeta_j,\partial S^*_{\alpha_j,\beta_j}) \ge dist(\zeta_j,U_j) = r_j.$$

For $\lambda\in\mathbb{R}$, we have

$$\zeta_j\notin \overline{S}^*_{\alpha_j,\beta_j} + z_j + 2(\zeta_j-w_j) + \lambda i(\zeta_j-w_j) \supset \overline{U}_j + 2(\zeta_j-w_j) + \lambda i(\zeta_j-w_j).$$

If $\pi-\beta_j+\alpha_j > \frac{\pi}{2}$, then $\zeta_j - w_j\in S^*_{\alpha_j,\beta_j}$. If $\pi-\beta_j+\alpha_j\le\frac{\pi}{2}$, then we can choose $\lambda\in\mathbb{R}$ such that $\zeta_j-w_j+\lambda i(\zeta_j-w_j)\in\overline{S}^*_{\alpha_j,\beta_j}$ and such that $|\zeta_j-w_j+\lambda i(\zeta_j-w_j)| = \frac{|\zeta_j-w_j|}{\cos\left(\frac{\beta_j-\alpha_j}{2}\right)}$. So there exists in both cases $\epsilon_j\in S^*_{\alpha_j,\beta_j}$ such that $\zeta_j\notin\overline{U}_j+\epsilon_j$ and $|\epsilon_j| = \frac{2dist(\zeta_j,\partial S^*_{\alpha_j,\beta_j})}{\cos\left(\frac{\beta_j-\alpha_j}{2}\right)}$.

Using (40) and (i), we obtain

$$
\begin{aligned}
|F(\zeta)| &\leq \frac{1}{\pi^k r_1^2 \dots r_k^2} \int_{\Pi_{j\leq k}\left(U_j \setminus (\overline{U}_j+\epsilon_j)\right)} |F(\eta)| dm(\eta) \\
&\leq \frac{2^k}{\pi^k \cos\left(\frac{\beta_1-\alpha_1}{2}\right)\dots\cos\left(\frac{\beta_k-\alpha_k}{2}\right)} \frac{\Pi_{1\leq j\leq k} dist(\zeta_j, \partial S^*_{\alpha_j,\beta_j})}{[\Pi_{1\leq j\leq k} dist(\zeta_j, \partial U_j)]^2} \|F\|_{H^{(1)}(U)},
\end{aligned}
$$

which proves (ii). □

COROLLARY 12.4. *$(H^{(1)}(U), \|.\|_{H^{(1)}(U)})$ is a Banach space, $F_{\overline{U}+\epsilon}$ is bounded on $\overline{U}+\epsilon$, and $\lim_{\substack{dist(\zeta,\partial U)\to+\infty \\ \zeta\in\overline{U}+\epsilon}} F(\zeta) = 0$ for every $F \in H^{(1)}(U)$ and every $\epsilon \in S^*_{\alpha,\beta}$.*

PROOF. It follows from (ii) that for every compact set $K \subset U$ there exists $m_K > 0$ such that $\max_{\zeta\in K}|F(\zeta)| \leq m_K\|F\|_{H^{(1)}(U)}$. So every Cauchy sequence $(F_n)_{n\geq1}$ in $(H^{(1)}(U), \|.\|_{H^{(1)}(U)})$ is a normal family which converges uniformly on every compact subset of U to a holomorphic function $F : U \to \mathbb{C}$. Since $\int_{\tilde{\partial} U+\epsilon} |F(\sigma) - F_n(\sigma)||d\sigma| = \lim_{R\to+\infty} \int_{B(0,R)\cap(\tilde{\partial}U+\epsilon)} |F(\sigma) - F_n(\sigma)|d\sigma|$, an easy argument shows that $F \in H^{(1)}(U)$ and that $\lim_{n\to+\infty} \|F - F_n\|_{H^{(1)}(U)} = 0$.

Now let $\epsilon > 0$. For $1 \leq j \leq k$, there exists $m_j > 0$ such that

$$dist(\zeta_j, \partial U) \geq m_j dist(\zeta_j, \partial S^*_{\alpha_j,\beta_j})$$

for every $\zeta_j \in \overline{U}_j + \epsilon_j$, which gives, for $\zeta \in \overline{U} + \epsilon$,

$$|F(\zeta)| \leq \frac{2^k}{\pi^k m_1 \dots m_k \cos\left(\frac{\beta_1-\alpha_1}{2}\right)\dots\cos\left(\frac{\beta_k-\alpha_k}{2}\right) \Pi_{1\leq j\leq k} dist(\zeta_j, \partial U_j)} \|F\|_{H^{(1)}(U)}.$$

Since $\inf_{\zeta_j\in\overline{U}_j+\epsilon_j} dist(\zeta_j, \partial U_j) > 0$ for $j \leq k$, this shows that F is bounded on $\overline{U} + \epsilon$, and that $\lim_{\substack{dist(\zeta,\partial U)\to+\infty \\ \zeta\in\overline{U}+\epsilon}} F(\zeta) = 0$. □

THEOREM 12.5. *Let U be an admissible open set with respect to $(\alpha, \beta) \in M_{a,b}$, and let $F \in H^{(1)}(U)$. Then*

$$\int_{\tilde{\partial} U+\epsilon} F(\sigma)d\sigma = 0$$

for every $\epsilon \in S^*_{\alpha,\beta}$, and

$$F(\zeta) = \frac{1}{(2\pi i)^k} \int_{\tilde{\partial} U+\epsilon} \frac{F(\sigma)d\sigma}{\Pi_{1\leq j\leq k}(\zeta_j - \sigma_j)}$$

for every $\epsilon \in S^*_{\alpha,\beta}$ and for every $\zeta \in U+\epsilon$, where ∂U_j is oriented from $e^{i(-\frac{\pi}{2}-\alpha_j)}.\infty$ to $e^{i(\frac{\pi}{2}-\beta j)}.\infty$ for $j \leq k$.

PROOF. Let $z \in \mathbb{C}^k$ satisfying the conditions of Definition 12.1 with respect to U, let $\epsilon \in S^*_{\alpha,\beta}$, let $L > 1$ such that $(z_j + e^{i\alpha_j}.\infty, z_j + Le^{i\alpha_j}] \subset \partial U_j$ and $[z_j + Le^{i\beta_j}, z_j + e^{i\beta_j}.\infty) \subset \partial U_j$ for $j \leq k$, and let $M > 1$. Set

$$
\begin{aligned}
M_j &:= \left((z_j + \epsilon_j + Le^{i\alpha_j}, z_j + \epsilon_j + e^{i\alpha_j}.\infty) \cup (z_j + \epsilon_j + Le^{i\beta_j}, z_j + \epsilon_j + e^{i\beta_j}.\infty)\right), \\
N_j &:= \left((z_j + L\epsilon_j + Le^{i\alpha_j}.\infty, z_j + L\epsilon_j + Le^{i\alpha_j}) \cup (z_j + L\epsilon_j + Le^{i\beta}, z_j + \epsilon_j + e^{i\beta_j}.\infty)\right),
\end{aligned}
$$

$$\begin{aligned}
\Gamma_{L,j,1} &= (\epsilon_j + \partial U_j) \setminus M_j, \\
\Gamma_{L,j,2} &= [z_j + \epsilon_j + Le^{i\beta_j}, z_j + L\epsilon_j + Le^{i\beta j}], \\
\Gamma_{L,j,3} &= (L\epsilon_j + \partial U_j) \setminus N_j, \\
\Gamma_{L,j,4} &= [z_j + L\epsilon_j + Le^{i\alpha_j}, z_j + \epsilon_j + Le^{i\alpha_j}], \\
\Gamma_{L,j} &= \cup_{1\le s\le 4}\Gamma_{L,j,s},
\end{aligned}$$

where the Jordan curve $\Gamma_{L,j}$ is oriented clockwise.

For $n \ge 1, \zeta_j \in \overline{S}^*_{\alpha_j,\beta_j}$, set $W_{j,n}(\zeta_j) = \frac{n^2}{\left(n+e^{\frac{\alpha_j+\beta_j}{2}i}\zeta_j\right)^2}$, and set

$$W_n(\zeta) = \Pi_{j\le k} W_{n,j}(\zeta_j),\ (\zeta \in \overline{S}^*_{\alpha,\beta}).$$

Then $|W_{n,j}(\zeta_j)| \le 1$ for $\zeta_j \in \overline{S}^*_{\alpha_j,\beta_j}$, $W_n(\zeta) \to 1$ as $n \to \infty$ uniformly on compact sets of $\overline{S}^*_{\alpha,\beta}$, and $\lim_{\substack{|\zeta|\to\infty \\ \zeta\in S^*_{\alpha,\beta}}} W_n(\zeta) = 0$.

Denote by $V_{L,j}$ the interior of $\Gamma_{L,j}$ and set $V_L = \Pi_{1\le j\le k} V_{L,j}$. If $\zeta \in V_L$, it follows from Cauchy's theorem that we have

$$\sum_{l\in\{1,2,3,4\}^k} \int_{\Pi_{j\le k}\Gamma_{L,j,l(j)}} W_n(\sigma - z - \epsilon)F(\sigma)d\sigma = \int_{\bar\partial V_L} W_n(\sigma - z - \epsilon)F(\sigma)d\sigma = 0.$$

Set $l_0(j) = 1$ for $j \le k$. It follows from the corollary that there exists $M > 0$ such that $|F(\zeta)| \le M$ for $\zeta \in \overline{U} + \epsilon$, and there exists $R_n > 0$ such that, for every L satisfying the conditions given above, we have $\int_{\Gamma_{L,j}} |W_n(\sigma_j - z_j - \epsilon_j)||d\sigma| \le R_n$. Also $\limsup_{L\to+\infty} \int_{\Gamma_{L,j,s}} |W_{n,j}(\sigma_j - z_j - \epsilon_j)||d(\sigma_j)| = 0$ for $s \ge 2, j \le k$.

Let $l \ne l_0$, and let $j_l \le k$ such that $j_l \ge 2$. We have

$$\begin{aligned}
&\limsup_{L\to+\infty} \left| \int_{\Pi_{j\le k}\Gamma_{L,j,l(j)}} |W_n(\sigma - z - \epsilon)F(\sigma)d\sigma \right| \\
\le\ & MR_n^{k-1} \int_{\Gamma_{L,j_l,l(j_l)}} |W_{n,j_l}(\zeta_{j_l} - z_{j_l} - \epsilon_{j_l})||d\sigma_{j_l}| = 0.
\end{aligned}$$

This gives

$$\begin{aligned}
\int_{\partial U+\epsilon} W_n(\sigma - z - \epsilon)F(\sigma)d\sigma &= \lim_{L\to+\infty} \int_{\Pi_{j\le k}\Gamma_{L,j,l_0(j)}} W_n(\sigma - z - \epsilon)F(\sigma)d\sigma \\
&= \lim_{L\to+\infty} \sum_{l\in\{1,2,3,4\}^k} \int_{\Pi_{j\le k}\Gamma_{L,j,l(j)}} W_n(\sigma - z - \epsilon)F(\sigma)d\sigma \\
&= 0.
\end{aligned}$$

It follows then from the Lebesgue dominated convergence theorem that $\int_{\partial U+\epsilon} F(\sigma)d\sigma = 0$.

Similarly, applying Cauchy's formula when $\zeta \in U + \epsilon$ is contained in V_L, we obtain

$$
\begin{aligned}
&\frac{1}{(2\pi i)^k}\int_{\partial U+\epsilon}\frac{W_n(\sigma-z-\epsilon)F(\sigma)}{(\zeta_1-\sigma_1)\dots(\zeta_k-\sigma_k)}d\sigma\\
= &\lim_{L\to+\infty}\frac{1}{(2\pi i)^k}\int_{\Gamma_{L,j,l_0(j)}}\frac{W_n(\sigma-z-\epsilon)F(\sigma)}{(\zeta_1-\sigma_1)\dots(\zeta_k-\sigma_k)}d\sigma\\
= &\lim_{L\to+\infty}\frac{1}{(2\pi i)^k}\sum_{l\in\{1,2,3,4\}^k}\int_{\Pi_{j\le k}\Gamma_{L,j,l(j)}}\frac{W_n(\sigma-z-\epsilon)F(\sigma)}{(\zeta_1-\sigma_1)\dots(\zeta_k-\sigma_k)}d\sigma\\
= &\lim_{L\to+\infty}\frac{1}{(2\pi i)^k}\int_{\tilde\partial V_L}\frac{W_n(\sigma-z-\epsilon)F(\sigma)}{(\zeta_1-\sigma_1)\dots(\zeta_k-\sigma_k)}d\sigma\\
= &W_n(\zeta-z-\epsilon)F(\zeta).
\end{aligned}
$$

It follows then again from the Lebesgue dominated convergence theorem that we have

$$
\begin{aligned}
F(\zeta) &= \lim_{n\to+\infty}W_n(\zeta-z-\epsilon)F(\zeta)\\
&= \lim_{n\to+\infty}\frac{1}{(2\pi i)^k}\int_{\partial U+\epsilon}\frac{W_n(\sigma-z-\epsilon)F(\sigma)}{(\zeta_1-\sigma_1)\dots(\zeta_k-\sigma_k)}d\sigma\\
&= \frac{1}{(2\pi i)^k}\int_{\partial U+\epsilon}\frac{F(\sigma)}{(\zeta_1-\sigma_1)\dots(\zeta_k-\sigma_k)}d\sigma.
\end{aligned}
$$

□

Let $\zeta\in U$, and let $\epsilon\in S^*_{\alpha,\beta}$. It follows from the theorem that there exists $\rho>0$ such that

$$|F(\zeta)|\le\frac{1}{(2\pi)^k}\frac{\|F\|_{H^{(1)}(U)}}{\Pi_{j\le k}dist(\zeta_j,\partial U_j+t\epsilon_j)}$$

for $t\in(0,\rho]$. Since

$$\lim_{t\to0^+}dist(\zeta_j,\partial U_j+t\epsilon_j)=\lim_{t\to0^+}dist(\zeta_j-t\epsilon_j,\partial U_j)=dist(\zeta_j,\partial U_j),$$

we obtain, for $F\in H^{(1)}(U)$, $\zeta\in U$,

$$|F(\zeta)|\le\frac{1}{(2\pi)^k}\frac{\|F\|_{H^{(1)}(U)}}{\Pi_{j\le k}dist(\zeta_j,\partial U_j)},\tag{36}$$

which improves inequality (ii) of Lemma 12.3.

If $\alpha_j=\beta_j$ for $j\le k$, then every (α,β) admissible open set U is a product of open half-planes and the space $H^{(1)}(U)$ is the usual Hardy space $H^1(U)$. The standard conformal mappings of the open unit disc D onto half planes induce an isometry from the Hardy space $H^1(D^k)$ onto $H^1(U)$. It follows then from standard results about $H^1(D^k)$, see Theorems 3.3.3 and 3.3.4 of **[31]**, that F admits a.e. a nontangential limit F^* on $\tilde\partial U$, and that

$$\lim_{\epsilon\to0}|\int_{\tilde\partial U}|F^*(\sigma)-F(\sigma+\epsilon)||d\sigma|=0.$$

This gives the formula

$$F(\zeta)=\frac{1}{(2\pi i)^k}\int_{\tilde\partial U}\frac{F^*(\sigma)d\sigma}{(\zeta_1-\sigma_1)\dots(\zeta_k-\sigma_k)}\text{ for every }\zeta\in U.\tag{37}$$

We did not investigate whether such nontangential limits of F on $\tilde\partial U$ exist in the general case.

Recall that the Smirnov class $\mathcal{N}^+(P^+)$ on the right-hand open half-plane P^+ consists in those functions F holomorphic on P^+ which can be written under the form $F = G/H$ where $G \in H^\infty(P^+)$ and where $H \in H^\infty(P^+)$ is outer, which means that we have, for $Re(\zeta) > 0$,

$$F(\zeta) = exp\left(\frac{1}{\pi}\int_{-\infty}^{+\infty}\frac{1-iy\zeta}{(\zeta-iy)(1+y^2)}\log|F^*(iy)|dy\right),$$

where $F^*(iy) = \lim_{x\to 0^+} F(x+iy)$ if defined a.e. on the vertical axis and satisfies $\int_{-\infty}^{+\infty}\frac{|\log|F^*(iy)||}{1+y^2}dy < +\infty$.

Set , for $Re(\zeta) > 0$,

$$F_n(\zeta) = exp\left(\frac{1}{\pi}\int_{-\infty}^{+\infty}\frac{1-iy\zeta}{(\zeta-iy)(1+y^2)}\sup(\log|F^*(iy)|, -n)dy\right).$$

It follows from the positivity of the Poisson kernel on the real line that we have $|F(\zeta)| \le |F_n(\zeta)|$ and $\lim_{n\to+\infty} F_n(\zeta) = F(\zeta)$ for $Re(\zeta) > 0$. Also the nontangential limit $F_n^*(iy)$ of F at iy exists a.e. on the imaginary axis and $|F_n^*(iy)| = \sup(e^{-n}, |F^*(iy)|)$ a.e., which shows that $\sup_{\zeta\in P^+}|F_n(\zeta)| = \sup_{\zeta\in P^+}|F(\zeta)|$ when n is sufficiently large. Hence $\lim_{n\to+\infty} F(\zeta)F_n^{-1}(\zeta) = 1$ for $\zeta \in P^+$.

This suggests the following notion.

Definition 12.6. *Let $U \subset \mathbb{C}^k$ be a connected open set.*

A holomorphic function $F \in H^\infty(U)$ is said to be strongly outer on U if there exists a sequence $(F_n)_{n\ge 1}$ of invertible elements of $H^\infty(U)$ satisfying the following conditions

(i) $|F(\zeta)| \le |F_n(\zeta)|$ $(\zeta \in U, n \ge 1)$,

(ii) $\lim_{n\to+\infty} F(\zeta)F_n^{-1}(\zeta) = 1$ $(\zeta \in U)$.

The Smirnov class $S(U)$ consists of those holomorphic functions F on U such that $FG \in H^\infty(U)$ for some strongly outer function $G \in H^\infty(U)$.

It follows from (ii) that $F(\zeta) \neq 0$ for every $\zeta \in U$ if F is strongly outer on U, and $F_{|V}$ is strongly outer on V if $V \subset U$. Similarly if $F \in \mathcal{S}(U)$ then $F_{|V} \in \mathcal{S}(V)$. Also it follows immediately from the definition that the set of bounded strongly outer functions on U is stable under products, and that if there is a conformal mapping θ from an open set $V \subset \mathbb{C}^k$ onto U then $F \in H^\infty(U)$ is strongly outer on U if and only if $F\circ\theta$ is strongly outer on V, and if G is holomorphic on U then $G \in \mathcal{S}(U)$ if and only $F\circ\theta \in \mathcal{S}(V)$.

Now let $(\alpha,\beta) \in M_{a,b}$ and let $U = \Pi_{j\le k}U_j$ be an admissible open set with respect to (α,β). Then each set U_j is conformally equivalent to the open unit disc $\mathbb{D}$, and so there exists a conformal mapping θ from $\mathbb{D}^k$ onto U, and the study of the class of bounded strongly outer functions on U (resp. the Smirnov class on U) reduces to the study of the class bounded strongly outer functions (resp. the Smirnov class) on $\mathbb{D}^k$.

Let $F \in H^\infty(\mathbb{D}^k)$ be strongly outer, and let $(F_n)_{n\ge1}$ be a sequence of invertible elements of $H^\infty(\mathbb{D}^k)$ satisfying the conditions of Definition 12.6 with respect to F. Denote by $\mathbb{T} = \partial\mathbb{D}$ the unit circle. Then $H^\infty(\mathbb{D}^k)$ can be identified to a w^*-closed subspace of $L^\infty(\mathbb{T}^k)$ with respect to the w^*-topology $\sigma(L^1(\mathbb{T}^k), L^\infty(\mathbb{T}^k))$.

Let $L \in H^\infty(\mathbb{D}^k)$ be a w^*-cluster point of the sequence $(FF_n^{-1})_{n\ge1}$. Since the map $G \to G(\zeta)$ is w^*-continuous on $H^\infty(\mathbb{D}^k)$ for $\zeta \in \mathbb{D}^k$, $L = 1$, and so $FH^\infty(\mathbb{D}^k)$ is w^*-dense in H^∞. When $k = 1$, this implies as well-known that F is outer, and the argument used for the half-plane shows that, conversely, every bounded outer function on $\mathbb{D}$ is strongly outer, and so $\mathcal{S}(\mathbb{D}) = \mathcal{N}^+(\mathbb{D})$.

Recall that a function $G \in H^\infty(\mathbb{D}^k)$ is said to be outer if

$$\log(|G(0,\dots,0)|) = \frac{1}{(2\pi)^k}\int_{\mathbb{T}^k} \log|G^*(e^{it_1},\dots,e^{it_k})|dt_1\dots dt_k,$$

where $G^*(e^{it_1},\dots,e^{it_k})$ denotes a.e. the nontangential limit of G at $(e^{it_1},\dots,e^{it_k})$, see **[31]**, Definition 4.4.3, and G is outer if and only if almost every slice function G_ω is outer on $\mathbb{D}$, where $G_\omega(\zeta) = G(\omega\zeta)$ for $\omega \in \mathbb{T}^k, \zeta \in \mathbb{D}$, see **[31]**, Lemma 4.4.4. If follows from Definition 12.6 that every slice function F_ω is strongly outer on $\mathbb{D}$ if $F \in H^\infty(\mathbb{D}^k)$ is strongly outer on $\mathbb{D}^k$, and so every strongly outer bounded function on $\mathbb{D}^k$ is outer. It follows from an example from **[31]** that the converse is false if $k \geq 2$.

PROPOSITION 12.7. *Let $k \geq 2$, and set $F(\zeta_1,\dots,\zeta_k) = e^{\frac{\zeta_1+\zeta_2+2}{\zeta_1+\zeta_2-2}}$ for $(\zeta_1,\dots,\zeta_k) \in \mathbb{D}^k$. Then F is outer on $\mathbb{D}^k$, but F is not strongly outer on $\mathbb{D}^k$.*

PROOF. Set $f(\zeta) = e^{\frac{\zeta+1}{\zeta-1}}$ for $\zeta \in \mathbb{D}$. Then $f \in H^\infty(\mathbb{D})$ is a singular inner function. Since $f(\zeta) \neq 0$ for $\zeta \in \mathbb{D}$, it follows from **[31]**, Lemma 4.4.4b that the function

$$\tilde{f} : (\zeta_1,\zeta_2) \to f(\frac{\zeta_1+\zeta_2}{2}, \frac{\zeta_1+\zeta_2}{2}) = e^{\frac{\zeta_1+\zeta_2+2}{\zeta_1+\zeta_2-2}}$$

is outer on $\mathbb{D}^2$. Hence we have

$$\begin{aligned}\log|F(0,\dots,0)| &= \log|\tilde{f}(0,0)| \\ &= \frac{1}{(2\pi)^2}\int_{\mathbb{T}^2} \tilde{f}(e^{it_1},e^{it_2})dt_1dt_2 \\ &= \frac{1}{(2\pi)^k}\int_{\mathbb{T}^k} F(e^{it_1},\dots,e^{it_k})dt_1\dots dt_k,\end{aligned}$$

and so F is outer on $\mathbb{D}^k$.

Now set $\omega = (1,\dots,1)$. Then $F_\omega = f$ is not outer on $\mathbb{D}$, and so F is not strongly outer on $\mathbb{D}^k$. □

The fact that some bounded outer functions on $\mathbb{D}$ are not strongly outer is not surprising: The Poisson integral of a real valued integrable function on $\mathbb{T}^k$ is the real part of some holomorphic function on $\mathbb{D}^k$ if an only if its Fourier coefficients vanish on $\mathbb{Z}^k \setminus (\mathbb{Z}^+)^k \cup (\mathbb{Z}^-)^k$, see **[31]**, Theorem 2.4.1, and so the construction of the sequence $(F_n)_{n\geq 1}$ satisfying the conditions of Definition 12.6 with respect to a bounded outer function F on $\mathbb{D}^k$ breaks down when $k \geq 2$. We conclude this appendix with the following trivial observations.

PROPOSITION 12.8. *Let $U = \Pi_{j\leq k}U_j \subset \mathbb{C}^k$ be an admissible open set with respect to some $(\alpha,\beta) \in M_{a,b}$.*

(i) Let $\theta_j : U_j \to \mathbb{D}$ be a conformal map and let $\pi_j : (\zeta_1,\dots,\zeta_k) \to \zeta_j$ be the j-th coordinate projection. If $f \in H^\infty(\mathbb{D})$ is outer, then $f \circ \theta_j \circ \pi_j$ is strongly outer on U.

(ii) The Smirnov class $\mathcal{S}(U)$ contains all holomorphic functions on U having polynomial growth at infinity.

PROOF. (i) Since f is strongly outer on $\mathbb{D}$, there exists a sequence $(f_n)_{n\geq 1}$ of invertible elements of $H^\infty(\mathbb{D})$ satisfying the conditions of Definition 12.6 with respect to f. Then the sequence $(f_n \circ \theta_j \circ \pi_j)_{n\geq 1}$ satisfies the conditions of Definition 12.6 with respect to $f \circ \theta_j \circ \pi_j$, and so $f \circ \theta_j \circ \pi_j$ is strongly outer on U.

(ii) For $j \leq k$ there exists $\gamma_j \in [-\pi, \pi)$ and $m_j \in \mathbb{R}$ such that open set U_j is contained in the open half plane $P_j := \{\zeta_j \in \mathbb{C} \mid Re(\zeta_j e^{i\gamma_j}) \geq m_j\}$. The function $\sigma \to \frac{1-\sigma}{2}$ is outer on $\mathbb{D}$, since $|\frac{1-\sigma}{2}| \leq \left|\frac{1+1/n-\sigma}{2}\right|$ for $\sigma \in \mathbb{D}$, and the function $\zeta_j \to \frac{\zeta_j e^{i\gamma_j}-m_j-1}{\zeta_j e^{i\gamma_j}-m_j+1}$ maps conformally U_j onto $\mathbb{D}$. Set $F_j(\zeta_1, \ldots, \zeta_k) = \frac{1-\frac{\zeta_j e^{i\gamma_j}-m_j-1}{\zeta_j e^{i\gamma_j}-m_j+1}}{2} = \frac{1}{\zeta_j e^{i\gamma_j}-m_j+1}$. It follows from (i) that F_j is strongly outer on $\Pi_{j\leq k} P_j$, hence strongly outer on U.

Now assume that a function F holomorphic on U has polynomial growth at infinity. Then there exists $p \geq 1$ such that $F\Pi_{j\leq k} F_j^p$ is bounded on U, and so $F \in \mathcal{S}(U)$. □

References

[1] G. R. Allan, H. G. Dales, and J. P. McClure, *Pseudo-Banach algebras*, Studia Math. **40** (1971), 55–69, DOI 10.4064/sm-40-1-55-69. MR0333737

[2] W. Arveson, *On groups of automorphisms of operator algebras*, J. Functional Analysis **15** (1974), 217–243, DOI 10.1016/0022-1236(74)90034-2. MR0348518

[3] V. Avanissian and R. Gay, *Sur une transformation des fonctionnelles analytiques et ses applications aux fonctions entières de plusieurs variables*, Bull. Soc. Math. France **103** (1975), no. 3, 341–384. MR0396988

[4] F. S. Baher, *On a basis in the space of continuous functions defined on a compactum* (Russian), Dokl. Akad. Nauk SSSR (N.S.) **101** (1955), 589–592. MR0069394

[5] Z. Bendouad, I. Chalendar, J. Esterle, and J. R. Partington, *Distances between elements of a semigroup and estimates for derivatives*, Acta Math. Sin. (Engl. Ser.) **26** (2010), no. 12, 2239–2254, DOI 10.1007/s10114-010-8569-6. MR2737295

[6] N. Bourbaki, *Éléments de mathématique. Fasc. XIII. Livre VI: Intégration. Chapitres 1, 2, 3 et 4: Inégalités de convexité, Espaces de Riesz, Mesures sur les espaces localement compacts, Prolongement d'une mesure, Espaces L^p* (French), Deuxième édition revue et augmentée. Actualités Scientifiques et Industrielles, No. 1175, Hermann, Paris, 1965. MR0219684

[7] N. Bourbaki, *Éléments de mathématique. Première partie. (Fascicule II.) Livre III: Topologie générale. Chapitre 1: Structures topologiques. Chapitre. 2: Structures uniformes* (French), Troisiéme édition entiérement refondue, Actualités Sci. Indust., No. 1142. Hermann, Paris, 1961. MR0141067

[8] I. Chalendar, J. Esterle, and J. R. Partington, *Boundary values of analytic semigroups and associated norm estimates*, Banach algebras 2009, Banach Center Publ., vol. 91, Polish Acad. Sci. Inst. Math., Warsaw, 2010, pp. 87–103, DOI 10.4064/bc91-0-5. MR2777505

[9] I. Chalendar, J. Esterle, and J. R. Partington, *Lower estimates near the origin for functional calculus on operator semigroups*, J. Funct. Anal. **269** (2015), no. 6, 1620–1635, DOI 10.1016/j.jfa.2015.07.005. MR3373429

[10] I. Chalendar, J. Esterle, and J. R. Partington, *Estimates near the origin for functional calculus on analytic semigroups*, J. Funct. Anal. **275** (2018), no. 3, 698–711, DOI 10.1016/j.jfa.2018.03.012. MR3806822

[11] P. J. Cohen, *Factorization in group algebras*, Duke Math. J. **26** (1959), 199–205. MR104982

[12] C. D'Antoni, R. Longo, and L. Zsidó, *A spectral mapping theorem for locally compact groups of operators*, Pacific J. Math. **103** (1982), no. 1, 17–24. MR687960

[13] H. G. Dales, *Banach algebras and automatic continuity*, London Mathematical Society Monographs. New Series, vol. 24, The Clarendon Press, Oxford University Press, New York, 2000. Oxford Science Publications. MR1816726

[14] J. Esterle, *Elements for a classification of commutative radical Banach algebras*, Radical Banach algebras and automatic continuity (Long Beach, Calif., 1981), Lecture Notes in Math., vol. 975, Springer, Berlin-New York, 1983, pp. 4–65. MR697578

[15] J. Esterle, *Quasimultipliers, representations of H^∞, and the closed ideal problem for commutative Banach algebras*, Radical Banach algebras and automatic continuity (Long Beach, Calif., 1981), Lecture Notes in Math., vol. 975, Springer, Berlin, 1983, pp. 66–162, DOI 10.1007/BFb0064548. MR697579

[16] J. Esterle, *Distance near the origin between elements of a strongly continuous semigroup*, Ark. Mat. **43** (2005), no. 2, 365–382, DOI 10.1007/BF02384785. MR2173957

[17] J. Esterle, *On the generation of Arveson weakly continuous semigroups*, Banach algebras and applications, Proceedings of the International Conference held at Oulu, July 3-11-2017, De Gruyter Proceedings in Mathematics 2020, ISBN 978-3-11-0602418.

[18] J. Esterle and E. Fašangová, *A Banach algebra approach to the weak spectral mapping theorem for locally compact abelian groups*, Operator semigroups meet complex analysis, harmonic analysis and mathematical physics, Oper. Theory Adv. Appl., vol. 250, Birkhäuser/Springer, Cham, 2015, pp. 155–170, DOI 10.1007/978-3-319-18494-4_11. MR3468215

[19] J. Esterle and A. Mokhtari, *Distance entre éléments d'un semi-groupe dans une algèbre de Banach* (French, with English and French summaries), J. Funct. Anal. **195** (2002), no. 1, 167–189, DOI 10.1006/jfan.2002.3963. MR1934356

[20] W. Feller, *Semi-groups of transformations in general weak topologies*, Ann. of Math. (2) **57** (1953), 287–308, DOI 10.2307/1969859. MR0054165

[21] W. Feller, *On the generation of unbounded semi-groups of bounded linear operators*, Ann. of Math. (2) **58** (1953), 166–174, DOI 10.2307/1969826. MR0055572

[22] J. E. Galé, *A notion of analytic generator for groups of unbounded operators*, Topological algebras, their applications, and related topics, Banach Center Publ., vol. 67, Polish Acad. Sci. Inst. Math., Warsaw, 2005, pp. 185–197, DOI 10.4064/bc67-0-15. MR2143925

[23] J. E. Galé and P. J. Miana, *One-parameter groups of regular quasimultipliers*, J. Funct. Anal. **237** (2006), no. 1, 1–53, DOI 10.1016/j.jfa.2006.03.021. MR2239257

[24] M. Haase, *The functional calculus for sectorial operators*, Operator Theory: Advances and Applications, vol. 169, Birkhäuser Verlag, Basel, 2006. MR2244037

[25] A. Martineau, *Sur les fonctionnnelles analytiques et la transformation de Fourier-Borel* (French), J. Analyse Math. **11** (1963), 1–164, DOI 10.1007/BF02789982. MR0159220

[26] A. Mokhtari, *Distance entre éléments d'un semi-groupe continu dans une algèbre de Banach* (French), J. Operator Theory **20** (1988), no. 2, 375–380. MR1004129

[27] A. Mukherjea and N. A. Tserpes, *Measures on topological semigroups: convolution products and random walks*, Lecture Notes in Mathematics, Vol. 547, Springer-Verlag, Berlin-New York, 1976. MR0467871

[28] B. Nyman, *On the one-dimensional translation group and semi-group in certain function spaces*, Thesis, University of Uppsala, 1950. MR0036444

[29] M. M. Rajoelina, *Elements de Cohen de $L^1(\mathbf{R}_+)$* (French), Math. Ann. **276** (1987), no. 2, 303–310, DOI 10.1007/BF01450744. MR870968

[30] F. Ricci, Hardy spaces in one variable, `homepage.sns.it/fricci/papers/hardy.pdf`

[31] W. Rudin, *Function theory in polydiscs*, W. A. Benjamin, Inc., New York-Amsterdam, 1969. MR0255841

[32] M. Rosenblum and J. Rovnyak, *Hardy classes and operator theory*, Dover Publications, Inc., Mineola, NY, 1997. Corrected reprint of the 1985 original. MR1435287

[33] Z. Semadeni, *Schauder bases in Banach spaces of continuous functions*, Lecture Notes in Mathematics, vol. 918, Springer-Verlag, Berlin-New York, 1982. MR653986

IMB, UMR 5251, Université de Bordeaux, 351, cours de la Libération, 33405 - Talence

Email address: `esterle@math.u-bordeaux.fr`

Contemporary Mathematics
Volume **743**, 2020
https://doi.org/10.1090/conm/743/14957

An integral Hankel operator on $H^1(\mathbb{D})$

Miron B. Bekker and Joseph A. Cima

Abstract. An integral Hankel operator generated by a Carleson measure with support on $[0,1)$ is investigated. We show that this operator boundedly maps the Hardy space $H^1(\mathbb{D})$ into the space of Cauchy Transforms. We also study some properties of the corresponding measure on the unit circle $\mathbb{T}$.

1. Introduction

Perhaps the best place to begin this article is a significant result of Ch. Pommerenke [**17**]. He asked (and successfully answered) the following question: for which analytic functions g in the unit disk $\mathbb{D}$ is the linear operator

$$f \to \int_0^z f(w)g'(w)dw$$

continuous on the Hardy space $H^2(\mathbb{D})$? His answer was that this operator is continuous on $H^2(\mathbb{D})$ if and only if $g \in BMOA$. This lead to others asking this question in the more general setting of the Hardy space $H^p(\mathbb{D})$, $0 < p < \infty$. The answers in this setting were settled in papers by A. Aleman and A. Siskakis [**3**] and A. Aleman and J. Cima [**2**]. This lead to a great flurry of activity for introducing several different types of linear, integral operators and investigating their behaviour on many other spaces, such as Bergman spaces, Dirichlet spaces, Analytic Morrey spaces, Fock spaces, etc .... The references given at the end of this paper give an abbreviated selection of some of the papers written over last 15 or so years and the interested reader will find a fairly complete background list of pertinent publications in the references to the listed papers.

It is the goal of this print to consider one such operator on the Hardy space $H^1(\mathbb{D})$. This will neccesiate introducing some material from the setting of the Cauchy transform of finite Borel measures on the unit circle $\mathbb{T}$, and some tools from that discipline. We will make the definitions and supply necessary background material in the next section.

2. Definitions and pertinent background material

There are many good references available for theory of Hardy spaces and we suggest a book by J. Cima, A. Matheson and W. Ross [**6**] as one such. In that text there is a detailed discussion of the Banach space known as the space of Cauchy

2010 *Mathematics Subject Classification.* Primary 30E20, 30H10, 30K99.

transforms (on the unit circle), denoted as CT. Namely, a function f analytic in the unit disk D belongs to CT if it can be represented in the form

$$f(z) = \int_{\mathbb{T}} \frac{d\nu(\zeta)}{1 - z\bar{\zeta}} \equiv (K\nu)(z), \tag{2.1}$$

where ν is a finite Borel measure on the unit circle $\mathbb{T}$, and $z \in \mathbb{D}$.

It is easy to see that the measure ν involved is not unique and adding the complex conjugate of any analytic polynomial against Lebesgue measure $dm(\zeta)$ to ν will yield the same analytic function f. The space CT can be realized as the dual of the Banach space A of all functions analytic in $\mathbb{D}$ and continuous on $\overline{\mathbb{D}}$. This requires that the norm be a coset norm, namely

$$A^* = \left(L^1(\mathbb{T})/\overline{H_0^1}\right) \oplus M_s, \tag{2.2}$$

where M_s denotes the singular measures. The norm in question for our f is given as

$$\|f\|_{CT} = \inf\{\|\mu\| : f(z) = (K\mu)(z)\}, \tag{2.3}$$

where $\|\mu\|$ is the total variation norm of the measure μ. In this situation there is a measure, say μ, such that $\|f\|_{CT} = \|\mu\|$. Further, if the measure giving f, say ν, is singular then it is unique among the singular measures, and $\|f\|_{CT} = \|\nu\|$.

For a given measurable function g on the unit circle the distribution function for g is defined as follows. Put

$$A(y, g) = A(y) = \{\zeta \in \mathbb{T} : |g(\zeta)| > y\}.$$

The distribution function $d(y)$, $y > 0$ is given as $d(y) = m(A(y))$, where m is the normalized Lebesgue measure on the unit circle T. Recall that the function d is a non-increasing and right-continuous function.

There is an important result due to Poltoratski [**16**] relating the size of the total variation of the singular part of the measure in question and a distribution function.

THEOREM 2.1. *Assume $y > 0$, $f(z) = (K\mu)(z)$, and that the lebesgue decomposition of μ is given as $d\mu(\zeta) = F(\zeta)dm(\zeta) + d\mu_s(\zeta)$, where μ_s is the singular part of the measure μ. Then the measures $\pi y \chi_{\{|K(\mu)>y|\}} dm$ converge weak* as y tends to infinity to $|\mu_s|$.*

Finally, in this section let us mention the motivation for the operator we wish to consider. The Hilbert matrix $H = (a_{ij})$, where $a_{ij} = 1/(i+j+1)$ is well studied and it maps the Hilbert space l^2 boundedly into itself. There is a natural way to lift this matricial operation to an operator on the Hardy space $H^2(\mathbb{D})$ as follows. Namely, if $f(z) = \sum_{n\geq 0} A_n z^n$ is an analytic function on the unit disk, then define a transformation of f by setting

$$(If)(z) = \sum_{n=0}^{\infty} \left(\sum_{k=0}^{\infty} \frac{A_k}{n+k+1} \right) z^n.$$

It is straightforward to check that for $f \in H^p(\mathbb{D})$ the following holds

$$(If)(z) = \int_0^1 \frac{f(t)}{1 - zt} dt. \tag{2.4}$$

We refer to [**14**] where some questions related to the transformation (2.4) are investigated. Since $\int_0^1 t^{i+j}dt = 1/(i+j+1)$ this easily generalized to Hankel matrices and the analogous integral operators where one uses the integral

$$(H_\mu f)(z) = \int_0^1 \frac{f(t)}{1-zt} d\mu(t),$$

where μ is a finite Borel measure on the interval $[0,1)$, and $f \in L^1(\mu)$. In the next section we will discuss one such extension for the Hardy space $H^1(\mathbb{D})$.

3. The embedding result

Assume μ is a positive Borel measure on the interval $[0,1)$ and $f \in L^1([0,1),\mu)$. The operator we wish to study is the integral Hankel operator defined for $f(z)$ analytic on $\mathbb{D}$ and is given by

$$(H_\mu f)(z) \equiv \int_{[0,1)} \frac{f(t)}{1-tz} d\mu(t), \quad z \in \mathbb{D}. \tag{3.1}$$

Even in the most basic case where μ is the Lebesgue measure on $[0,1)$ and $f \in H^1(\mathbb{D})$ (and consequently by the Fejer-Riesz theorem in $L^1((-1,1)dt)$) there is a class of examples given by Diamantopoulos and Siskakis ([**7**]), that shows that H_μ need not map $H^1(\mathbb{D})$ into itself. Namely, the functions

$$f_\epsilon(z) = \frac{1}{1-z}\left[\frac{1}{z}\log\frac{1}{1-z}\right]^{-1-\epsilon}$$

are in $H^1(\mathbb{D})$ for $\epsilon > 0$ (see [**8**]), but $H_\mu f_\epsilon$ are not in $H^1(\mathbb{D})$.

Hence one can ask for which of the classical Banach spaces the range of H_μ will be continuously embedded. The following statement answers this question.

THEOREM 3.1. *Assume the positive Borel measure μ is a Carleson measure with support in $[0,1)$. The Hankel operator H_μ is a continuous linear mapping from the Hardy space $H^1(\mathbb{D})$ into the space of Cauchy Transforms.*

Proof. It suffices to prove that for $f \in H^1(\mathbb{D})$ the function $H_\mu f$ defines a bounded linear functional on the disk algebra A.To this end we will show that for $g \in A$

$$\lim_{r\to 1}\int_{\mathbb{T}} (H_\mu f)(re^{i\phi})\overline{g(e^{i\phi})}dm(\phi)$$

exists. For the notational purposes set

$$M(r) = \int_{\mathbb{T}} (H_\mu f)(re^{i\phi})\overline{g(e^{i\phi})}dm(\phi).$$

Rewriting M we have

$$M(r) = \int_{[0,1)} f(t)\overline{g(rt)}d\mu(t).$$

Since g is uniformly continuous on $[0,1]$, we may use the assumption that μ is a Carleson measure to conclude that given $\epsilon > 0$ there is an $r' \in (0,1)$ so that for

$r' < r_1 < r_2 < 1$ we have

$$\begin{aligned}|M(r_1) - M(r_2)| &= \left| \int_{[0,1)} f(t)\overline{[g(r_1 t) - g(r_2 t)]} d\mu(t) \right| \\ &< \epsilon \int_{[0,1)} |f(t)| d\mu(t) \\ &\leq C\epsilon \|f\|_{H^1},\end{aligned}$$

from which it follows that

$$\lim_{r \to 1} M(r)$$

exists.

This proves that H_μ is in the CT space. Hence there eixsts a finite Borel measure, say ν, for which

$$(H_\mu f)(z) = \int_{\mathbb{T}} \frac{d\nu(\zeta)}{1 - z\bar{\zeta}} = (K\nu)(z). \tag{3.2}$$

We show that the mapping $f \to H_\mu f$ is continuous by using the closed graph theorem applied to the product space $H^1(\mathbb{D}) \times CT$ with the appropriate norms. First, assume for $f_n \in H^1(\mathbb{D})$ and $H_\mu f_n \equiv F_n$ there are two functions, $f \in H^1(\mathbb{D})$ and $F \in CT$ for which

$$(f_n, F_n) \to (f, F)$$

in the graph norm of the product space. In particular we have convergence of both sequences uniformly on compacta,

$$f_n(z) \rightrightarrows f(z) \qquad \text{and} \qquad F_n(z) \rightrightarrows F(z).$$

If ν_n denote the measure that corresponds to F_n $(n = 1, 2, \ldots)$ with $\|F_n\|_{CT} = \|\nu_n\|$, then they are bounded in total variation norm (choose $z = 0$ in the integral representation). Then the sequence $\{\nu_n\}$ contains a subsequence (again written as $\{\nu_n\}$) that converges weak* to a measure, say ν. Since, for each $z \in \mathbb{D}$ the function $\dfrac{1}{1 - z\bar{\zeta}}$ as a function of ζ, is continuous on the closed unit disk, we conclude

$$\lim_{n \to \infty} \int_{\mathbb{T}} \frac{d\nu_n(\zeta)}{1 - z\bar{\zeta}} = \int_{\mathbb{T}} \frac{d\nu(\zeta)}{1 - z\bar{\zeta}}.$$

But this implies that

$$F(z) = \int_{\mathbb{T}} \frac{d\nu(\zeta)}{1 - z\bar{\zeta}}$$

and this is what was to be proved. □

4. The corresponding measure

Now let us consider the equality, for $|z| < 1$, of

$$(H_\mu f)(re^{i\phi}) = \int_{\mathbb{T}} \frac{d\nu(\zeta)}{1 - re^{i\phi}\bar{\zeta}} = (K\nu)(re^{i\phi})$$

for $\phi \in (0, 2\pi)$.

THEOREM 4.1. *The singular part of the measure ν in the above representation is a point mass at the point $\zeta = 1$.*

Proof. By rewriting the denominator $(1 - tre^{i\phi}) = (1 - te^{i\phi}) - (tre^{i\phi} - te^{i\phi})$ one can expand about the point $z = e^{i\phi}$ and see that there is a neighborhood of this point where H_ν is analytic. Assuming this one has that $H_\nu f$ is analytic on $\mathbb{T} \setminus \{1\}$. Then the integral

$$\int_{\mathbb{T}} \frac{d\nu(\zeta)}{1 - re^{i\phi}\bar{\zeta}}$$

can be extended to be analytic on the closed disk near any point in $\mathbb{T} \setminus \{1\}$. Hence, the integral is bounded on any compact subset of $\mathbb{T}$ which does not contain the point $\zeta = 1$. Thus, given any measurable subset of $\mathbb{T}$, say E, such that there is an $\eta > 0$ for which $E \cap (|\zeta - 1| < 2\eta) = \emptyset$ there is an M such that $|(H_\mu f)(\zeta)| = |(K\nu)(\zeta)| < M$ for all $\zeta \in E$. Choose a non-negative continuous function g on $\mathbb{T}$ which is one on E and zero on $\mathbb{T} \cap (|\zeta - 1| < \eta)$. Consider the statement of the Poltoratski theorem [**16**] in this context. First, note that if $y > M$, the set $A(y) = \{\zeta |\, |(K\nu)(\zeta)| > y\} = \emptyset$ and hence, $\chi_{\{|(K\nu)| > y\}}(\zeta) = 0$ for $\zeta \in E$. Consider now

$$\begin{aligned} &\int_{\mathbb{T}} g(\zeta)\chi_{\{|(K\nu)|>y\}}(\zeta)dm(\zeta) \\ = &\int_{|\zeta-1|<\eta} g(\zeta)\chi_{\{|(K\nu)|>y\}}(\zeta)dm(\zeta) + \int_{\mathbb{T}\setminus\{|\zeta-1|<\eta\}} g(\zeta)\chi_{\{|(K\nu)|>y\}}(\zeta)dm(\zeta). \end{aligned}$$

Since $g(\zeta) = 0$ in the first integral and $\chi_{\{|(K\nu)|>y\}}(\zeta) = 0$ for $y > M$ in the second integral, we have

$$\begin{aligned} \lim_{y\to\infty} \pi y \int_{\mathbb{T}} g(\zeta)\chi_{\{|(K\nu)|>y\}}(\zeta)dm(\zeta) &= 0 \\ &= \int_{\mathbb{T}} g(\zeta) d|\nu_s|(\zeta) \\ &\geq |\nu_s|(E). \end{aligned}$$

Hence, $|\nu_s|$ is a singular measure with support at $\zeta = 1$, and so must be a point mass, $|\nu_s|(\zeta) = |c|\delta_1(\zeta)$. Consequently

$$\nu_s(\zeta) = c\delta_1(\zeta)$$

where c is a complex constant. This establishes the result. $\square$

From the above results we see that if the measure

$$d\nu(\zeta) = F(\zeta)dm(\zeta) + c\delta_1(\zeta)$$

corresponds to the integral operator then the function

$$F(z) = \int_{\mathbb{T}} \frac{F(\zeta)}{1 - z\bar{\zeta}} dm(\zeta) \tag{4.1}$$

is continuous on $\overline{\mathbb{D}} \setminus \{1\}$. We can make more remarks concerning this function.

PROPOSITION 4.2. *The function F defined by* (4.1) *has the following property:*

$$\lim_{r\uparrow 1}(1 - r)F(r) = -c. \tag{4.2}$$

Proof. For $0 < r < 1$ we have

$$(H_\mu f)(r) = \int_0^1 \frac{f(t)}{1 - rt} d\mu(t) = F(r) + \frac{c}{1 - r}.$$

Let $\epsilon > 0$ be given. For $0 < \delta < 1$ we have

$$|\chi_{(1-\delta,1)}(t)f(t)| \leq |f(t)|$$

and

$$\lim_{\delta\uparrow 1} |\chi_{(1-\delta,1)}(t)f(t)| = 0.$$

Using the assumption $f \in L^1(d\mu)$ for $f \in H^1$ and the Lebesgue dominated convergence theorem we conclude that there is a $\delta_0 \in (0,1)$ with

$$\int_{1-\delta_0}^{1} |f(t)|d\mu(t) < \epsilon.$$

Hence

$$(1-r)\left|\int_{1-\delta_0}^{1} \frac{f(t)}{1-rt} d\mu(t)\right| < \epsilon.$$

Also

$$(1-r)\left|\int_{0}^{1-\delta_0} \frac{f(t)}{1-rt} d\mu(t)\right| \leq \frac{1-r}{1-r(1-\delta_0)} C\|f\|_{H^1}$$

implying that

$$\lim_{r\uparrow 1}(1-r)\int_{0}^{1-\delta_0} \frac{f(t)}{1-rt} d\mu(t) = 0.$$

Thus

$$\lim_{r\uparrow 1}(1-r)(H_\mu f)(r) = 0 = \lim_{r\uparrow 1}(1-r)\left[F(r) + \frac{c}{1-r}\right]$$

for $f \in H^1$. This yealds the desired equality. □

REMARK 4.3. The Proposition implies that

$$\lim_{z\to 1}(1-z)F(z) = -c$$

for all $z \in \mathbb{D}$ for which $(1-z)/(1-zt)| \leq M$ with t near the point 1 (e.g. a truncated cone in D with vertex in 1 and $t > 1/2$).

REMARK 4.4. If $c \neq 0$ Proposition 4.2 implies that F can not be in H^1.

Finally, it is possible to make one more comment about the correspondence between the function $f \in H^1(\mathbb{D})$ and the Lebesgue decomposition of the measure in question.

PROPOSITION 4.5. *There is a choice of the measures in the Lebesgue decomposition of ν so that the map of $H^1(\mathbb{D})$ into M_s and $L^1(\mathbb{T})/\overline{H_0^1(\mathbb{D})}$ is continuous.*

Proof. It suffices to assume that $f_n \in H^1(\mathbb{D})$ and $f_n \to 0$ in $H^1(\mathbb{D})$ and then prove that the map $f_n \mapsto \nu_s^n$ is continuous, where ν_s^n is an appropriate singular measure corresponding to f_n in the integral decomposition. We choose $F_n \in L^1(\mathbb{T})$ and ν_s^n as mentioned above, so we have

$$\|H_\mu f\|_{CT} = \|F_n\|_{L^1(\mathbb{T})} + \|\nu_s^n\| \to 0$$

implying the continuity in question. □

The following fact may be worthwhile noting.

PROPOSITION 4.6. *Assume μ is a Carleson measure on $[0,1)$ and H_μ is the associated Hankel operator. Then H_μ maps polynomials into BMOA.*

Proof. It suffices to prove that $H_\mu z^k$, $k \in \mathbb{Z}_+$, is in BMOA. Let $p(z)$ be a polynomial. As above it suffices to prove that for $0 < r < 1$

$$\left|\int_{\mathbb{T}} \overline{H_\mu(z^k)(r\zeta)} p(\zeta) dm(\zeta)\right| \leq C\|p\|_{H^1}$$

where C is independent of p.

Write

$$\int_{\mathbb{T}} \overline{\left(\int_0^1 \frac{t^k}{1 - r\zeta t} d\mu(t)\right)} p(\zeta) dm(\zeta)$$

$$= \int_0^1 t^k d\mu(t) \int_{\mathbb{T}} \frac{p(\zeta)}{1 - r\bar{\zeta}t} dm(\zeta) = \int_0^1 t^k d\mu(t) \int_{\mathbb{T}} \frac{\zeta p(\zeta)}{\zeta - rt} dm(\zeta)$$

$$= \int_0^1 t^k rtp(rt) d\mu(t).$$

Taking absolute value one obtains

$$\left|\int_{\mathbb{T}} \overline{(H_\mu z^k)(r\zeta)} p(\zeta) dm(\zeta)\right| \leq Cr\|z^{k+1}p_r\|_{H^1} \leq C\|p\|_{H_1}.$$

References

[1] A. Aleman, *Some open problems on a class of integral operators on spaces of analytic functions*, Topics in complex analysis and operator theory, Univ. Málaga, Málaga, 2007, pp. 139–140. MR2394658

[2] A. Aleman and J. A. Cima, *An integral operator on H^p and Hardy's inequality*, J. Anal. Math. **85** (2001), 157–176, DOI 10.1007/BF02788078. MR1869606

[3] A. Aleman and A. G. Siskakis, *An integral operator on H^p*, Complex Variables Theory Appl. **28** (1995), no. 2, 149–158, DOI 10.1080/17476939508814844. MR1700079

[4] M. B. Bekker and J. A. Cima, *Some classical integral operators on the Hardy spaces H^p*, The first NEAM, Theta Ser. Adv. Math., vol. 22, Editura Fundaţiei Theta, Bucharest, 2018, pp. 1–17. MR3839028

[5] C. Chatzifountas, D. Girela, and J. Á. Peláez, *A generalized Hilbert matrix acting on Hardy spaces*, J. Math. Anal. Appl. **413** (2014), no. 1, 154–168, DOI 10.1016/j.jmaa.2013.11.046. MR3153575

[6] J. A. Cima, A. L. Matheson, and W. T. Ross, *The Cauchy transform*, Mathematical Surveys and Monographs, vol. 125, American Mathematical Society, Providence, RI, 2006. MR2215991

[7] E. Diamantopoulos and A. G. Siskakis, *Composition operators and the Hilbert matrix*, Studia Math. **140** (2000), no. 2, 191–198, DOI 10.4064/sm-140-2-191-198. MR1784632

[8] P. L. Duren, Theory of H^p Spaces, Dover Publications, Mineola NY, 2000, 292 p.

[9] P. Galanopoulos, D. Girela, J. Á. Peláez, and A. G. Siskakis, *Generalized Hilbert operators*, Ann. Acad. Sci. Fenn. Math. **39** (2014), no. 1, 231–258, DOI 10.5186/aasfm.2014.3912. MR3186815

[10] P. Galanopoulas, N. Merchan and A. G. Siskakis, *A family of Dirichlet-Morrey spaces.* arXiv 1704.05281v1.

[11] P. Galanopoulos and J. Á. Peláez, *A Hankel matrix acting on Hardy and Bergman spaces*, Studia Math. **200** (2010), no. 3, 201–220, DOI 10.4064/sm200-3-1. MR2733266

[12] H. D. Hung, L. D. Ky and T. T. Quang, *Hausdorff operators on holomorphic Hardy spaces and applications.* arXiv 1703.01015v4.

[13] N. Merchán, *Mean Lipschitz spaces and a generalized Hilbert operator*, Collect. Math. **70** (2019), no. 1, 59–69, DOI 10.1007/s13348-018-0217-y. MR3902918

[14] A. R. Mirotin and I. S. Kovalyova. *The Markov-Stieltjes transform as an operator.* arXiv 1611.06584v1.

[15] J. Pau, *Integration operators between Hardy spaces on the unit ball of $\mathbb{C}^n$*, J. Funct. Anal. **270** (2016), no. 1, 134–176, DOI 10.1016/j.jfa.2015.10.009. MR3419758

[16] A. G. Poltoratski, *On the distributions of boundary values of Cauchy integrals*, Proc. Amer. Math. Soc. **124** (1996), no. 8, 2455–2463, DOI 10.1090/S0002-9939-96-03363-1. MR1327037

[17] Ch. Pommerenke, *Schlichte Funktionen und analytische Funktionen von beschränkter mittlerer Oszillation* (German), Comment. Math. Helv. **52** (1977), no. 4, 591–602, DOI 10.1007/BF02567392. MR0454017

[18] A. G. Siskakis, *Volterra operators on spaces of analytic functions—a survey*, Proceedings of the First Advanced Course in Operator Theory and Complex Analysis, Univ. Sevilla Secr. Publ., Seville, 2006, pp. 51–68. MR2290748

[19] A. G. Siskakis and R. Zhao, *A Volterra type operator on spaces of analytic functions*, Function spaces (Edwardsville, IL, 1998), Contemp. Math., vol. 232, Amer. Math. Soc., Providence, RI, 1999, pp. 299–311, DOI 10.1090/conm/232/03406. MR1678342

Department of Mathematics, University of Pittsburgh at Johnstown, Johnstown, PA 15905, US

Email address: `bekker@pitt.edu`

The University of North Carolina at Chapell Hill, CB 3250, 329 Phillips Hall, Chapel Hill, NC 27599, USA

Email address: `cima@email.unc.edu`

Contemporary Mathematics
Volume **743**, 2020
https://doi.org/10.1090/conm/743/14958

A panorama of positivity. II: Fixed dimension

Alexander Belton, Dominique Guillot, Apoorva Khare, and Mihai Putinar

Abstract. This survey contains a selection of topics unified by the concept of positive semidefiniteness (of matrices or kernels), reflecting natural constraints imposed on discrete data (graphs or networks) or continuous objects (probability or mass distributions). We put emphasis on entrywise operations which preserve positivity, in a variety of guises. Techniques from harmonic analysis, function theory, operator theory, statistics, combinatorics, and group representations are invoked. Some partially forgotten classical roots in metric geometry and distance transforms are presented with comments and full bibliographical references. Modern applications to high-dimensional covariance estimation and regularization are included.

Contents

2010 *Mathematics Subject Classification.* 15-02, 26-02, 15B48, 51F99, 15B05, 05E05, 44A60, 15A24, 15A15, 15A45, 15A83, 47B35, 05C50, 30E05, 62J10.

Key words and phrases. metric geometry, positive semidefinite matrix, Toeplitz matrix, Hankel matrix, positive definite function, completely monotone functions, absolutely monotonic functions, entrywise calculus, generalized Vandermonde matrix, Schur polynomials, symmetric function identities, totally positive matrices, totally non-negative matrices, totally positive completion problem, sample covariance, covariance estimation, hard / soft thresholding, sparsity pattern, critical exponent of a graph, chordal graph, Loewner monotonicity, convexity, and super-additivity.

D.G. is partially supported by a University of Delaware Research Foundation grant, by a Simons Foundation collaboration grant for mathematicians, and by a University of Delaware Research Foundation Strategic Initiative grant. A.K. is partially supported by Ramanujan Fellowship SB/S2/RJN-121/2017 and MATRICS grant MTR/2017/000295 from SERB (Govt. of India), by grant F.510/25/CAS-II/2018(SAP-I) from UGC (Govt. of India), and by a Young Investigator Award from the Infosys Foundation.

This is the second part of a two-part survey; we include on p. 146 the table of contents for the first part [**9**]. The survey in its unified form may be found online; see [**8**]. The abstract, keywords, MSC codes, and introduction are the same for both parts.

1. Introduction

Matrix positivity, or positive semidefiniteness, is one of the most wide-reaching concepts in mathematics, old and new. Positivity of a matrix is as natural as positivity of mass in statics or positivity of a probability distribution. It is a notion which has attracted the attention of many great minds. Yet, after at least two centuries of research, positive matrices still hide enigmas and raise challenges for the working mathematician.

The vitality of matrix positivity comes from its breadth, having many theoretical facets and also deep links to mathematical modelling. It is not our aim here to pay homage to matrix positivity in the large. Rather, the present survey, split for technical reasons into two parts, has a limited but carefully chosen scope.

Our panorama focuses on entrywise transforms of matrices which preserve their positive character. In itself, this is a rather bold departure from the dogma that canonical transformations of matrices are not those that operate entry by entry. Still, this apparently esoteric topic reveals a fascinating history, abundant characteristic phenomena and numerous open problems. Each class of positive matrices or kernels (regarding the latter as continuous matrices) carries a specific toolbox of internal transforms. Positive Hankel forms or Toeplitz kernels, totally positive matrices, and group-invariant positive definite functions all possess specific *positivity preservers*. As we see below, these have been thoroughly studied for at least a century.

One conclusion of our survey is that the classification of positivity preservers is accessible in the dimension-free setting, that is, when the sizes of matrices are unconstrained. In stark contrast, precise descriptions of positivity preservers in fixed dimension are elusive, if not unattainable with the techniques of modern mathematics. Furthermore, the world of applications cares much more about matrices of fixed size than in the free case. The accessibility of the latter was by no means a sequence of isolated, simple observations. Rather, it grew organically out of distance geometry, and spread rapidly through harmonic analysis on groups, special functions, and probability theory. The more recent and highly challenging path through fixed dimensions requires novel methods of algebraic combinatorics and symmetric functions, group representations, and function theory.

As well as its beautiful theoretical aspects, our interest in these topics is also motivated by the statistics of big data. In this setting, functions are often applied entrywise to covariance matrices, in order to induce sparsity and improve the quality of statistical estimators (see [**33, 34, 53**]). Entrywise techniques have recently increased in popularity in this area, largely because of their low computational complexity, which makes them ideal to handle the ultra high-dimensional datasets arising in modern applications. In this context, the dimensions of the matrices are fixed, and correspond to the number of underlying random variables. Ensuring that positivity is preserved by these entrywise methods is critical, as covariance matrices must be positive semidefinite. Thus, there is a clear need to produce characterizations of entrywise preservers, so that these techniques are widely applicable and mathematically justified. We elaborate further on this in the second part of the survey.

We conclude by remarking that, while we have tried to be comprehensive in our coverage of the field of matrix positivity and the entrywise calculus, there are very likely to be some inadvertent omissions. Even if our survey is not complete in terms of results and connections, we hope that it serves to impress upon the reader the depth and breadth, the classical history and modern applications, and the influence and beauty of the many facets of positivity.

2. A selection of classical results on entrywise positivity preservers

We begin by mentioning some results from the first part of the survey ([**9**] or [**8**]), which are used or referred to in this part.

2.1. From metric geometry to matrix positivity. As discussed in the first part of the survey, the study of entrywise positivity preservers naturally emerged out of considerations of metric geometry. We recall here some early results of Schoenberg, beginning with the following connection between metric geometry and matrix positivity.

THEOREM 2.1 (Schoenberg [**55**]). *Let $d \geq 1$ be an integer and let (X, ρ) be a metric space. An $(n+1)$-tuple of points $x_0, x_1, \ldots, x_n$ in X can be isometrically embedded into Euclidean space $\mathbb{R}^d$, but not into $\mathbb{R}^{d-1}$, if and only if the matrix*

$$[\rho(x_0, x_j)^2 + \rho(x_0, x_k)^2 - \rho(x_j, x_k)^2]_{j,k=1}^n \tag{2.1}$$

is positive semidefinite with rank equal to d.

The positivity of the matrix (2.1) is equivalent to the statement that the associated $(n+1) \times (n+1)$ matrix

$$[-\rho(x_j, x_k)^2]_{j,k=0}^n$$

is *conditionally positive semidefinite*: recall that a real symmetric matrix A is conditionally positive semidefinite if $\mathbf{u}^T A \mathbf{u} \geq 0$ whenever the coordinates of the real vector $\mathbf{u}$ sum to zero.

Schoenberg's Theorem 2.1 marks an early appearance of positive and conditionally positive matrices in the analysis literature. It says that applying the function $-x^2$ entrywise transforms Euclidean-distance matrices into conditionally positive semidefinite matrices. A natural next step is to remove the word "conditionally" and ask which functions transform distance matrices, from a given metric space (X, ρ), into positive matrices. This is precisely the definition of *positive definite functions* on (X, ρ).

Schoenberg showed [**56**] that Euclidean spaces are characterized by the property that Gaussian kernels with arbitrary variances are positive definite on them. He similarly showed [**55**] that among metric spaces of diameter no more than π, the unit spheres $S^{d-1} \subset \mathbb{R}^d$ and $S^\infty \subset \ell^2_{\mathbb{R}}$ admit a similar characterization in terms of just one function, cosine. Following this result, and the work of Bochner [**14, 15**] in classifying positive definite functions on Euclidean and compact homogeneous spaces, Schoenberg was interested in understanding classes of positive definite functions on these spheres.

THEOREM 2.2 (Schoenberg [**57**]). *Let $f : [-1,1] \to \mathbb{R}$ be continuous.*

(1) *For a given dimension $d \geq 2$, the function $f \circ \cos$ is positive definite on the unit sphere S^{d-1} if and only if it has a distinguished Fourier-series decomposition with non-negative coefficients. That is,*

$$f(\cos\theta) = \sum_{k=0}^{\infty} c_k P_k^{(\lambda)}(\cos\theta) \qquad (\theta \in \mathbb{R}), \tag{2.2}$$

where $P_k^{(\lambda)}$ are the ultraspherical orthogonal polynomials with $\lambda = (d-2)/2$ and the coefficients $c_k \geq 0$ for all $k \geq 0$ with $\sum_{k=0}^\infty c_k < \infty$.

(2) *The function $f(\cos\theta)$ is positive definite on all finite-dimensional spheres, or, equivalently, is positive definite on S^∞, if and only if*

$$f(\cos\theta) = \sum_{k=0}^{\infty} c_k \cos^k\theta, \tag{2.3}$$

where $c_k \geq 0$ for all $k \geq 0$ and $\sum_{k=0}^\infty c_k < \infty$.

By freeing the previous result from the spherical context, Schoenberg obtained his celebrated result on positivity preservers.

THEOREM 2.3 (Schoenberg [**57**]). *Let $f : [-1,1] \to \mathbb{R}$ be continuous. If the matrix $[f(a_{jk})]_{j,k=1}^n$ is positive semidefinite for all $n \geq 1$ and all positive semidefinite matrices $[a_{jk}]_{j,k=1}^n$ with entries in $[-1,1]$, then, and only then,*

$$f(x) = \sum_{k=0}^{\infty} c_k x^k \qquad (x \in [-1,1]),$$

where $c_k \geq 0$ for all $k \geq 0$ and $\sum_{k=0}^\infty c_k < \infty$.

2.2. Entrywise functions preserving positivity in all dimensions. Theorem 2.3 provides a definitive answer to one version of the following central question, which is the driving idea throughout this survey.

Which functions, when applied entrywise to certain classes of matrices, preserve positive semidefiniteness?

The fundamental result for answering this question is the Schur product theorem [**58**]: if A and B are positive semidefinite matrices of the same size, then their entrywise product is positive semidefinite too. As observed by Pólya and Szegö [**51**], the fact that the set of positive matrices forms a closed convex cone immediately implies, by Schur's result, that every power series with non-negative Maclaurin coefficients is a positivity preserver; they asked if there are any other functions with this property. It follows from Schoenberg's Theorem 2.3 that there are no additional continuous functions, and Rudin [**54**] subsequently removed the continuity hypothesis for real-valued functions on $(-1, 1)$.

A similar variant was proved by Vasudeva [**60**], for a different domain. To state this result, and for later, we recall some notation from the first part of the survey [**9**].

DEFINITION 2.4. Fix a domain $I \subset \mathbb{C}$ and integers $m, n \geq 1$. Let $\mathcal{P}_n(I)$ denote the set of $n \times n$ Hermitian positive semidefinite matrices with entries in I, with $\mathcal{P}_n(\mathbb{C})$ abbreviated to $\mathcal{P}_n$. A function $f : I \to \mathbb{C}$ acts *entrywise* on a matrix

$$A = [a_{jk}]_{1 \leq j \leq m,\ 1 \leq k \leq n} \in I^{m \times n}$$

by setting

$$f[A] := [f(a_{jk})]_{1 \leq j \leq m,\ 1 \leq k \leq n} \in \mathbb{C}^{m \times n}.$$

Below, we allow the dimensions m and n to vary, while keeping the uniform notation $f[-]$. We also let $\mathbf{1}_{m \times n}$ denote the $m \times n$ matrix with each entry equal to one. Note that $\mathbf{1}_{n \times n} \in \mathcal{P}_n(\mathbb{R})$.

Now we can state Vasudeva's result.

THEOREM 2.5 (Vasudeva [**60**]). *Let $f : (0, \infty) \to \mathbb{R}$. Then $f[-]$ preserves positivity on $\mathcal{P}_n\big((0, \infty)\big)$ for all $n \geq 1$, if and only if $f(x) = \sum_{k=0}^{\infty} c_k x^k$ on $(0, \infty)$, where $c_k \geq 0$ for all $k \geq 0$.*

A final variant is for matrices with possibly complex entries. This result was conjectured by Rudin in [**54**] and proved four years later.

THEOREM 2.6 (Herz [**37**]). *Let $D(0,1)$ denote the open unit disc in $\mathbb{C}$, and suppose $f : D(0,1) \to \mathbb{C}$. The entrywise map $f[-]$ preserves positivity on $\mathcal{P}_n\big(D(0,1)\big)$ for all $n \geq 1$, if and only if*

$$f(z) = \sum_{j,k \geq 0} c_{jk} z^j \overline{z}^k \qquad \text{for all } z \in D(0,1),$$

where $c_{jk} \geq 0$ for all j, $k \geq 0$.

2.3. The Horn–Loewner theorem and its variants. The first part of this survey [**9**] focuses on various refinements of our central question when the matrices under consideration are of arbitrary dimension (the "dimension-free" setting). Here, we consider the situation where the dimension N of the test matrices is fixed. This turns out to be highly challenging, and remains open to date for each $N \geq 3$. The following necessary condition was first published by R. Horn (who in [**40**] attributes it to his PhD advisor C. Loewner), and is essentially the only general result known.

THEOREM 2.7 (Horn–Loewner [**40**]). *Let $f : (0, \infty) \to \mathbb{R}$ be continuous. Fix a positive integer n and suppose $f[-]$ preserves positivity on $\mathcal{P}_n\big((0, \infty)\big)$. Then $f \in C^{n-3}\big((0, \infty)\big)$,*

$$f^{(k)}(x) \geq 0 \qquad \text{whenever } x \in (0, \infty) \text{ and } 0 \leq k \leq n-3,$$

and $f^{(n-3)}$ *is a convex non-decreasing function on* $(0,\infty)$. *Furthermore, if* $f \in C^{n-1}((0,\infty))$, *then* $f^{(k)}(x) \geq 0$ *whenever* $x \in (0,\infty)$ *and* $0 \leq k \leq n-1$.

This theorem has produced several variants: the arguments are purely local, they involve low-rank matrices, and continuity need not be assumed. Another possibility involves working with real-analytic functions, and we use this below.

LEMMA 2.8 (Belton–Guillot–Khare–Putinar [**6**] and Khare–Tao [**43**]). *Let* n *be a positive integer, suppose* $0 < \rho \leq \infty$ *and let* $f(x) = \sum_{k\geq 0} c_k x^k$ *be a convergent power series on* $I = [0,\rho)$ *that preserves positivity entrywise for all rank-one matrices in* $\mathcal{P}_n(I)$. *Suppose further that* $c_{m'} < 0$ *for some* m'.

(1) *If* $\rho < \infty$, *then* $c_m > 0$ *for at least* n *values of* $m < m'$. *(Thus, the first* n *non-zero Maclaurin coefficients of* f, *if they exist, must be positive.)*
(2) *If* $\rho = \infty$, *then* $c_m > 0$ *for at least* n *values of* $m < m'$ *and at least* n *values of* $m > m'$. *(Thus, if* f *is a polynomial, then the first* n *non-zero coefficients and the last* n *non-zero coefficients of* f, *if they exist, are all positive.)*

These results, and others in the literature for smooth functions, admit a common generalization that was recently obtained.

THEOREM 2.9 (Khare [**42**]). *Let* $a \in \mathbb{R}_+$ *and* $\epsilon \in (0,\infty)$, *and suppose* $f : [a, a+\epsilon) \to \mathbb{R}$ *is smooth. Fix integers* n, p, q *such that* $n \geq 1$ *and* $0 \leq p \leq q \leq n$, *with* $p = 0$ *if* $a = 0$, *and such that* f *has* $q-p$ *non-zero derivatives at* a *of order at least* p; *let*

$$m_p < m_{p+1} < \cdots < m_{q-1}$$

be the orders of these derivatives.

If there exists $\mathbf{u} := (u_1, \ldots, u_n)^T \in (0,1)^n$ *with distinct entries and such that* $f[a\mathbf{1}_{n\times n} + t\mathbf{u}\mathbf{u}^T] \in \mathcal{P}_n(\mathbb{R})$ *for all* $t \in [0,\epsilon)$, *then the derivative* $f^{(k)}(a)$ *is non-negative whenever* $0 \leq k \leq m_{q-1}$.

The proof of Theorem 2.9 involves a determinant computation that generalizes one by Horn and Loewner, and leads to an unexpected connection to symmetric function theory. See Theorem 3.22 for more details.

2.4. Preservers of positive Hankel matrices. Finally for this chapter, we consider entrywise maps preserving the set of positive Hankel matrices. A distinguished subset of these matrices arise as moment matrices for measures on the real line; we collect some concepts from the first part of the survey.

DEFINITION 2.10. A measure μ with support in $\mathbb{R}$ is said to be *admissible* if μ is non-negative and all its moments are finite:

$$s_k(\mu) := \int_{\mathbb{R}} x^k \, \mathrm{d}\mu(x) < \infty \qquad (k \in \mathbb{Z}_+).$$

The sequence $\mathbf{s}(\mu) := (s_k(\mu))_{k=0}^{\infty}$ is the *moment sequence* of μ, and the *moment matrix* of μ is the semi-infinite Hankel matrix

$$H_\mu := \begin{bmatrix} s_0(\mu) & s_1(\mu) & s_2(\mu) & \cdots \\ s_1(\mu) & s_2(\mu) & s_3(\mu) & \cdots \\ s_2(\mu) & s_3(\mu) & s_4(\mu) & \cdots \\ \vdots & \vdots & \vdots & \ddots \end{bmatrix}.$$

A function $f : \mathbb{R} \to \mathbb{R}$ acts entrywise on moment sequences, so that

$$f[\mathbf{s}(\mu)] := (f(s_0(\mu)), \ldots, f(s_k(\mu)), \ldots),$$

and $f[H_\mu] = H_\sigma$ if $f[\mathbf{s}(\mu)] = \mathbf{s}(\sigma)$ for some admissible measure σ.

Working with positive moment matrices and their entrywise preservers provides a route to proving stronger versions of Vasudeva's and Schoenberg's theorems. We conclude this section by stating these results.

THEOREM 2.11 (Belton–Guillot–Khare–Putinar [**7**]). *Suppose $I = (0, \infty)$ and $f : I \to \mathbb{R}$. The following are equivalent.*

(1) *The entrywise map $f[-]$ preserves positivity on $\mathcal{P}_n(I)$ for all $n \geq 1$.*
(2) *There exists $u_0 \in (0,1)$ such that $f[-]$ preserves positivity for all moment matrices of the form H_μ, where $\mu = a\delta_1 + b\delta_{u_0}$ and a, $b \in I$.*
(3) *The function f has a power-series representation $\sum_{k=0}^{\infty} c_k x^k$ valid for all $x \in I$, where the Maclaurin coefficients $c_k \geq 0$ for all $k \geq 0$.*

THEOREM 2.12 (Belton–Guillot–Khare–Putinar [**7**]). *Suppose $0 < \rho \leq \infty$, let $I = (-\rho, \rho)$ and suppose $f : I \to \mathbb{R}$. The following are equivalent.*

(1) *The entrywise map $f[-]$ preserves positivity on $\mathcal{P}_n(I)$, for all $n \geq 1$.*
(2) *The entrywise map $f[-]$ preserves positivity on the set of Hankel matrices in $\mathcal{P}_n(I)$ of rank at most 3, for all $n \geq 1$.*
(3) *The function f is real analytic, and absolutely monotonic on $(0, \rho)$, so that $f(x) = \sum_{k=0}^{\infty} c_k x^k$ for all $x \in I$, with $c_k \geq 0$ for all $k \geq 0$.*

3. Entrywise polynomials preserving positivity in fixed dimension

Having discussed at length the dimension-free setting, we now turn our attention to functions that preserve positivity in a fixed dimension $N \geq 2$. This is a natural question from the standpoint of both theory as well as applications. This latter connection to applied fields and to high-dimensional covariance estimation will be explained below in Chapter 5.

Mathematically, understanding the functions f such that $f[-] : \mathcal{P}_N \to \mathcal{P}_N$ for fixed $N \geq 2$, is a non-trivial and challenging refinement of Schoenberg's 1942 theorem. A complete characterization was found for $N = 2$ by Vasudeva [**60**]:

THEOREM 3.1 (Vasudeva [**60**]). *Given a function $f : (0, \infty) \to \mathbb{R}$, the entrywise map $f[-]$ preserves positivity on $\mathcal{P}_2\big((0, \infty)\big)$ if and only f is non-negative, non-decreasing, and multiplicatively mid-convex:*

$$f(x)f(y) \geq f\big(\sqrt{xy}\big)^2 \qquad \text{for all } x, y > 0.$$

In particular, f is either identically zero or never zero on $(0, \infty)$, and f is also continuous.

On the other hand, if $N \geq 3$, then such a characterization remains open to date. As mentioned above, perhaps the only known result for general entrywise preservers is the Horn–Loewner theorem 2.7 or its more general variants, some of which are stated above.

In light of this challenging scarcity of results in fixed dimension, a strategy adopted in the literature has been to further refine the problem, in one of several ways:

(1) Restrict the class of functions, while operating entrywise on all of $\mathcal{P}_N$ (over some given domain I, say $(0,\rho)$ or $(-\rho,\rho)$ for $0<\rho\le\infty$). For example, in this survey we consider possibly non-integer power functions, polynomials and power series, and even linear combinations of real powers.
(2) Restrict the class of matrices and study entrywise functions over this class in a fixed dimension. For instance, popular sub-classes of matrices include positive matrices with rank bounded above, or with a given sparsity pattern (zero entries), or classes such as Hankel or Toeplitz matrices; or intersections of these classes. For instance, in discussing the Horn–Loewner and Schoenberg–Rudin results, we encountered Toeplitz and Hankel matrices of low rank.
(3) Study the problem under both of the above restrictions.

In this chapter we begin with the first of these restrictions. Specifically, we will study polynomial maps that preserve positivity, when applied entrywise to $\mathcal{P}_N$. Recall from the Schur product theorem that if the polynomial f has only non-negative coefficients then $f[-]$ preserves positivity on $\mathcal{P}_N$ for every dimension $N\ge 1$. It is natural to expect that if one reduces the test set, from all dimensions to a fixed dimension, then the class of polynomial preservers should be larger. Remarkably, until 2016 not a single example was known of a polynomial positivity preserver with a negative coefficient. Then, in quick succession, the two papers [**6**, **43**] provided a complete understanding of the sign patterns of entrywise polynomial preservers of $\mathcal{P}_N$. The goal of this chapter is to discuss some of the results in these works.

3.1. Characterizations of sign patterns. Until further notice, we work with entrywise polynomial or power-series maps of the form

$$f(x)=c_0x^{n_0}+c_1x^{n_1}+\cdots,\quad \text{with } 0\le n_0<n_1<\cdots, \tag{3.1}$$

and $c_j\in\mathbb{R}$ typically non-zero, which preserve $\mathcal{P}_N(I)$ for various I. Our goal is to try and understand their sign patterns, that is, which c_j can be negative. The first observation is that as soon as I contains the interval $(0,\rho)$ for any $\rho>0$, by the Horn–Loewner type necessary conditions in Lemma 2.8, the lowest N non-zero coefficients of $f(x)$ must be positive.

The next observation is that if $I\not\subset\mathbb{R}_+$, then, in general, there is no structured classification of the sign patterns of the power series preservers on $\mathcal{P}_N(I)$. For example, let k be a non-negative integer; the polynomials

$$f_{k,t}(x):=t(1+x^2+\cdots+x^{2k})-x^{2k+1}\qquad(t>0)$$

do not preserve positivity entrywise on $\mathcal{P}_N\big((-\rho,\rho)\big)$ for any $N\ge 2$. This may be seen by taking $\mathbf{u}:=(1,-1,0,\ldots,0)^T$ and $A:=\eta\mathbf{u}\mathbf{u}^T$ for some $0<\eta<\rho$, and noting that

$$\mathbf{u}^Tf_{k,t}[A]\mathbf{u}=-4\eta^{2k+1}<0.$$

Similarly, if one allows complex entries and uses higher-order roots of unity, such negative results (vis-à-vis Lemma 2.8) are obtained for complex matrices.

Given this, in the rest of the chapter we will focus on $I=(0,\rho)$ for $0<\rho\le\infty$.[1] As mentioned above, if f as in (3.1) entrywise preserves positivity even on rank-one matrices in $\mathcal{P}_N\big((0,\rho)\big)$ then its first N non-zero Maclaurin coefficients are positive.

---

[1]That said, we also briefly discuss the one situation in which our results do apply more generally, even to $I=D(0,\rho)\subset\mathbb{C}$ (an open complex disc).

Our goal is to understand if any other coefficient can be negative (and if so, which of them). This has at least two ramifications:

(1) It would yield the first example of a polynomial entrywise map (for a fixed dimension) with at least one negative Maclaurin coefficient. Recall the contrast to Schoenberg's theorem in the dimension-free setting.

(2) This also yields the first example of a polynomial (or power series) that entrywise preserves positivity on $\mathcal{P}_N(I)$ but not $\mathcal{P}_{N+1}(I)$. In particular it would imply that the Horn–Loewner type necessary condition in Lemma 2.8(1) is "sharp".

These goals are indeed achieved in the particular case $n_0 = 0, \dots, n_{N-1} = N-1$ in [**6**], and subsequently, for arbitrary $n_0 < \cdots < n_{N-1}$ in [**43**]. (In fact, in the latter work the n_j need not even be integers; this is discussed below.) Here is a 'first' result along these lines. Henceforth we assume that $\rho < \infty$; we will relax this assumption midway through Section 3.5 below.

THEOREM 3.2 (Belton–Guillot–Khare–Putinar [**6**] and Khare–Tao [**43**]). *Suppose $N \geq 2$ and $n_0 < \cdots < n_{N-1}$ are non-negative integers, and ρ, c_0, ..., c_{N-1} are positive scalars. Given $\epsilon_M \in \{0, \pm 1\}$ for all $M > n_{N-1}$, there exists a power series*

$$f(x) = c_0 x^{n_0} + \cdots + c_{N-1} x^{n_{N-1}} + \sum_{M > n_{N-1}} d_M x^M$$

such that f is convergent on $(0, \rho)$, the entrywise map $f[-]$ preserves positivity on $\mathcal{P}_N\big((0,\rho)\big)$ and d_M has the same sign (positive, negative or zero) as ϵ_M for all $M > n_{N-1}$.

OUTLINE OF PROOF. The claim is such that it suffices to show the result for exactly one $\epsilon_M = -1$. Indeed, given the claim, for each $M > n_{N-1}$ there exists $\delta_M \in (0, 1/M!)$ such that $\sum_{j=0}^{N-1} c_j x^{n_j} + dx^M$ preserves positivity entrywise on $\mathcal{P}_N\big((0,\rho)\big)$ whenever $|d| \leq \delta_M$. Now let $d_M := \epsilon_M \delta_M$ for all $M > n_{N-1}$, and define

$$f_M(x) := \sum_{j=0}^{N_1} c_j x^{n_j} + d_M x^M \quad \text{and} \quad f(x) := \sum_{M > n_{N-1}} 2^{n_{N-1} - M} f_M(x).$$

Then it may be verified that $|f(x)| \leq \sum_{j=0}^{N-1} c_j x^{n_j} + 2^{n_{N-1}} e^{x/2}$, and hence f has the desired properties. $\square$

Thus it suffices to show the existence of a polynomial positivity preserver on $\mathcal{P}_N\big((0,\rho)\big)$ with precisely one negative Maclaurin coefficient, the leading term. In the next few sections we explain how to achieve this goal. In fact, one can show a more general result, for real powers as well.

THEOREM 3.3 (Khare–Tao [**43**]). *Fix an integer $N \geq 2$ and real exponents $n_0 < \cdots < n_{N-1} < M$ in the set $\mathbb{Z}_+ \cup [N-2, \infty)$. Suppose ρ, c_0, ..., $c_{N-1} > 0$ as above. Then there exists $c' < 0$ such that the function*

$$f(x) = c_0 x^{n_0} + \cdots + c_{N-1} x^{n_{N-1}} + c' x^M \qquad (x \in (0, \rho))$$

preserves positivity entrywise on $\mathcal{P}_N\big((0,\rho)\big)$. [Here and below, we set $0^0 := 1$.]

The restriction of the n_j lying in $\mathbb{Z}_+ \cup [N-2, \infty)$ is a technical one that is explained in a later chapter on the study of entrywise powers preserving positivity on $\mathcal{P}_N\big((0,\infty)\big)$; see Theorem 4.1.

REMARK 3.4. A stronger result, Theorem 3.15, which also applies to real powers, is stated below. We mention numerous ramifications of the results in this chapter following that result.

The proofs of the preceding two theorems crucially use type-A representation theory (specifically, a family of symmetric functions) that naturally emerges here via generalized Vandermonde determinants. These symmetric homogeneous polynomials are introduced and used in the next section.

For now, we explain how Theorem 3.3 helps achieve a complete classification of the sign patterns of a family of generalised power series, of the form

$$f(x) = \sum_{j=0}^{\infty} c_j x^{n_j}, \qquad n_j \in \mathbb{Z}_+ \cup [N-2, \infty) \text{ for all } j \geq 0,$$

but without the requirement that that exponents are non-decreasing. In this generality, one first notes that the Horn–Loewner-type Lemma 2.8 still applies: if some coefficient $c_{j_0} < 0$, then there must be at least N indices j such that $n_j < n_{j_0}$ and $c_j > 0$. The following result shows that once again, this necessary condition is the best possible.

THEOREM 3.5 (Classification of sign patterns for real-power series preservers, Khare–Tao [**43**]). *Fix an integer $N \geq 2$, and distinct real exponents n_0, n_1, ... in $\mathbb{Z}_+ \cup [N-2, \infty)$. Suppose $\epsilon_j \in \{0, \pm 1\}$ is a choice of sign for each $j \geq 0$, such that if $\epsilon_{j_0} = -1$ then $\epsilon_j = +1$ for at least N choices of j such that $n_j < n_{j_0}$. Given any $\rho > 0$, there exists a choice of coefficients c_j with sign ϵ_j such that*

$$f(x) := \sum_{j=0}^{\infty} c_j x^{n_j}$$

is convergent on $(0, \rho)$ and preserves positivity entrywise on $\mathcal{P}_N\big((0, \rho)\big)$.

Notice this result is strictly more general than Theorem 3.2, because the sequence $n_0, n_1, \ldots$ can contain an infinite decreasing sequence of positive non-integer powers, for example, all rational elements of $[N-2, \infty)$. Thus Theorem 3.5 covers a larger class of functions than even Hahn or Puiseux series.

Theorem 3.5 is derived from Theorem 3.3 in a similar fashion to the proof of Theorem 3.2, and we refer the reader to [**43**, Section 1] for the details.

3.2. Schur polynomials; the sharp threshold bound for a single matrix. We now explain how to prove Theorem 3.3. The present section will discuss the case of integer powers, and end by proving the theorem for a single 'generic' rank-one matrix. In the following section we show how to extend the results to all rank-one matrices for integer powers. The subsequent section will complete the proof for real powers, and then for matrices of all ranks.

The key new tool that is indispensable to the following analysis is that of *Schur polynomials.* These can be defined in a number of equivalent ways; we refer the reader to [**16**] for more details, including the equivalence of these definitions shown using ideas of Karlin–Macgregor, Lindström, and Gessel–Viennot. For our purposes the definition of Cauchy is the most useful:

DEFINITION 3.6. Given non-negative integers $N \geq 1$ and $n_0 < \cdots < n_{N-1}$, let

$$\mathbf{n} := (n_0, \ldots, n_{N-1})^T, \qquad \text{and} \quad \mathbf{n}_{\min} := (0, 1, \ldots, N-1)^T,$$

and define $V(\mathbf{n}) := \prod_{0 \leq i < j \leq N-1}(n_j - n_i)$.

Given a vector $\mathbf{u} = (u_1, \dots, u_N)^T$ and a non-negative integer k, let $\mathbf{u}^{\circ k} := (u_1^k, \dots, u_N^k)^T$, and let $\mathbf{u}^{\circ \mathbf{n}}$ be the $N \times N$ matrix with (j,k) entry $\mathbf{u}_j^{n_{k-1}}$.

The *Schur polynomial* in variables $u_1, \dots, u_N$ of *degree* $\mathbf{n}$ is given by

$$s_{\mathbf{n}}(\mathbf{u}) := \frac{\det \mathbf{u}^{\circ \mathbf{n}}}{\det \mathbf{u}^{\circ \mathbf{n}_{\min}}}. \tag{3.2}$$

Notice that the numerator is a generalized Vandermonde determinant, so a homogeneous and alternating polynomial, while the denominator is the usual Vandermonde determinant in the indeterminates $u_1, \dots, u_N$. Hence their ratio $s_{\mathbf{n}}(\mathbf{u})$ is a homogeneous symmetric polynomial in $\mathbb{Z}[u_1, \dots, u_N]$. It follows that Schur polynomials are well defined when working over any commutative unital ring.

Schur polynomials are an extremely well-studied family of symmetric functions. Their appeal lies in the important observation that they are the characters of all irreducible (finite-dimensional) polynomial representations of the complex Lie group $GL_n(\mathbb{C})$ (or of the Lie algebra $\mathfrak{sl}_{n+1}(\mathbb{C})$). In this setting, the definition of Cauchy is a special case of the Weyl character formula. Thus, its specialization yields the corresponding Weyl dimension formula, which will be of use below:

$$s_{\mathbf{n}}((1, \dots, 1)^T) = \prod_{0 \leq i < j \leq N-1} \frac{n_j - n_i}{j - i} = \frac{V(\mathbf{n})}{V(\mathbf{n}_{\min})}. \tag{3.3}$$

An alternate proof of (3.3) comes from the *principal specialization formula*: for a variable q, one has that

$$s_{\mathbf{n}}\big((1, q, \dots, q^{N-1})^T\big) = \prod_{0 \leq i < j \leq N-1} \frac{q^{n_j} - q^{n_i}}{q^j - q^i}; \tag{3.4}$$

this follows from (3.2) because now the numerator is also a standard Vandermonde determinant. We also refer the reader to [**48**] for many more results and properties of Schur polynomials.

Returning to polynomial positivity preservers, we wish to consider functions of the form

$$f(x) = c_0 x^{n_0} + \cdots + c_{N-1} x^{n_{N-1}} + c' x^M,$$

with non-negative integers $n_0 < \cdots < n_{N-1} < M$ and positive coefficients $c_0, \dots, c_{N-1}$. We are interested in characterizing those $c' \in \mathbb{R}$ for which the entrywise map $f[-]$ preserves positivity on $\mathcal{P}_N\big((0, \rho)\big)$. By the Schur product theorem, this is equivalent to finding the smallest c' such that $f[-]$ is a preserver. We may assume that $c' < 0$, so we rescale by $t := |c'|^{-1}$ and define

$$p_t(x) := t \sum_{j=0}^{N-1} c_j x^{n_j} - x^M. \tag{3.5}$$

The goal now is to find the smallest $t > 0$ such that $p_t[-]$ preserves positivity on $\mathcal{P}_N\big((0, \rho)\big)$. We next achieve this goal for a single rank-one matrix.

PROPOSITION 3.7. *With notation as above, define*

$$\mathbf{n}_j = (n_0, \dots, n_{j-1}, \widehat{n_j}, n_{j+1}, \dots, n_{N-1}, M)^T$$

for $0 \leq j \leq N-1$. Given a vector $\mathbf{u} \in (0, \infty)^N$ with distinct coordinates, the following are equivalent.

(1) *The matrix $p_t[\mathbf{u}\mathbf{u}^T]$ is positive semidefinite.*

(2) $\det p_t[\mathbf{u}\mathbf{u}^T] \geq 0$.

(3) $t \geq \sum_{j=0}^{N-1} \frac{s_{\mathbf{n}_j}(\mathbf{u})^2}{c_j s_{\mathbf{n}}(\mathbf{u})^2}$.

In particular, this shows that for a generic rank-one matrix in $\mathcal{P}_N\big((0,\rho)\big)$, there does exist a positivity-preserving polynomial with a negative leading term.

In essence, the equivalences in Proposition 3.7 hold more generally; this is distilled into the following lemma.

LEMMA 3.8 (Khare–Tao [**44**][2]). *Fix $\mathbf{w} \in \mathbb{R}^N$ and a positive-definite matrix H. Fix $t > 0$ and define $P_t := tH - \mathbf{w}\mathbf{w}^T$. The following are equivalent.*

(1) P_t is positive semidefinite.

(2) $\det P_t \geq 0$.

(3) $t \geq \mathbf{w}^T H^{-1} \mathbf{w} = 1 - \frac{\det(H - \mathbf{w}\mathbf{w}^T)}{\det H}$.

We refer the reader to [**44**] for the detailed proof of Lemma 3.8, remarking only that the equality in assertion (3) follows by using Schur complements in two different ways to expand the determinant of the matrix $\begin{bmatrix} H & \mathbf{w} \\ \mathbf{w}^T & 1 \end{bmatrix}$.

Now Proposition 3.7 follows directly from Lemma 3.8, by setting

$$H = \sum_{j=0}^{N-1} c_j \mathbf{u}^{\circ n_j} (\mathbf{u}^{\circ n_j})^T \quad \text{and} \quad \mathbf{w} = \mathbf{u}^{\circ M},$$

where H is positive definite because of the following general matrix factorization (which is also used below).

PROPOSITION 3.9. *Let $f(x) = \sum_{k=0}^M f_k x^k$ be a polynomial with coefficients in a commutative ring R. For any integer $N \geq 1$ and any vectors $\mathbf{u} = (u_1, \dots, u_N)^T$ and $\mathbf{v} = (v_1, \dots, v_N)^T \in R^N$, it holds that*

$$f[t\mathbf{u}\mathbf{v}^T] = \sum_{k=0}^{M} f_k t^k \mathbf{u}^{\circ k} (\mathbf{v}^{\circ k})^T \tag{3.6}$$

$$= \begin{bmatrix} 1 & u_1 & \cdots & u_1^M \\ 1 & u_2 & \cdots & u_2^M \\ \vdots & \vdots & \ddots & \vdots \\ 1 & u_N & \cdots & u_N^M \end{bmatrix} \begin{bmatrix} f_0 & 0 & \cdots & 0 \\ 0 & f_1 t & \cdots & 0 \\ \vdots & \vdots & \ddots & \vdots \\ 0 & 0 & \cdots & f_M t^M \end{bmatrix} \begin{bmatrix} 1 & v_1 & \cdots & v_1^M \\ 1 & v_2 & \cdots & v_2^M \\ \vdots & \vdots & \ddots & \vdots \\ 1 & v_N & \cdots & v_N^M \end{bmatrix}^T,$$

where 1 is a multiplicative identity which is adjoined to R if necessary.

Now to adopt Lemma 3.8(3), this same equation and the Cauchy–Binet formula allow one to compute $\det(H - \mathbf{w}\mathbf{w}^T)$ in the present situation, and this yields precisely that $t \geq \sum_{j=0}^{N-1} \frac{s_{\mathbf{n}_j}(\mathbf{u})^2}{c_j s_{\mathbf{n}}(\mathbf{u})^2}$, as desired.

---

[2]The work [**44**] is an extended abstract of the paper [**43**], but some of the results in it have different proofs from [**43**].

3.3. The threshold for all rank-one matrices: a Schur positivity result. We continue toward a proof of Theorem 3.3. The next step is to use Proposition 3.7 to achieve an intermediate goal: a threshold bound for c' that works for all rank-one matrices in $\mathcal{P}_N\big((0,\rho)\big)$, still working with integer powers. Clearly, to do so one has to understand the supremum of each ratio $R_j := s_{\mathbf{n}_j}(\mathbf{u})^2/s_{\mathbf{n}}(\mathbf{u})^2$, as $\mathbf{u}$ runs over vectors in $(0,\sqrt{\rho})^N$ with distinct coordinates. More precisely, one has to understand the supremum of the weighted sum $\sum_j R_j/c_j$.

This observation was first made in the work [**6**] for the case $n_j = j$, that is, $\mathbf{n} = \mathbf{n}_{\min}$. It led to the first proof of Theorem 3.3, with all of the denominators being the same: $s_{\mathbf{n}_{\min}}(\mathbf{u}) = 1$. We now use another equivalent definition of Schur polynomials, by Littlewood, realizing them as sums of monomials corresponding to certain Young tableaux. Every monomial has a non-negative integer coefficient. It follows by the continuity and homogeneity of $s_{\mathbf{n}_j}$ and the Weyl Dimension Formula (3.3), that the supremum in the previous paragraph equals the value at $(\sqrt{\rho},\dots,\sqrt{\rho})^T$, namely

$$\sup_{\mathbf{u}\in(0,\sqrt{\rho})^N} s_{\mathbf{n}_j}(\mathbf{u})^2 = \frac{V(\mathbf{n}_j)^2}{V(\mathbf{n}_{\min})^2}\rho^{M-n_j}.$$

Since all of these suprema are attained at the same point $\sqrt{\rho}(1,\dots,1)^T$, the weighted sum in Proposition 3.7(3) also attains its supremum at the same point. Thus, we conclude using Proposition 3.7 that

$$f(x) = \sum_{j=0}^{N-1} c_j x^{n_j} + c' x^M$$

preserves positivity entrywise on all rank-one matrices $\mathbf{u}\mathbf{u}^T \in \mathcal{P}_N\big((0,\rho)\big)$ if and only if

$$c' \geq -\left(\sum_{j=0}^{N-1} \frac{V(\mathbf{n}_j)^2}{c_j V(\mathbf{n}_{\min})^2}\rho^{M-n_j}\right)^{-1}.$$

In fact, if $\mathbf{n} = \mathbf{n}_{\min}$ then the entire argument above goes through even when one changes the domain to the open complex disc $D(0,\rho)$, or any intermediate domain $(0,\rho) \subset D \subset D(0,\rho)$. This is precisely the content of the main result in [**6**].

THEOREM 3.10 (Belton–Guillot–Khare–Putinar [**6**]). *Fix $\rho > 0$ and integers $M \geq N \geq 2$. Let*

$$f(z) = \sum_{j=0}^{N-1} c_j z^j + c' z^M, \qquad \text{where } c_0,\dots,c_{N-1},c' \in \mathbb{R},$$

and let $I := \overline{D}(0,\rho)$ be the closed disc in the complex plane with centre 0 and radius ρ. The following are equivalent.

1. *The entrywise map $f[-]$ preserves positivity on $\mathcal{P}_N(I)$.*
2. *The entrywise map $f[-]$ preserves positivity on rank-one matrices in $\mathcal{P}_N\big((0,\rho)\big)$.*
3. *Either $c_0, \dots, c_{N-1}, c'$ are all non-negative, or $c_0, \dots, c_{N-1}$ are positive and*

$$c' \geq -\left(\sum_{j=0}^{N-1} \frac{V(\mathbf{n}_j)^2}{c_j V(\mathbf{n}_{\min})^2}\rho^{M-j}\right)^{-1},$$

where $\mathbf{n}_j := (0, 1, \dots, j-1, \widehat{j}, j+1, \dots, N-1, M)^T$ *for* $0 \leq j \leq N-1$.

This theorem provides a complete understanding of which polynomials of degree at most N preserve positivity entrywise on $\mathcal{P}_N\big((0,\rho)\big)$ and, more generally, on any subset of $\mathcal{P}_N\big(\overline{D}(0,\rho)\big)$ that contains the rank-one matrices in $\mathcal{P}_N\big((0,\rho)\big)$.

REMARK 3.11. Clearly (1) $\implies$ (2) here, and the proof of (2) $\iff$ (3) was outlined above via Proposition 3.7. We defer mentioning the proof strategy for (2) $\implies$ (1), because we will later see a similar theorem over $I = (0,\rho)$ for more general powers n_j. The proof of that result, Theorem 3.15, will be outlined in some detail.

Having dealt with the base case of $\mathbf{n} = \mathbf{n}_{\min}$, as well as $\mathbf{n} = (k, k+1, \dots, k+N-1)$ for any $k \in \mathbb{Z}_+$, which holds by the Schur product theorem, we now turn to the general case. In general, $s_{\mathbf{n}}(\mathbf{u})$ is no longer a monomial, and so it is no longer clear if and where the supremum of each ratio $s_{\mathbf{n}_j}(\mathbf{u})^2/s_{\mathbf{n}}(\mathbf{u})^2$, or of their weighted sum, is attained for $\mathbf{u} \in (0, \sqrt{\rho})^N$. The threshold bound for all rank-one matrices itself is not apparent, and the bound for all matrices in $\mathcal{P}_N\big((0,\rho)\big)$ is even more inaccessible.

By a mathematical miracle, it turns out that the same phenomenon as in the base case holds in general. Namely, the ratio of each $s_{\mathbf{n}_j}$ and $s_{\mathbf{n}}$ attains its supremum at $\sqrt{\rho}(1,\dots,1)^T$. Hence one can proceed as above to obtain a uniform threshold for c', which works for all rank-one matrices in $\mathcal{P}_N\big((0,\rho)\big)$.

EXAMPLE 3.12. To explain the ideas of the preceding paragraph, we present an example. Suppose

$$N = 3, \qquad \mathbf{n} = (0,2,3), \qquad M = 4, \quad \text{and} \quad \mathbf{u} = (u_1, u_2, u_3)^T.$$

Then

$$\begin{aligned} \mathbf{n}_3 &= (0,2,4), \\ s_{\mathbf{n}}(\mathbf{u}) &= u_1u_2 + u_2u_3 + u_3u_1, \\ \text{and} \quad s_{\mathbf{n}_3}(\mathbf{u}) &= (u_1+u_2)(u_2+u_3)(u_3+u_1). \end{aligned}$$

The claim is that $s_{\mathbf{n}_3}(\mathbf{u})/s_{\mathbf{n}}(\mathbf{u})$ is coordinatewise non-decreasing for $\mathbf{u} \in (0,\infty)^3$; the assertion about its supremum on $(0,\sqrt{\rho})^N$ immediately follows from this. It suffices by symmetry to show the claim only for one variable, say u_3. By the quotient rule,

$$s_{\mathbf{n}}(\mathbf{u})\partial_{u_3}s_{\mathbf{m}}(\mathbf{u}) - s_{\mathbf{m}}(\mathbf{u})\partial_{u_3}s_{\mathbf{n}}(\mathbf{u}) = (u_1+u_2)(u_1u_3 + 2u_1u_2 + u_2u_3)u_3,$$

and this is clearly non-negative on the positive orthant, proving the claim. As we see, the above expression is, in fact, monomial positive, from which numerical positivity follows immediately.

In fact, an even stronger fact holds. Viewed as a polynomial in u_3, every coefficient in the above expression is in fact *Schur positive*. In other words, the coefficient of each u_3^j is a non-negative combination of Schur polynomials in u_1 and u_2:

$$(u_1+u_2)(u_1u_3 + 2u_1u_2 + u_2u_3)u_3 = \sum_{j\geq 0} p_j(u_1,u_2)u_3^j,$$

where

$$p_j(u_1,u_2) = \begin{cases} 2s_{(1,3)}(u_1,u_2) & \text{if } j=1, \\ s_{(0,3)}(u_1,u_2) + s_{(1,2)}(u_1,u_2) & \text{if } j=2, \\ 0 & \text{otherwise.} \end{cases}$$

In particular, this implies that each coefficient is monomial positive, whence numerically positive. We recall here that the monomial positivity of Schur polynomials follows from the definition of $s_{\mathbf{n}}(\mathbf{u})$ using Young tableaux.

The miracle to which we alluded above, is that the Schur positivity in the preceding example in fact holds in general.

THEOREM 3.13 (Khare–Tao [**43**]). *If $n_0 < \cdots < n_{N-1}$ and $m_0 < \cdots < m_{N-1}$ are N-tuples of non-negative integers such that $m_j \geq n_j$ for $j = 0, \ldots, N-1$, then the function*

$$f_{\mathbf{m},\mathbf{n}} : (0,\infty)^N \to \mathbb{R}; \ \mathbf{u} \mapsto \frac{s_{\mathbf{m}}(\mathbf{u})}{s_{\mathbf{n}}(\mathbf{u})}$$

is non-decreasing in each coordinate. Furthermore, if

$$s_{\mathbf{n}}(\mathbf{u})\partial_{u_N} s_{\mathbf{m}}(\mathbf{u}) - s_{\mathbf{m}}(\mathbf{u})\partial_{u_N} s_{\mathbf{n}}(\mathbf{u}) \tag{3.7}$$

is considered as a polynomial in u_N, then the coefficient of every monomial u_N^j is a Schur-positive polynomial in $u_1, \ldots, u_{N-1}$.

The second, stronger part of Theorem 3.13 follows from a deep and highly non-trivial result in symmetric function theory (or type-A representation theory) by Lam, Postnikov, and Pylyavskyy [**45**], following earlier results by Skandera. We refer the reader to this paper and to [**43**] for more details. Notice also that the first assertion in Theorem 3.13 only requires the numerical positivity of the expression (3.7). This is given a separate proof in [**43**], using the method of condensation due to Charles Lutwidge Dodgson [**18**].[3] In this context, we add for completeness that in [**43**] the authors also show a log-supermodularity (or FKG, or MTP_2) phenomenon for determinants of totally positive matrices.

3.4. Real powers; the threshold works for all matrices. We now return to the proof of Theorem 3.3, which holds for real powers. Our next step is to observe that the first part of Theorem 3.13 now holds for all real powers. Since one can no longer define Schur polynomials in this case, we work with generalized Vandermonde determinants instead:

COROLLARY 3.14. *Fix N-tuples of real powers $\mathbf{n} = (n_0 < \cdots < n_{N-1})$ and $\mathbf{m} = (m_0 < \cdots < m_{N-1})$, such that $n_j \leq m_j$ for all j. Letting $\mathbf{u}^{\circ \mathbf{n}} := [u_j^{n_{k-1}}]_{j,k=1}^N$ as above, the function*

$$f : \{\mathbf{u} \in (0,\infty)^N : u_i \neq u_j \text{ if } i \neq j\} \to \mathbb{R}; \ \mathbf{u} \mapsto \frac{\det \mathbf{u}^{\circ \mathbf{m}}}{\det \mathbf{u}^{\circ \mathbf{n}}}$$

is non-decreasing in each coordinate.

We sketch here one proof. The version for integer powers, Theorem 3.13, gives the version for rational powers, by taking a "common denominator" $L \in \mathbb{Z}$ such that Lm_j and Ln_j are all integers, and using a change of variables $y_j := u_j^{1/L}$. The

[3]This article by Dodgson immediately follows his better-known 1865 publication, *Alice's Adventures in Wonderland.*

general version for real powers then follows by considering rational approximations and taking limits.

Corollary 3.14 helps prove the real-power version of Theorem 3.3, just as Theorem 3.13 would have shown the integer-powers case of Theorem 3.3. Namely, first note that Proposition 3.7 holds even when the n_j are real powers; the only changes are (a) to assume that the coordinates of $\mathbf{u}$ are distinct, and (b) to rephrase the last assertion (3) to the following:

$$t \geq \sum_{j=0}^{N-1} \frac{(\det \mathbf{u}^{\circ \mathbf{n}_j})^2}{c_j (\det \mathbf{u}^{\circ \mathbf{n}})^2}.$$

These arguments help prove the first part of the following result, which is the culmination of these ideas.

THEOREM 3.15 (Khare–Tao [**43**]). *Fix an integer $N \geq 1$ and real exponents $n_0 < \cdots < n_{N-1} < M$, as well as scalars $\rho > 0$ and $c_0, \ldots, c_{N-1}, c'$. Let*

$$f(x) := \sum_{j=0}^{N-1} c_j x^{n_j} + c' x^M.$$

The following are equivalent.

(1) *The function f preserves positivity entrywise on all rank-one matrices in $\mathcal{P}_N\big((0,\rho)\big)$.*
(2) *The function f preserves positivity entrywise on all Hankel rank-one matrices in $\mathcal{P}_N\big((0,\rho)\big)$.*
(3) *Either the coefficients $c_0, \ldots, c_{N-1}$ and c' are non-negative, or $c_0, \ldots, c_{N-1}$ are positive and*

$$c' \geq -\left(\sum_{j=0}^{N-1} \frac{V(\mathbf{n}_j)^2}{c_j V(\mathbf{n})^2} \rho^{M-n_j} \right)^{-1},$$

where $V(\mathbf{n})$ and $\mathbf{n}_j$ are as defined above.

If, moreover, the exponents n_j all lie in $\mathbb{Z}_+ \cup [N-2, \infty)$, then these assertions are also equivalent to the following.

(4) *The function f preserves positivity entrywise on $\mathcal{P}_N\big((0,\rho)\big)$.*

Before sketching the proof, we note several ramifications of this result.

(1) The theorem completely characterizes linear combinations of up to $N+1$ powers that entrywise preserve positivity on $\mathcal{P}_N\big((0,\rho)\big)$. The same is true for any subset of $\mathcal{P}_N\big((0,\rho)\big)$ that contains all rank-one positive semidefinite Hankel matrices.
(2) As discussed above, Theorem 3.15 implies Theorem 3.5, which helps in understanding which sign patterns correspond to countable sums of real powers that preserve positivity entrywise on $\mathcal{P}_N\big((0,\rho)\big)$ (or on the subset of rank-one matrices). In particular, the existence of sign patterns which are not all non-negative shows the existence of functions which preserve positivity on $\mathcal{P}_N$ but not on $\mathcal{P}_{N+1}$.
(3) Theorem 3.15 bounds $A^{\circ M}$ in terms of a multiple of $\sum_{j=0}^{N-1} c_j A^{\circ n_j}$. More generally, one can do this for an arbitrary convergent power series instead of a monomial, in the spirit of Theorem 3.2. Even more generally, one may work with Laplace transforms of measures; see Corollary 3.17 below.

For completeness, we also mention two developments related (somewhat more distantly) to the above results.

- A refinement of a conjecture of Cuttler, Greene, and Skandera (2011) and its proof; see [**43**] for more details. In particular, this approach assists with a novel characterization of weak majorization, using Schur polynomials.
- A related "Schubert cell-type" stratification of the cone $\mathcal{P}_N(\mathbb{C})$; see [**6**] for further details.

We conclude this section by outlining the proof of Theorem 3.15.

PROOF. Clearly, (4) $\implies$ (1) $\implies$ (2). If (2) holds, then, by Lemma 2.8, either all the c_j and c' are non-negative, or c_j is positive for all j. Thus, we suppose that $c_j > 0 > c'$.

Note that if $\mathbf{u}(u_0) := (1, u_0, \dots, u_0^{N-1})^T$ for some $u_0 \in (0,1)$, then

$$A(u_0) := \rho u_0^2 \mathbf{u}(u_0)\mathbf{u}(u_0)^T$$

is a rank-one Hankel matrix and hence in our test set. Repeating the analysis in Section 3.2, using generalized Vandermonde determinants instead of Schur polynomials and rank-one Hankel matrices of the form $A(u_0)$,

$$\begin{aligned}|c'|^{-1} &\geq \sup_{u_0 \in (0,1)} \sum_{j=0}^{N-1} \frac{(\det[\sqrt{\rho} u_0 \mathbf{u}(u_0)]^{\circ \mathbf{n}_j})^2}{c_j (\det[\sqrt{\rho} u_0 \mathbf{u}(u_0)]^{\circ \mathbf{n}})^2} \\ &= \sum_{j=0}^{N-1} \lim_{u_0 \to 1^-} \sum_{j=0}^{N-1} \frac{(\det \mathbf{u}(u_0)^{\circ \mathbf{n}_j})^2}{c_j (\det \mathbf{u}(u_0)^{\circ \mathbf{n}})^2} (\rho u_0^2)^{M - n_j},\end{aligned}$$

where the equality follows from Corollary 3.14 above. The real-exponent version of (3.4) holds if $q \in (0,\infty) \setminus \{1\}$ and the exponents n_j are real and non-decreasing:

$$\det \mathbf{u}(q)^{\circ \mathbf{n}} = \prod_{0 \leq i < k \leq N-1} (q^{n_k} - q^{n_i}) = V(q^{\circ \mathbf{n}}).$$

Applying this identity, the above computation yields

$$|c'|^{-1} \geq \lim_{u_0 \to 1^-} \sum_{j=0}^{N-1} \frac{V(u_0^{\circ \mathbf{n}_j})^2}{V(u_0^{\circ \mathbf{n}})^2} \frac{(\rho u_0^2)^{M-n_j}}{c_j} = \sum_{j=0}^{N-1} \frac{V(\mathbf{n}_j)^2}{c_j V(\mathbf{n})^2} \rho^{M-n_j}.$$

Thus (2) $\implies$ (3). Conversely, that (3) $\implies$ (1) follows by a similar analysis to that given above, using Corollary 3.14 and the density of matrices $\mathbf{u}\mathbf{u}^T$, where $\mathbf{u} \in \left(0, \sqrt{\rho}\right)^N$ has distinct entries, in the set of all rank-one matrices in $\mathcal{P}_N\big((0,\rho)\big)$.

It remains to show that (1) $\implies$ (4) if all the exponents $n_j \in \mathbb{Z}_+ \cup [N-2, \infty)$. We proceed by induction on N. The case $N = 1$ is immediate. For the inductive step, we apply the extension principle of the following Proposition 3.16 with $h = f$, which requires verification that $f'[-]$ preserves positivity on $\mathcal{P}_{N-1}$. This is a straightforward calculation via the induction hypothesis. □

The following extension principle was inspired by work of FitzGerald and Horn [**25**].

PROPOSITION 3.16 (Khare–Tao [**43**]). *Suppose* $0 < \rho \leq \infty$, *and* $I = (0,\rho)$, $(-\rho, \rho)$ *or the closure of one of these sets. Let* $h : I \to \mathbb{R}$ *be a continuously differentiable function on the interior of* I. *If* $h'[-]$ *preserves positivity entrywise* *on $\mathcal{P}_{N-1}(I)$ and $h[-]$ does so on the rank-one matrices in $\mathcal{P}_N(I)$, then $h[-]$ in fact preserves positivity on all of $\mathcal{P}_N(I)$.*[4]

Proposition 3.16 relies on two arguments found in [**25**]: (a) every matrix in $\mathcal{P}_N$ may be written as the sum of a rank-one matrix in $\mathcal{P}_N$, and a matrix in $\mathcal{P}_{N-1}$ with its last row and column both zero, and (b) applying the integral identity

$$h(x) - h(y) = \int_x^y h'(t)\,\mathrm{d}t = \int_0^1 (x-y)h'(\lambda x + (1-\lambda)y)\,\mathrm{d}\lambda$$

entrywise to this decomposition. See [**43**, Section 3] for more details. The original use of these arguments was when h is a power function; this is explained in Chapter 4 below.

3.5. Power series preservers and beyond; unbounded domains. In the remainder of this chapter, we use Theorem 3.15 to derive several corollaries; thus, we retain and use the notation of that theorem. As discussed following Theorem 3.15, the first consequence extends the theorem from bounding monomials $A^{\circ M} = (x^M)[A]$ by a multiple of $\sum_{j=0}^{N-1} c_j A^{\circ n_j}$, to bounding $f[A]$ for more general power series. Even more generally, one can work with Laplace transforms of real measures on $\mathbb{R}$.

COROLLARY 3.17 (Khare–Tao [**43**]). *Let the notation be as for Theorem 3.15, with $c_j > 0$ for all j. Suppose μ is a real measure supported on $[n_{N-1}+\epsilon, \infty)$ for some $\epsilon > 0$, and let*

$$g_\mu(x) := \int_{n_{N-1}+\epsilon}^{\infty} x^t\,\mathrm{d}\mu(t). \tag{3.8}$$

If g_μ is absolutely convergent at ρ, then there exists a finite threshold $t_\mu > 0$ such that, for all $A \in \mathcal{P}_N\big((0,\rho)\big)$, the matrix

$$t_\mu \sum_{j=0}^{N-1} c_j A^{\circ n_j} - g_\mu[A]$$

is positive semidefinite.

PROOF. By Theorem 3.15 and the fact that $\mathcal{P}_N(\mathbb{R})$ is a closed convex cone, it suffices to show the finiteness of the quantity

$$\int_{n_{N-1}+\epsilon}^{\infty} \sum_{j=0}^{N-1} \frac{V(\mathbf{n}_j)^2}{c_j V(\mathbf{n})^2} \rho^{M-n_j}\,\mathrm{d}\mu_+(M),$$

where μ_+ is the positive part of μ. This follows from the hypotheses. $\square$

We now turn to the $\rho = \infty$ case, which was briefly alluded to above. In other words, the domain is now unbounded: $I = (0, \infty)$. As in the bounded-domain case, the question of interest is to classify all possible sign patterns of polynomial or power-series preservers on $\mathcal{P}_N(I)$ for a fixed integer N.

Similar to the above discussion for bounded I, the crucial step in classifying sign patterns of power series (or more general functions, as in Theorem 3.5) is to work with integer powers and precisely one coefficient that can be negative. Thus, one

[4] An analogous version of this results holds for $I = D(0,\rho)$ or its closure in $\mathbb{C}$, with $h : I \to \mathbb{C}$ analytic. This is used to prove the corresponding implication in Theorem 3.10 above.

first observes that Lemma 2.8(2) holds in the unbounded-domain case $I = (0, \infty)$. Hence given a polynomial

$$f(x) = \sum_{j=0}^{2N-1} c_j x^{n_j} + c' x^M,$$

where

$$0 \leq n_0 < \cdots < n_{N-1} < M < n_N < n_{N+1} \cdots < n_{2N-1},$$

if $f[-]$ preserves positivity on $\mathcal{P}_N\big((0,\infty)\big)$, then either all the coefficients $c_0, \ldots, c_{2N-1}, c'$ are non-negative, or $c_0, \ldots, c_{2N-1}$ are positive and c' can be negative. In this case, an explicit threshold is not known as it is in Theorem 3.15, but we now explain why such a threshold exists.

We start from (3.6) and repeat the subsequent analysis via the Cauchy–Binet formula. To find a uniform threshold for c' that works for all rank-one matrices in $\mathcal{P}_N\big((0,\infty)\big)$, it suffices to bound, uniformly from above, certain ratios of sums of squares of Schur polynomials. This may be done because of the following tight bounds.

PROPOSITION 3.18 (Khare–Tao [**43**]). *If* $\mathbf{n} := (n_0, ..., n_{N-1})$ *and*

$$\mathbf{u} := (u_1, ..., u_N),$$

where $n_0 < \cdots < n_{N-1}$ *are non-negative integers and* $u_1 \leq \cdots \leq u_N$ *are non-negative real numbers, then*

$$\mathbf{u}^{\mathbf{n}-\mathbf{n}_{\min}} \leq s_{\mathbf{n}}(\mathbf{u}) \leq \frac{V(\mathbf{n})}{V(\mathbf{n}_{\min})} \mathbf{u}^{\mathbf{n}-\mathbf{n}_{\min}}, \tag{3.9}$$

where $\mathbf{n}_{\min} := (0, \ldots, n_{N-1})$. *The constants* 1 *and* $V(\mathbf{n})/V(\mathbf{n}_{\min})$ *on each side of (*3.9*) cannot be improved.*

We refer the reader to [**43**, Section 4] for further details, including how Proposition 3.18 implies the existence of preservers f as above for rank-one matrices with $c' < 0$. The extension from rank-one matrices to all of $\mathcal{P}_N\big((0,\infty)\big)$ is carried out using the extension principle in Proposition 3.16.

In a sense, Proposition 3.18 isolates the 'leading term' of every Schur polynomial. This calculation can be generalized to the case of non-integer powers,[5] which helps extend the above results for the unbounded domain $I = (0, \infty)$ to real powers. This yields the desired classification, similar to Theorem 3.5 in the bounded-domain case.

THEOREM 3.19 (Khare–Tao [**43**]). *Let* $N \geq 2$, *and let*

$$\{\alpha_j : j \geq 0\} \subset \mathbb{Z}_+ \cup [N-2, \infty)$$

be a set of distinct real numbers. For each $j \geq 0$, *let* $\epsilon_j \in \{0, \pm 1\}$ *be a sign and suppose that, whenever* $\epsilon_{j_0} = -1$, *then* $\epsilon_j = +1$ *for at least* N *choices of* j *such that* $\alpha_j < \alpha_{i_0}$ *and also for at least* N *choices of* j *such that* $\alpha_j > \alpha_{i_0}$. *There exists a series with real coefficients,*

$$f(x) = \sum_{j=0}^{\infty} c_j x^{\alpha_j}$$

[5] We refer the reader again to [**43**, Section 5] for the details, which use additional concepts from type-A representation theory: the Harish-Chandra–Itzykson–Zuber integral and Gelfand–Tsetlin patterns.

which converges on $(0,\infty)$, *preserves positivity entrywise on* $\mathcal{P}_N\big((0,\infty)\big)$, *and is such that* c_j *has the same sign as* ϵ_j *for all* $j \geq 0$.

Note that, in particular, Theorem 3.19 reaffirms that the Horn–Loewner-type conditions in Lemma 2.8(2) are sharp.

3.6. Digression: Schur polynomials from smooth functions, and new symmetric function identities. Before proceeding to additional applications of Theorem 3.15 and related results, we take a brief detour to explain how Schur polynomials arise naturally from any sufficiently differentiable function.

THEOREM 3.20 (Khare [**42**]). *Fix non-negative integers* $m_0 < m_1 < \ldots < m_{N-1}$, *as well as scalars* $\epsilon > 0$ *and* $a \in \mathbb{R}$. *Let* $M := m_0 + \cdots + m_{N-1}$ *and suppose the function* $f : [a, a+\epsilon) \to \mathbb{R}$ *is* M*-times differentiable at* a. *Given vectors* $\mathbf{u}$, $\mathbf{v} \in \mathbb{R}^N$, *define* $\Delta : [0, \epsilon') \to \mathbb{R}$ *for a sufficiently small* $\epsilon' \in (0, \epsilon)$ *by setting*

$$\Delta(t) := \det f[a\mathbf{1}_{N\times N} + t\mathbf{u}\mathbf{v}^T].$$

Then,

(3.10)

$$\Delta^{(M)}(0) = \sum_{\mathbf{m} \vdash M} \binom{M}{m_0, m_1, \ldots, m_{N-1}} V(\mathbf{u})V(\mathbf{v}) s_{\mathbf{m}}(\mathbf{u}) s_{\mathbf{m}}(\mathbf{v}) \prod_{k=0}^{N-1} f^{(m_k)}(a),$$

where the first factor in the summand is a multinomial coefficient, and we sum over all partitions $\mathbf{m} = (m_0, \ldots, m_{N-1})$ *of* M *with unequal parts, that is,* $M = m_0 + \cdots + m_{N-1}$ *and* $0 \leq m_0 < \cdots < m_{N-1}$.

In particular, $\Delta(0) = \Delta'(0) = \cdots = \Delta^{(\binom{N}{2}-1)}(0) = 0$.

REMARK 3.21. As a special case, if $f : \mathbb{R} \to \mathbb{R}$ is smooth at a, and $\mathbf{u}$, $\mathbf{v} \in \mathbb{R}^N$, then defining $\Delta(t) := \det f[a\mathbf{1}_{N\times N} + t\mathbf{u}\mathbf{v}^T]$ gives a function Δ which is smooth at 0, and Theorem 3.20 gives all of these derivatives via the formula (3.10). The general version of Theorem 3.20 is a key ingredient in showing Theorem 2.9, which subsumes all known variants of Horn–Loewner-type necessary conditions in fixed dimension.

The key determinant computation required to prove the original Horn–Loewner necessary condition in fixed dimension (see Theorem 2.7) is the special case of Theorem 3.20 where $\mathbf{u} = \mathbf{v}$ and $m_j = j$ for all j. In this situation, $s_{\mathbf{m}}(\mathbf{u}) = s_{\mathbf{m}}(\mathbf{v}) = 1$, so Schur polynomials do not appear. The general version of Theorem 3.20 decouples the vectors $\mathbf{u}$ and $\mathbf{v}$, and holds for all $M > 0$ if f is smooth (as in Loewner's setting). Moreover, it reveals the presence of Schur polynomials in every case other than the ones studied by Loewner, that is, when $M > \binom{N}{2}$.

While Theorem 3.20 involves derivatives of a smooth function, the result and its proof are, in fact, completely algebraic, and valid over any commutative ring. To show this, an algebraic analogue of the differential operator is required, with more structure than is given by a derivation. The precise statement and its proof may be found in [**42**, Section 2].

We conclude this section by applying Theorem 3.20 and its algebraic avatar to symmetric function theory. We begin by recalling the famous *Cauchy summation identity* [**48**, Example I.4.6]: if $f_0(x) := 1 + x + x^2 + \cdots$ is the geometric series, viewed as a formal power series over a commutative unital ring R, and $u_1, \ldots, u_N$,

$v_1, \ldots, v_N$ are commuting variables, then

$$\det f_0[\mathbf{u}\mathbf{v}^T] = V(\mathbf{u})V(\mathbf{v}) \sum_{\mathbf{m}} s_{\mathbf{m}}(\mathbf{u}) s_{\mathbf{m}}(\mathbf{v}), \tag{3.11}$$

where the sum runs over all partitions $\mathbf{m}$ with at most N parts.[6]

A natural question is whether similar formulae hold when f_0 is replaced by other formal power series. Very few such results were known; this includes one due to Frobenius [**26**], for the function $f_c(x) := (1 - cx)/(1 - x)$ with c an scalar. (This is also connected to theta functions and elliptic Frobenius–Stickelberger–Cauchy determinant identities.) For this function,

$$\begin{aligned}\det f_c[\mathbf{u}\mathbf{v}^T] &= \det \left[\frac{1 - cu_j v_k}{1 - u_j v_k} \right]_{j,k=1}^N \\ &= V(\mathbf{u})V(\mathbf{v})(1 - c)^{N-1} \\ &\quad \times \Big(\sum_{\mathbf{m}: m_0 = 0} s_{\mathbf{m}}(\mathbf{u}) s_{\mathbf{m}}(\mathbf{v}) + (1 - c) \sum_{\mathbf{m}: m_0 > 0} s_{\mathbf{m}}(\mathbf{u}) s_{\mathbf{m}}(\mathbf{v}) \Big).\end{aligned} \tag{3.12}$$

A third, obvious identity is if f is a 'fewnomial' with at most $N - 1$ terms. In this case, $f[\mathbf{u}\mathbf{v}^T]$ is a sum of at most $N-1$ rank-one matrices, and so its determinant vanishes.

The following result extends all three of these cases to an arbitrary formal power series over an arbitrary commutative ring R, and with an additional $\mathbb{Z}_+$-grading.

THEOREM 3.22 (Khare [**42**]). *Fix a commutative unital ring R and let t be an indeterminate. Let $f(t) := \sum_{M \geq 0} f_M t^M \in R[[t]]$ be an arbitrary formal power series. Given vectors $\mathbf{u}$, $\mathbf{v} \in R^N$, where $N \geq 1$, we have that*

(3.13)

$$\det f[t\mathbf{u}\mathbf{v}^T] = V(\mathbf{u})V(\mathbf{v}) \sum_{M \geq \binom{N}{2}} t^M \sum_{\mathbf{m} = (m_{N-1}, \ldots, m_0) \vdash M} s_{\mathbf{m}}(\mathbf{u}) s_{\mathbf{m}}(\mathbf{v}) \prod_{k=0}^{N-1} f_{m_k}.$$

The heart of the proof involves first computing, for each $M \geq 0$, the coefficient of t^M in $\det f[t\mathbf{u}\mathbf{v}^T]$, over the "universal ring"

$$R' := \mathbb{Q}[u_1, \ldots, u_N, v_1, \ldots, v_N, f_0, f_1, \ldots],$$

where u_j, v_k and f_m are algebraically independent over $\mathbb{Q}$. These coefficients are seen to equal $\Delta^{(M)}(0)/M!$, by the algebraic version of Theorem 3.20. Thus, (3.13) holds over R'. Then note that both sides of (3.13) lie in the subring $R_0 := \mathbb{Z}[u_1, \ldots, u_N, v_1, \ldots, v_N, f_0, f_1, \ldots]$, so the identity holds in R_0. Finally, it holds as claimed by specializing from R_0 to R.

An alternate approach to proving Theorem 3.22 is also provided in [**42**]. The identity (3.6) is applied, along with the Cauchy–Binet formula, to each truncated Taylor–Maclaurin polynomial $f_{\leq M}$ of $f(x)$. The result follows by taking limits in the t-adic topology, using the t-adic continuity of the determinant function.

[6]Usually one uses infinitely many indeterminates in symmetric function theory, but given the connection to the entrywise calculus in a fixed dimension, we will restrict our attention to u_j and v_j for $1 \leq j \leq N$.

3.7. Further applications: linear matrix inequalities, Rayleigh quotients, and the cube problem. This chapter ends with further ramifications and applications of the above results. First, notice that Theorem 3.15 implies the following linear matrix inequality version that is 'sharp' in more than one sense:

COROLLARY 3.23. *Fix $\rho > 0$, real exponents $n_0 < \cdots < n_{N-1} < M$ for some integer $N \geq 1$, and scalars $c_j > 0$ for all j. Then,*

$$A^{\circ M} \leq \mathcal{C}\Big(c_0 A^{\circ n_0} + \cdots + c_{N-1} A^{\circ n_{N-1}}\Big),$$
$$\text{where } \mathcal{C} = \sum_{j=0}^{N-1} \frac{V(\mathbf{n}_j)^2}{c_j V(\mathbf{n})^2} \rho^{M-n_j},$$

for all $A \in \mathcal{P}_N\big((0,\rho)\big)$ of rank one, or of all ranks if $n_0, \ldots, n_{N-1} \in \mathbb{Z}_+ \cup [N-2,\infty)$. Moreover, the constant $\mathcal{C}$ is the smallest possible, as is the number of terms N on the right-hand side.

In the above Corollary and henceforth, the notations $A \leq B$ and $B \geq A$ mean that the matrix $B - A$ is positive semidefinite.

Seeking a uniform threshold such as $\mathcal{C}$ in the preceding inequality can also be achieved (as explained above) by first working with a single positive matrix, then optimizing over all matrices. The first step here can be recast as an extremal problem that involves Rayleigh quotients:

PROPOSITION 3.24 (see [**6, 43**]). *Fix an integer $N \geq 2$ and real exponents $n_0 < \cdots < n_{N-1} < M$, where each $n_j \in \mathbb{Z}_+ \cup [N-2,\infty)$. Given positive scalars $c_0, \ldots, c_{N-1}$, let*

$$h(x) := \sum_{j=0}^{N-1} c_j x^{n_j} \qquad (x \in (0,\infty)).$$

Then, for $0 < \rho < \infty$ and $A \in \mathcal{P}_N\big([0,\rho]\big)$,

$$t\, h[A] \geq A^{\circ M} \qquad \text{if and only if} \quad t \geq \varrho(h[A]^{\dagger/2} A^{\circ M} h[A]^{\dagger/2}), \tag{3.14}$$

where $\varrho[B]$ and $B^\dagger$ denote the spectral radius and the Moore–Penrose pseudo-inverse of a square matrix B, respectively. Moreover, for every non-zero matrix $A \in \mathcal{P}_N\big([0,\rho]\big)$, the following variational formula holds:

$$\varrho(h[A]^{\dagger/2} A^{\circ M} h[A]^{\dagger/2}) = \sup_{\mathbf{u} \in (\ker h[A])^\perp \setminus \{\mathbf{0}\}} \frac{\mathbf{u}^T A^{\circ M} \mathbf{u}}{\mathbf{u}^T h[\mathbf{u}\mathbf{u}^T] \mathbf{u}} \leq \sum_{j=0}^{N-1} \frac{V(\mathbf{n}_j)^2}{V(\mathbf{n})^2} \frac{\rho^{M-n_j}}{c_j}.$$

Proposition 3.24 is shown using the Kronecker normal form for matrix pencils; see the treatment in [**27**, Section X.6]. When the matrix A is a generic rank-one matrix, the above generalized Rayleigh quotient has a closed-form expression, which features Schur polynomials for integer powers. This reveals connections between Rayleigh quotients, spectral radii, and symmetric functions.

PROPOSITION 3.25. *Let the notation be as in Proposition 3.24, but now with n_j not necessarily in $\mathbb{Z}_+ \cup [N-2,\infty)$. If $A = \mathbf{u}\mathbf{u}^T$, where $\mathbf{u} \in (0,\infty)^N$ has distinct coordinates, then $h[A]$ is invertible, and the threshold bound*

$$\varrho(h[A]^{\dagger/2} A^{\circ M} h[A]^{\dagger/2}) = (\mathbf{u}^{\circ M})^T h[\mathbf{u}\mathbf{u}^T]^{-1} \mathbf{u}^{\circ M} = \sum_{j=0}^{N-1} \frac{(\det \mathbf{u}^{\circ \mathbf{n}_j})^2}{c_j (\det \mathbf{u}^{\circ \mathbf{n}})^2}. \tag{3.15}$$

In fact, the proof of the final equality in (3.15) is completely algebraic, and reveals new determinantal identities that hold over any field $\mathbb{F}$ with at least N elements.

PROPOSITION 3.26 (Khare–Tao [**43**]). *Suppose $N \geq 1$ and*

$$0 \leq n_0 < \ldots < n_{N-1} < M$$

are integers, and $\mathbf{u}, \mathbf{v} \in \mathbb{F}^N$ each have distinct coordinates. Let $c_j \in \mathbb{F}^\times$ and define $h(t) := \sum_{j=0}^{N-1} c_j t^{n_j}$. Then $h[\mathbf{u}\mathbf{v}^T]$ is invertible, and

$$(\mathbf{v}^{\circ M})^T h[\mathbf{u}\mathbf{v}^T]^{-1} \mathbf{u}^{\circ M} = \sum_{j=0}^{N-1} \frac{\det \mathbf{u}^{\circ \mathbf{n}_j} \det \mathbf{v}^{\circ \mathbf{n}_j}}{c_j \det \mathbf{u}^{\circ \mathbf{n}} \det \mathbf{v}^{\circ \mathbf{n}}}.$$

The final result is a variant of the matrix-cube problem [**49**], and connects to spectrahedra [**13, 61**] and modern optimization theory. Given two or more real symmetric $N \times N$ matrices $A_0, \ldots, A_{M+1}$, the corresponding matrix cube of size $2\eta > 0$ is

$$\mathcal{U}[\eta] := \{A_0 + \sum_{m=1}^{M+1} u_m A_m : u_m \in [-\eta, \eta]\}.$$

The matrix-cube problem is to find the largest $\eta > 0$ such that $\mathcal{U}[\eta] \subset \mathcal{P}_N(\mathbb{R})$. In the present setting of the entrywise calculus, the above results imply asymptotically matching upper and lower bounds for the size of the matrix cube.

THEOREM 3.27 (see [**6, 43**]). *Suppose $M \geq 0$ and $0 \leq n_0 < n_1 < \cdots$ are integers. Fix positive scalars $\rho > 0$, $0 < \alpha_1 < \cdots < \alpha_{M+1}$, and $c_j > 0$ for all $j \geq 0$, and define for each $N \geq 1$ and each matrix $A \in \mathcal{P}_N([0, \rho])$, the cube*

$$\mathcal{U}_A[\eta] := \left\{ \sum_{j=0}^{N-1} c_j A^{\circ n_j} + \sum_{m=1}^{M+1} u_m A^{\circ (n_{N-1} + \alpha_m)} : u_m \in [-\eta, \eta] \right\}. \tag{3.16}$$

Also define for $N \geq 1$ and $\alpha > 0$:

$$\mathcal{K}_\alpha(N) := \sum_{j=0}^{N-1} \frac{V(\mathbf{n}_j(\alpha, N))^2}{V(\mathbf{n}(N))^2} \frac{\rho^{\alpha - n_j}}{c_j}, \tag{3.17}$$

where $\mathbf{n}(N) := (n_0, \ldots, n_{N-1})^T$, and

$$\mathbf{n}_j(\alpha, N) := (n_0, \ldots, n_{j-1}, n_{j+1}, \ldots, n_{N-1}, n_{N-1} + \alpha).$$

Then for each fixed $N \geq 1$, we have the uniform upper and lower bounds:

(3.18)

$$\begin{aligned} \eta \leq \left(\mathcal{K}_{\alpha_1}(N) + \cdots + \mathcal{K}_{\alpha_{M+1}}(N)\right)^{-1} &\implies \mathcal{U}_A[\eta] \subset \mathcal{P}_N \text{ for all } A \in \mathcal{P}_N([0, \rho]) \\ &\implies \eta \leq \mathcal{K}_{\alpha_{M+1}}(N)^{-1}. \end{aligned}$$

Moreover, if the n_j grow linearly, in that

$$\alpha_{M+1} - \alpha_M \geq n_{j+1} - n_j \qquad \text{for all } j \geq 0,$$

then the lower and upper bounds for $\eta = \eta_N$ in (3.18) are asymptotically equal as $N \to \infty$:

$$\lim_{N \to \infty} \mathcal{K}_{\alpha_{M+1}}(N)^{-1} \sum_{m=1}^{M+1} \mathcal{K}_{\alpha_m}(N) = 1.$$

3.8. Entrywise preservers of totally non-negative Hankel matrices. The first part of this survey discusses entrywise preservers of totally positive and totally non-negative matrices; these turn out to be very rigid in nature. If, instead, we consider the subfamily of totally non-negative matrices which are Hankel, then a richer class of preservers emerges, as well as a parallel story to that of entrywise positivity preservers on all matrices.

DEFINITION 3.28. A real matrix A is said to be *totally non-negative* or *totally positive* if every minor of A is non-negative or positive, respectively. We will denote these matrices, as well as the property, by TN and TP.

In the recent article [**22**] by Fallat, Johnson, and Sokal, the authors study when various classes of totally non-negative (TN) matrices are closed under taking sums or Schur products. As they observe, the set of all TN matrices is not closed under these operations; for example, the 3×3 identity matrix and the all-ones matrix $\mathbf{1}_{3\times 3}$ are both TN but their sum is not.

It is of interest to isolate a class of TN matrices that is a closed convex cone, and is furthermore closed under taking Schur products. Indeed, it is under these conditions that the observation of Pólya–Szegö (see Section 2.2) holds, leading to large classes of TN preservers.

Such a class of matrices has been identified in both the dimension-free as well as fixed-dimension settings. It consists of the *TN Hankel matrices*. In a fixed dimension, there is the following classical result from 1912.

LEMMA 3.29 (Fekete [**24**]). *Let A be a possibly rectangular real Hankel matrix such that all of its contiguous minors are positive. Then A is totally positive.*

Recall that a minor is said to be *contiguous* if it is obtained from successive rows and successive columns of A.

If A is a square Hankel matrix, let $A^{(1)}$ be the square submatrix of A obtained by removing the first row and the last column. Notice that every contiguous minor of A is a principal minor of either A or $A^{(1)}$. Combined with Fekete's lemma, these observations help show another folklore result.

THEOREM 3.30. *Let A be a square real Hankel matrix. Then A is TN or TP if and only if both A and $A^{(1)}$ are positive semidefinite or positive definite, respectively.*

Theorem 3.30 is a very useful bridge between matrix positivity and total non-negativity. A related dimension-free variant (see [**2, 28**]) concerns the Stieltjes moment problem: a sequence $(s_0, s_1, \ldots,)$ is the moment sequence of an admissible measure on $\mathbb{R}_+$ (see Definition 2.10) if and only if the Hankel matrices $H := (s_{j+k})_{j,k\geq 0}$ and $H^{(1)}$ (obtained by excising the first row of H, or equivalently, the first column) are both positive semidefinite. By Theorem 3.30, this is equivalent to saying that H is totally non-negative.

With Theorem 3.30 in hand, one can easily show several basic facts about TN Hankel matrices; we collect these in the following result for convenience.

LEMMA 3.31. *For an integer $N \geq 1$ and a set $I \subset \mathbb{R}_+$, let $HTN_N(I)$ denote the set of $N \times N$ TN Hankel matrices with entries in I. For brevity, we let $HTN_N := HTN_N(\mathbb{R}_+)$.*

(1) *The family HTN_N is closed under taking sums and non-negative scalar multiples, or more generally, integrals against non-negative measures (as long as these exist).*

(2) *In particular, if* μ *is an admissible measure supported on* $\mathbb{R}_+$*, then its moment matrix* $H_\mu := (s_{j+k}(\mu))_{j,k=0}^{\infty}$ *is totally non-negative.*
(3) HTN_N *is closed under taking entrywise products.*
(4) *If the power series* $f(x) = \sum_{k\geq 0} c_k x^k$ *is convergent on* $I \subset \mathbb{R}_+$*, with* $c_k \geq 0$ *for all* $k \geq 0$*, then the entrywise map* $f[-]$ *preserves total non-negativity on* $HTN_N(I)$*, for all* $N \geq 1$*.*

Given Lemma 3.31(4), which is identical to the start of the story for positivity preservers, it is natural to expect parallels between the two settings. This does in fact occur, in both the dimension-free and the fixed-dimension settings, and we now elaborate on both of these. For example, one can ask if a Schoenberg-type phenomenon also holds for preservers of total non-negativity on $\bigcup_{N\geq 1} HTN_N([0,\rho))$ with $0 < \rho \leq \infty$. This is indeed the case; we set $\rho = \infty$ for ease of exposition. From Theorem 2.12 and the subsequent discussion, it follows via Hamburger's theorem that the class of functions $\sum_{k\geq 0} c_k x^k$ with all $c_k \geq 0$ characterizes the entrywise maps preserving the set of moment sequences of admissible measures supported on $[-1,1]$. By the above discussion, in considering the family of matrices HTN_N for all $N \geq 1$, we are studying moment sequences of admissible measures supported on $I = \mathbb{R}_+$, or the related Hausdorff moment problem for $I = [0,1]$. In this case, one also has a Schoenberg-like characterization, outside of the origin.

THEOREM 3.32 (Belton–Guillot–Khare–Putinar [**7**]). *Let* $f : \mathbb{R}_+ \to \mathbb{R}$*. The following are equivalent.*

(1) *Applied entrywise, the map* f *preserves the set* HTN_N *for all* $N \geq 1$*.*
(2) *Applied entrywise, the map* f *preserves positive semidefiniteness on* HTN_N *for all* $N \geq 1$*.*
(3) *Applied entrywise, the map* f *preserves the set of moment sequences of admissible measures supported on* $\mathbb{R}_+$*.*
(4) *Applied entrywise, the map* f *preserves the set of moment sequences of admissible measures supported on* $[0,1]$*.*
(5) *The function* f *agrees on* $(0,\infty)$ *with an absolutely monotonic entire function, hence is non-decreasing, and* $0 \leq f(0) \leq \lim_{\epsilon\to 0^+} f(\epsilon)$*.*

REMARK 3.33. If we work only with $f : (0,\infty) \to \mathbb{R}$, then we are interested in matrices in HTN_N with positive entries. Since the only matrices in HTN_N with a zero entry are scalar multiples of the elementary square matrices E_{11} or E_{NN} (equivalently, the only admissible measures supported in $\mathbb{R}_+$ with a zero moment are of the form $c\delta_0$), the test set does not really reduce, and hence the preceding theorem still holds in essence: we must replace HTN_N by $HTN_N((0,\infty))$ in (1) and (2), reduce the class of admissible measures to those that are not of the form $c\delta_0$ in (3) and (4), and end (5) at 'entire function'. These five modified statements are, once again, equivalent, and provide further equivalent conditions to those of Vasudeva (Theorems 2.5 and 2.11).

In a similar vein, we now present the classification of sign patterns of polynomial or power-series functions that preserve TN entrywise in a fixed dimension on Hankel matrices. This too turns out to be exactly the same as for positivity preservers.

THEOREM 3.34 (Khare–Tao [**43**]). *Fix $\rho > 0$ and real exponents $n_0 < \cdots < n_{N-1} < M$. For any real coefficients $c_0, \ldots, c_{N-1}, c'$, let*

$$f(x) := \sum_{j=0}^{N-1} c_j x^{n_j} + c' x^M. \tag{3.19}$$

The following are equivalent.

(1) *The entrywise map $f[-]$ preserves TN on the rank-one matrices in $HTN_N((0,\rho))$.*
(2) *The entrywise map $f[-]$ preserves positivity on the rank-one matrices in $HTN_N((0,\rho))$.*
(3) *Either all the coefficients $c_0, \ldots, c_{N-1}, c'$ are non-negative, or $c_0, \ldots, c_{N-1}$ are positive and $c' \geq -\mathcal{C}^{-1}$, where*

$$\mathcal{C} = \sum_{j=0}^{N-1} \frac{V(\mathbf{n}_j)^2}{V(\mathbf{n})^2} \frac{\rho^{M-n_j}}{c_j}. \tag{3.20}$$

If $n_j \in \mathbb{Z}_+ \cup [N-2, \infty)$ for $j = 0, \ldots, N-1$, then conditions (1), (2) and (3) are further equivalent to the following.

(4) *The entrywise map $f[-]$ preserves TN on $HTN_N([0,\rho])$.*

In particular, this produces further equivalent conditions to Theorem 3.15. Notice that assertion (2) here is valid because the rank-one matrices used in proving Theorem 3.15 are of the form $c\mathbf{u}\mathbf{u}^T$, where $\mathbf{u} = (1, u_0, \ldots, u_0^{N-1})^T$, $u_0 \in (0,1)$, and $c \in (0,\rho)$, so that $c\mathbf{u}\mathbf{u}^T \in HTN_N((0,\rho))$.

The consequences of Theorem 3.15 also carry over for TN preservers. For instance, one can bound Laplace transforms analogously to Corollary 3.17, by replacing the words "positive semidefinite" by "totally non-negative" and the set $\mathcal{P}_N((0,\rho))$ by $HTN_N((0,\rho))$. Similarly, one can completely classify the sign patterns of power series that preserve TN entrywise on Hankel matrices of a fixed size:

THEOREM 3.35 (Khare–Tao [**43**]). *Theorems 3.5 and 3.19 hold upon replacing the phrase "preserves positivity entrywise on $\mathcal{P}_N((0,\rho))$" with "preserves TN entrywise on $HTN_N((0,\rho))$", for both $\rho < \infty$ and for $\rho = \infty$.*

We point the reader to [**43**, End of Section 9] for details.

To conclude, it is natural to seek a general result that relates the positivity preservers on $\mathcal{P}_N(I)$ and TN preservers on the set $HTN_N(I)$ for domains $I \subset \mathbb{R}_+$. Here is one variant which helps prove the above theorems, and which essentially follows from Theorem 3.30.

PROPOSITION 3.36 (Khare–Tao [**43**]). *Fix integers $1 \leq k \leq N$ and a scalar $0 < \rho \leq \infty$. Suppose $f : [0,\rho) \to \mathbb{R}$ is such that the entrywise map $f[-]$ preserves positivity on $\mathcal{P}_N^k([0,\rho))$, the set of matrices in $\mathcal{P}_N([0,\rho))$ with rank no more than k. Then $f[-]$ preserves total non-negativity on $HTN_N([0,\rho)) \cap \mathcal{P}_N^k([0,\rho))$.*

4. Power functions

A natural approach to tackle the problem of characterizing entrywise preservers in fixed dimension is to examine if some natural simple functions preserve positivity. One such family is the collection of power functions, $f(x) = x^\alpha$ for $\alpha > 0$.

Characterizing which fractional powers preserve positivity entrywise has recently received much attention in the literature. One of the first results in this area reads as follows.

THEOREM 4.1 (FitzGerald and Horn [**25**, Theorem 2.2]). *Let $N \geq 2$ and let $A = [a_{jk}] \in \mathcal{P}_N(\mathbb{R}_+)$. For any real number $\alpha \geq N-2$, the matrix $A^{\circ\alpha} := [a_{jk}^\alpha]$ is positive semidefinite. If $0 < \alpha < N-2$ and α is not an integer, then there exists a matrix $A \in \mathcal{P}_N\big((0,\infty)\big)$ such that $A^{\circ\alpha}$ is not positive semidefinite.*

Theorem 4.1 shows that every real power $\alpha \geq N-2$ entrywise preserves positivity, while no non-integers in $(0, N-2)$ do. This surprising "phase transition" phenomenon at the integer $N-2$ is referred to as the "critical exponent" for preserving positivity. Studying which powers entrywise preserve positivity is a very natural and interesting problem. It also often provides insights to determine which general functions preserve positivity. For example, Theorem 4.1 suggests that functions that entrywise preserve positivity on $\mathcal{P}_N$ should have a certain number of non-negative derivatives, which is indeed the case by Theorem 2.7.

OUTLINE OF THE PROOF. The first part of Theorem 4.1 relies on an ingenious idea that we now sketch. The result is obvious for $N = 2$. Let us assume it holds for some $N-1 \geq 2$, let $A \in \mathcal{P}_N(\mathbb{R}_+)$, and let $\alpha \geq N-2$. Write A in block form,

$$A = \begin{bmatrix} B & \xi \\ \xi^T & a_{NN} \end{bmatrix},$$

where B has dimension $(N-1)\times(N-1)$ and $\xi \in \mathbb{R}^{N-1}$. Assume without loss of generality that $a_{NN} \neq 0$ (as the case where $a_{NN} = 0$ follows from the induction hypothesis) and let $\zeta := (\xi^T, a_{NN})^T/\sqrt{a_{NN}}$. Then $A - \zeta\zeta^T = (B - \xi\xi^T)/a_{NN} \oplus 0$, where $(B-\xi\xi^T)/a_{NN}$ is the Schur complement of a_{NN} in A. Hence $A - \zeta\zeta^T$ is positive semidefinite. By the fundamental theorem of calculus, for any x, $y \in \mathbb{R}$,

$$x^\alpha = y^\alpha + \alpha \int_0^1 (x-y)(\lambda x + (1-\lambda)y)^{\alpha-1}\,\mathrm{d}\lambda.$$

Using the above expression entrywise, we obtain

$$A^{\circ\alpha} = \zeta^{\circ\alpha}(\zeta^{\circ\alpha})^T + \int_0^1 (A - \zeta\zeta^T) \circ (\lambda A + (1-\lambda)\zeta\zeta^T)^{\circ(\alpha-1)}\,\mathrm{d}\lambda.$$

Observe that the entries of the last row and column of the matrix $A - \zeta\zeta^T$ are all zero. Using the induction hypothesis and the Schur product theorem, it follows that the integrand is positive semidefinite, and therefore so is $A^{\circ\alpha}$.

The converse implication in Theorem 4.1 is shown by considering a matrix of the form $a\mathbf{1}_{N\times N} + t\mathbf{u}\mathbf{u}^T$, where a, $t > 0$, the coordinates of $\mathbf{u}$ are distinct, and $t1$ is small. Recall this is the exact same class of matrices that was useful in proving the Horn–Loewner theorem 2.7 as well as its strengthening in Theorem 2.9. The original proof, by FitzGerald and Horn [**25**], used $\mathbf{u} = (1, 2, \ldots, N)^T$, while a later proof by Fallat, Johnson and Sokal [**22**] used the same argument, now with $\mathbf{u} = (1, u_0, \ldots, u_0^{N-1})^T$; the motivation in [**22**] was to work with Hankel matrices, and the matrix $a\mathbf{1}_{N\times N} + t\mathbf{u}\mathbf{u}^T$ is indeed Hankel. That said, the argument of FitzGerald and Horn works more generally than both of these proofs, to show that, for any non-integral power $\alpha \in (0, N-2)$, $a > 0$, and vector $\mathbf{u} \in (0,\infty)^N$ with distinct coordinates, there exists $t > 0$ such that $(a\mathbf{1}_{N\times N} + t\mathbf{u}\mathbf{u}^T)^{\circ\alpha}$ is not positive semidefinite. $\square$

In her 2017 paper [**41**], Jain provided a remarkable strengthening of the result mentioned at the end of the previous proof, which removes the dependence on t entirely.

THEOREM 4.2 (Jain [**41**]). *Let*

$$A := [1 + u_j u_k]_{j,k=1}^N = \mathbf{1}_{N \times N} + \mathbf{u}\mathbf{u}^T,$$

where $N \geq 2$ and $\mathbf{u} = (u_1, \dots, u_N)^T \in (0,\infty)^N$ has distinct entries. Then $A^{\circ\alpha}$ is positive semidefinite for $\alpha \in \mathbb{R}$ if and only if $\alpha \in \mathbb{Z}_+ \cup [N-2, \infty)$.

Jain's result identifies a family of rank-two positive semidefinite matrices, every one of which encodes the classification of powers preserving positivity over all of $\mathcal{P}_N\big((0,\infty)\big)$. In a sense, her rank-two family is the culmination of previous work on positivity preserving powers for $\mathcal{P}_N\big((0,\infty)\big)$, since for rank-one matrices, every entrywise power preserves positivity: $(\mathbf{u}\mathbf{u}^T)^{\circ\alpha} = \mathbf{u}^{\circ\alpha}(\mathbf{u}^{\circ\alpha})^T$.

An immediate consequence of these results is the classification of the entrywise powers preserving positivity on the $N \times N$ TN Hankel matrices. Recall from the results in Section 3.8 (including Lemma 3.31(4)) that there is to be expected a strong correlation between this classification and the one in Theorem 4.1.

COROLLARY 4.3. *Given $N \geq 2$, the following are equivalent for an exponent $\alpha \in \mathbb{R}$.*

1. *The entrywise power function $x \mapsto x^\alpha$ preserves total non-negativity on HTN_N (see Lemma 3.31).*
2. *The entrywise map $x \mapsto x^\alpha$ preserves positivity on HTN_N.*
3. *The entrywise map $x \mapsto x^\alpha$ preserves positivity on the matrices in $HTN_N\big((0,\infty)\big)$ of rank at most two.*
4. *The exponent $\alpha \in \mathbb{Z}_+ \cup [N-2, \infty)$.*

PROOF. That (4) $\implies$ (2) and (2) $\implies$ (1) follow from Theorems 4.1 and 3.30, respectively. That (1) $\implies$ (2) and (2) $\implies$ (3) are obvious, and Jain's Theorem 4.2 shows that (3) $\implies$ (4). $\square$

A problem related to the above study of entrywise powers preserving positivity, is to characterize infinitely divisible matrices. This problem was also considered by Horn in [**40**]. Recall that a complex $N \times N$ matrix is said to be *infinitely divisible* if $A^{\circ\alpha} \in \mathcal{P}_N$ for all $\alpha \in \mathbb{R}_+$. Denote the *incidence matrix* of A by $M(A)$:

$$M(A)_{jk} = m_{jk} := \begin{cases} 0 & \text{if } a_{jk} = 0 \\ 1 & \text{otherwise.} \end{cases}$$

Also, let

$$L(A) := \{\mathbf{x} \in \mathbb{C}^N : \sum_{j,k=1}^N m_{jk} x_j \overline{x_k} = 0\},$$

and note that $L(A)$ is the kernel of $M(A)$ if $M(A)$ is positive semidefinite.

Assuming the arguments of the entries are chosen in a *consistent way* [**40**], we let

$$\log^\# A := M(A) \circ \log[A] = [\mu_{jk} \log a_{jk}]_{j,k=1}^N,$$

with the usual convention $0 \log 0 = 0$.

THEOREM 4.4 (Horn [**40**, Theorem 1.4]). *An $N \times N$ matrix A is infinitely divisible if and only if (a) A is Hermitian, with $a_{jj} \geq 0$ for all j, (b) $M(A) \in \mathcal{P}_N$, and (c) $\log^{\#} A$ is positive semidefinite on $L(A)$.*

4.1. Sparsity constraints. Theorem 4.1 was recently extended to more structured matrices. Given $I \subset \mathbb{R}$ and a graph $G = (V, E)$ on the finite vertex set $V = \{1, \ldots, N\}$, we define the cone of positive-semidefinite matrices with *zeros according to* G:

$$\mathcal{P}_G(I) := \{A = [a_{jk}] \in \mathcal{P}_N(I) : a_{jk} = 0 \text{ if } (j,k) \notin E \text{ and } i \neq j\}. \tag{4.1}$$

Note that if $(j,k) \in E$, then the entry a_{jk} is unconstrained; in particular, it is allowed to be 0. Consequently, the cone $\mathcal{P}_G := \mathcal{P}_G(\mathbb{R})$ is a closed subset of $\mathcal{P}_N$.

A natural refinement of Theorem 4.1 involves studying powers that entrywise preserve positivity on $\mathcal{P}_G$. In that case, the flavor of the problem changes significantly, with the discrete structure of the graph playing a prominent role.

DEFINITION 4.5 (Guillot–Khare–Rajaratnam [**30**]). Given a simple graph $G = (V, E)$, let

$$\mathcal{H}_G := \{\alpha \in \mathbb{R} : A^{\circ\alpha} \in \mathcal{P}_G \text{ for all } A \in \mathcal{P}_G(\mathbb{R}_+)\}. \tag{4.2}$$

Define the *Hadamard critical exponent of* G to be

$$CE(G) := \min\{\alpha \in \mathbb{R} : [\alpha, \infty) \subset \mathcal{H}_G\}. \tag{4.3}$$

Notice that, by Theorem 4.1, for every graph $G = (V, E)$, the critical exponent $CE(G)$ exists, and lies in $[\omega(G) - 2, |V| - 2]$, where $\omega(G)$ is the size of the largest complete subgraph of G, that is, the clique number. To compute such critical exponents is natural and highly non-trivial.

FitzGerald and Horn proved that $CE(K_n) = n - 2$ for all $n \geq 2$ (Theorem 4.1), while it follows from [**31**, Proposition 4.2] that $CE(T) = 1$ for every tree T. For a general graph, it is not *a priori* clear what the critical exponent is or how to compute it. A natural family of graphs that encompasses both complete graphs and trees is that of chordal graphs. Recall that a graph is *chordal* if it does not contain an induced cycle of length 4 or more. Chordal graphs feature extensively in many areas, such as the theory of graphical models [**46**], and in problems involving positive-definite completions (see [**59**]). Examples of important chordal graphs include trees, complete graphs, Apollonian graphs, band graphs, and split graphs.

Recently, Guillot, Khare, and Rajaratnam [**30**] were able to compute the complete set of entrywise powers preserving positivity on $\mathcal{P}_G$ for all chordal graphs G. Here, the critical exponent can be described purely combinatorially.

THEOREM 4.6 (Guillot–Khare–Rajaratnam [**30**]). *Let $K_r^{(1)}$ denote the complete graph with one edge removed, and let G be a finite simple connected chordal graph. The critical exponent for entrywise powers preserving positivity on $\mathcal{P}_G$ is $r - 2$, where r is the largest integer such that K_r or $K_r^{(1)}$ is an induced subgraph of G. More precisely, the set of entrywise powers preserving $\mathcal{P}_G$ is $\mathcal{H}_G = \mathbb{Z}_+ \cup [r-2, \infty)$, with r as before.*

The set of entrywise powers preserving positivity was also computed in [**30**] for cycles and bipartite graphs.

THEOREM 4.7 (Guillot–Khare–Rajaratnam [**30**]). *The critical exponent of cycles and bipartite graphs is* 1.

Surprisingly, the critical exponent does not depend on the size of the graph for cycles and bipartite graphs. In particular, it is striking that any power greater than 1 preserves positivity for families of dense graphs such as bipartite graphs. Such a result is in sharp contrast to the general case, where there is no underlying structure of zeros. That small powers can preserve positivity is important for applications, since such entrywise procedures are often used to regularize positive definite matrices, such as covariance or correlation matrices, where the goal is to minimally modify the entries of the original matrix (see [**47**, **63**] and Chapter 5 below).

For a general graph, the problem of computing the set $\mathcal{H}_G$ or the critical exponent $CE(G)$ remains open. We now outline some other natural open problems in the area.

Problems.

(1) In every currently known case (Theorems 4.6, 4.7), $CE(G)$ is equal to $r-2$, where r is the largest integer such that K_r or $K_r^{(1)}$ is an induced subgraph of G. Is the same true for every graph G?
(2) Is $CE(G)$ always an integer? Can this be proved without computing $CE(G)$ explicitly?
(3) Recall that every chordal graph is perfect. Can the critical exponent be calculated for other broad families of graphs such as the family of perfect graphs?

4.2. Rank constraints and other Loewner properties. Another approach to generalize Theorem 4.1 is to examine other properties of entrywise functions such as monotonicity, convexity, and super-additivity (with respect to the Loewner semidefinite ordering) [**29**, **38**]. Given a set $V \subset \mathcal{P}_N(I)$, recall that a function $f : I \to \mathbb{R}$ is

- *positive on* V with respect to the Loewner ordering if $f[A] \geq 0$ for all $0 \leq A \in V$;
- *monotone on* V with respect to the Loewner ordering if $f[A] \geq f[B]$ for all A, $B \in V$ such that $A \geq B \geq 0$;
- *convex on* V with respect to the Loewner ordering if
$$f[\lambda A + (1-\lambda)B] \leq \lambda f[A] + (1-\lambda)f[B]$$
for all $\lambda \in [0,1]$ and all A, $B \in V$ such that $A \geq B \geq 0$;
- *super-additive on* V with respect to the Loewner ordering if
$$f[A+B] \geq f[A] + f[B]$$
for all A, $B \in V$ for which $f[A+B]$ is defined.

The following relations between the first three notions were obtained by Hiai.

THEOREM 4.8 (Hiai [**38**, Theorem 3.2]). *Let $I = (-\rho, \rho)$ for some $\rho > 0$.*

(1) *For each $N \geq 3$, the function f is monotone on $\mathcal{P}_N(I)$ if and only if f is differentiable on I and f' is positive on $\mathcal{P}_N(I)$.*
(2) *For each $N \geq 2$, the function f is convex on $\mathcal{P}_N(I)$ if and only if f is differentiable on I and f' is monotone on $\mathcal{P}_N(I)$.*

Power functions satisfying any of the above four properties have been characterized by various authors. In recent work, Hiai [**38**] has extended Theorem 4.1 by considering the odd and even extensions of the power functions to $\mathbb{R}$. For $\alpha > 0$, the even and odd extensions to $\mathbb{R}$ of the power function $f_\alpha(x) := x^\alpha$ are defined to be $\phi_\alpha(x) := |x|^\alpha$ and $\psi_\alpha(x) := \operatorname{sign}(x)|x|^\alpha$. The first study of powers $\alpha > 0$ for which ϕ_α preserves positivity entrywise on $\mathcal{P}_N(\mathbb{R})$ was carried out by Bhatia and Elsner [**10**]. Subsequently, Hiai studied the power functions ϕ_α and ψ_α that preserve Loewner positivity, monotonicity, and convexity entrywise, and showed for positivity preservers that the same phase transition occurs at $n-2$ for ϕ_α and ψ_α, as demonstrated in [**25**]. The work was generalized in [**29**] to matrices satisfying rank constraints.

DEFINITION 4.9. Fix non-negative integers $n \geq 2$ and $n \geq k$, and a set $I \subset \mathbb{R}$. Let $\mathcal{P}_n^k(I)$ denote the subset of matrices in $\mathcal{P}_n(I)$ that have rank at most k, and let

$$\begin{aligned} \mathcal{H}_{\text{pos}}(n,k) &:= \{\alpha > 0 : x^\alpha \text{ preserves positivity on } \mathcal{P}_n^k(\mathbb{R}_+)\}, \\ \mathcal{H}^\phi_{\text{pos}}(n,k) &:= \{\alpha > 0 : \phi_\alpha \text{ preserves positivity on } \mathcal{P}_n^k(\mathbb{R})\}, \\ \mathcal{H}^\psi_{\text{pos}}(n,k) &:= \{\alpha > 0 : \psi_\alpha \text{ preserves positivity on } \mathcal{P}_n^k(\mathbb{R})\}. \end{aligned} \tag{4.4}$$

Similarly, let $\mathcal{H}_J(n,k)$, $\mathcal{H}^\phi_J(n,k)$ and $\mathcal{H}^\psi_J(n,k)$ denote sets of the entrywise powers preserving Loewner properties on $\mathcal{P}_n^k(\mathbb{R}_+)$ or $\mathcal{P}_n^k(\mathbb{R})$, where

$$J \in \{\text{monotonicity, convexity, super-additivity}\}.$$

The set of entrywise powers preserving the above notions are given in the table below (see [**29**, Theorem 1.2]).

J	$\mathcal{H}_J(n,k)$	$\mathcal{H}^\phi_J(n,k)$	$\mathcal{H}^\psi_J(n,k)$
Positivity			
$k=1$	$\mathbb{R}$ G–K–R	$\mathbb{R}$ G–K–R	$\mathbb{R}$ G–K–R
$2 \leq k \leq n$	$\mathbb{N} \cup [n-2,\infty)$ FitzGerald–Horn	$2\mathbb{N} \cup [n-2,\infty)$ Hiai, Bhatia–Elsner, G–K–R	$(-1+2\mathbb{N}) \cup [n-2,\infty)$ Hiai, G–K–R
Monotonicity			
$k=1$	$\mathbb{R}_+$ G–K–R	$\mathbb{R}_+$ G–K–R	$\mathbb{R}_+$ G–K–R
$2 \leq k \leq n$	$\mathbb{N} \cup [n-1,\infty)$ FitzGerald–Horn	$2\mathbb{N} \cup [n-1,\infty)$ Hiai, G–K–R	$(-1+2\mathbb{N}) \cup [n-1,\infty)$ Hiai, G–K–R
Convexity			
$k=1$	$[1,\infty)$ G–K–R	$[1,\infty)$ G–K–R	$[1,\infty)$ G–K–R
$2 \leq k \leq n$	$\mathbb{N} \cup [n,\infty)$ Hiai, G–K–R	$2\mathbb{N} \cup [n,\infty)$ Hiai, G–K–R	$(-1+2\mathbb{N}) \cup [n,\infty)$ Hiai, G–K–R
Super-additivity			
$1 \leq k \leq n$	$\mathbb{N} \cup [n,\infty)$ G–K–R	$2\mathbb{N} \cup [n,\infty)$ G–K–R	$(-1+2\mathbb{N}) \cup [n,\infty)$ G–K–R

TABLE 1. Summary of real Hadamard powers preserving Loewner properties, with additional rank constraints. See Bhatia–Elsner [**10**], FitzGerald–Horn [**25**], Guillot–Khare–Rajaratnam [**29**], and Hiai [**38**].

5. Motivation from statistics

The study of entrywise functions preserving positivity has recently attracted renewed attraction due to its importance in the estimation and regularization of covariance/correlation matrices. Recall that the covariance between two random variables X_j and X_k is given by

$$\sigma_{jk} = \operatorname{Cov}(X_j, X_k) = E\big[(X_j - E[X_j])(X_k - E[X_k])\big],$$

where $E[X_j]$ denotes the expectation of X_j. In particular, $\operatorname{Cov}(X_j, X_j) = \operatorname{Var}(X_j)$, the variance of X_j. The *covariance matrix* of a random vector $\mathbf{X} := (X_1, \dots, X_m)$, is the matrix $\Sigma := [\operatorname{Cov}(X_j, X_k)]_{j,k=1}^m$. Covariance matrices are a fundamental tool that measure linear dependencies between random variables. In order to discover relations between variables in data, statisticians and applied scientists need to obtain estimates of the covariance matrix Σ from observations $\mathbf{x}_1, \dots, \mathbf{x}_n \in \mathbb{R}^m$ of $\mathbf{X}$. A traditional estimator of Σ is the *sample covariance matrix* S given by

$$S = [s_{jk}]_{j,k=1}^m = \frac{1}{n-1}\sum_{i=1}^n (\mathbf{x}_i - \overline{\mathbf{x}})(\mathbf{x}_i - \overline{\mathbf{x}})^T, \tag{5.1}$$

where $\overline{\mathbf{x}} := \frac{1}{n}\sum_{i=1}^n \mathbf{x}_i$ is the average of the observations. In the case where the random vector $\mathbf{X}$ has a multivariate normal distribution with mean μ and covariance matrix Σ, one can show that $\overline{\mathbf{x}}$ and $\frac{n-1}{n}S$ are the maximum likelihood estimators of μ and Σ, respectively [**3**, Chapter 3]. It is not difficult to show that S is an unbiased estimator of Σ. More generally, under weak assumptions, one can show that the distribution of $\sqrt{n}(S - \Sigma)$ is asymptotically normal as $n \to \infty$. The exact description of the limiting distribution depends on the moments and the cumulants of $\mathbf{X}$ (see [**12**, Chapter 6.3]). For example, in the two-dimensional case, we have the following result.

Let $N_m(\mu, \Sigma)$ denote the m-dimensional normal distribution with mean μ and covariance matrix Σ.

PROPOSITION 5.1 (see [**12**, Example 6.4]). *Let $\mathbf{x}_1, \dots, \mathbf{x}_n \in \mathbb{R}^2$ be an independent and identically distributed sample from a bivariate vector $\mathbf{X} = (X_1, X_2)$ with mean $\mu = (\mu_1, \mu_2)$ and finite fourth-order moments, and let S be as in Equation (5.1). Then*

$$\sqrt{n}\left[\begin{bmatrix} s_1^2 \\ s_{12} \\ s_2^2 \end{bmatrix} - \begin{bmatrix} \sigma_1^2 \\ \sigma_{12} \\ \sigma_2^2 \end{bmatrix}\right] \xrightarrow{\mathrm{d}} N_3(\mathbf{0}, \Omega),$$

where Ω is the symmetric 3×3 matrix

$$\Omega = \begin{bmatrix} \mu_4^1 - (\mu_2^1)^2 & \mu_{31}^{12} - \mu_{11}^{12}\mu_2^1 & \mu_{22}^{12} - \mu_2^1\mu_2^2 \\ \mu_{31}^{12} - \mu_{11}^{12}\mu_2^1 & \mu_{22}^{12} - (\mu_{11}^{12})^2 & \mu_{13}^{12} - \mu_{11}^{12}\mu_2^2 \\ \mu_{22}^{12} - \mu_2^1\mu_2^2 & \mu_{31}^{12} - \mu_{11}^{12}\mu_2^1 & \mu_4^2 - (\mu_2^2)^2 \end{bmatrix},$$

and $\mu_k^i = E[(X_i - \mu_i)^k]$ and $\mu_{kl}^{ij} = E[(X_i - \mu_i)^k(X_j - \mu_j)^l]$.

In traditional statistics, one usually assumes the number of samples n is large enough for asymptotic results such as the one above to apply. In covariance estimation, one typically requires a sample size at least a few times the number of variables m for that to apply. In such a case, the sample covariance matrix provides a good approximation of the true covariance matrix Σ. However, this ideal setting is rarely seen nowadays. Indeed, our systematic and automated way of collecting data today yields datasets where the number of variables is often orders of magnitude larger than the number of instances available for study [**19**]. Classical statistical methods were not designed and are not suitable for analyzing data in such settings. Developing new methodologies that are adapted to modern high-dimensional problems is the object of active research. In the case of covariance estimation, several strategies have been proposed to replace the traditional sample covariance matrix estimator S. These approaches typically leverage low-dimensional structures in the data (low rank, sparsity, ...) to obtain reasonable covariance estimates, even when the sample size is small compared to the dimension of the problem (see [**52**] for a detailed description of such techniques). One such approach involves applying functions to the entries of sample covariance matrices to improve their properties (see [**5, 11, 21, 35, 36, 47, 53, 63**]). For example, *hard thresholding* a matrix entails setting to zero the entries of the matrix that are smaller in absolute value than a prescribed value $\epsilon > 0$. Letting

$$f_\epsilon^H(x) = \begin{cases} x & \text{if } |x| > \epsilon, \\ 0 & \text{otherwise,} \end{cases} \tag{5.2}$$

hard thresholding is equivalent to applying the function f_ϵ^H entrywise to the entries of the matrix. Another popular example that was first studied in the context of wavelet shrinkage [**20**] is *soft thresholding*, where f_ϵ^H is replaced by

$$f_\epsilon^S : x \mapsto \operatorname{sign}(x)\big(|x| - \epsilon\big)_+ \qquad \text{with } y_+ := \max\{y, 0\}.$$

Soft thresholding not only sets small entries to zero, it also shrinks all the other entries continuously towards zero. Several other thresholding and shrinkage procedures were also recently proposed in the context of covariance estimation (see [**23**] and the references therein).

Compared to other techniques, the above procedure has several advantages. Firstly, the resulting estimators are often significantly more precise than the sample covariance matrices. Secondly, applying a function to the entries of a matrix is very simple and not computationally intensive. The procedure can therefore be performed in very high dimensions and in real-time applications. This is in contrast to several other techniques that require solving optimization problems and often become too intensive to be used in modern applications. A downside of the entrywise calculus, however, is that the positive definiteness of the resulting matrices is not guaranteed. As the parameter space of covariance matrices is the cone of positive definite matrices, it is critical that the resulting matrices be positive definite for the technique to be useful and widely applicable. The problem of characterizing positivity preservers thus has an immediate impact in the area of covariance estimation by providing useful functions that can be applied entrywise to covariance estimates in order to regularize them.

Several characterizations of when thresholding procedures preserve positivity have recently been obtained.

5.1. Thresholding with respect to a graph. In [**33**], the concept of thresholding with respect to a graph was examined. In this context, the elements to threshold are encoded in a graph $G = (V, E)$ with $V = \{1, \ldots, p\}$. If $A = (a_{jk})$ is a $p \times p$ matrix, we denote by A_G the matrix with entries

$$(A_G)_{jk} = \begin{cases} a_{jk} & \text{if } (j,k) \in E \text{ or } j = k, \\ 0 & \text{otherwise.} \end{cases}$$

We say that A_G is the matrix obtain by thresholding A with respect to the graph G. The main result of [**33**] characterizes the graphs G for which the corresponding thresholding procedure preserves positivity. Denote by $\mathcal{P}_N^+$ the set of real symmetric $N \times N$ positive definite matrices and by $\mathcal{P}_G^+$ the subset of positive definite matrices contained in $\mathcal{P}_G$ (see Equation 4.1).

THEOREM 5.2 (Guillot–Rajaratnam [**33**, Theorem 3.1]). *The following are equivalent:*

(1) $A_G \in \mathcal{P}_N^+$ *for all* $A \in \mathcal{P}_N^+$*;*
(2) $G = \bigcup_{i=1}^d G_i$*, where* $G_1, \ldots, G_d$ *are disconnected and complete components of* G*.*

The implication (2) $\implies$ (1) of the theorem is intuitive and straightforward, since principal submatrices of positive definite matrices are positive definite. That (1) $\implies$ (2) may come as a surprise though, and shows that indiscriminate or arbitrary thresholding of a positive definite matrix can quickly lead to loss of positive definiteness.

Theorem 5.2 also generalizes to matrices that already have zero entries. In that case, the characterization of the positivity preservers remains essentially the same.

THEOREM 5.3 (Guillot–Rajaratnam [**33**, Theorem 3.3]). *Let* $G = (V, E)$ *be an undirected graph and let* $H = (V, E')$ *be a subgraph of* G*, so that* $E' \subset E$*. Then* A_H *is positive definite for every* $A \in \mathcal{P}_G^+$ *if and only if* $H = G_1 \cup \cdots \cup G_k$*, where* $G_1, \ldots, G_k$ *are disconnected induced subgraphs of* G*.*

5.2. Hard and soft thresholding. Theorems 5.2 and 5.3 address the case where positive definite matrices are thresholded with respect to a given pattern of entries, regardless of the magnitude of the entries of the original matrix. The more natural case where the entries are hard or soft thresholded was studied in [**33, 34**]. In applications, it is uncommon to threshold the diagonal entries of estimated covariance matrices, as the diagonal contains the variance of the underlying variables. Hence, for a given function $f : \mathbb{R} \to \mathbb{R}$ and a real matrix $A = [a_{jk}]$, we let the matrix $f^*[A]$ be defined by setting

$$f^*[A]_{jk} := \begin{cases} f(a_{jk}) & \text{if } j \neq k, \\ a_{jk} & \text{otherwise.} \end{cases}$$

THEOREM 5.4 (Guillot–Rajaratnam [**33**, Theorem 3.6]). *Let* G *be a connected undirected graph with* $n \geq 3$ *vertices. The following are equivalent.*

(1) *There exists* $\epsilon > 0$ *such that, for every* $A \in \mathcal{P}_G^+$*, we have* $(f_\epsilon^H)^*[A] \in \mathcal{P}_n^+$*.*
(2) *For every* $\epsilon > 0$ *and every* $A \in \mathcal{P}_G^+$*, we have* $f_\epsilon^H[A] \in \mathcal{P}_n^+$*.*
(3) G *is a tree.*

The case of soft thresholding was considered in [**34**]. Surprisingly, the characterization of the thresholding levels that preserve positivity is exactly the same as in the case of hard thresholding.

THEOREM 5.5 (Guillot–Rajaratnam [**34**, Theorem 3.2]). *Let $G = (V, E)$ be a connected graph with $n \geq 3$ vertices. Then the following are equivalent:*

(1) *There exists $\epsilon > 0$ such that for every $A \in \mathcal{P}_G^+$, we have $(f_\epsilon^S)^*[A] \in \mathcal{P}_n^+$.*
(2) *For every $\epsilon > 0$ and every $A \in \mathcal{P}_G^+$, we have $f_\epsilon^S[A] \in \mathcal{P}_n^+$.*
(3) *G is a tree.*

An extension of Schoenberg's theorem (Theorem 2.3) to the case where the function f is only applied to the off-diagonal entries of the matrix was also obtained in [**34**].

THEOREM 5.6 (Guillot–Rajaratnam [**34**, Theorem 4.21]). *Let $0 < \rho \leq \infty$ and $f : (-\rho, \rho) \to \mathbb{R}$. The matrix $f^*[A]$ is positive semidefinite for all $A \in \mathcal{P}_n\big((-\rho, \rho)\big)$ and all $n \geq 1$ if and only if $f(x) = xg(x)$, where*

(1) *g is analytic on the disc $D(0, \rho)$;*
(2) *$\|g\|_\infty \leq 1$;*
(3) *g is absolutely monotonic on $(0, \rho)$.*

When $\rho = \infty$, the only functions satisfying the above conditions are the affine functions $f(x) = ax$ for $0 \leq a \leq 1$.

5.3. Rank and sparsity constraints. An explicit and useful characterization of entrywise functions preserving positivity on $\mathcal{P}_N$ for a fixed N still remains out of reach as of today. Motivated by applications in statistics, the authors in [**31, 32**] examined the cases where the matrices in $\mathcal{P}_N$ satisfy supplementary rank and sparsity constraints that are common in applications.

Observe that the sample covariance matrix (Equation (5.1)) has rank at most n, where n is the number of samples used to compute it. Moreover, as explained at the start of this Chapter, it is common in modern applications that n is much smaller than the dimension p. Hence, when studying the regularization approach described above, it is natural to consider positive semidefinite matrices with bounded rank.

An immediate application of Schoenberg's theorem on spheres (see Equation (2.2)) provides a characterization of entrywise positivity preservers of correlation matrices of all dimensions, with rank bounded by n. Recall that a correlation matrix is the covariance matrix of a random vector where each variable has variance 1, so is a positive semidefinite matrix with diagonal entries equal to 1. As in Equation (2.2), we denote the ultraspherical orthogonal polynomials by $P_k^{(\lambda)}$.

THEOREM 5.7 (Reformulation of [**57**, Theorem 1]). *Let $n \in \mathbb{N}$ and let $f : [-1, 1] \to \mathbb{R}$. The following are equivalent.*

(1) *$f[A] \in \mathcal{P}_N$ for all correlation matrices $A \in \mathcal{P}_N\big([-1, 1]\big)$ with rank no more than n and all $N \geq 1$.*
(2) *$f(x) = \sum_{j=0}^\infty a_j P_j^{(\lambda)}(x)$ with $a_j \geq 0$ for all $j \geq 0$ and $\lambda = (n-1)/2$.*

PROOF. The result follows from [**57**, Theorem 1] and the observation that correlation matrices of rank at most n are in correspondence with Gram matrices of vectors in S^{n-1}. □

In order to approach the case of matrices of a fixed dimension, we introduce some notation.

DEFINITION 5.8. Let $I \subset \mathbb{R}$. Define $\mathcal{S}_n(I)$ to be the set of $n \times n$ symmetric matrices with entries in I. Let $\operatorname{rank} A$ denote the rank of a matrix A. We define:

$$\mathcal{S}_n^k(I) := \{A \in \mathcal{S}_n(I) : \operatorname{rank} A \leq k\},$$
$$\mathcal{P}_n^k(I) := \{A \in \mathcal{P}_n(I) : \operatorname{rank} A \leq k\}.$$

The main result in [**32**] provides a characterization of entrywise functions mapping $\mathcal{P}_n^l$ into $\mathcal{P}_n^k$.

THEOREM 5.9 (Guillot–Khare–Rajaratnam [**32**, Theorem B]). *Let $0 < R \leq \infty$ and $I = [0, R)$ or $(-R, R)$. Fix integers $n \geq 2$, $1 \leq k < n-1$, and $2 \leq l \leq n$. Suppose $f \in C^k(I)$. The following are equivalent.*

(1) *$f[A] \in \mathcal{S}_n^k$ for all $A \in \mathcal{P}_n^l(I)$;*
(2) *$f(x) = \sum_{t=1}^r c_t x^{i_t}$ for some $c_t \in \mathbb{R}$ and some $i_t \in \mathbb{N}$ such that*

$$\sum_{t=1}^{r} \binom{i_t + l - 1}{l-1} \leq k. \tag{5.3}$$

Similarly, $f[-] : \mathcal{P}_n^l(I) \to \mathcal{P}_n^k$ if and only if f satisfies (2) and $c_t \geq 0$ for all t. Moreover, if $I = [0, R)$ and $k \leq n-3$, then the assumption that $f \in C^k(I)$ is not required.

Notice that Theorem 5.9 is a fixed-dimension result with rank constraints. This may be considered a refinement of a similar, dimension-free result with rank constraints shown in [**4**], in which the authors arrive at the same conclusion as in part (2) above. We compare the two settings: in [**4**], (a) the hypotheses held for all dimensions N rather than in a fixed dimension; (b) the test matrices were a larger set in each dimension, compared to just the positive matrices considered in Theorem 5.9; (c) the test matrices did not consist only of rank-one matrices, similar to Theorem 5.9; and (d) the test functions f in the dimension-free case were assumed to be measurable, rather than C^k as in the fixed-dimension case. Thus, Theorem 5.9 is (a refinement of) the fixed-dimension case of the first main result in [**4**].[7]

The (2) $\implies$ (1) implication in Theorem 5.9 is clear. Indeed, let $i \geq 0$ and $A = \sum_{j=1}^l u_j u_j^T \in \mathcal{P}_n^l(I)$. Then

$$A^{\circ i} = \sum_{m_1 + \cdots + m_l = i} \binom{i}{m_1, \ldots, m_l} \mathbf{w_m} \mathbf{w_m}^T \qquad \text{where } \mathbf{w_m} := u_1^{\circ m_1} \circ \cdots \circ u_l^{\circ m_l}$$

and $\binom{i}{m_1, \ldots, m_l}$ is a multinomial coefficient. Note that there are exactly $\binom{i+l-1}{l-1}$ terms in the previous summation. Therefore $\operatorname{rank} A^{\circ i} \leq \binom{i+l-1}{l-1}$, and so (1) easily follows from (2). The proof that (1) $\implies$ (2) is much more challenging; see [**32**] for details.

In [**31**], the authors focus on the case where sparsity constraints are imposed to the matrices instead of rank constraints. Positive semidefinite matrices with

[7] We also point out the second main result in *loc. cit.*, that is, [**4**, Theorem 2], which classifies all continuous entrywise maps $f : \mathbb{C} \to \mathbb{C}$ that obey similar rank constraints in all dimensions. Such maps are necessarily of the form $g(z) = \sum_{j=1}^p \beta_j z^{m_j} (\overline{z})^{n_j}$, where the exponents m_j and n_j are non-negative integers. This should immediately remind the reader of Rudin's conjecture in the 'dimension-free' case, and its resolution by Herz; see Theorem 2.6.

zeros according to graphs arise naturally in many applications. For example, in the theory of Markov random fields in probability theory ([**46**, **62**]), the nodes of a graph G represent components of a random vector, and edges represent the dependency structure between nodes. Thus, absence of an edge implies marginal or conditional independence between the corresponding random variables, and leads to zeros in the associated covariance or correlation matrix (or its inverse). Such models therefore yield parsimonious representations of dependency structures. Characterizing entrywise functions preserving positivity for matrices with zeros according to a graph is thus of tremendous interest for modern applications. Obtaining such characterizations is, however, much more involved than the original problem considered by Schoenberg as one has to enforce and maintain the sparsity constraint. The problem of characterizing functions preserving positivity for sparse matrices is also intimately linked to problems in spectral graph theory and many other problems (see e.g. [**1**, **17**, **39**, **50**]).

As before, for a given graph $G = (V, E)$ on the finite vertex set $V = \{1, \dots, N\}$, we denote by $\mathcal{P}_G(I)$ the set of positive-semidefinite matrices with entries in I and zeros according to G, as in (4.1). Given a function $f : \mathbb{R} \to \mathbb{R}$ and $A \in \mathcal{S}_{|G|}(\mathbb{R})$, denote by $f_G[A]$ the matrix such that

$$f_G[A]_{jk} := \begin{cases} f(a_{jk}) & \text{if } (j,k) \in E \text{ or } j = k, \\ 0 & \text{otherwise.} \end{cases}$$

The first main result in [**31**] is an explicit characterization of the entrywise positive preservers of $\mathcal{P}_G$ for any collection of trees (other than copies of K_2). Following Vasudeva's classification for $\mathcal{P}_{K_2}$ in Theorem 3.1, trees are the only other graphs for which such a classification is currently known.

THEOREM 5.10 (Guillot–Khare–Rajaratnam [**31**, Theorem A]). *Suppose $I = [0, R)$ for some $0 < R \leq \infty$, and $f : I \to \mathbb{R}_+$. Let G be a tree with at least 3 vertices, and let A_3 denote the path graph on 3 vertices. The following are equivalent.*

(1) *$f_G[A] \in \mathcal{P}_G$ for every $A \in \mathcal{P}_G(I)$;*
(2) *$f_T[A] \in \mathcal{P}_T$ for all trees T and all matrices $A \in \mathcal{P}_T(I)$;*
(3) *$f_{A_3}[A] \in \mathcal{P}_{A_3}$ for every $A \in \mathcal{P}_{A_3}(I)$;*
(4) *The function f satisfies*

$$f\left(\sqrt{xy}\right)^2 \leq f(x)f(y) \qquad \text{for all } x, y \in I \tag{5.4}$$

and is super-additive on I, that is,

$$f(x+y) \geq f(x) + f(y) \qquad \text{whenever } x, y, x+y \in I. \tag{5.5}$$

The implication (4) $\implies$ (1) was further extended to all chordal graphs: it is the following result with $c = 2$ and $d = 1$.

THEOREM 5.11 (Guillot–Khare–Rajaratnam [**30**]). *Let G be a chordal graph with a perfect elimination ordering of its vertices $\{v_1, \dots, v_n\}$. For all $1 \leq k \leq n$, denote by G_k the induced subgraph on G formed by $\{v_1, \dots, v_k\}$, so that the neighbors of v_k in G_k form a clique. Define $c = \omega(G)$ to be the clique number of G, and let*

$$d := \max\{\deg_{G_k}(v_k) : k = 1, \dots, n\}.$$

If $f : \mathbb{R} \to \mathbb{R}$ is any function such that $f[-]$ preserves positivity on $\mathcal{P}_c^1(\mathbb{R})$ and $f[M + N] \geq f[M] + f[N]$ for all $M \in \mathcal{P}_d$ and $N \in \mathcal{P}_d^1$, then $f[-]$ preserves positivity on $\mathcal{P}_G(\mathbb{R})$. [Here, $\mathcal{P}_d^1$ denotes the matrices in $\mathcal{P}_d$ of rank at most one.]

See [**30**] for other sufficient conditions for a general entrywise function to preserve positivity on $\mathcal{P}_G$ for G chordal.

To state the final result in this section, recall that Schoenberg's theorem (Theorem 2.3) shows that entrywise functions preserving positivity for all matrices (that is, according to the family of complete graphs K_n for $n \geq 1$) are absolutely monotonic on the positive axis. It is not clear if functions satisfying (5.4) and (5.5) in Theorem 5.10 are necessarily absolutely monotonic, or even analytic. As shown in [**31**, Proposition 4.2], the critical exponent (see Definition 4.5) of every tree is 1. Hence, functions satisfying (5.4) and (5.5) do not need to be analytic. The second main result in [**31**] demonstrates that even if the function is analytic, it can in fact have arbitrarily long strings of negative Taylor coefficients.

THEOREM 5.12 (Guillot–Khare–Rajaratnam [**31**, Theorem B]). *There exists an entire function $f(z) = \sum_{n=0}^{\infty} a_n z^n$ such that*

(1) *$a_n \in [-1, 1]$ for every $n \geq 0$;*
(2) *The sequence $(a_n)_{n\geq 0}$ contains arbitrarily long strings of negative numbers;*
(3) *For every tree G, $f_G[A] \in \mathcal{P}_G$ for every $A \in \mathcal{P}_G(\mathbb{R}_+)$.*

In particular, if $\Delta(G)$ denotes the maximum degree of the vertices of G, then there exists a family G_n of graphs and an entire function f that is not absolutely monotonic, such that

(1) *$\sup_{n\geq 1} \Delta(G_n) = \infty$;*
(2) *$f_{G_n}[A] \in \mathcal{P}_{G_n}$ for every $A \in \mathcal{P}_{G_n}(\mathbb{R}_+)$.*

Table of contents from Part I of the survey

Table of contents from Part II of the survey

References

[1] Jim Agler, J. William Helton, Scott McCullough, and Leiba Rodman, *Positive semidefinite matrices with a given sparsity pattern*, Proceedings of the Victoria Conference on Combinatorial Matrix Analysis (Victoria, BC, 1987), Linear Algebra Appl. **107** (1988), 101–149, DOI 10.1016/0024-3795(88)90240-6. MR960140

[2] Naum Ilyich Akhiezer, *The classical moment problem and some related questions in analysis*, Translated by N. Kemmer, Hafner Publishing Co., New York, 1965. MR0184042

[3] Theodore W. Anderson, *An introduction to multivariate statistical analysis*, 3rd ed., Wiley Series in Probability and Statistics, Wiley-Interscience [John Wiley & Sons], Hoboken, NJ, 2003. MR1990662

[4] Aharon Atzmon and Allan Pinkus, *Rank restricting functions*, Linear Algebra Appl. **372** (2003), 305–323, DOI 10.1016/S0024-3795(03)00549-4. MR1999153

[5] Zhi Dong Bai and Li-Xin Zhang, *Semicircle law for Hadamard products*, SIAM J. Matrix Anal. Appl. **29** (2007), no. 2, 473–495, DOI 10.1137/050640424. MR2318359

[6] Alexander Belton, Dominique Guillot, Apoorva Khare, and Mihai Putinar, *Matrix positivity preservers in fixed dimension. I*, Adv. Math. **298** (2016), 325–368, DOI 10.1016/j.aim.2016.04.016. MR3505743

[7] ———, *Moment-sequence transforms*, Preprint (available at http://arxiv.org/abs/1610.05740) (2016).

[8] ———, *A panorama of positivity*, Preprint (available at http://arxiv.org/abs/1812.05482) (2018).

[9] ———, *A panorama of positivity. I: Dimension free*, Analysis of Operators on Function Spaces, The Serguei Shimorin memorial volume (Alexandru Aleman, Håkan Hedenmalm, Dmitry Khavinson, and Mihai Putinar, eds.), Trends in Mathematics, Birkhauser, Basel, to appear.

[10] Rajendra Bhatia and Ludwig Elsner, *Positivity preserving Hadamard matrix functions*, Positivity **11** (2007), no. 4, 583–588, DOI 10.1007/s11117-007-2104-8. MR2346444

[11] Peter J. Bickel and Elizaveta Levina, *Covariance regularization by thresholding*, Ann. Statist. **36** (2008), no. 6, 2577–2604, DOI 10.1214/08-AOS600. MR2485008

[12] Martin Bilodeau and David Brenner, *Theory of multivariate statistics*, Springer Texts in Statistics, Springer-Verlag, New York, 1999. MR1705291

[13] Grigoriy Blekherman, Pablo A. Parrilo, and Rekha R. Thomas (eds.), *Semidefinite optimization and convex algebraic geometry*, MOS-SIAM Series on Optimization, vol. 13, Society for Industrial and Applied Mathematics (SIAM), Philadelphia, PA; Mathematical Optimization Society, Philadelphia, PA, 2013. MR3075433

[14] Salomon Bochner, *Monotone Funktionen, Stieltjessche Integrale und harmonische Analyse* (German), Math. Ann. **108** (1933), no. 1, 378–410, DOI 10.1007/BF01452844. MR1512856

[15] Salomon Bochner, *Hilbert distances and positive definite functions*, Ann. of Math. (2) **42** (1941), 647–656, DOI 10.2307/1969252. MR0005782

[16] David M. Bressoud and Shi-Yuan Wei, *Combinatorial equivalence of definitions of the Schur function*, A tribute to Emil Grosswald: number theory and related analysis, Contemp. Math., vol. 143, Amer. Math. Soc., Providence, RI, 1993, pp. 59–64, DOI 10.1090/conm/143/00990. MR1210512

[17] Richard A. Brualdi, *The mutually beneficial relationship of graphs and matrices*, CBMS Regional Conference Series in Mathematics, vol. 115, American Mathematical Society, Providence, RI, 2011. MR2808017

[18] Charles Lutwidge Dodgson, *Condensation of determinants, being a new and brief method for computing their arithmetical values*, Proc. R. Soc. Lond. **15** (1866), 150–155.

[19] David L. Donoho, *High-dimensional data analysis: The curses and blessings of dimensionality*, AMS Math Challenges Lecture **1** (2000), 32 pp.

[20] David L. Donoho and Iain M. Johnstone, *Ideal spatial adaptation by wavelet shrinkage*, Biometrika **81** (1994), no. 3, 425–455, DOI 10.1093/biomet/81.3.425. MR1311089

[21] Noureddine El Karoui, *Operator norm consistent estimation of large-dimensional sparse covariance matrices*, Ann. Statist. **36** (2008), no. 6, 2717–2756, DOI 10.1214/07-AOS559. MR2485011

[22] Shaun Fallat, Charles R. Johnson, and Alan D. Sokal, *Total positivity of sums, Hadamard products and Hadamard powers: results and counterexamples*, Linear Algebra Appl. **520** (2017), 242–259, DOI 10.1016/j.laa.2017.01.013. MR3611466

[23] Jianqing Fan, Yuan Liao, and Han Liu, *An overview of the estimation of large covariance and precision matrices*, Econom. J. **19** (2016), no. 1, C1–C32, DOI 10.1111/ectj.12061. MR3501529

[24] Michael Fekete, *Über ein problem von Laguerre*, Rend. Circ. Math. Palermo **34** (1912), 89–120.

[25] Carl H. FitzGerald and Roger A. Horn, *On fractional Hadamard powers of positive definite matrices*, J. Math. Anal. Appl. **61** (1977), no. 3, 633–642, DOI 10.1016/0022-247X(77)90167-6. MR0506356

[26] Ferdinand Georg Frobenius, *Ueber die elliptischen Functionen zweiter Art* (German), J. Reine Angew. Math. **93** (1882), 53–68, DOI 10.1515/crll.1882.93.53. MR1579913

[27] Felix Ruvimovich Gantmacher, *The theory of matrices. Vols. 1, 2*, Translated by K. A. Hirsch, Chelsea Publishing Co., New York, 1959. MR0107649

[28] Felix Ruvimovich Gantmacher and Mark Grigorievich Krein, *Sur les matrices complètement non négatives et oscillatoires* (French), Compositio Math. **4** (1937), 445–476. MR1556987

[29] Dominique Guillot, Apoorva Khare, and Bala Rajaratnam, *Complete characterization of Hadamard powers preserving Loewner positivity, monotonicity, and convexity*, J. Math. Anal. Appl. **425** (2015), no. 1, 489–507, DOI 10.1016/j.jmaa.2014.12.048. MR3299675

[30] Dominique Guillot, Apoorva Khare, and Bala Rajaratnam, *Critical exponents of graphs*, J. Combin. Theory Ser. A **139** (2016), 30–58, DOI 10.1016/j.jcta.2015.11.003. MR3436051

[31] Dominique Guillot, Apoorva Khare, and Bala Rajaratnam, *Preserving positivity for matrices with sparsity constraints*, Trans. Amer. Math. Soc. **368** (2016), no. 12, 8929–8953, DOI 10.1090/tran6669. MR3551594

[32] Dominique Guillot, Apoorva Khare, and Bala Rajaratnam, *Preserving positivity for rank-constrained matrices*, Trans. Amer. Math. Soc. **369** (2017), no. 9, 6105–6145, DOI 10.1090/tran/6826. MR3660215

[33] Dominique Guillot and Bala Rajaratnam, *Retaining positive definiteness in thresholded matrices*, Linear Algebra Appl. **436** (2012), no. 11, 4143–4160, DOI 10.1016/j.laa.2012.01.013. MR2915274

[34] Dominique Guillot and Bala Rajaratnam, *Functions preserving positive definiteness for sparse matrices*, Trans. Amer. Math. Soc. **367** (2015), no. 1, 627–649, DOI 10.1090/S0002-9947-2014-06183-7. MR3271272

[35] Alfred Hero and Bala Rajaratnam, *Large-scale correlation screening*, J. Amer. Statist. Assoc. **106** (2011), no. 496, 1540–1552, DOI 10.1198/jasa.2011.tm11015. MR2896855

[36] Alfred Hero and Bala Rajaratnam, *Hub discovery in partial correlation graphs*, IEEE Trans. Inform. Theory **58** (2012), no. 9, 6064–6078, DOI 10.1109/TIT.2012.2200825. MR2966078

[37] Carl S. Herz, *Fonctions opérant sur les fonctions définies-positives* (French), Ann. Inst. Fourier (Grenoble) **13** (1963), 161–180. MR0152832

[38] Fumio Hiai, *Monotonicity for entrywise functions of matrices*, Linear Algebra Appl. **431** (2009), no. 8, 1125–1146, DOI 10.1016/j.laa.2009.04.001. MR2547900

[39] Leslie Hogben, *Spectral graph theory and the inverse eigenvalue problem of a graph*, Electron. J. Linear Algebra **14** (2005), 12–31, DOI 10.13001/1081-3810.1174. MR2202430

[40] Roger A. Horn, *The theory of infinitely divisible matrices and kernels*, Trans. Amer. Math. Soc. **136** (1969), 269–286, DOI 10.2307/1994714. MR264736

[41] Tanvi Jain, *Hadamard powers of some positive matrices*, Linear Algebra Appl. **528** (2017), 147–158, DOI 10.1016/j.laa.2016.06.030. MR3652842

[42] Apoorva Khare, *Smooth entrywise positivity preservers, a Horn–Loewner master theorem, and Schur polynomials*, Preprint (available at http://arxiv.org/abs/1809.01823) (2018).

[43] Apoorva Khare and Terence Tao, *On the sign patterns of entrywise positivity preservers in fixed dimension*, Preprint (available at http://arxiv.org/abs/1708.05197) (2017).

[44] ———, *Schur polynomials, entrywise positivity preservers, and weak majorization*, Sem. Lothar. Combin. **80B** (2018), #14 (12 pp.).

[45] Thomas Lam, Alexander Postnikov, and Pavlo Pylyavskyy, *Schur positivity and Schur log-concavity*, Amer. J. Math. **129** (2007), no. 6, 1611–1622, DOI 10.1353/ajm.2007.0045. MR2369890

[46] Steffen L. Lauritzen, *Graphical models*, Oxford Statistical Science Series, vol. 17, The Clarendon Press, Oxford University Press, New York, 1996. Oxford Science Publications. MR1419991

[47] Ai Li and Steve Horvath, *Network neighborhood analysis with the multi-node topological overlap measure*, Bioinformatics **23** (2007), no. 2, 222–231.

[48] Ian G. Macdonald, *Symmetric functions and Hall polynomials*, 2nd ed., Oxford Mathematical Monographs, The Clarendon Press, Oxford University Press, New York, 1995. With contributions by A. Zelevinsky; Oxford Science Publications. MR1354144

[49] Arkadi Nemirovski, *Advances in convex optimization: conic programming*, International Congress of Mathematicians. Vol. I, Eur. Math. Soc., Zürich, 2007, pp. 413–444, DOI 10.4171/022-1/17. MR2334199

[50] Allan Pinkus, *Strictly positive definite functions on a real inner product space*, Adv. Comput. Math. **20** (2004), no. 4, 263–271, DOI 10.1023/A:1027362918283. MR2032278

[51] Georg Pólya and Gabor Szegő, *Aufgaben und Lehrsätze aus der Analysis. Band II: Funktionentheorie, Nullstellen, Polynome Determinanten, Zahlentheorie* (German), Springer-Verlag, Berlin-New York, 1971. Vierte Auflage; Heidelberger Taschenbücher, Band 74. MR0344041

[52] Mohsen Pourahmadi, *High-dimensional covariance estimation*, Wiley Series in Probability and Statistics, John Wiley & Sons, Inc., Hoboken, NJ, 2013. MR3235948

[53] Adam J. Rothman, Elizaveta Levina, and Ji Zhu, *Generalized thresholding of large covariance matrices*, J. Amer. Statist. Assoc. **104** (2009), no. 485, 177–186, DOI 10.1198/jasa.2009.0101. MR2504372

[54] Walter Rudin, *Positive definite sequences and absolutely monotonic functions*, Duke Math. J. **26** (1959), 617–622. MR109204

[55] Isaac Jacob Schoenberg, *Remarks to Maurice Fréchet's article "Sur la définition axiomatique d'une classe d'espace distanciés vectoriellement applicable sur l'espace de Hilbert" [MR1503246]*, Ann. of Math. (2) **36** (1935), no. 3, 724–732, DOI 10.2307/1968654. MR1503248

[56] Isaac Jacob Schoenberg, *Metric spaces and positive definite functions*, Trans. Amer. Math. Soc. **44** (1938), no. 3, 522–536, DOI 10.2307/1989894. MR1501980

[57] Isaac Jacob Schoenberg, *Positive definite functions on spheres*, Duke Math. J. **9** (1942), 96–108. MR5922

[58] Issai Schur, *Bemerkungen zur Theorie der beschränkten Bilinearformen mit unendlich vielen Veränderlichen* (German), J. Reine Angew. Math. **140** (1911), 1–28, DOI 10.1515/crll.1911.140.1. MR1580823

[59] Ronald L. Smith, *The positive definite completion problem revisited*, Linear Algebra Appl. **429** (2008), no. 7, 1442–1452, DOI 10.1016/j.laa.2008.04.020. MR2444334

[60] Harkrishan Vasudeva, *Positive definite matrices and absolutely monotonic functions*, Indian J. Pure Appl. Math. **10** (1979), no. 7, 854–858. MR537245

[61] Cynthia Vinzant, *What is ... a spectrahedron?*, Notices Amer. Math. Soc. **61** (2014), no. 5, 492–494, DOI 10.1090/noti1116. MR3203240

[62] Joe Whittaker, *Graphical models in applied multivariate statistics*, Wiley Series in Probability and Mathematical Statistics: Probability and Mathematical Statistics, John Wiley & Sons, Ltd., Chichester, 1990. MR1112133

[63] Bin Zhang and Steve Horvath, *A general framework for weighted gene co-expression network analysis*, Stat. Appl. Genet. Mol. Biol. **4** (2005), Art. 17, 45, DOI 10.2202/1544-6115.1128. MR2170433

(A. Belton) Department of Mathematics and Statistics, Lancaster University, Lancaster, UK

Email address: a.belton@lancaster.ac.uk

(D. Guillot) Department of Mathematical Sciences, University of Delaware, Newark, DE, USA

Email address: dguillot@udel.edu

(A. Khare) Department of Mathematics, Indian Institute of Science, Analysis and Probability Research Group, Bangalore, India

Email address: khare@iisc.ac.in

(M. Putinar) Department of Mathematics, University of California at Santa Barbara, CA, USA and School of Mathematics, Statistics and Physics, Newcastle University, Newcastle Upon Tyne, UK

Email address: mputinar@math.ucsb.edu, mihai.putinar@ncl.ac.uk

Contemporary Mathematics
Volume **743**, 2020
https://doi.org/10.1090/conm/743/14959

Boundary values of holomorphic distributions in negative Lipschitz classes

Anthony G. O'Farrell

ABSTRACT. We consider the behaviour at a boundary point of an open subset $U \subset \mathbb{C}$ of distributions that are holomorphic on U and belong to what are called negative Lipschitz classes. The result explains the significance for holomorphic functions of series of Wiener type involving Hausdorff contents of dimension between 0 and 1. We begin with a survey about function spaces and capacities that sets the problem in context and reviews the relevant general theory.

1. Introduction

1.1. Boundary values. It may happen that all bounded holomorphic functions on an open set $U \subset \mathbb{C}$ admit a 'reasonable boundary value' at some boundary point. This was first noted by Gamelin and Garnett [**18**]. The condition for the existence of such a boundary value is expressed using a series of 'Wiener type', and involves the Ahlfors analytic capacity, γ. The condition is

$$\sum_{n=1}^{\infty} 2^n \gamma(A_n \setminus U) < +\infty.$$

Here, if b is the boundary point in question, A_n denotes the annulus

$$A_n(b) := \left\{ z \in \mathbb{C} : \frac{1}{2^{n+1}} \leq |z-b| \leq \frac{1}{2^n} \right\}.$$

This condition says that in an appropriate sense the complement of U is very thin at b; in particular it implies that U has full area density at b, i.e.

$$\lim_{r \downarrow 0} \frac{|\mathbb{B}(b,r) \cap U|}{\pi r^2} = 1,$$

where we denote the area of a set $E \subset \mathbb{C}$ by $|E|$. When the series converges, it is emphatically *not* the case that the limit

$$\lim_{z \to b, z \in U} f(z)$$

exists for all functions f bounded and holomorphic on U (unless all such functions extend holomorphically to a neighbourhood of b). But for each such function there

2010 *Mathematics Subject Classification.* Primary 30E25; 46F20.

is a (unique) value which we may call $f(b)$, with the property that for some set $E \subset U$ having full area density at the point b

$$\lim_{z \to b, z \in U} f(z) = f(a).$$

1.2. Peak points. This result is one of many about the boundary behaviour of analytic and harmonic functions on arbitrary open sets. The original Wiener series (cf. [**42**] or [**1**]) involved logarithmic capacity in dimension two, Newtonian capacity in dimension three, and other Riesz capacities in higher dimensions, and characterised boundary points that are regular for the Dirichlet problem. Later these points were recognised as *peak points* for the space of functions harmonic on an open set U and continuous on its closure. The first person to use such a series with holomorphic functions was Melnikov [**17**, Theorem VIII.4.5]. He characterised the peak points for the uniform closure on a compact set $X \subset \mathbb{C}$ of the algebra of all rational functions having poles off X. He used the Ahlfors capacity, and he showed that a point $b \in X$ is a peak point if and only if

$$\sum_{n=1}^{\infty} 2^n \gamma(\mathring{A}_n \setminus X) < +\infty.$$

(This was used by Gamelin and Garnett to obtain their above-quoted result.)

For a bounded open set $U \subset \mathbb{C}$, and a point $b \in \partial U$, the condition

$$\sum_{n=1}^{\infty} 2^n \alpha(A_n \setminus U) < +\infty,$$

where α denotes the so-called *continuous* analytic capacity (introduced by Dolzhenko) is necessary and sufficient for b to be a peak point for the algebra of all continuous functions on $\overline{U}$, holomorphic on U [**17**].

1.3. Capacities. The vague idea that *there is a capacity for every problem* has gathered momentum over time. A *capacity* is a function c that assigns nonnegative extended real numbers to sets, and is nondecreasing:

$$E_1 \subset E_2 \implies c(E_1) \le c(E_2).$$

Keldysh [**24**] used Newtonian capacity to solve the problem of stability for the Dirichlet problem. Vitushkin used analytic capacity to solve the problem of uniform rational approximation [**17**, Chapter VIII]. Vitushkin's theorem is completely analogous to Keldysh's: harmonic functions have been replaced by holomorphic functions, and Newtonian capacity by analytic capacity. The same switch relates Wiener's regularity criterion and Melnikov's peak-point criterion.

In an influential little book [**7**], Carleson explained how other capacities (particularly kernel capacities) could be used to solve problems about boundary values, convergence of Fourier series, and removable singularities, and in an appendix (prepared by Wallin) he listed over a thousand articles from Mathematical Reviews up to 1965 that involve some combination of these ideas.

1.4. Continuous point evaluations. In relation to L^p holomorphic approximation, the appropriate capacity is a condenser capacity. The groundwork on condenser capacities and (generalized) extremal length had already been laid down by Fuglede [**16**]. Hedberg [**22**] (see also [**2**]) worked out the analogue of Vitushkin's theorem for L^p approximation on compact $X \subset \mathbb{C}$, and [**21**] proved the analogue of Melnikov's theorem. Hedberg's result is about *continuous point evaluations.* To explain this concept, consider a Banach space F of 'functions' on some set $E \subset \mathbb{C}$, where each element $f \in F$ is defined almost-everywhere on E with respect to area measure m. Suppose $b \in \overline{E}$ and the subspace F_b, consisting of those $f \in F$ that extend holomorphically to some neighbourhood of b, is a dense subset of F. Then we say that F *admits a continuous point evaluation at* b if there exists $\kappa > 0$ such that

$$|f(b)| \le \kappa \|f\|_F, \ \forall f \in F_b.$$

This means that the functional $f \mapsto f(b)$ has a continuous extension from F_b to the whole of F. Taking the case where F is the closure $R^p(X)$ in $L^p(X, m)$ of the rational functions with poles off a compact $X \subset \mathbb{C}$, Hedberg showed that if $2 < p < +\infty$, then $R^p(X)$ admits a continuous point evaluation at b if and only if

$$\sum_{n=1}^{\infty} 2^{nq} \Gamma_q(A_n(b) \setminus X) < +\infty.$$

Here $q = p/(p-1)$ is the conjugate index, and Γ_q is a certain condenser capacity. When $p < 2$, $R^p(X)$ never admits a continuous point evaluation at b, unless b is an interior point of X. In the case $p = 2$, Hedberg left an interesting gap between the sharpest known necessary condition and the sharpest known sufficient condition, and this gap was closed by Fernström [**14**]. Historically, the existence of continuous point evaluations in the L^2 case attracted considerable attention, because of hopes that it might provide a way to attack the invariant subspace problem for operators on Hilbert space, and hopes that it might provide a way to attack the L^2 rational approximation problem [**3**–**5**].

The existence problem for continuous point evaluations at boundary points has also been studied for harmonic functions in the Sobolev space $W^{1,2}$, and Kolsrud [**25**] gave a solution in terms of Wiener series.

In the literature, continuous point evaluations are often referred to as *bounded* point evaluations.

1.5. Continuous point derivations. There are similar results about the possibility that the k-th complex derivative $f \mapsto f^{(k)}$ may have a continuous extension from F_b to all of F. These involve the same Wiener series as continuous point evaluations, except that the base 2 is replaced by 2^{k+1}. For instance, the $R^p(X)$ result (also due to Hedberg) involves the condition

$$\sum_{n=1}^{\infty} 2^{(k+1)qn} \Gamma_q(A_n(b) \setminus X) < +\infty.$$

The earliest such result was for the uniform closure of the rationals, and was due to Hallstrom [**20**].

1.6. Intrinsic capacities. The present author began to formalize the pairing of problems and capacities in his thesis [**30**]. He considered the limited context of uniform algebras on compact subsets of the plane. To each functor $X \mapsto F(X)$ that associates a uniform algebra to each compact $X \subset \mathbb{C}$, and subject to certain coherence assumptions, he associated a capacity

$$\alpha(F, \cdot) : \mathcal{O} \to [0, +\infty),$$

where $\mathcal{O}$ is the topology of $\mathbb{C}$. He then proved a *Capacity Uniqueness Theorem*, which stated that the map $F \mapsto \alpha(F,\cdot)$ is injective on the set of such functors, i.e. the capacity determines the functor. The *Local Capacity Uniqueness Theorem* states that two functors F and G have $F(X) = G(X)$ for a given compact set X if and only if the capacities $\alpha(F,\cdot)$ and $\alpha(G,\cdot)$ agree on all open subsets of the complement of X. Thus $F(X) \stackrel{?}{=} G(X)$ is a problem for which there are *two* capacities, not one! Vitushkin's theorem on rational approximation is the case when $F(X)$ is the uniform closure of the rational functions having poles off X and $G(X)$ is the algebra of all functions continuous on X and holomorphic on $\mathring{X}$. This part of the thesis is unpublished, mainly because the main result is essentially equivalent to a theorem of Davie [**9**], obtained independently. In another unpublished chapter, the author established that the results of Melnikov and Hallstrom extended to all these $F(X)$, replacing the analytic capacity by $\alpha(F,\cdot)$.

Other work by Wang [**41**] and the author [**31, 32**] established a link between equicontinuous pointwise Hölder conditions at a boundary point and series in which 2 is replaced by 2^λ for a nonintegral $\lambda > 1$. For instance, Hölder conditions of order α are related to the convergence of series such as

$$\sum_{n=1}^{\infty} 2^{(1+\alpha)n}\gamma(A_n \setminus X).$$

Moving on from the uniform norm, the author considered parallel questions for Lipschitz or Hölder norms. Building on a result of Dolzhenko, he established that the equivalent of continuous analytic capacity is the *lower β-dimensional Hausdorff content* M_*^β, with $\beta = \alpha + 1$. (For $\beta > 0$, the *β-dimensional Hausdorff content* $M^\beta(E)$ of a set $E \subset \mathbb{R}^d$ is defined to be the infimum of the sums $\sum_{n=0}^{\infty} r_n^\beta$, taken over all countable coverings of E by closed balls $(\mathbb{B}(a_n, r_n))_n$. If we replace r_n^β by $h(r_n)$ for an increasing function $h : [0,+\infty) \to [0,+\infty)$ we get the Hausdorff h-content $M_h(E)$. The *lower β-dimensional Hausdorff content* $M_*^\beta(E)$ is defined to be the supremum of $M_h(E)$, taken over all h such that $0 \le h(r) \le r^\beta$ for all $r > 0$, and $r^{-\beta}h(r) \to 0$ as $r \downarrow 0$.) He proved [**29**] the analogue of Vitushkin's theorem for rational approximation. Later, Lord and he [**26**] proved the analogue of Hallstrom's theorem for boundary derivatives. For the k-th derivative, this involved the series condition

$$\sum_{n=1}^{\infty} 2^{(k+1)n} M_*^{\alpha+1}(A_n(b) \setminus X) < +\infty.$$

1.7. SCS. Moving to a more general context, the author introduced the notion of a *Symmetric Concrete Space* F on $\mathbb{R}^d$, and considered the relation between problems about a given such space F, in combination with an elliptic operator L, and an appropriate associated capacity, the L-F-cap. A Symmetric Concrete Space (SCS) on $\mathbb{R}^d$ is a complete locally-convex topological vector space F over the field $\mathbb{C}$, such that

- $\mathcal{D} \hookrightarrow F \hookrightarrow \mathcal{D}^*$;
- F is a topological $\mathcal{D}$-module under the usual product $\phi\cdot f$ of a test function and a distribution;
- F is closed under complex conjugation;
- The affine group of $\mathbb{R}^d$ acts by composition on F, and each compact set of affine maps gives an equicontinuous family of composition operators.

Here $\mathcal{D} = C^\infty_{\rm cs}(\mathbb{R}^d, \mathbb{C})$, is the space of *test functions* and $\mathcal{D}^*$ is its dual, the space of distributions, $A \hookrightarrow B$ stands for "*$A \subset B$ and the inclusion map is continuous*". (In fact, it is elementary that if A and B are metrizable SCS, then $A \subset B$ implies $A \hookrightarrow B$.)

A SCS is a Symmetric Concrete Banach Space (SCBS) when it is normable and is equipped with a norm.

We shall be concerned only with the case $d = 2$, and we identify $\mathbb{R}^2$ with $\mathbb{C}$.

The various analytic capacities are $\frac{\partial}{\partial \bar{z}}$-$F$-cap for particular F. The author planned a book about this subject, but this project has never been completed. Some extracts with useful ideas and results were published. The most useful ideas concern localness. For SCS F and G, we define

$$F_{\rm loc} := \{f \in \mathcal{D}^* : \phi \cdot f \in F,\ \forall \phi \in \mathcal{D}\},$$

$$F_{\rm cs} := \{\phi \cdot f : f \in F \text{ and } \phi \in \mathcal{D}\},$$

$$F \overset{\rm loc}{\hookrightarrow} G \iff F_{\rm loc} \hookrightarrow G_{\rm loc},$$

$$F \overset{\rm loc}{=} G \iff F_{\rm loc} = G_{\rm loc},$$

and observe that

$$F \overset{\rm loc}{=} F_{\rm loc} \overset{\rm loc}{=} F_{\rm cs}.$$

Published results include the following:
(1) A *Fundamental Theorem of Calculus for SCS* that are weakly-locally invariant under Calderon-Zygmund operators [**35**, Lemma 12]. This says that

$$D{\textstyle\int} F \overset{\rm loc}{=} {\textstyle\int} DF \overset{\rm loc}{=} F,$$

where

$$DF := \mathcal{D} + \operatorname{span}\left\{\frac{\partial f}{\partial x_j} : 1 \le j \le d, f \in F\right\}$$

and

$${\textstyle\int} F := \left\{f \in \mathcal{D}^* : \frac{\partial f}{\partial x_j} \in F, \text{ for } 1 \le j \le d\right\}.$$

(2) A *1-reduction principle* that allows us to establish equivalences between problems for different operators L [**35**, Theorem 1]. The identity operator $\mathbb{1} : f \mapsto f$ is elliptic. If U is open, then the equation $\mathbb{1} f = 0$ on U just means that $U \cap \operatorname{supp}(f) = \emptyset$. The idea is to reduce questions about L and some space F to equivalent problems about $\mathbb{1}$ and the space $LF := \{Lf : f \in F\}$.
(3) A general Sobolev-type embedding theorem [**34**] involving the concept of the *order* of an SCS, and
(4) A theorem that says that in dimension two all SCS are essentially (technically, weakly-locally-) T-invariant [**36**], i.e. invariant under the Vitushkin localization operators (see Section 6 below).

In 1990 the author circulated a set of notes on the concept of SCS and the main examples. Some ideas from these papers were expounded by Tarkhanov in his book on the *Cauchy Problem for Solutions of Elliptic Equations* [**39**, Chapter 1].

The general point of view raised many particular questions, and some of these have been solved, while other loose ends remain.

2. The Problem

Our objective in the present paper is to address a loose end connected to the results on boundary behaviour of holomorphic functions mentioned above. For $0 < \alpha < 1$, the $\frac{\partial}{\partial \bar{z}}$-$F$-cap associated to the Lipschitz class Lipα is $M^{\alpha+1}$, and that associated to the little Lipschitz class lipα is $M_*^{\alpha+1}$. Kaufmann [**23**] showed that M^1 is the $\frac{\partial}{\partial \bar{z}}$-BMO-cap, the capacity associated to the space of functions of bounded mean oscillation, and Verdera [**40**] established that M_*^1 is the $\frac{\partial}{\partial \bar{z}}$-VMO-cap, the capacity associated to the space of functions of vanishing mean oscillation. Verdera proved the Vitushkin theorem for VMO.

The question is, *what do M^β and M_*^β have to do with the boundary behaviour of analytic functions when* $0 < \beta < 1$?. What is the significance of the condition

$$\sum_{n=1}^{\infty} 2^n M^\beta(A_n \setminus U) < +\infty,$$

when $0 < \beta < 1$, where U is a bounded open subset of $\mathbb{C}$ and $b \in \partial U$?

We are considering a local problem, and it is worth noting that there are several different meanings commonly attached to the global Lipschitz classes, and the little Lipschitz classes. For $0 < \alpha < 1$, we define Lip$\alpha(\mathbb{R}^d)$ to be the space of bounded functions $f : \mathbb{R}^d \to \mathbb{C}$ that satisfy a Lipschitz-alpha condition:

$$|f(x) - f(y)| \leq \kappa_f |x - y|^\alpha, \ \forall x, y \in \mathbb{R}^d.$$

We would obtain a locally-equivalent Banach SCS if we omit the word 'bounded'. We would also obtain a locally-equivalent SCBS if we just require the Lipschitz condition for $|x - y| \leq 1$. Another locally-equivalent space is obtained by requiring the Lipschitz condition with repect to the spherical metric (associated to the stereographic projection $\mathbb{S}^d \to \mathbb{R}^d$). We shall shortly meet another locally-equivalent space, defined in terms of the Poisson transform. It makes no difference for our problem which of these versions is used, and we can exploit this fact by choosing whatever version is easiest to use in each context. For this paper, we define lipα to be the closure of $\mathcal{D}$ in Lipα. This space is locally-equivalent to the space of functions that have restriction in lip(α, X) for each compact $X \subset \mathbb{R}^d$, but it has an additional property 'near ∞', irrelevant for our purposes.

3. Results

3.1. The spaces T_s and C_s. The answer to the problem will not surprise anyone who has studied the paper [**35**], but may be regarded as rather strange by others.

The first step in trying to identify the L-F-cap for given L and F is based on the principle that the compact sets $X \subset \mathbb{R}^d$ that have (L-F-cap)$(X) = 0$ should be the *sets of removable singularities* for solutions of $Lf = 0$ of class F. This means that (L-F-cap)$(X) = 0$ should be the necessary and sufficient condition that the restriction map

$$\{f \in F : Lf = 0 \text{ on } U\} \to \{f \in F : Lf = 0 \text{ on } U \setminus X\}$$

be surjective for each open set $U \subset \mathbb{R}^d$.

In [**35**, p.140] it was established (as a special case of the 1-reduction principle) that for nonintegral β the set function M^β is zero on the sets of removable singularities for holomorphic functions of a Lipschitz class, but when $0 < \beta < 1$ this is a *negative Lipschitz class*, there denoted $T_{\beta-1}$.

The negative Lipschitz classes can be described in a number of equivalent ways. In informal terms, the basic idea is that the T_s for $s \in \mathbb{R}$ form a one-dimensional 'scale' of spaces of distributions, i.e. a family of spaces totally-ordered under local inclusion. When $0 < s < 1$, T_s is locally-equal to Lip$s$. Differentiation takes T_s down to T_{s-1}, and DT_s is locally-equivalent to T_{s-1}. Thus if $s < 0$ and $k \in \mathbb{N}$ has $s + k > 0$, then T_s is locally-equal to $D^k\mathrm{Lip}(s + k)$. The elements of T_s having compact support may also be characterised by the growth of the Poisson transform as we approach the plane from the upper half of $\mathbb{R}^3$, or by the growth of the convolution with the heat kernel. This idea originated in the work of Littlewood and Paley and was fully developed by Taibleson [**38**, Chapter 5]. The Poisson kernel is

$$P_t(z) := \frac{t}{\pi(t^2 + |z|^2)^{\frac{3}{2}}}, \ (t > 0, z \in \mathbb{C}).$$

It is real-analytic in z and t, and is harmonic in (z, t) in the upper half-space

$$\mathbb{H}^3 := \mathbb{C} \times (0, +\infty).$$

For a distribution $f \in \mathcal{E}^* := (C^\infty(\mathbb{C}, \mathbb{C}))^*$ having compact support, the Poisson transform of f is the convolution

$$F(z, t) := (P_t * f)(z)$$

where $P_t * f$ denotes the convolution on $\mathbb{C} = \mathbb{R}^2$. For $s < 0$ we say (following [**35**]) that f belongs to the 'negative Lipschitz space' T_s if

$$\|f\|_s := \sup\{t^{-s}|F(z, t)| : z \in \mathbb{C}, t > 0\} < +\infty,$$

and belongs to the 'small negative Lipschitz space C_s if, in addition,

$$\lim_{t \downarrow 0} t^{-s} \sup\{|F(z, t)| : z \in \mathbb{C}\} = 0.$$

For $s \geq 0$, we define T_s and C_s by requiring that for $f \in \mathcal{E}^*$, $f \in T_s$ (respectively C_s) if and only if all k-th order partial derivatives of f belong to T_{s-k} (respectively C_{s-k}) for each (or, equivalently, for some) integer $k > s$.

The Riesz transforms[1], convolution with $|z|^{t-d}$, map T_s locally into T_{s+t}, so behave like 'fractional integrals'.

The scale corresponding to the little Lipschitz class C_s may be described as the closure of the space $\mathcal{D}$ in T_s.

Delicate questions arise at integral values s, and we shall not consider such s in this paper.

3.2. Statements. For an open set $U \subset \mathbb{C}$, and $s \in \mathbb{R}$, let

$$A^s(U) := \{f \in C_s : f \text{ is holomorphic on } U\},$$

and

$$B^s(U) := \{f \in T_s : f \text{ is holomorphic on } U\}.$$

[1]Strictly speaking the order t Riesz transform is the operator $(-\Delta)^{-t/2}$, which for $t > 0$ is convolution with $c_t|z|^{t-d}$, for a certain constant c_t [**38**, p 117].

We are interested in the range $-1 < s < 0$, and for such s the elements of $A^s(U)$ and $B^s(U)$ are distributions on $\mathbb{C}$ that may fail to be representable by integration against a locally-L^1 function, so the definition of continuous point evaluation given above does not apply. However, we can make a straightforward adjustment. We shall prove the following lemma:

LEMMA 3.1. *For each* $s \in \mathbb{R}$, *each open set* $U \subset \mathbb{C}$ *and each* $b \in \mathbb{C}$, *the set* $\{f \in A^s(U) : f \text{ is holomorphic on some neighbourhood of } b\}$ *is dense in* $A^s(U)$.

Here, when we say that the distribution f on $\mathbb{C}$ is holomorphic on an open set V, we mean that its distributional $\bar{\partial}$-derivative has support off V, i.e.

$$\left\langle \phi, \frac{\partial f}{\partial \bar{z}} \right\rangle := -\left\langle \frac{\partial \phi}{\partial \bar{z}}, f \right\rangle = 0$$

whenever the test function ϕ has support in V. Recall that by Weyl's Lemma this means that the restriction $f|V$ is represented by an ordinary holomorphic function, so that it and all its derivatives have well-defined values throughout V.

Let us denote

$$A^s_b(U) := \{f \in A^s(U) : f \text{ is holomorphic on some neighbourhood of } b\}.$$

DEFINITION 3.2. We say that $A^s(U)$ admits a continuous point evaluation at a point $b \in \mathbb{C}$ if the functional $f \mapsto f(b)$ extends continuously from $A^s_b(U)$ to the whole of $A^s(U)$.

Our main result is this:

THEOREM 3.3. *Let* $0 < \beta < 1$ *and* $s = \beta - 1$. *Let* $U \subset \mathbb{C}$ *be a bounded open set, and* $b \in \partial U$. *Then* $A^s(U)$ *admits a continuous point evaluation at* b *if and only if*

$$\sum_{n=1}^{\infty} 2^n M_*^{\beta}(A_n \setminus U) < +\infty.$$

3.3. Weak-star continuous evaluations. We can also give a result about the big Lipschitz class $B^s(U)$. We cannot replace $A^s(U)$ by $B^s(U)$ in Lemma 3.1 as it stands. However, the T_s spaces are dual spaces, and so have a weak-star topology (see Subsection 6.2 for details), and restricting this topology gives us a second useful topology on $B^s(U)$. We have the following:

LEMMA 3.4. *For each* $s \in \mathbb{R}$, *each open set* $U \subset \mathbb{C}$ *and each* $b \in \mathbb{C}$, *The set* $\{f \in B^s(U) : f \text{ is holomorphic on some neighbourhood of } b\}$ *is weak-star dense in* $B^s(U)$.

Denoting

$$B^s_b(U) := \{f \in B^s(U) : f \text{ is holomorphic on some neighbourhood of } b\},$$

we can then give the following definition:

DEFINITION 3.5. We say that $B^s(U)$ admits a weak-star continuous point evaluation at a point $b \in \mathbb{C}$ if the functional $f \mapsto f(b)$ extends weak-star continuously from $B^s_b(U)$ to the whole of $B^s(U)$.

Our result for $B^s(U)$ is this:

THEOREM 3.6. *Let $0 < \beta < 1$ and $s = \beta - 1$. Let $U \subset \mathbb{C}$ be a bounded open set, and $b \in \partial U$. Then $B^s(U)$ admits a weak-star continuous point evaluation at b if and only if*

$$\sum_{n=1}^{\infty} 2^n M^\beta(A_n \setminus U) < +\infty.$$

3.4. Boundary derivatives. In the same spirit, we get results about boundary derivatives. We denote the set of positive integers by $\mathbb{N}$.

THEOREM 3.7. *Let $0 < \beta < 1$, $s = \beta - 1$, and let $k \in \mathbb{N}$. Let $U \subset \mathbb{C}$ be a bounded open set, and $b \in \partial U$. Then the functional $f \mapsto f^{(k)}(b)$ has a continuous extension from $A_b^s(U)$ to the whole of $A^s(U)$ if and only if*

$$\sum_{n=1}^{\infty} 2^{(k+1)n} M_*^\beta(A_n \setminus U) < +\infty.$$

THEOREM 3.8. *Let $0 < \beta < 1$, $s = \beta - 1$, and let $k \in \mathbb{N}$. Let $U \subset \mathbb{C}$ be a bounded open set, and $b \in \partial U$. Then the functional $f \mapsto f^{(k)}(b)$ has a weak-star continuous extension from $B_b^s(U)$ to the whole of $B^s(U)$ if and only if*

$$\sum_{n=1}^{\infty} 2^{(k+1)n} M^\beta(A_n \setminus U) < +\infty.$$

The spaces $A^s(U)$ are not algebras — essentially SCS are only algebras when they are locally-included in C^0 — so we avoid using the term *derivation*, lest we confuse people.

3.5. Harmonic functions. The foregoing results concern objects f that are not 'proper functions'. Using the ideas related to 1-reduction, we may derive a theorem about ordinary harmonic functions:

For $0 < \alpha < 1$, let $H^\alpha(U)$ denote the space of (complex-valued) functions that are harmonic on U and belong to the little Lipschitz α class on the closure of U (or, equivalently, have an extension belonging to the global little Lipschitz class). For $b \in \partial U$, let

$$H_b^\alpha(U) := \{h \in H^\alpha(U) : h \text{ is harmonic on a neigbourhood of } b\}.$$

If we denote, as is usual,

$$\frac{\partial}{\partial z} := \frac{1}{2}\left(\frac{\partial}{\partial x} - i\frac{\partial}{\partial y}\right)$$

and

$$\frac{\partial}{\partial \bar{z}} := \frac{1}{2}\left(\frac{\partial}{\partial x} + i\frac{\partial}{\partial y}\right),$$

then $\Delta = 4\dfrac{\partial}{\partial z}\dfrac{\partial}{\partial \bar{z}}$.

THEOREM 3.9. *Let $0 < \alpha < 1$, let $U \subset \mathbb{C}$ be bounded and open, and $b \in \partial U$. Then*

(1) $H_b^\alpha(U)$ is dense in $H^\alpha(U)$.

(2) The functional $h \mapsto \dfrac{\partial h}{\partial z}(b)$ *extends continuously from* $H_b^\alpha(U)$ *to* $H^\alpha(U)$ *if and only if*

$$\sum_{n=1}^{\infty} 2^n M_*^\alpha(A_n \setminus U) < +\infty.$$

(3) The $\mathbb{C}^2$*-valued function* $h \mapsto (\nabla h)(b)$ *extends continuously from* $H_b^\alpha(U)$ *to* $H^\alpha(U)$ *if and only if*

$$\sum_{n=1}^{\infty} 2^n M_*^\alpha(A_n \setminus U) < +\infty.$$

4. Examples

4.1. Smooth boundary. If U is smoothly-bounded, then there are no continuous point evaluations at any boundary point on $A^s(U)$ for any $s < 0$. Indeed, if b belongs any to any nontrivial continuum $K \subset \mathbb{C} \setminus U$, then no such continuous point evaluation exists at b.

4.2. Multiple components. If b belongs to the boundary of two (or more) connected components of the open set U, then no such continuous point evaluation exists at b.

To see this, note that the assumptions imply that for all small enough r, the circle $|z - b| = r$ meets the complement of U, and this implies that for large enough n, the M^β content of $A_n \setminus U$ is at least $2^{-n\beta}$. Hence the series in Theorem 3.3 diverges for all $s \in (-1, 0)$.

This contrasts with the behaviour found in [**26**] for ordinary Lipschitz classes, for which interesting behaviour is possible at the boundary of Jordan domains with piecewise-smooth boundary, and at common boundary points of two components.

4.3. Slits. Let $a_n \downarrow 0$, $r_n \downarrow 0$ and

$$a_{n+1} + r_{n+1} < a_n - r_n, \ \forall n \in \mathbb{N}.$$

Then 0 is a boundary point of the slit domain

$$U := \mathring{\mathbb{B}}(0, a_1 + r_1) \setminus \bigcup_{n=1}^{\infty} [a_n - r_n, a_n + r_n].$$

For a line segment I of length d, we have

$$M^\beta(I) = M_*^\beta(I) = d^\beta,$$

for $0 < \beta < 1$. Then for $-1 < s < 0$, $A^s(U)$ admits a continuous point evaluation at 0 if and only if $B^s(U)$ admits a weak-star continuous point evaluation at 0, and if and only if

$$\sum_{n=1}^{\infty} \frac{r_n^{s+1}}{a_n} < +\infty. \tag{4.1}$$

This follows at once from Theorems 3.3 and 3.6 in case $a_n = 2^{-n}$. In the general case, one obtains it by imitating the proofs, using contours that pass between the slits. The corresponding condition for the existence of a k-th order continuous point derivation is

$$\sum_{n=1}^{\infty} \frac{r_n^{(k+1)(s+1)}}{a_n} < +\infty.$$

For example, if $a_n = 2^{-n}$ and $r_n = 4^{-n}$, then there is a continuous point evaluation at 0 on $A^s(U)$ if and only if $s > -\frac{1}{2}$.

4.4. Road-runner sets. The M^β content of a disc and of one of its diameters are both fixed multiples of (radius)$^\beta$, and the lower content is the same, so the same condition (4.1) is necessary and sufficient for the existence of a continuous point evaluation on $A^s(U)$ at 0 on the so-called road-runner set

$$U := \mathring{\mathbb{B}}(0, a_1 + r_1) \setminus \bigcup_{n=1}^{\infty} \mathbb{B}(a_n, r_n),$$

when the a_n and r_n are as in the last subsection.

4.5. Below minus 1. The L-F-cap capacity of a singleton is positive as soon as there are distributions f of class F having $Lf = 0$ on a deleted neighbourhood of 0. In the case $L = \frac{\partial}{\partial \bar{z}}$, this happens when the distribution represented by the L^1_{loc} function $\frac{1}{z}$ belongs to F_{loc}. That explains why, in the case of L^p holomorphic functions, there is a major transition at $p = 2$; the function $\frac{1}{z}$ belongs to L^p_{loc} when $1 \le p < 2$. (The 'smoothness' of $L^p(\mathbb{R}^d)$ is $-d/p$. This can be extended below $p = 1$ by using Hardy spaces H^p instead of L^p.)

The 'delta-function', δ_0, the unit point mass at the origin, is the d-bar (distributional) derivative of $1/z$. More precisely

$$\frac{\partial \frac{1}{\pi z}}{\partial \bar{z}} = \delta_0.$$

The Poisson transform of δ_0 is just the Poisson kernel $t/\pi(|z|^2 + t^2)^{\frac{3}{2}}$, so grows no faster than $1/t^2$ as $t \downarrow 0$. Thus δ_0 belongs to T_{-2} and $\frac{1}{z} \in T_{-1}$.

5. Tools

We abbreviate $\|f\|_{T_s}$ to $\|f\|_s$. We use K to denote a positive constant which is independent of everything but the value of the parameter s, and which may be different at each occurrence.

5.1. The strong module property. For a nonnegative integer k, and $\phi \in \mathcal{D}$, we use the notation

$$N_k(\phi) := d(\phi)^k \cdot \sup |\nabla^k(\phi)|.$$

where $d(\phi)$ denotes the diameter of the support of ϕ. Here, we take the norm $|\nabla^k \phi|$ to be the maximum of all the k-th order partial derivatives of ϕ.

Note that for $\kappa > 0$, $N_k(\kappa \cdot \phi) = \kappa \cdot N_k(\phi)$, that

$$N_0(\phi) \le N_1(\phi) \le N_2(\phi) \le \cdots,$$

and that (by Leibnitz' formula)

$$N_k(\phi \cdot \psi) \le 2^k N_k(\phi) N_k(\psi)$$

whenever $\phi, \psi \in \mathcal{D}$ and $k \in \mathbb{N}$.

Also, $N_k(\phi)$ is invariant under rescaling: If $\phi \in \mathcal{D}$, $r > 0$, and $\psi(x) := \phi(r \cdot x)$, then $N_k(\psi) = N_k(\phi)$ for each k.

An SCBS F has the *(order k-) strong module property* if there exists $k \in \mathbb{Z}_+$ and $K > 0$ such that

$$\|\phi \cdot f\|_F \leq K \cdot N_k(\phi) \cdot \|f\|_F, \ \forall f \in F.$$

Most common SCBS have this property. Note that for every SCBS and each $\phi \in \mathcal{D}$ the fact that F is a topological $\mathcal{D}$-module tells us that there is some constant $K(\phi)$ such that

$$\|\phi \cdot f\|_F \leq K(\phi)\|f\|_F, \ \forall f \in F.$$

Thus the strong module property amounts to saying that the least $K(\phi)$ are dominated by some $N_k(\phi)$, up to a fixed multiplicative constant.

It is readily seen that the ordinary Lipschitz spaces have the order 1 strong module property.

5.2. Standard pinchers. If $b \in \mathbb{R}^d$, a *standard pincher at b* is a sequence of nonnegative test functions $(\phi_n)_n$, such that $\phi_n = 1$ on a neighbourhood of b, the diameter of $\mathrm{supp}(\phi_n)$ tends to zero, and for each $k \in \mathbb{N}$, the sequence $(N_k(\phi_n))_n$ is bounded.

The elements ϕ_n of a standard pincher make the transition from the value 1 at b down to zero in a reasonably gentle way, so that the various derivatives are not greatly larger than they have to be in order to achieve the transition.

It is easy to see that such sequences exist. For instance, they may be constructed in the form $\phi_n(x) = \psi(n|x - b|)$, where $\psi : [0, +\infty) \to [0, 1]$ is C^∞, has $\psi = 1$ near 0 and $\psi = 0$ off $[0, 1]$.

5.3. The Cauchy transform. The Cauchy transform is the convolution operator

$$\mathfrak{C}f := \frac{1}{\pi z} * f.$$

It acts (at least) on distributions having compact support, and it almost inverts the d-bar operator $\frac{\partial}{\partial \bar{z}}$:

$$\frac{\partial}{\partial \bar{z}}\mathfrak{C}f = f = \mathfrak{C}\frac{\partial f}{\partial \bar{z}},$$

whenever $f \in \mathcal{E}^*$. Recall from Subsection 1.7 that if F is an SCS we have the associated spaces F_{cs} and F_{loc}. The Cauchy transform maps $(T_s)_{\mathrm{cs}}$ continuously into T_{s+1} and $(C_s)_{\mathrm{cs}}$ into C_{s+1}, so in combination with $\frac{\partial}{\partial \bar{z}}$ it can be used to relate properties of T_s to properties of T_{s+1}. For our present purposes, this allows us to move from our spaces of distributions corrresponding to $-1 < s < 0$ to spaces of ordinary $\mathrm{Lip}(s+1)$ functions.

The Cauchy kernel $\frac{1}{\pi \bar{z}}$ does not belong to L^1, but it does belong to L^1_{loc}, and indeed there is a uniform bound on its norm on discs of fixed radius:

$$\left\|\frac{1}{\pi \bar{z}}\right\|_{L^1(\mathbb{B}(a,r))} \leq 2r, \ \forall a \in \mathbb{C}, \forall r > 0.$$

So if F is a SCBS, and translation acts isometrically on F, then

$$\|\mathfrak{C}f\|_F \leq d\|f\|_F, \tag{5.1}$$

whenever $f \in F_{\mathrm{cs}}$ is supported in a disc of radius d.

5.4. Evaluating the Cauchy transform. The value of $(\mathfrak{C}f)(b)$ at a point off the support of the distribution f may be evaluated in the obvious way:

LEMMA 5.1. *Let $f \in \mathcal{E}^*$, and $b \in \mathbb{C} \setminus \operatorname{supp}(f)$. Let $\chi \in \mathcal{D}$ be any test function having $\chi(z) = 1/(z-b)$ near* $\operatorname{supp}(f)$. *Then*

$$\mathfrak{C}(f)(b) = \left\langle \frac{\chi}{\pi}, f \right\rangle.$$

PROOF. We take $b = 0$, without loss in generality.

For any $\psi \in \mathcal{D}$ with $\int \psi dm = 1$ and $\operatorname{supp}(\psi) \cap \operatorname{supp}(f) = \emptyset$, we have

$$\langle \psi, \mathfrak{C}(f) \rangle = -\left\langle \psi * \frac{1}{\pi z}, f \right\rangle.$$

Let $(\phi_n)_n$ be a standard pincher at 0 and take

$$\psi_n = \frac{\phi_n}{\int \phi_n dm}.$$

Then $\psi_n * \frac{1}{\pi z} \to \frac{1}{\pi z} = \chi/\pi$ in C^∞ topology on a neighbourhood of $\operatorname{supp}(f)$, so

$$\left\langle \frac{\chi}{\pi}, f \right\rangle = \lim_n \left\langle \psi_n * \frac{1}{\pi z}, f \right\rangle = -\lim_n \langle \psi_n, \mathfrak{C}(f) \rangle = \mathfrak{C}(f)(0).$$

□

5.5. The Vitushkin localization operator. The Vitushkin localization operator is defined by

$$\mathfrak{T}_\phi(f) := \mathfrak{C}\left(\phi \cdot \frac{\partial f}{\partial \bar{z}}\right).$$

Here $\phi \in \mathcal{D}$ and $f \in \mathcal{D}^*$.

In view of the distributional equation

$$\frac{\partial}{\partial \bar{z}} \mathfrak{T}_\phi(f) = \phi \cdot \frac{\partial f}{\partial \bar{z}},$$

$\mathfrak{T}_\phi(f)$ is holomorphic wherever f is holomorphic and off the support of ϕ.

It was established in [**36**] (using soft general arguments) that whenever F is an SCS, $\mathfrak{T}_\phi$ maps F continuously into F_{loc}, and that when F is an SCBS, we actually get a continuous map into the Banach subspace $F_\infty \subset F_{\text{loc}}$ normed by

$$\|f\|_{F_\infty} := \sup\{\|f|B\|_{F(B)} : B \text{ is a ball of radius } 1\},$$

where $f|B$ denotes the restriction coset $f + J(F, B)$, with $J(F, B)$ equal to the space of all elements $g \in F$ that vanish near B, and the $F(B)$ norm of a restriction is the infimum of the F norms of all its extensions in F, i.e.

$$\|f|B\|_{F(B)} := \inf\{\|h\|_F : h \in F, h - f = 0 \text{ near } B\}.$$

When F is an SCBS with the strong module property (of order k), and translation acts isometrically on F, the identity

$$\mathfrak{T}_\phi(f) = \phi \cdot f - \mathfrak{C}\left(\frac{\partial \phi}{\partial \bar{z}} \cdot f\right)$$

together with equation (5.1) yields the more precise estimate

$$\|\mathfrak{T}_\phi(f)\|_F \leq K N_{k+1}(\phi) \cdot \|f\|_F, \ \forall \phi \in \mathcal{D}, \forall f \in F. \tag{5.2}$$

5.6. Our spaces T_s. We denote the $(T_s)_\infty$ norm of a distribution f by $\|f\|_{s,\infty}$. Notice that if the support of f has diameter at most 1, then $\|f\|_s$ and $\|f\|_{s,\infty}$ are comparable, i.e. stay within constant multiplicative bounds of one another. In fact, using only the translation-invariance of the norm and the (ordinary) $\mathcal{D}$-module property, it is easy to see that, for such f,

$$\|f\|_{s,\infty} \leq \|f\|_s \leq K\|f\|_{s,\infty},$$

where K is independent of f.

LEMMA 5.2. *Let $k \in \mathbb{N}$ and $-k-1 < s < -k$. Then $(T_s)_\infty$ has the order $(k+2)$ strong module property, and in fact*

$$\|\phi \cdot f\|_s \leq K \cdot N_{k+2}(\phi) \cdot \|f\|_s,$$

for all $\phi \in \mathcal{D}$ with $d(\phi) \leq 1$ and all $f \in T_s$, where $K > 0$ is independent of ϕ and f.

PROOF. We use induction on k, *starting with $k = -1$.*

For $0 < s < 1$, Lip(s) has the strong module property of order 1, and since T_s is locally-equal to Lip(s), we have the result in this case.

Now suppose it holds for some k, and fix $s \in (-k-2, -k-1)$.

It suffices to prove the estimate for ϕ supported in $\mathbb{B}(0,2)$, and multiplying f by a fixed test function ψ that equals 1 on $\mathbb{B}(0,2)$, we may assume that f has compact support (without changing $\phi \cdot f$ or increasing $\|f\|_s$ by more than a fixed constant that depends only on ψ and s). Then $g := \mathfrak{C}f \in T_{s+1}$ has $\dfrac{\partial g}{\partial \bar{z}} = f$ and $\|g\|_{s+1} \leq K\|f\|_s$. Also $\frac{\partial}{\partial \bar{z}}$ maps T_{s+1} continuously into T_s, and $\frac{\partial}{\partial \bar{z}}\mathfrak{C}(\phi \cdot f) = \phi \cdot f$, so using (5.2) with k replaced by $k+2$, we have

$$\|\phi \cdot f\|_s \leq K\|\mathfrak{C}(\phi \cdot f)\|_{s+1} = K\|\mathfrak{T}_\phi(g)\|_{s+1} \leq K \cdot N_{k+3}(\phi)\|g\|_{s+1,\infty},$$

since $(T_{s+1})_\infty$ has the strong module property of order $k+2$, by the induction hypothesis. Then

$$\|\phi \cdot f\|_s \leq K \cdot N_{k+3}(\phi) \cdot \|g\|_{s+1} \leq K \cdot N_{k+3}(\phi) \cdot \|f\|_s$$

Hence the result holds for $k+1$, completing the induction step. □

REMARK 5.3. A similar result holds for all s, but for positive s one has to replace $\|g\|$ by $\|g - p\|$, where p is the degree $\lfloor s \rfloor$ Taylor polynomial of g about a.

5.7. The C_s norm on small discs. Since $\mathfrak{C}$ maps $(C_s)_{\text{cs}}$ into C_{s+1}, induction also gives the following:

LEMMA 5.4. *Let $s < 0$, $f \in C_s$ and $a \in \mathbb{C}$. Then for each $\epsilon > 0$ there exists $r > 0$ and $g \in C_s$ such that $f = g$ on $\mathring{\mathbb{B}}(a,r)$ and $\|g\|_{T_s} < \epsilon$.* □

REMARK 5.5. We note that since $\mathfrak{T}_\phi(f)$ is holomorphic off the support of ϕ and has a zero at ∞, $\mathfrak{T}_\phi$ maps C_s into C_s.

5.8. Estimate for $\langle \phi, f \rangle$. The strong module property gives an estimate for the action of $f \in F$ on a given $\phi \in \mathcal{D}$:

LEMMA 5.6. *Suppose F is an SCBS with the order k strong module property. Then for each compact $X \subset \mathbb{C}$, there exists $K > 0$ such that*

$$|\langle \phi, f \rangle| \leq K \cdot N_k(\phi) \cdot \|f\|_F,$$

whenever $\phi \in \mathcal{D}$, $f \in F$ and $\text{supp}(\phi \cdot f) \subset X$.

PROOF. Fix $\chi \in \mathcal{D}$ with $\chi = 1$ near X. Then since $f \mapsto \langle \chi, f \rangle$ is continuous there exists $K > 0$ such that

$$|\langle \chi, f \rangle| \leq K \cdot \|f\|_F.$$

Thus

$$|\langle \phi, f \rangle| = |\langle \chi, \phi \cdot f| \leq K \cdot N_k(\phi) \cdot \|f\|_F.$$

□

Putting it another way, the order k strong module property says that the operator $f \mapsto \phi \cdot f$ on F has operator norm dominated by $N_k(\phi)$, and this last lemma says that the functional $f \mapsto \langle \phi, f \rangle$ has $F(X)^*$ norm dominated by $N_k(\phi)$.

It turns out that we can improve substantially on this estimate, for our particular spaces. The trick is to pay close attention to the support of $\phi \cdot f$.

5.9. Scaling: better estimate. The action of an affine map $A : \mathbb{R}^d \to \mathbb{R}^d$ on distributions is defined by

$$\langle \phi, f \circ A \rangle := |A|^{-1} \langle \phi \circ A^{-1}, f \rangle, \ \forall \phi \in \mathcal{D}.$$

In the case of a dilation $A(z) = r \cdot z$ on $\mathbb{C}$, this means that

$$\langle \phi, f \rangle = r^2 \langle \phi \circ A, f \circ A \rangle.$$

Taking into account the fact that $N_k(\phi) = N_k(\phi \circ A)$ whereas the identity

$$(P_t * f)(z) = (P_{t/r} * (f \circ A))(rz)$$

gives

$$\|f \circ A\|_s = r^s \|f\|_s,$$

we obtain:

LEMMA 5.7. *Let* $-2 < s < 0$. *Then*

$$|\langle \phi, f \rangle| \leq K \cdot N_3(\phi) \cdot \|f\|_s \cdot r^{s+2},$$

whenever $\phi \in \mathcal{D}$ *and* $\operatorname{supp}(\phi \cdot f) \in \mathbb{B}(0, r)$. □

Note that for $f \in C_s$, the norm of f in $(T_s)_{\mathbb{B}(a,r)}$ tends to zero as $r \downarrow 0$, so we can replace the constant K in the estimate by $\eta(r)$, where η depends on f, and $\eta(r) \to 0$ as $r \downarrow 0$.

5.10. Hausdorff content estimate. Next, using a covering argument, we can bootstrap the estimate to:

LEMMA 5.8. *Let* $-2 < s < 0$. *Then*

$$|\langle \phi, f \rangle| \leq K \cdot N_3(\phi) \cdot \|f\|_s \cdot M^{s+2}(\operatorname{supp}(\phi \cdot f)),$$

whenever $\phi \in \mathcal{D}$ *and* $f \in T_s$. □

Before giving the proof of this lemma, we need some preliminaries.

A *closed dyadic square* is a set of the form $I_{m,n} \times I_{r,n}$ (for integers n, m, r) where

$$I_{m,n} := \left\{ x \in \mathbb{R} : \frac{m}{2^n} \leq x \leq \frac{m+1}{2^n} \right\}.$$

Let $\mathcal{S}_2$ denote the family of closed dyadic squares. For $\beta > 0$, the *β-dimensional dyadic content* of a set $E \subset \mathbb{R}^2$ is

$$M_2^\beta(E) := \inf\left\{\sum_{n=1}^\infty \text{side}(S_n)^\beta : E \subset \bigcup_{n=1}^\infty S_n, \text{ and } S_n \in \mathcal{S}_2\right\}.$$

This content is comparable to M^β:

$$M^\beta(E) \le 2^{\beta/2} M_2^\beta(E), \quad M_2^\beta(E) \le 2^{\beta+2} M^\beta(E),$$

for all bounded sets E.

Thus it suffices to prove Lemma 5.8 with M^{s+2} replaced by M_2^{s+2} in the statement.

Also, since both sides change by a factor r^2 when f and ϕ are replaced by $f(r\cdot)$ and $\phi(r\cdot)$, it suffices to consider the case when $E := \text{supp}(\phi \cdot f)$ has diameter at most 1. Given $\epsilon > 0$, we may cover such an E by a countable sequence $(S_n)_n$ of dyadic squares, of side at most 1, with

$$\sum_{n=1}^\infty \text{side}(S_n)^\beta < M_2^\beta + \epsilon.$$

We now state the key partition-of-identity lemma:

LEMMA 5.9. *Let $k \in \mathbb{N}$ be given. There exist positive constants K and Λ such that whenever $E \in \mathbb{R}^2$ is compact, and*

$$E \subset \bigcup_{n=1}^\infty S_n,$$

where the S_n are dyadic squares of side at most 1, there exists a sequence of test functions $(\phi_n)_n$ such that
(1) $\phi_n = 0$ except for finitely many n;
(2) $\sum_n \phi_n = 1$ on a neighbourhood of E;
(3) $N_k(\phi_n) \le K$ for all n, and
(4) $\text{supp}(\phi_n) \subset \Lambda S_n$.

Here ΛS denotes the square with the same centre as S and Λ times the side. We shall show that Λ may be taken equal to 5, although we do not claim this is sharp.

PROOF. Rearrange the S_n in nonincreasing order of size. The interiors of the $\frac{5}{4}S_n$ form an open covering of E, so we may select a finite subcover,

$$\mathcal{F} := \left\{\tfrac{5}{4}\mathring{S}_n : 1 \le n \le N\right\}.$$

Remove all squares from the sequence $(S_n)_n$ that are contained in $(\frac{5}{4}S_1) \setminus S_1$, renumber the remaining squares, and adjust n; then remove all in $(\frac{5}{4}S_2) \setminus S_2$, and so on. Now no element $S_n \in \mathcal{F}$ is contained in any square 'cordon' $(\frac{5}{4}S_m) \setminus S_m$.

Group the squares of $\mathcal{F}$ into generations

$$\mathcal{G}_m := \{S \in \mathcal{F} : \text{side}(S) = 2^{-m}\}$$

for $m = 0, 1, 2, \ldots$.

Each (finite) generation $\mathcal{G}_m$ forms part of the tesselation $\mathcal{T}_m$ of the whole plane by dyadic squares of side 2^{-m}. We can construct a uniform partition of unity on the whole plane subordinate to the covering by the open squares $5\mathring{S}$, with $S \in \mathcal{T}_m$, as follows:

Choose $\rho \in C^\infty([0,+\infty))$ such that ρ is nonincreasing, $\rho(r) = 1$ for $0 \le r \le \frac{5}{8}$ and $\rho(r) = 0$ for $r \ge \frac{3}{4}$. Then define

$$\theta(x,y) := \rho(x)\rho(y).$$

For a dyadic square S having centre (a,b) and side 1, define $\theta_S(x,y) := \theta(x-a, y-b)$, and

$$\tau := \sum_{S \in \mathcal{T}_1} \theta_S.$$

Then $\theta_S = 1$ on $\frac{5}{4}S$ and is supported on $\frac{3}{2}S$, so that $1 \le \tau \le 4$. Let

$$\psi_S := \frac{\theta_S}{\tau}.$$

Then the test functions ψ_S, for $S \in \mathcal{T}_1$ form a partition of unity, and $c_k := N_k(\psi_S)$ is independent of S. (This partition is invariant under translation by Gaussian integers.)

For general $m \in \mathbb{N}$, and a dyadic square S of side 2^{-m}, define $\psi_S(z) := \psi_{2^m S}(2^m z)$. Then the ψ_S, for $S \in \mathcal{T}_m$ also form a nonnegative smooth partition of unity, $N_k(\psi_S) = c_k$ is independent of S (and m), the support of ψ_S is contained in $\frac{3}{2}S$, hence at most 4 ψ_S are nonzero at any given point.

Note that

$$|\nabla^k \psi_S| \le \left(\mathrm{diam}\tfrac{3}{2}S\right)^k c_k = \left(\tfrac{3}{\sqrt{2}}\right)^k \cdot (\mathrm{side} S)^k \cdot c_k$$

for each $k \in \mathbb{N}$.

For a dyadic square S, let S^+ denote the set of 9 dyadic squares of the same size that meet S. For any family $\mathcal{H}$ of dyadic squares, let

$$\mathcal{H}^+ := \bigcup \{S^+ : S \in \mathcal{H}\}.$$

Thus $S^{++} := (S^+)^+$ is the family of 25 squares, consisting of S, the 8 other dyadic squares of the same size that meet S, and the 16 other dyadic squares of the same size that meet at least one of those 8 squares. Observe that the smooth function

$$\sum_{T \in S^+} \psi_T$$

is supported in $\bigcup S^{++} = 5S$ and has sum identically 1 on $\frac{3}{2}S$.

We now proceed to construct the desired collection of functions (ϕ_n).

Let

$$\sigma_m := \sum_{S \in \mathcal{G}_m^+} \psi_S.$$

Then σ_m is supported in $\bigcup \mathcal{G}_m^{++}$ and

$$\sigma_m = 1 \text{ on } K_m := \bigcup_{S \in \mathcal{G}_m} \tfrac{3}{2}S.$$

Since at most 4 ψ_S are nonzero at any one point, we have

$$|\nabla^k \sigma_m| \le 4c_k \left(\tfrac{3}{\sqrt{2}}\right)^k \cdot 2^{km}, \ \forall k \in \mathbb{N}. \tag{5.3}$$

Now take the squares $S \in \mathcal{G}_0^+$, and allocate each one to a 'nearest' square $n(S) \in \mathcal{G}_0$ so that:
(1) if $S \in \mathcal{G}_0$, take $n(S) = S$;
(2) if $S \notin \mathcal{G}_0$, pick $n(S)$ with $S \in n(S)^+$. (There may be up to eight ways to pick $n(S)$. It does not matter which you choose.)

Next, let

$$\phi_T := \sum_{n(S)=T} \psi_S, \ \forall T \in \mathcal{G}_0.$$

Then each ϕ_T is supported on $5T$, and

$$\sigma_0 = \sum_{T \in \mathcal{G}_0} \phi_T.$$

Since the sum defining ϕ_T has at most 9 terms, and its support has diameter at most 5 times that of T, we have

$$N_k(\phi_T) \le 9 \cdot 5^k \cdot c_k.$$

Let $\tau_0 := \sigma_0$.

Next, consider $\mathcal{G}_1$. As before, allocate each square $S \in \mathcal{G}_1^+$ to a nearest square $n(S) \in \mathcal{G}_1$, but this time let

$$\phi_T := (1-\tau_0) \sum_{n(S)=T} \psi_S, \ \forall T \in \mathcal{G}_1.$$

Then

$$\sum_{T \in \mathcal{G}_1} \phi_T = (1-\tau_0)\sigma_1.$$

and

$$\tau_1 := \tau_0 + (1-\tau_0)\sigma_1$$

is supported in $\bigcup(\mathcal{G}_0^{++} \cup \mathcal{G}_1^{++})$ and is identically equal to 1 on $K_0 \cup K_1$.

Continuing in this way, for $m \ge 1$ we allocate each square $S \in \mathcal{G}_{m+1}^+$ to a nearest square $n(S) \in \mathcal{G}_{m+1}$, and let

$$\phi_T := (1-\tau_m) \sum_{n(S)=T} \psi_S, \ \forall T \in \mathcal{G}_{m+1},$$

and

$$\tau_{m+1} := \tau_m + (1-\tau_m)\sigma_{m+1}.$$

Then

$$\sum_{T \in \mathcal{G}_{m+1}} \phi_T = (1-\tau_m)\sigma_{m+1}$$

and $\tau_{m+1} = 1$ on $K_0 \cup \cdots \cup K_{m+1}$.

When we have worked through all the nonempty generations $\mathcal{G}_m$, we will have defined ϕ_{S_n} for each S_n, and (renaming ϕ_{S_n} as ϕ_n) we have $\sum_n \phi_n = 1$ on the union of all the $\frac{3}{2}S_n$, and hence on a neighbourhood of E. Since ϕ_n is supported on $5S_n$, it remains to prove the estimate (3) of the statement, i.e. to prove that $\sup_n N_k(\phi_n) < +\infty$.

This amounts to showing that there is a constant $K > 0$ (depending on k) such that for $0 \le m \in \mathbb{Z}$ and $S \in \mathcal{G}_m$, we have

$$|\nabla^k \phi_S| \le K 2^{km}.$$

To do this, we start by proving that for each k there exists $C = C_k > 0$ such that

$$|\nabla^k \tau_m| \le C \cdot 2^{km}, \ \forall k \in \mathbb{N}. \tag{5.4}$$

for all $m \ge 0$.

To see this, we use induction on k and on m, and the identity

$$\tau_{m+1} = \sigma_{m+1} + (1 - \sigma_{m+1})\tau_m, \tag{5.5}$$

together with the bound (5.3).

Let us write

$$d_k := 4c_k \left(\tfrac{3}{\sqrt{2}}\right)^k,$$

so that (5.3) becomes $|\nabla^k \sigma_m| \le d_k 2^{km}$. Since $\tau_0 = \sigma_0$, then for any k, we know that (5.4) holds for $m = 0$ as long as C is at least d_k.

Take the case $k = 1$, and proceed by induction on m. If (5.4) holds for $k = 1$ and some m, then the identity (5.5) gives

$$|\nabla \tau_{m+1}| \le d_1 2^{m+1} + |\nabla \tau_m| + |\nabla \sigma_{m+1}| \le \left(d_1 + \frac{C}{2} + d_1\right) 2^{m+1}.$$

Thus we get (5.4) with m replaced by $m+1$, as long as $C \ge 4d_1$. This proves the case $k = 1$, with $C_1 = 4d_1$.

Now suppose that $k > 1$, and we have (5.4) with k replaced by any number r from 1 to $k-1$ and C replaced by some C_r. We proceed by induction on m. We have the case $m = 0$, with any constant $C \ge d_k$. Suppose we have the case m, with a constant C.

Using the identity, we can estimate $|\nabla^k \tau_{m+1}|$ by

$$d_k 2^{k(m+1)} + C \cdot 2^{km} + \sum_{j=1}^{k} \binom{k}{j} |\nabla^j \sigma_{m+1}| \cdot |\nabla^{k-j} \tau_m|.$$

This is no greater than

$$\left(\frac{C}{2^k} + R\right) \cdot 2^{k(m+1)},$$

where R is an expression involving $d_1, \ldots, d_k$ and $C_1, \ldots, C_{k-1}$. So as long as $C > 2R$, we get (5.4) with m replaced by $m+1$, and the induction goes through.

So we have (5.4) for all k and m. It follows easily that for some $C > 0$ (depending on k) and for each $S \in \mathcal{G}_m^+$ we have

$$|\nabla^k (1 - \tau_m) \cdot \psi_S| \le C \cdot 2^{km},$$

and this gives

$$|\nabla^k \phi_S| \le 9C \cdot 2^{km}$$

whenever $S \in \mathcal{G}_m$, as required. □

PROOF OF LEMMA 5.8. With $E = \mathrm{supp}(\phi \cdot f)$, take the partition of the identity (ϕ_n) constructed in Lemma 5.9, and note that $\phi = \sum_n \phi \cdot \phi_n$ on a neighbourhood of E. Thus

$$\langle \phi, f \rangle = \sum_{n=1}^{N} \langle \phi \cdot \phi_n, f \rangle.$$

Now apply Lemma 5.7 with ϕ replace by $\phi \cdot \phi_n$. The fact that N_k is submultiplicative implies that $N_3(\phi \cdot \phi_n) \le K N_3(\phi)$, so we get the stated result at once. □

Further, using the remark about $\eta(r) \downarrow 0$, we get a stronger statement for elements of C_s:

LEMMA 5.10. *Let* $-2 < s < 0$. *Then*

$$|\langle \phi, f\rangle| \le K \cdot N_3(\phi) \cdot \|f\|_s \cdot M_*^{s+2}(\text{supp}(\phi \cdot f),$$

whenever $\phi \in \mathcal{D}$ *and* $f \in C_s$. □

6. Proofs of preliminary lemmas

6.1. Proof of Lemma 3.1.

PROOF. Fix $f \in A^s(U)$. Take some $\psi \in \mathcal{D}$ having $\psi = 1$ near $\overline{U}$. Then $f_1 := \psi \cdot f \in A^s(U)$, so we may write $f = f_1 + f_2$, where $f_2 \in C_s$ vanishes near $\overline{U}$, and hence is holomorphic near b. So it remains to show that we can approximate f_1 by elements of $A_b^s(U)$.

Now f_1 has compact support. Take a standard pincher $(\phi_n)_n$ at b.

Take $g_n := T_{\phi_n}(f_1)$. Then $\|g_n\|_s \le K\|f_1\|_s$. Since T_{ϕ_n} depends only on the restriction of f_1 to the support of ϕ_n, an application of Lemmas 5.4 and 5.2 shows that $\|g_n\|_s \to 0$ as $n \uparrow \infty$. Thus $f_1 - g_n \to f_1$ in T_s norm. Finally,

$$\frac{\partial}{\partial \bar{z}}(f_1 - g_n) = (1 - \phi_n)\frac{\partial f_1}{\partial \bar{z}},$$

so $f_1 - g_n$ is holomorphic on a neighbourhood of b, and so belongs to $A_b^s(U)$. □

6.2. Proof of Lemma 3.4. First, we have to explain the weak-star topology in question, by specifying a specific predual for T_s.

The fact is that T_s is essentially the double dual of C_s. More, it is a *concrete dual*: An SCS F is called *small* if it is the closure of $\mathcal{D}$. If F is a small SCS, then its dual F^* is naturally isomorphic to an SCS, where the isomorphism is the restriction map $L \mapsto L|\mathcal{D}$. We call this SCS the concrete dual of F, and denote it by the same symbol F^*. Also F_{loc} and F_{cs} are also small,

$$(F_{\text{loc}})^* = (F^*)_{\text{cs}},$$

$$(F_{\text{cs}})^* = (F^*)_{\text{loc}},$$

and so

$$(F^*) \stackrel{\text{loc}}{=} (F_{\text{loc}})^* \stackrel{\text{loc}}{=} (F_{\text{cs}})^*.$$

In the case of C_s, for $0 < s < 1$, the concrete dual C_s^* is also small, and we have

$$(C_s^*)^* \stackrel{\text{loc}}{=} T_s.$$

This fact is basically due to Sherbert, who observed the isomorphism

$$\text{lip}(\alpha, K)^{**} = \text{Lip}(\alpha, K)$$

for all compact metric spaces K. The key to this is the fact that for each $L \in \text{lip}(\alpha, K)^*$ that annihilates constants there exists a measure μ on $K \times K$, having no mass on the diagonal, such that

$$Lf = \int_{K \times K} \frac{f(x) - f(y)}{\text{dist}(x, y)^\alpha} d\mu(x, y),$$

whenever $f \in \text{lip}(\alpha, K)$. In particular, each point-mass at an off-diagonal point $(z, w) \in \mathbb{C} \times \mathbb{C}$ gives an element of the dual of $\text{lip}\alpha(\mathbb{C})$:

$$Q(z,w)(f) := \frac{f(z) - f(w)}{|z-w|^\alpha}, \quad \forall f \in \text{lip}\alpha.$$

This might lead one to suspect that the dual is non-separable, but the norm topology on these functionals is not discrete. In fact, the map

$$P : \mathbb{C}^2 \setminus \text{diagonal} \to \text{lip}\alpha^*$$

is continuous, and indeed locally Hölder-continuous: one may show that

$$\|Q(z,w) - Q(z',w)\|_{\text{lip}\alpha^*} \le \frac{4|z-z'|^\alpha}{|z-w|}$$

whenever $|z-z'| < \frac{1}{2}|z-w|$. Hence the functional L on $\text{lip}\alpha_{\text{loc}}$ represented by a given measure μ may be approximated in the dual norm by finite linear combinations of elements from

$$\mathcal{P} := \{Q(z,w) : z \neq w\}.$$

By smearing the point masses, each functional $Q(z,w)$ may be approximated in the dual norm by functionals $\int_{\mathbb{C}} Q(z+\zeta, w+\zeta)\phi(\zeta)dm(\zeta)$, where $\phi \in \mathcal{D}$ has $\int \phi\, dm = 1$, which send an element $f \in \text{lip}(\alpha)$ to

$$\begin{aligned}&\int_{\mathbb{C}} Q(z+\zeta, w+\zeta)(f)\cdot\phi(\zeta)\,dm(\zeta)\\ &= \int_{\mathbb{C}} f(\omega)\left(\frac{\phi(\omega-z)-\phi(\omega-w)}{|z-w|^\alpha}\right)dm(\omega),\end{aligned}$$

and the function

$$\omega \mapsto \frac{\phi(\omega-z)-\phi(\omega-w)}{|z-w|^\alpha}$$

is a test function, so the functional L may be approximated by test functions. Thus $\text{lip}\alpha^*_{\text{loc}}$ has a concrete dual, and by Sherbert's result this can only be $\text{Lip}\alpha_{cs}$.

Moreover, it follows that a sequence in $\text{Lip}\alpha$ is weak-star convergent to zero if and only if it is bounded in $\text{Lip}\alpha$ norm and converges pointwise to zero on the span of $\mathcal{P}$. So in fact it suffices to show that it is bounded in norm and converges pointwise on $\mathbb{C}$. But we already know that if (ϕ_n) is a standard pincher, then, for $f \in \text{Lip}\alpha$, $T_{\phi_n} f$ is bounded in $\text{Lip}\alpha$ norm and converges uniformly to zero, hence we conclude that $T_{\phi_n} f$ is weak-star convergent to zero. This proves the lemma in case $0 < s < 1$.

For other nonintegral s, we obtain it by applying the Fundamental Theorem of Calculus. In particular, for the case of immediate interest, $-1 < s < 0$, we have that

$$((C_s)_{\text{loc}})^{**} \overset{\text{loc}}{=} ((DC_{s+1})_{\text{loc}})^{**} \overset{\text{loc}}{=} \int T_{s+1} \overset{\text{loc}}{=} T_s,$$

so to show that, for $f \in T_s$, the sequence $(T_{\phi_n} f)$ converges weak-star in T_s, it suffices to show that $(\mathfrak{C}T_{\phi_n} f)$ converges weak-star in T_{s+1}. We may assume that f has compact support, since $(T_{\phi_n} f)$ depends only on the restriction of f to a neighbourhood of $\text{supp}\phi$, and then taking $g = \mathfrak{C}f \in T_{s+1}$, it suffices to show that $\mathfrak{C}T_{\phi_n}\dfrac{\partial g}{\partial \bar{z}}$ converges weak-star to zero. But

$$\mathfrak{C}T_{\phi_n}\frac{\partial g}{\partial \bar{z}} = \mathfrak{C}^2\left(\phi_n \cdot \frac{\partial^2 g}{\partial \bar{z}^2}\right),$$

so we are just dealing with the equivalent of $\mathfrak{T}_\phi$ for the d-bar-squared operator instead of the d-bar operator, so it is bounded on Lipα and on C^0, independently of n, and thus we have the desired weak-star convergence.

REMARK 6.1. We expect that the argument of Subsection 6.2 may be used more generally, i.e we *conjecture* the following:

Let F be a small SCBS, such that F^* is also small, $F^{**} \overset{\text{loc}}{\hookrightarrow} C^0$, and the span of the point evaluations is dense in F^*, and F^{**} has the strong module property. Then whenever (ϕ_n) is a standard pincher, and L is an elliptic operator with smooth coefficients,

$$L^{-1}(\phi_n \cdot Lf) \to 0 \text{ weak-star } \forall f \in F^{**}.$$

Here, L^{-1} denotes some suitably-chosen parametrix for L.

7. Proofs of Theorems

7.1. Proof of Theorem 3.3. We fix $\beta \in (0,1)$ and $s = \beta - 1$, and without loss in generality we assume that the boundary point $b = 0$.

First, consider the 'only if' direction. Suppose the series diverges:

$$\sum_{n=1}^{\infty} 2^n M_*^\beta(A_n \setminus U) = +\infty.$$

We wish to show that there exist $f \in A_0^s(U)$ having $\|f\|_s \leq 1$ and $|f(0)|$ arbitrarily large.

Since M_*^β is subadditive, there exists at least one of the four right-angle sectors

$$S_r := \left\{ z \in \mathbb{C} : |\arg(i^r z)| < \frac{\pi}{4} \right\}$$

(for $r \in \{0,1,2,3\}$) such that

$$\sum_{n=1}^{\infty} 2^n M_*^\beta((S_r \cap A_n) \setminus U) = +\infty.$$

We may assume that this happens for $r = 0$, and we may assume further that U contains the whole complement of S_0 and the whole exterior of the unit disc. So we may select closed sets $E_n \subset S_0 \cap A_n$ such that $U \cap E_n = \emptyset$ and

$$\sum_{n=1}^{\infty} 2^n M_*^\beta(E_n) = +\infty.$$

We may select numbers $\lambda_n > 0$ such that the individual terms $\lambda_n 2^n M_*^\beta(E_n) \leq 1$, and yet

$$\sum_{n=1}^{\infty} \lambda_n 2^n M_*^\beta(E_n) = +\infty.$$

For each n, by Frostman's Lemma, we may select a positive Radon measure supported on E_n such that (1) $\mu_n(\mathbb{B}(a,r)) \leq r^\beta$ for all $a \in \mathbb{C}$ and all $r > 0$ (i.e. μ_n 'has growth β'), (2) the total variation $\|\mu_n\| \geq K \cdot M_*^\beta(E_n)$, and (3) $\mu_n(\mathbb{B}(a,r))/r^\beta \to 0$

uniformly in a as $r \downarrow 0$. Then taking $r \mapsto h(r)$ to be the upper concave envelope of $r \mapsto \sup_a \mu_n(\mathbb{B}(a,r)$ on $[0,+\infty)$, we have $\|\mu_n\| \leq M_h(E_n) \leq M_*^\beta(E_n)$, so $\lambda_n 2^n\|\mu_n\| \leq 1$ and

$$\sum_{n=1}^{\infty} \lambda_n 2^n \|\mu_n\| = +\infty.$$

Let $h_n := \lambda_n \mathfrak{C}(\mu_n)$. Then $h_n \in C_s$, h_n is holomorphic off $\operatorname{supp}(\mu_n)$, hence $h_n \in A_0^s(U)$. Also $\Re(h_n(0)) \geq \lambda_n \frac{2^n}{\sqrt{2}}\|\mu_n\|$. Hence

$$\left|\sum_{n=1}^{N} h_n(0)\right| \geq \frac{1}{\sqrt{2}} \sum_{n=1}^{N} \lambda_n 2^n \|\mu_n\| \to +\infty$$

as $N \uparrow \infty$. So now it suffices to show that $f_N := \sum_{n=1}^N h_n$ is bounded in T_s norm, independently of $N \in \mathbb{N}$.

For this, it suffices to show that the $\frac{\partial}{\partial \bar z}$-derivatives

$$g_N := \frac{\partial}{\partial \bar z} f_N = \sum_{n=1}^{N} \lambda_n \mu_n$$

are bounded in $T_{s-1} = T_{\beta-2}$, i.e. that for some $K > 0$ we have

$$\sum_{n=1}^{N} \frac{\lambda_n}{\pi} \cdot \int \frac{t\, d\mu_n(\zeta)}{(t^2 + |z-\zeta|^2)^{\frac{3}{2}}} \leq K t^{\beta-2},$$

whenever $z \in \mathbb{C}$ and $t > 0$.

When $t \geq 1$, we have the trivial estimate (independent of z)

$$\lambda_n (P_t * \mu_n)(z,t) \leq \frac{\lambda_n}{\pi} \frac{M_*^\beta(E_n)}{t^2} \leq \frac{1}{\pi 2^n t^2},$$

so this gives $|P_t * g_N| \leq K t^{-2} \leq K t^{\beta-2}$.

So to finish, fix $t \in (0,1)$, and choose $m \in \mathbb{N}$ such that $2^{-m-1} \leq t \leq 2^{-m}$, take the n-th term in the sum, and consider separately the ranges of n:
case 1°: $n > m-2$, and case 2°: $n < m-2$,
and the possible positions of z in relation to A_n.

Case 1°:
The trivial estimate also gives

$$\lambda_n \cdot (P_t * \mu_n)(z,t) \leq \frac{M^\beta(A_n)}{\pi t^2} \leq \frac{(2^{-n})^\beta}{\pi t^2},$$

so we get an estimate for the total contribution from all the Case 1° terms:

$$t^{2-\beta} \sum_{n=m-2}^{\infty} \lambda_n \cdot (P_t * \mu_n)(z,t) \leq \frac{1}{\pi} \sum_{n=m-2}^{\infty} 2^{(m+1-n)\beta} = \frac{8^\beta}{\pi(1-2^{-\beta})},$$

Case 2°:
To deal with this we have to consider the position of z in relation to A_n.

There are at most three n such that the distance from z to A_n is less than 2^{-n-1}. For these we can use the uniform estimate

$$t^{2-\beta} \cdot (P_t * \mu_n)(z,t) \leq K,$$

which follows from the fact that μ_n has growth β. (just write the value of $P_t * \mu_n(w)$ as a sum of the integrals over the annuli

$$\{z \in \mathbb{C} : 2^m t < |z - w| \leq 2^{m+1} t\}$$

from 0 to $-\log_2 t$ plus the integral over the disc $\mathbb{B}(w,t)$).

For the remaining $n \in \{1, \dots, m-3\}$, the estimate

$$(P_t * \mu_n)(z,t) \leq \frac{t \cdot M^\beta(A_n)}{\pi \cdot \mathrm{dist}(z, A_n)^3}$$

gives

$$t^{2-\beta} \cdot (P_t * \mu_n)(z,t) \leq (2^{\beta-3})^{m+1-n},$$

so

$$t^{2-\beta} \sum_{n=1}^{m-3} \lambda_n \cdot (P_t * \mu_n)(z,t) \leq 3K + \sum_{n=1}^{m-3} (2^{\beta-3})^{m+1-n} = K,$$

another constant (depending on β), and we are done.

Now consider the converse. Suppose $\sum_n 2^n M_*^\beta(A_n \setminus U) < +\infty$. We want to show that $A^s(U)$ admits a continuous point evaluation at 0.

If V is an open subset of U, then $A^s(U)$ is a subset of $A^s(V)$, so it suffices to prove the result for U that are contained in $\mathring{\mathbb{B}}(0, \frac{1}{2})$. We assume this is the case.

We may choose radial functions $\psi_n \in \mathcal{D}$ such that $\psi_n = 1$ on A_n, $\psi_n = 0$ off $A_{n-1} \cup A_n \cup A_{n+1}$, and for each k the sequence $(N_k(\psi_n))_n$ is bounded. Let

$$\phi_n := \frac{\psi_n}{\sum_{m=1}^\infty \psi_m}$$

on the complement of $\{0\}$, and $\phi_n(0) = 0$. Then each $\phi_n \in \mathcal{D}$, is zero off $A_{n-1} \cup A_n \cup A_{n+1}$, the sequences $(N_k(\phi_n))_n$ are all bounded, and $\sum_n \phi_n = 1$ on the union of all the A_n.

Fix a test function χ that equals 1 on $\mathbb{B}(0, \frac{1}{2})$ and is supported on $\mathbb{B}(0,1)$.

Fix $f \in A_0^s(U)$. We want to prove that $|f(0)| \leq K\|f\|_s$, where $K > 0$ does not depend on f.

We have $f(0) = (\chi \cdot f)(0)$, $\chi \cdot f \in A^s(U)$, and $\|\chi \cdot f\|_s \leq K\|f\|_s$, so it suffices to prove the estimate for $f \in A_0^s(U)$ having support in $\mathbb{B}(0,1)$.

Choose $N \in \mathbb{N}$ such that $f(z)$ is holomorphic for $|z| < 2^{2-N}$. Define $\phi_0(z)$ to be $1 - \phi_N(z)$ when $|z| < 2^{-N-1}$ and 0 otherwise. Then $\phi_0 \in \mathcal{D}$, $N_k(\phi_0) = N_k(\phi_N)$, and the test function

$$\phi := \phi_0 + \sum_{n=1}^N \phi_n$$

is equal to 1 near $\mathbb{B}(0,1)$.

We have

$$f = \phi \cdot f = \mathfrak{C}\left(\frac{\partial}{\partial \bar{z}}(\phi \cdot f)\right).$$

Since $\dfrac{\partial}{\partial \bar{z}}\phi = 0$ on the support of f, this equals

$$\mathfrak{C}\left(\phi \frac{\partial f}{\partial \bar{z}}\right) = \mathfrak{C}\left(\phi_0 \frac{\partial f}{\partial \bar{z}}\right) + \sum_{n=1}^N \mathfrak{C}\left(\phi_n \frac{\partial f}{\partial \bar{z}}\right) = \sum_{n=1}^N \mathfrak{C}(\phi_n \frac{\partial f}{\partial \bar{z}}),$$

since f is holomorphic on $\mathrm{supp}(\phi_0)$.

Take a test function ψ that equals $1/z$ for $2^{-N} < |z| < 2$.

Applying Lemma 5.1, we have

$$\mathfrak{C}\left(\phi_n \cdot \frac{\partial f}{\partial \bar{z}}\right)(0) = -\left\langle \frac{\psi}{\pi}, \phi_n \cdot \frac{\partial f}{\partial \bar{z}}\right\rangle = -\left\langle \frac{\psi \cdot \phi_n}{\pi}, \frac{\partial f}{\partial \bar{z}}\right\rangle.$$

Thus

$$f(0) = -\sum_{n=1}^{N} \left\langle \frac{\phi_n}{\pi z}, \frac{\partial f}{\partial \bar{z}}\right\rangle$$

(Here, by ϕ_n/z we understand the test function that equals 0 at the origin and ϕ_n/z everywhere else in $\mathbb{C}$.)

Applying the Hausdorff content estimate from Lemma 5.10, we have

$$\begin{aligned} |f(0)| &\leq K \cdot \sum_{n=1}^{N} N_3\left(\frac{\phi_n}{z}\right) \cdot M_*^\beta\left(\operatorname{supp}(\phi_n \cdot \frac{\partial f}{\partial \bar{z}})\right) \cdot \left\|\frac{\partial f}{\partial \bar{z}}\right\|_{s-1} \\ &\leq K \cdot \sum_{n=1}^{\infty} 2^n M_*^\beta((A_{n-1} \cup A_n \cup A_{n+1}) \setminus U) \cdot \left\|\frac{\partial f}{\partial \bar{z}}\right\|_{s-1}, \end{aligned}$$

since $N_3(\phi_n/z) \leq 2^{n+1} N_3(\phi_n) \leq K2^n$.

Since M_*^β is subadditive and $\left\|\frac{\partial f}{\partial \bar{z}}\right\|_{s-1} \leq K\|f\|_s \leq K$, we get

$$|f(0)| \leq K \cdot \sum_{n=1}^{\infty} 2^n M_*^\beta(A_n \setminus U) \cdot \|f\|_s.$$

This completes the proof.

REMARK 7.1. The proof actually shows that the sum of the series is the dual norm of the point evaluation $f \mapsto f(b)$, up to multiplicative constants that depend only on β.

7.2. Proof of Theorems 3.6, 3.7 and 3.8. To prove Theorem 3.6, one can use exactly the same argument, just replacing M_*^β by M^β, and using Lemma 5.8 instead of Lemma 5.10.

To prove the other two theorems, one just uses the corresponding Cauchy-Pompeiu formulas for derivatives:

$$\mathfrak{C}(\mu_n)^{(k)}(0) = \frac{k!}{\pi} \int \frac{d\mu_n(z)}{z^{k+1}}$$

for the 'only if' direction, and

$$f^{(k)}(0) = -\sum_{n=1}^{N} \left\langle \frac{k!\phi_n}{\pi z^{k+1}}, \frac{\partial f}{\partial \bar{z}}\right\rangle$$

for the 'if' direction.

7.3. Proof of Theorem 3.9. The point is that a distribution $f \in C_s$ satisfies $\Delta f = 0$ on the open set U if and only if $\frac{\partial f}{\partial z}$ is holomorphic on U, and the operator $\frac{\partial}{\partial z}$ maps T_s into T_{s-1}, and is inverted on $(T_{s-1})_{\text{cs}}$ by the 'anti-Cauchy' transform. So the results are just reformulations of Theorem 3.3.

REMARK 7.2. This is an example of 1-reduction, and one could also formulate equivalent results about other elliptic operators. In particular, M_*^β is also the capacity for T_β and the operator $\left(\frac{\partial}{\partial \bar{z}}\right)^2$, which is associated to complex elastic potentials, and it is the capacity for $T_{\beta+2}$ (a space of functions that are twice differentiable, but may have discontinuities in the third derivative) and the operator Δ^2, associated to elastic plates.

8. Concluding remarks

8.1. In [**6**, p.311] Carleson proved that for $0 < \beta < 1$ the M^β-null sets are the removable singularities for the class of Lipβ "multiple-valued holomorphic functions having single-valued real part." The expression in quotation marks is really code for "harmonic functions", so this is really the first version of the fact that M^β is the Δ-Lipβ-cap.

In the same paper, Carleson proved a precursor to Dolzhenko's theorem [**13**] about removable singularities for Lipβ holomorphic functions. He left a little gap, between the Hausdorff content and the nearby Riesz capacity, and Dolzhenko closed the gap.

8.2. Conjecture. Recently [**37**] the author showed that the existence of a continuous point derivation on $A^s(U)$ at b, for some positive $s < 1$, implies that the value of the derivation may be calculated by taking limits of difference quotients from a subset $E \subset U$ having full area density at b. In case U also satisfies an interior cone condition at b, the value may be calculated by taking limits along the midline of the cone. It seems reasonable to hope that for negative s, if $A^s(U)$ admits a continuous point evaluation at b, then the value can be calculated in a similar way, as

$$\lim_{z\to b, z\in E} f(z)$$

for some $E \subset U$ having full area density at b, and for segments $E \subset U$ (if any) along which nontangential approach to b is possible. In the case of L^p spaces, results along these lines have also been obtained by Wolf [**44**] and Deterding [**10–12**]. See also [**14, 27, 43**].

8.3. Question. Suppose F is an SCBS on $\mathbb{R}^d$ having the strong module property

$$\|\phi \cdot f\|_F \le K \cdot N_k(\phi) \cdot \|f\|_F, \quad \forall \phi \in \mathcal{D}\ \forall f \in F,$$

for some positive constant K and some nonnegative integer k. Define an inner capacity $c_{F,k}$ by the rule that for each compact $E \subset \mathbb{R}^d$ the value $c_{F,k}(E)$ is the least nonnegative number c such that

$$|\langle \phi, f\rangle| \le N_k(\phi) \cdot \|f\|_F \cdot c$$

whenever $\phi \in \mathcal{D}$, $f \in F$, and $\text{supp}(\phi \cdot f) \subset E$. For example, if $F = L^\infty$, it is easy to see that $c_{F,0}(E)$ is the d-dimensional Lebesgue measure of E, whereas for $F = L^1$, $c_{F,0}(E) = 1$ for all E.

The question is this: *For which F and k is it the case that*
$c_{F,k} \le K \cdot (1\text{-}F\text{-cap})$ *for some constant K?*

Recall that for compact $E \subset \mathbb{R}^d$,

$$1\text{-}F\text{-cap}(E) := \inf\{|\langle \chi, f\rangle| : \|f\|_F \le 1, \text{supp}(f) \subset E\},$$

where $\chi \in \mathcal{D}$ is any fixed test function such that $\chi = 1$ on E.

We have seen that this holds for $F = T_s$, $s \in \mathbb{R}$. Does it hold for all SCBS having the strong module property?

8.4. Question. If an SCBS F has the order k strong module property, when is there an SCBS locally-equal to DF that has the strong order $k+1$ module property? And what about $\int F$?

DF_{loc} is the Frechet space topologised by the seminorms defined by

$$\|f\|_n = \inf\left\{ \|g_1 + \cdots + g_d\|_{F(\mathbb{B}(0,n))} : g_j \in F, f = \frac{\partial g_1}{\partial x_1} + \cdots + \frac{\partial g_d}{\partial x_d} \right\}.$$

Note that if $d = 2$, and F is weakly-locally invariant under Calderon-Zygmund operators (or just under the Beurling transform), then DF_{loc} is topologised by the seminorms

$$\|f\|_n = \inf\left\{ \|g\|_{F(\mathbb{B}(0,n))} : g \in F_{\text{cs}}, f = \frac{\partial g}{\partial \bar{z}} \right\},$$

and this implies that each $\|\phi \cdot f\|_n$ is dominated by $N_{k+1}(\phi) \cdot \|f\|_n$, because

$$\phi \cdot \frac{\partial g}{\partial \bar{z}} = \frac{\partial}{\partial \bar{z}}(\phi \cdot g) - \left(\frac{\partial \phi}{\partial \bar{z}}\right) \cdot g.$$

This property is a kind of local version of the strong module property.

8.5. Acknowledgment. The author is grateful to the referee for a careful reading of the typescript and for corrections and suggestions that materially improved the exposition.

References

[1] D. H. Armitage and S. J. Gardiner, *Classical potential theory*, Springer Monographs in Mathematics, Springer-Verlag London, Ltd., London, 2001. MR1801253

[2] T. Bagby, *Quasi topologies and rational approximation*, J. Functional Analysis **10** (1972), 259–268, DOI 10.1016/0022-1236(72)90025-0. MR0355058

[3] J. E. Brennan, *Invariant subspaces and rational approximation*, J. Functional Analysis **7** (1971), 285–310, DOI 10.1016/0022-1236(71)90036-x. MR0423059

[4] J. E. Brennan, *Point evaluations and invariant subspaces*, Proceedings of an International Symposium on Operator Theory (Indiana Univ., Bloomington, Ind., 1970), Indiana Univ. Math. J. **20** (1971), no. 10, 879–881, DOI 10.1512/iumj.1971.20.20070. MR0407640

[5] J. E. Brennan, *Bounded point derivations on certain function algebras*, 50 years with Hardy spaces, Oper. Theory Adv. Appl., vol. 261, Birkhäuser/Springer, Cham, 2018, pp. 173–190. MR3792096

[6] L. Carleson, *On null-sets for continuous analytic functions*, Ark. Mat. **1** (1951), 311–318, DOI 10.1007/BF02591368. MR0041924

[7] L. Carleson, *Selected problems on exceptional sets*, Van Nostrand Mathematical Studies, No. 13, D. Van Nostrand Co., Inc., Princeton, N.J.-Toronto, Ont.-London, 1967. MR0225986

[8] B. E. J. Dahlberg, *On exceptional sets at the boundary for subharmonic functions*, Ark. Mat. **15** (1977), no. 2, 305–312, DOI 10.1007/BF02386046. MR0463467

[9] A. M. Davie, *Analytic capacity and approximation problems*, Trans. Amer. Math. Soc. **171** (1972), 409–444, DOI 10.2307/1996389. MR350009

[10] S. Deterding. *Bounded point derivations on certain function spaces*, Thesis, University of Kentucky. 2018.

[11] S. Deterding. *A formula for a bounded point derivation on $R^p(X)$*. Preprint 2017. arXiv:1709.04050.

[12] S. Deterding, *Bounded point derivations on $R^p(X)$ and approximate derivatives*, Math. Scand. **124** (2019), no. 1, 132–148, DOI 10.7146/math.scand.a-109998. MR3904752

[13] E. P. Dolženko, *The removability of singularities of analytic functions* (Russian), Uspekhi Mat. Nauk **18** (1963), no. 4 (112), 135–142. MR0155965

[14] C. Fernstrøm, *Bounded point evaluations and approximation in L^p by analytic functions*, Spaces of analytic functions (Sem. Functional Anal. and Function Theory, Kristiansand, 1975), Springer, Berlin, 1976, pp. 65–68. Lecture Notes in Math., Vol. 512. MR0486560

[15] C. Fernström and J. C. Polking, *Bounded point evaluations and approximation in L^p by solutions of elliptic partial differential equations*, J. Functional Analysis **28** (1978), no. 1, 1–20, DOI 10.1016/0022-1236(78)90077-0. MR0481468

[16] B. Fuglede, *Extremal length and functional completion*, Acta Math. **98** (1957), 171–219, DOI 10.1007/BF02404474. MR0097720

[17] T. W. Gamelin, *Uniform algebras*, Prentice-Hall, Inc., Englewood Cliffs, N. J., 1969. MR0410387

[18] T. W. Gamelin and J. Garnett, *Distinguished homomorphisms and fiber algebras*, Amer. J. Math. **92** (1970), 455–474, DOI 10.2307/2373334. MR303296

[19] J. Garnett, *Analytic capacity and measure*, Lecture Notes in Mathematics, Vol. 297, Springer-Verlag, Berlin-New York, 1972. MR0454006

[20] A. P. Hallstrom, *On bounded point derivations and analytic capacity*, J. Functional Analysis **4** (1969), 153–165, DOI 10.1016/0022-1236(69)90028-7. MR0243358

[21] L. I. Hedberg, *Bounded point evaluations and capacity*, J. Functional Analysis **10** (1972), 269–280, DOI 10.1016/0022-1236(72)90026-2. MR0352493

[22] L. I. Hedberg, *Approximation in the mean by analytic functions*, Trans. Amer. Math. Soc. **163** (1972), 157–171, DOI 10.2307/1995714. MR432886

[23] R. Kaufman, *Hausdorff measure, BMO, and analytic functions*, Pacific J. Math. **102** (1982), no. 2, 369–371. MR686557

[24] M. V. Keldych, *On the solubility and the stability of Dirichlet's problem* (Russian), Uspekhi Matem. Nauk **8** (1941), 171–231. MR0005249

[25] T. Kolsrud, *Bounded point evaluations and balayage*, Ark. Mat. **20** (1982), no. 1, 137–146, DOI 10.1007/BF02390505. MR660132

[26] D. J. Lord and A. G. O'Farrell, *Boundary smoothness properties of* Lip α *analytic functions*, J. Anal. Math. **63** (1994), 103–119, DOI 10.1007/BF03008420. MR1269216

[27] J. E. McCarthy and L. M. Yang, *Bounded point evaluations on the boundaries of L regions*, Indiana Univ. Math. J. **43** (1994), no. 3, 857–883, DOI 10.1512/iumj.1994.43.43037. MR1305950

[28] M. S. Mel′nikov, *Estimate of the Cauchy integral over an analytic curve* (Russian), Mat. Sb. (N.S.) **71 (113)** (1966), 503–514. MR0206301

[29] A. G. O'Farrell, *Hausdorff content and rational approximation in fractional Lipschitz norms*, Trans. Amer. Math. Soc. **228** (1977), 187–206, DOI 10.2307/1998526. MR432887

[30] A. G. O'Farrell, *Capacities in Uniform Approximation*, ProQuest LLC, Ann Arbor, MI, 1973. Thesis (Ph.D.)–Brown University. MR2623760

[31] A. G. O'Farrell, *Analytic capacity, Hölder conditions and τ-spikes*, Trans. Amer. Math. Soc. **196** (1974), 415–424, DOI 10.2307/1997036. MR361116

[32] A. G. O'Farrell, *Analytic capacity and equicontinuity*, Bull. London Math. Soc. **10** (1978), no. 3, 276–279, DOI 10.1112/blms/10.3.276. MR519908

[33] A. O'Farrell. *Symmetric Concrete Spaces*. Maynooth College. 1993.

[34] A. G. O'Farrell, *The order of a symmetric concrete space*, Proc. Roy. Irish Acad. Sect. A **88** (1988), no. 1, 39–48. MR974282

[35] A. G. O'Farrell, *The* 1*-reduction for removable singularities, and the negative Hölder spaces*, Proc. Roy. Irish Acad. Sect. A **88** (1988), no. 2, 133–151. MR986220

[36] A. G. O'Farrell, *T-invariance*, Proc. Roy. Irish Acad. Sect. A **92** (1992), no. 2, 185–203. MR1204219

[37] A. G. O'Farrell, *Derivatives at the boundary for analytic Lipschitz functions* (Russian, with Russian summary), Mat. Sb. **207** (2016), no. 10, 119–140, DOI 10.4213/sm8629; English transl., Sb. Math. **207** (2016), no. 9-10, 1471–1490. MR3588975

[38] E. M. Stein, *Singular integrals and differentiability properties of functions*, Princeton Mathematical Series, No. 30, Princeton University Press, Princeton, N.J., 1970. MR0290095

[39] N. N. Tarkhanov, *The Cauchy problem for solutions of elliptic equations*, Mathematical Topics, vol. 7, Akademie Verlag, Berlin, 1995. MR1334094

[40] J. Verdera, *BMO rational approximation and one-dimensional Hausdorff content*, Trans. Amer. Math. Soc. **297** (1986), no. 1, 283–304, DOI 10.2307/2000469. MR849480

[41] J. L. M. Wang, *An approximate Taylor's theorem for* $R(X)$, Math. Scand. **33** (1973), 343–358 (1974), DOI 10.7146/math.scand.a-11496. MR0348124

[42] N. Wiener. The Dirichlet Problem. MIT J Math Phys 3 (1924) 127-146.

[43] E. Wolf, *Bounded point evaluations and smoothness properties of functions in* $R^p(X)$, Trans. Amer. Math. Soc. **238** (1978), 71–88, DOI 10.2307/1997798. MR470679

[44] E. Wolf, *Smoothness properties of functions in* $R^2(X)$ *at certain boundary points*, Internat. J. Math. Math. Sci. **2** (1979), no. 3, 415–426, DOI 10.1155/S0161171279000326. MR542956

Mathematics and Statistics, NUI, Maynooth, Co Kildare, W23 HW31, Ireland
Email address: anthony.ofarrell@mu.ie

Contemporary Mathematics
Volume **743**, 2020
https://doi.org/10.1090/conm/743/14960

Cyclicity in Dirichlet type spaces

K. Kellay, F. Le Manach, and M. Zarrabi

Dedicated to Thomas Ransford on the occasion of his 60th birthday.

ABSTRACT. We study cyclicity in the Dirichlet type spaces for outer functions whose zero set is countable.

1. Introduction and main result

Let X be a Banach space of functions holomorphic in the open unit disk $\mathbb{D}$, such that the shift operator $S : f(z) \to zf(z)$ is a continuous map from X into itself. Given $f \in X$, we denote by $[f]_X$ the smallest closed S–invariant subspace of X containing f, namely

$$[f]_X = \overline{\{pf : p \text{ is a polynomial}\}}.$$

We say that f is cyclic in X if $[f]_X = X$.

The problem of cyclic vectors in the Dirichlet spaces goes back to the work of Beurling and Carleson (see [**B**, **C**]). The classical Dirichlet space $\mathcal{D}$ consists of holomorphic functions on the unit disc whose derivatives are square integrable. While Beurling characterizes cyclic vectors in the Hardy space H^2, the problem of characterizing the cyclic vectors in the Dirichlet space $\mathcal{D}$ is much more difficult. Beurling's theorem says that the cyclic vectors in H^2 are the outer functions. On the other hand we know that there are outer functions in the Dirichlet space which are not cyclic in $\mathcal{D}$. In fact, the cyclicity of such a function depends on the distribution of the zeros of the radial limit f^* of f on the unit circle. The Brown–Shields conjecture [**HS**] claims that $f \in \mathcal{D}$ is cyclic iff f is an outer function and the set of all zeros of f^* is a set of logarithmic capacity zero. A partial (positive) answer to this conjecture was given in [**EKR2**, **EKR1**]. We mention the results of Beurling [**B**] about the boundary behavior for the functions of the Dirichlet spaces: if $f \in \mathcal{D}$ we write $f^*(\zeta) = \lim_{r\to 1-} f(r\zeta)$, then the radial limit f^* exists –q.e on $\mathbb{T}$, that is f^* exists outside a set of capacity logarithmic zero. As a consequence of a weak-type inequality the invariant subspace $\mathcal{D}_E$ defined by

$$\mathcal{D}_E = \{f \in \mathcal{D},\ f^*|_E = 0 \text{ q.e.}\}$$

2010 *Mathematics Subject Classification.* Primary 46E22; Secondary 31A05, 31A15, 31A20, 47B32.

Key words and phrases. Dirichlet spaces, Bergman spaces, Smirnov space, cyclic vectors.

The research of the first author is partially supported by by the project ANR-18-CE40-0035 and by the Joint French-Russian Research Project PRC CNRS/RFBR,2017-2019.

is closed in $\mathcal{D}$. Carleson in [**C**] proved that for every closed subset E of the unit circle which has zero logarithmic capacity, there exists a cyclic function in $\mathcal{D}$ which vanishes on E.

We denote by $A(\mathbb{D})$ the disc algebra. Hedenmalm and Shields showed in [**HS**] that if $f \in \mathcal{D} \cap A(\mathbb{D})$ is an outer function and $\mathcal{Z}(f^*) = \{\zeta \in \mathbb{T} : f^*(\zeta) = 0\}$, the zero set of f^*, is countable then f is cyclic in $\mathcal{D}$. Richter and Sundberg in [**RS1**] improve this result by showing that if $f \in \mathcal{D}$ is outer and $\underline{\mathcal{Z}}(f) = \{\zeta \in \mathbb{T} : \liminf_{z\to\zeta} |f(z)| = 0\}$ is countable then f is cyclic in $\mathcal{D}$. When the set of zeros of f^* is not countable, see [**EKMR, EKR1, EKR2**] in the case of the classical Dirichlet space $\mathcal{D}^2_0$ and [**EKR3**] in the case of $\mathcal{D}^2_\alpha$, $0 < \alpha < 1$ for further results on cyclicity in that context.

In this paper we are interested in cyclicity, in more general Dirichlet spaces, of outer functions such that the zero set is countable. We now introduce some notations. The Dirichlet/Besov space $\mathcal{D}^p_\alpha$ with $p \geq 1$ and $\alpha > -1$ is given by

$$\mathcal{D}^p_\alpha = \left\{ f \in \mathrm{Hol}(\mathbb{D}) : \|f\|^p_{\mathcal{D}^p_\alpha} = |f(0)|^p + \int_{\mathbb{D}} |f'(z)|^p \mathrm{dA}_\alpha(z) < \infty \right\}.$$

where dA_α denotes the finite measure on the unit disc $\mathbb{D}$ given by

$$\mathrm{dA}_\alpha(z) := (1+\alpha)(1-|z|^2)^\alpha \mathrm{dA}(z),$$

and $\mathrm{dA}(z) = \mathrm{d}x\mathrm{d}y/\pi$ stands for the normalized area measure on $\mathbb{D}$. If $p = 2$ and $\alpha = 1$, then $\mathcal{D}^2_1$ is the Hardy space H^2 and the classical Dirichlet space corresponds to $p = 2$ and $\alpha = 0$, $\mathcal{D}^2_0 = \mathcal{D}$. The following theorem is the main result of this paper.

THEOREM. *Let $p > 1$ be such that $\alpha + 1 < p \leq \alpha + 2$ and let $f \in \mathcal{D}^p_\alpha \cap A(\mathbb{D})$. If f is outer and $\mathcal{Z}(f)$ is countable, then f is cyclic in $\mathcal{D}^p_\alpha$.*

Notice that when $1 < p < \alpha + 1$, $H^p(\mathbb{D})$ is continuously embedded in $\mathcal{D}^p_\alpha$ and every outer function $f \in H^p(\mathbb{D})$ is cyclic for $\mathcal{D}^p_\alpha$ (Proposition 3.1). On the other hand when $p > \alpha + 2$ then every function which vanishes at least at one point is not cyclic in $\mathcal{D}^p_\alpha$.

The method used for the proof of Theorem 3.10 is inspired by that of the Hedenmalm and Shields [**HS**] in the case of the classical Dirichlet space and the paper [**EKR2**] .

Throughout the paper, we use the following notations:

- $A \lesssim B$ means that there is an absolute constant C such that $A \leq CB$.
- $A \asymp B$ if both $A \lesssim B$ and $B \lesssim A$ hold.

2. Dirichlet space and duality

The Bergman spaces $\mathcal{A}^p_\alpha$ with $p \geq 1$, $\alpha > -1$ are given by

$$\mathcal{A}^p_\alpha(\mathbb{D}) = \left\{ f \in \mathrm{Hol}(\mathbb{D}),\ \|f\|^p_{\mathcal{A}^p_\alpha} = \int_{\mathbb{D}} |f(z)|^p \mathrm{dA}_\alpha(z) < \infty \right\}.$$

We define the Bergman spaces $\mathcal{A}^p_\alpha(\mathbb{D}_e)$ on the exterior disk $\mathbb{D}_e = (\mathbb{C}\cup\{\infty\})\setminus\overline{\mathbb{D}}$ with $p \geq 1$ and $\alpha > -1$ by

$$\mathcal{A}^p_\alpha(\mathbb{D}_e) = \left\{ g \in \mathrm{Hol}(\mathbb{D}_e),\ g(\infty) = 0 \text{ and } \|g\|^p_{\mathcal{A}^p_\alpha} = \int_{\mathbb{D}_e} |g(z)|^p \frac{(|z|^2-1)^\alpha}{|z|^{4-p+2\alpha}} \mathrm{dA}(z) < \infty \right\}.$$

Note that $\mathcal{A}^p_\alpha(\mathbb{D})$ and $\mathcal{A}^p_\alpha(\mathbb{D}_e)$ are isometrically isomorphic via the isometry $R : f \mapsto Rf$ defined on $\mathcal{A}^p_\alpha(\mathbb{D})$ by

$$Rf(z) = \frac{1}{z}\overline{f\left(\frac{1}{\overline{z}}\right)}, \quad z \in \mathbb{D}_e. \tag{2.1}$$

Indeed, by the variable change $z \mapsto 1/\overline{z}$,

$$\int_{\mathbb{D}} |f(z)|^p \mathrm{dA}_\alpha(z) = \int_{\mathbb{D}_e} |\overline{f(1/\overline{z})}/z|^p \frac{(|z|^2-1)^\alpha}{|z|^{4-p+2\alpha}} \mathrm{dA}(z)$$

Futhermore if $f = \sum_{n\geq 0} a_n z^n \in \mathcal{A}^p_\alpha(\mathbb{D})$ then by (2.1)

$$Rf(z) = \sum_{n=0}^{\infty} \frac{\overline{a_n}}{z^{n+1}}, \quad z \in \mathbb{D}_e. \tag{2.2}$$

Denote by S the shift operator on $\mathcal{A}^p_\alpha(\mathbb{D})$ for $p \geq 1$ and $\alpha > -1$, that is the multiplication by z on $\mathcal{A}^p_\alpha(\mathbb{D})$. Let S^* denote the backward shift, that is

$$S^* f(z) = \frac{f(z) - f(0)}{z}.$$

Notice that S^* is continuous on $\mathcal{A}^p_\alpha(\mathbb{D})$ for $p \geq 1$ and $\alpha > -1$. Indeed, for $f \in \mathcal{A}^p_\alpha(\mathbb{D})$ we get by subharmonicity ([**HKZ**, proposition 1.1]) that

$$\left| \frac{f(z) - f(0)}{z} \right| \leq \sup_{|w|\leq 1/2} |f'(w)| \lesssim \|f\|_{\mathcal{A}^p_\alpha(\mathbb{D})}, \quad |z| < 1/2.$$

Since $f \mapsto f(0)$ is continuous on $\mathcal{A}^p_\alpha(\mathbb{D})$ ([**HKZ**, proposition 1.1]), we have

$$\begin{aligned} \|S^* f\|^p_{\mathcal{A}^p_\alpha} &\leq \int_{|z|\leq 1/2} \|f\|^p_{\mathcal{A}^p_\alpha} \mathrm{dA}_\alpha(z) + 2^p \int_{1/2<|z|<1} |f(z) - f(0)|^p \mathrm{dA}_\alpha(z) \\ &\lesssim \|f\|^p_{\mathcal{A}^p_\alpha(\mathbb{D})} + \|f - f(0)\|^p_{\mathcal{A}^p_\alpha(\mathbb{D})} \\ &\lesssim \|f\|^p_{\mathcal{A}^p_\alpha(\mathbb{D})}. \end{aligned}$$

From now, we suppose that $p > 1$ and we denote by $q = \frac{p}{p-1}$.

LEMMA 2.1. *Suppose that $-1 < \alpha < p - 1$. Then $\langle \cdot, \cdot \rangle$ defined on $\mathcal{D}^p_\alpha \times \mathcal{A}^q_{-\alpha q/p}(\mathbb{D})$ by*

$$\langle f, g \rangle = \int_{\mathbb{D}} f'(z)\overline{S^* g(z)} dA(z) + f(0)\overline{g(0)}, \qquad f \in \mathcal{D}^p_\alpha,\ g \in \mathcal{A}^q_{-\alpha q/p}(\mathbb{D}), \tag{2.3}$$

is linear on the left, anti-linear on the right and

$$|\langle f, g \rangle| \lesssim \|f\|_{\mathcal{D}^p_\alpha} \|g\|_{\mathcal{A}^q_{-\alpha q/p}(\mathbb{D})}.$$

PROOF. Since $-\alpha q/p > -1$, $(f, g) \to \langle f, g \rangle$ is well defined. Clearly this map is linear on the left and antilinear on the right. It is therefore sufficient to show that

$$|\langle f, g \rangle| \lesssim \|f\|_{\mathcal{D}^p_\alpha} \|g\|_{\mathcal{A}^q_{-\alpha q/p}(\mathbb{D})}.$$

Using Hölder's inequality and the fact that the maps S^* and $f \mapsto f(0)$ are continuous on the space $\mathcal{A}^q_{-\alpha q/p}(\mathbb{D})$, we get

$$\begin{aligned}
|\langle f,g\rangle| &\le \int_{\mathbb{D}} |f'(z)|\frac{(1-|z|^2)^{\alpha/p}}{(1-|z|^2)^{\alpha/p}}|\overline{S^*g(z)}|dA(z) + |f(0)\overline{g(0)}| \\
&\le \left(\int_{\mathbb{D}} |f'(z)|^p \mathrm{dA}_\alpha(z)\right)^{1/p}\left(\int_{\mathbb{D}} |S^*g(z)|^q \mathrm{dA}_{-\alpha q/p}\right)^{1/q} + |f(0)||g(0)| \\
&\le \|f\|_{\mathcal{D}^p_\alpha}\ \|S^*g\|_{\mathcal{A}^q_{-\alpha q/p}(\mathbb{D})} + |f(0)||g(0)| \\
&\lesssim \|f\|_{\mathcal{D}^p_\alpha}\ \|g\|_{\mathcal{A}^q_{-\alpha q/p}(\mathbb{D})}.
\end{aligned}$$

□

The previous lemma shows that $\langle\cdot,\cdot\rangle$ defines a duality between $\mathcal{D}^p_\alpha$ and $\mathcal{A}^q_{-\alpha q/p}(\mathbb{D})$. The following result shows that $\mathcal{A}^q_{-\alpha q/p}(\mathbb{D})$ can be identified as the dual of $\mathcal{D}^p_\alpha$.

PROPOSITION 2.2. *Let $p > 1$ and $-1 < \alpha < p-1$. The dual of $\mathcal{D}^p_\alpha$, noted by $\mathcal{D}^{p\,\prime}_\alpha$, is isomorphic to $\mathcal{A}^q_{-\alpha q/p}(\mathbb{D})$.*

PROOF. We will show that the mapping $g \mapsto \langle\cdot,g\rangle$ is an isomorphism of $\mathcal{A}^q_{-\alpha q/p}(\mathbb{D})$ in $\mathcal{D}^{p\,\prime}_\alpha$, the dual of $\mathcal{D}^p_\alpha$. This mapping is well defined, antilinear, continuous and injective. Let's show that it's surjective. Take L in $\mathcal{D}^{p\,\prime}_\alpha$. For all $f \in \mathcal{A}^p_\alpha(\mathbb{D})$, we consider F the primitive of f on $\mathbb{D}$ such that $F(0) = 0$. It's easy to see that $F \in \mathcal{D}^p_\alpha$. We define the mapping L_0 on $\mathcal{A}^p_\alpha(\mathbb{D})$ by $L_0(f) = L(F)$. Thus L_0 belong to the dual of $\mathcal{A}^p_\alpha(\mathbb{D})$, since

$$|L_0(f)| = |L(F)| \le \|L\|\|F\|_{\mathcal{D}^p_\alpha} = \|L\|\|f\|_{\mathcal{A}^p_\alpha}.$$

By the Hahn-Banach theorem, L_0 extends to $L^p_\alpha(\mathbb{D}) = L^p(\mathbb{D}, \mathrm{dA}_\alpha)$ in a continuous linear form $\widetilde{L_0}$. By the Riesz representation theorem, there exists $\psi_0 \in L^p_{-\alpha q/p}(\mathbb{D}) = L^p_\alpha(\mathbb{D})'$ such that for any $g \in L^p_\alpha(\mathbb{D})$,

$$\widetilde{L_0}(g) = \int_{\mathbb{D}} g(z)\overline{\psi_0}(z)dA(z).$$

Let P be the linear map defined by

$$P : f \mapsto \left(z \mapsto \int_{\mathbb{D}} \frac{f(w)}{(1-z\overline{w})^2}dA(w)\right).$$

According to [**HKZ**, Theorem 1.10], P is a bounded projection from $L^s_\gamma(\mathbb{D})$ onto $\mathcal{A}^s_\gamma(\mathbb{D})$ for $\gamma < s-1$ which is the case when $(s,\gamma) = (p,\alpha)$ and $(s,\gamma) = (q, -\alpha q/p)$. Set $\psi = P(\psi_0) \in \mathcal{A}^q_{-\alpha q/p}(\mathbb{D})$. So for $f \in \mathcal{A}^p_\alpha(\mathbb{D})$, we get

$$\begin{aligned}
L_0(f) = \widetilde{L_0}(f) &= \int_{\mathbb{D}} f(z)\overline{\psi_0}(z)dA(z) \\
&= \int_{\mathbb{D}}\int_{\mathbb{D}} \frac{f(w)}{(1-z\overline{w})^2}\overline{\psi_0}(z)dA(w)dA(z) \\
&= \int_{\mathbb{D}} f(w)\overline{\int_{\mathbb{D}} \frac{\psi_0(z)}{(1-w\overline{z})^2}dA(z)}dA(w) \\
&= \int_{\mathbb{D}} f(w)\overline{\psi(w)}dA(w).
\end{aligned}$$

Thus we showed that there is $\psi \in \mathcal{A}^q_{-\alpha q/p}(\mathbb{D})$ such that for any $F \in \mathcal{D}^p_\alpha$ with $F(0) = 0$, we have

$$L(F) = \int_{\mathbb{D}} F'(z)\overline{\psi(z)}dA(z).$$

Set $\varphi(z) = z\psi(z) + \overline{L(1)} \in \mathcal{A}^q_{-\alpha q/p}(\mathbb{D})$. We have $S^*\varphi = \psi$. Hence for $h \in \mathcal{D}^p_\alpha$

$$\begin{aligned} L(h) &= L(h - h(0)) + L(h(0)) \\ &= \int_{\mathbb{D}} h'(z)\overline{\psi(z)}dA(z) + h(0)L(1) \\ &= \int_{\mathbb{D}} h'(z)\overline{S^*\varphi(z)}dA(z) + h(0)\overline{\varphi(0)} = \langle h, \varphi\rangle. \end{aligned}$$

This shows that the mapping $g \mapsto \langle \cdot, g\rangle$ is surjective and defines an isomorphism from $\mathcal{A}^q_{-\alpha q/p}(\mathbb{D})$ onto $\mathcal{D}^{p\,\prime}_\alpha$. □

Remarks. If $p > 1$ and $\alpha < p - 1$, the dual of $\mathcal{D}^p_\alpha$ is identified as $\mathcal{A}^q_{-\alpha q/p}(\mathbb{D})$. Also the spaces $\mathcal{A}^q_{-\alpha q/p}(\mathbb{D})$ and $\mathcal{A}^q_{-\alpha q/p}(\mathbb{D}_e)$ are isomorphic, so we can identify the dual of $\mathcal{D}^p_\alpha$ with $\mathcal{A}^q_{-\alpha q/p}(\mathbb{D}_e)$ by the duality

$$\langle f, g\rangle_e = \langle f, R^{-1}g\rangle, \quad f \in \mathcal{D}^p_\alpha,\ g \in \mathcal{A}^q_{-\alpha q/p}(\mathbb{D}_e).$$

In the following we will introduce the tools to use the Hedenmalm and Shields Theorem [**HS**, Theorem 1]. For all $\varphi \in \mathcal{D}^{p\,\prime}_\alpha$, we set

$$\widetilde{\varphi}(\lambda) = \langle f_\lambda, \varphi\rangle, \quad \lambda \in \mathbb{D}_e$$

where f_λ is given by

$$f_\lambda(z) = (\lambda - z)^{-1}, \qquad z \in \mathbb{D}.$$

We define then as in [**HS**]

$$\mathcal{D}^{p\,*}_\alpha = \left\{\widetilde{\varphi},\ \varphi \in \mathcal{D}^{p\,\prime}_\alpha\right\}.$$

Let $\varphi \in \mathcal{D}^{p\,\prime}_\alpha$, we have

$$\widetilde{\varphi}(\lambda) = \left\langle \sum_{n=0}^{\infty} \frac{z^n}{\lambda^{n+1}}, \varphi \right\rangle = \sum_{n=0}^{\infty} \frac{\langle z^n, \varphi\rangle}{\lambda^{n+1}}.$$

We identify φ as an element of $\mathcal{A}^q_{-\alpha q/p}(\mathbb{D})$ that we write

$$\varphi(z) = \sum_{n\geq 0} a_n z^n, \quad z \in \mathbb{D}.$$

So if $n = 0$, $\langle z^n, \varphi\rangle = \overline{\varphi(0)} = \overline{a_0}$ and if $n \geq 1$,

$$\begin{aligned} \langle z^n, \varphi\rangle &= \int_{\mathbb{D}} nz^{n-1}\overline{S^*\varphi(z)}dA(z) \\ &= \int_{\mathbb{D}} nz^{n-1} \sum_{m=1}^{\infty} \overline{a_m}\ \overline{z}^{m-1}dA(z) \\ &= \sum_{m=1}^{\infty} n\overline{a_m} \int_0^1 \int_0^{2\pi} r^{n+m-2}e^{i\theta(n-m)}d\theta/\pi\ rdr \\ &= \overline{a_n}\int_0^1 2nr^{2n-1}dr = \overline{a_n}. \qquad (2.4) \end{aligned}$$

Thus for $\lambda \in \mathbb{D}_e$,

$$\widetilde{\varphi}(\lambda) = \sum_{n=0}^{\infty} \frac{\overline{a_n}}{\lambda^{n+1}}.$$

Moreover, according to (2.2), we also have

$$R\varphi(\lambda) = \sum_{n=0}^{\infty} \frac{\overline{a_n}}{\lambda^{n+1}}, \quad \lambda \in \mathbb{D}_e.$$

So

$$\mathcal{D}_\alpha^{p\,*} = \mathcal{A}^q_{-\alpha q/p}(\mathbb{D}_e).$$

The following lemma will be useful for expressing duality (see [**HS**, Lemma 3]).

LEMMA 2.3. *Let $p > 1$ and $-1 < \alpha < p-1$. Let $f \in \mathcal{D}_\alpha^p$ and $g \in \mathcal{A}^q_{-\alpha q/p}(\mathbb{D}_e)$. For $0 \leq r < 1$, we set*

$$f_r(z) = f(rz), \quad z \in \mathbb{D} \quad \text{and} \quad g_{1/r}(z) = g(z/r), \quad z \in \mathbb{D}_e,$$

Then

$$\langle f, g\rangle_e = \lim_{r\to 1^-} \langle f_r, g_{1/r}\rangle_e = \lim_{r\to 1^-} \sum_{n=0}^{\infty} a_n b_n r^n = \lim_{r\to 1^-} \frac{1}{2\pi}\int_0^{2\pi} f(re^{i\theta})g(e^{i\theta}/r)e^{i\theta}d\theta,$$

where $f(z) = \sum_{n=0}^{\infty} a_n z^n$ and $g(1/z) = \sum_{n=0}^{\infty} b_n z^{n+1}$, $z \in \mathbb{D}$.

3. Cyclicity in $\mathcal{D}_\alpha^p$

We start this section by comparing the spaces $\mathcal{D}_\alpha^p$ and the Hardy spaces $H^p(\mathbb{D})$. We suppose $p \geq 1$ and $\alpha > -1$. Let $H^\infty(\mathbb{D})$ be the algebra of bounded analytic functions on the open unit disc $\mathbb{D}$ and let $H^p(\mathbb{D})$ be the Hardy space of analytic functions f on $\mathbb{D}$ such that

$$\|f\|_{H^p} = \sup_{r<1} M_p(f,r) < \infty,$$

where

$$M_p(f,r) = \left(\frac{1}{2\pi}\int_0^{2\pi} |f(re^{i\theta})|^p d\theta\right)^{1/p}.$$

Let $\mathcal{N}$ be the Nevanlinna class of analytic functions f on $\mathbb{D}$ for which

$$\sup_{r<1} \int_{\mathbb{T}} \ln^+ |f(r\zeta)||d\zeta| < \infty.$$

By Fatou's Theorem, the radial limit $f^*(\zeta) = \lim_{r\to 1-} f(r\zeta)$ exists a.e on $\mathbb{T}$ and $\ln|f^*| \in L^1(\mathbb{T})$. Recall that $f \in \mathcal{N}$ if and only if $f = \varphi/\psi$, where $\varphi, \psi \in H^\infty(\mathbb{D})$.

Let $\mathcal{N}^+$ be the Smirnov class of analytic functions $f \in \mathcal{N}$ such that

$$\sup_{r<1} \int_{\mathbb{T}} \ln^+ |f(r\zeta)||d\zeta| = \int_{\mathbb{T}} \ln^+ |f^*(\zeta)||d\zeta|.$$

The function $f \in \mathcal{N}^+$ if and only if $f = \varphi/\psi$ where $\varphi, \psi \in H^\infty(\mathbb{D})$ and ψ is an outer function, that is, ψ has the form

$$\psi(z) = \exp \int_{\mathbb{T}} \frac{\zeta + z}{\zeta - z} \log \psi^*(\zeta) \frac{|d\zeta|}{2\pi}, \qquad z \in \mathbb{D}.$$

A function $f \in H^p(\mathbb{D})$ is cyclic for $H^p(\mathbb{D})$ if and only if f is outer [**N**, 4.8.4]. We then study the different possible inclusions between the spaces $\mathcal{D}_\alpha^p$ and $H^p(\mathbb{D})$ to obtain first conditions on the cyclicity in the Dirichlet spaces.

PROPOSITION 3.1. *Let $p \geq 1$ and $\alpha > -1$. If $p < \alpha + 1$ then $H^p(\mathbb{D})$ is continuously embedded in $\mathcal{D}^p_\alpha$. Consequently, if $f \in H^p(\mathbb{D})$ is outer then f is cyclic for $\mathcal{D}^p_\alpha$.*

PROOF. Let $f \in H^p(\mathbb{D})$, $z = re^{it} \in \mathbb{D}$ and $r < \rho < 1$. By Cauchy's formula,

$$f'(z) = \frac{1}{2\pi}\int_0^{2\pi} \frac{f(\rho e^{i(\theta+t)})}{(\rho e^{i\theta} - r)^2}\rho e^{i(\theta-t)}d\theta.$$

Now, by Minkowski's inequality,

$$\begin{aligned} M_p(f', r) &= \left(\frac{1}{2\pi}\int_0^{2\pi}\left|\frac{1}{2\pi}\int_0^{2\pi}\frac{f(\rho e^{i(\theta+t)})}{(\rho e^{i\theta}-r)^2}\rho e^{i(\theta-t)}d\theta\right|^p dt\right)^{1/p} \\ &\leq \frac{\rho}{2\pi}\int_0^{2\pi}\left(\frac{1}{2\pi}\int_0^{2\pi}\frac{|f(\rho e^{i(\theta+t)})|^p}{|\rho e^{i\theta}-r|^{2p}}dt\right)^{1/p} d\theta \\ &\leq \frac{1}{2\pi}\int_0^{2\pi}\frac{\rho}{|\rho e^{i\theta}-r|^2}d\theta\ M_p(f,\rho) \\ &= \frac{\rho}{\rho^2-r^2}M_p(f,\rho) \leq \frac{1}{\rho - r}M_p(f,\rho). \end{aligned}$$

Now letting $\rho \to 1$, we get

$$\frac{1}{2\pi}\int_0^{2\pi} |f'(re^{i\theta})|^p d\theta \leq \frac{1}{(1-r)^p}\|f\|^p_{H^p},$$

Since $p < \alpha + 1$,

$$\begin{aligned} \int_{\mathbb{D}} |f'(z)|^p \mathrm{dA}_\alpha(z) &= \int_0^1\int_0^{2\pi} |f'(re^{i\theta})|^p d\theta (1-r^2)^\alpha r dr/\pi \\ &\leq 2^{\alpha+1}\int_0^1 \frac{(1-r)^\alpha}{(1-r)^p}dr\|f\|^p_{H^p} \\ &= \frac{2^{\alpha+1}}{\alpha+1-p}\|f\|^p_{H^p}. \end{aligned}$$

So $H^p(\mathbb{D})$ is continuously embedded in $\mathcal{D}^p_\alpha$. Now the result follows from the fact that an outer function is cyclic in $H^p(\mathbb{D})$. □

Remark. If $p < \alpha + 1$, the Dirichlet space $\mathcal{D}^p_\alpha = \mathcal{A}^p_{\alpha-p}(\mathbb{D})$, see [**Wu**]. Therefore, in this case, there exists an inner function which is cyclic in $\mathcal{D}^p_\alpha$, see [**Ro**]. If $p > \alpha + 1$ we have the following result.

PROPOSITION 3.2. *Let $p > 1$ and $p > \alpha + 1$. The Dirichlet space $\mathcal{D}^p_\alpha$ is continuously embedded in $H^p(\mathbb{D})$. Therefore if $f \in \mathcal{D}^p_\alpha$ is cyclic in $\mathcal{D}^p_\alpha$ then f is an outer function.*

PROOF. Let $f \in \mathcal{D}^p_\alpha$ and $r \in [1/2, 1[$. We have

$$f(re^{i\theta}) = \int_0^r f'(se^{i\theta})e^{i\theta}ds + f(0).$$

Note that $|f(0)| \leq \|f\|_{\mathcal{D}^p_\alpha}$ and by subharmonicity, there exists $C > 0$ such that $|f'(se^{i\theta})| \leq C\|f\|_{\mathcal{D}^p_\alpha}$, $0 \leq s \leq 1/2$. So

$$|f(re^{i\theta})| \leq \int_{1/2}^r |f'(se^{i\theta})|ds + (C/2+1)\|f\|_{\mathcal{D}^p_\alpha}.$$

By Hölder's inequality, and since $\alpha q/p = \alpha/(p-1) < 1$,

$$\begin{aligned}
&\Big(\int_0^{2\pi} |f(re^{i\theta})|^p d\theta\Big)^{1/p} \\
&\lesssim \left[\int_0^{2\pi} \Big(\int_{1/2}^r |f'(se^{i\theta})|ds\Big)^p d\theta\right]^{1/p} + \|f\|_{\mathcal{D}^p_\alpha} \\
&\lesssim \left[\int_0^{2\pi} \Big(\int_{1/2}^r |f'(se^{i\theta})|^p(1-s^2)^\alpha ds\Big)\Big(\int_{1/2}^r (1-s^2)^{-\alpha q/p} ds\Big)^{p/q} d\theta\right]^{1/p} + \|f\|_{\mathcal{D}^p_\alpha} \\
&\lesssim \left[\int_0^{2\pi}\int_{1/2}^1 |f'(se^{i\theta})|^p(1-s^2)^\alpha 2sdsd\theta\right]^{1/p}\left[\int_{1/2}^1 (1-s^2)^{-\alpha q/p} ds\right]^{1/q} + \|f\|_{\mathcal{D}^p_\alpha} \\
&\lesssim \Big(1-\frac{\alpha q}{p}\Big)\|f\|_{\mathcal{D}^p_\alpha} + \|f\|_{\mathcal{D}^p_\alpha}.
\end{aligned}$$

So $\|f\|_{H^p} \lesssim \|f\|_{\mathcal{D}^p_\alpha}$. Hence if f is cyclic for $\mathcal{D}^p_\alpha$ then f is also cyclic for $H^p(\mathbb{D})$ and f is then an outer function. □

Remark. We have $\mathscr{D}^2_1(\mathbb{D}) = H^2(\mathbb{D})$ and $\mathscr{D}^2_0(\mathbb{D}) = \mathcal{D}$. So if $1 \le p \le 2$ and $p = \alpha + 1$, we obtain by interpolation theorem [**Wu**, (3.8)]), that $\mathcal{D}^p_\alpha$ is continuously embedded in $H^p(\mathbb{D})$. Futhermore if $p > \alpha + 2$, then $\mathcal{D}^p_\alpha$ is continuously embedded in $H^\infty(\mathbb{D})$ (see the proof of [**Wu**, Theorem 4.2]).

We can summarize here all the inclusions obtained:

$$\begin{aligned}
p < \alpha+1 \quad &\Longrightarrow \quad H^p(\mathbb{D}) \subset \mathcal{D}^p_\alpha = \mathcal{A}^p_{\alpha-p}(\mathbb{D}) \\
1 \le p \le 2 \text{ and } p = \alpha+1 \quad &\Longrightarrow \quad \mathcal{D}^p_\alpha \subset H^p(\mathbb{D}) \\
p > \alpha+1 \quad &\Longrightarrow \quad \mathcal{D}^p_\alpha \subset H^p(\mathbb{D}) \\
p > \alpha+2 \quad &\Longrightarrow \quad \mathcal{D}^p_\alpha \subset H^\infty(\mathbb{D}).
\end{aligned}$$

We assume in the following that $p > \alpha + 1$. We will prove that any outer function of $A(\mathbb{D}) \cap \mathcal{D}^p_\alpha$ whose set of zeros is reduced to a single point is cyclic in $\mathcal{D}^p_\alpha$. For that we will use a Hedenmalm-Shields Theorem [**HS**, Theorem 1]. We first need to define the following notions. Let $X \subset \mathrm{Hol}(\mathbb{D})$ be a Banach space. The multiplier set of X, noted $M(X)$, is defined by

$$M(X) = \{\varphi \in \mathrm{Hol}(\mathbb{D}),\ \varphi f \in X,\ \forall f \in X\}.$$

If $X \subset \mathrm{Hol}(\mathbb{D}_e)$ we define in a similar way $M(X)$.
As in [**HS**] we identify the dual X' of X with a space X^* of holomorphic functions on $\mathbb{D}_e$. Finally for $E \subset \mathbb{T}$ a closed set of zero Lebesgue measure, we set

$$\mathscr{H}_E(\mathscr{N}^+, X^*) = \left\{\varphi \in \mathrm{Hol}(\mathbb{C}\cup\{\infty\}\setminus E),\ \varphi|_{\mathbb{D}} \in \mathscr{N}^+(\mathbb{D}),\ \varphi|_{\mathbb{D}_e} \in X^*\right\}.$$

We denote by $\mathrm{Hol}(\overline{\mathbb{D}})$, respectively $\mathrm{Hol}(\overline{\mathbb{D}_e})$, the space of all holomorphic functions in a neighborhood of $\overline{\mathbb{D}}$, respectively $\overline{\mathbb{D}}_e$.

THEOREM 3.3 (Hedenmalm-Shields [**HS**]). *Let $X \subset \mathrm{Hol}(\mathbb{D})$ be a Banach space. Assume that*

(1) *The embedding map of X into* $\mathrm{Hol}(\overline{\mathbb{D}})$ *is continuous and X contains* $\mathrm{Hol}(\overline{\mathbb{D}})$ *as a dense subspace*

(2) $X \cap A(\mathbb{D})$ *is a Banach algebra, containing* $\mathrm{Hol}(\overline{\mathbb{D}})$ *as a dense algebra.*
(3) $\mathrm{Hol}(\overline{\mathbb{D}}) \subset M(X)$.
(4) $\mathrm{Hol}(\overline{\mathbb{D}_e}) \subset M(X^*) = H^\infty(\mathbb{D}_e)$.

If $f \in X \cap A(\mathbb{D})$ *is an outer function and if*

$$\mathscr{H}_{\mathcal{Z}(f)}(\mathcal{N}^+, X^*) = \{0\}$$

then f *is cyclic in* X.

Hedenmalm and Shields show that if$f \in A(\mathbb{D}) \cap \mathscr{D}_0^2(\mathbb{D})$ is an outer function and $\mathcal{Z}(f) = \{1\}$ then $\mathscr{H}_{\mathcal{Z}(f)}(\mathcal{N}^+, \mathscr{D}_0^2(\mathbb{D})^*) = \{0\}$ and so f is cyclic (see also [**EKR2**, **EKMR**]). We will prove a similar result for $\mathcal{D}_\alpha^p$ where $\alpha+1 < p \leq \alpha+2$.

THEOREM 3.4. *Let* $p > 1$ *and* $p > \alpha + 1$. *If* $f \in A(\mathbb{D}) \cap \mathcal{D}_\alpha^p$ *is outer and if*

$$\mathscr{H}_{\mathcal{Z}(f)}(\mathcal{N}^+, \mathcal{D}_\alpha^{p*}) = \{0\}$$

then f *is cyclic on* $\mathcal{D}_\alpha^p$.

To prove this result, we will use Theorem 3.3. For that we need only to show the following lemma (see the proof of [**DRS**, lemma 11]).

LEMMA 3.5. *Let* $p > 1$ *and* $\alpha > -1$. *Then* $M(\mathcal{A}_\alpha^p(\mathbb{D}_e)) = H^\infty(\mathbb{D}_e)$.

PROOF. Let $f \in \mathcal{A}_\alpha^p(\mathbb{D}_e)$ and $g \in H^\infty(\mathbb{D}_e)$. We have

$$\int_{\mathbb{D}_e} |f(z)g(z)|^p \frac{(|z|^2-1)^\alpha}{|z|^{4-p+2\alpha}} \mathrm{dA}(z) \leq \|g\|_\infty^p \|f\|_{\mathcal{A}_\alpha^p}^p.$$

So $fg \in \mathcal{A}_\alpha^p(\mathbb{D}_e)$ and $H^\infty(\mathbb{D}_e) \subset M(\mathcal{A}_\alpha^p(\mathbb{D}_e))$.

Now let $g \in M(\mathcal{A}_\alpha^p(\mathbb{D}_e))$ and let $M_g : \mathcal{A}_\alpha^p(\mathbb{D}_e) \to \mathcal{A}_\alpha^p(\mathbb{D}_e)$ be the operator given by $M_g(f) = fg$. By the closed graph theorem, M_g is bounded. For $z \in \mathbb{D}_e$, the linear functional $\Lambda_z : \mathcal{A}_\alpha^p(\mathbb{D}_e) \to \mathbb{C}$ defined by $\Lambda_z(f) = f(z)$, is continuous ([**HKZ**, proposition 1.1]). So for $f \in \mathcal{A}_\alpha^p(\mathbb{D}_e)$ and $z \in \mathbb{D}_e$,

$$|f(z)g(z)| = |\Lambda_z(M_g f)| \leq \|\Lambda_z\| \|M_g\| \|f\|_{\mathcal{A}_\alpha^p}.$$

Hence

$$\|\Lambda_z\| |g(z)| \leq \|\Lambda_z\| \|M_g\|$$

and $g \in H^\infty(\mathbb{D}_e)$. So $M(\mathcal{A}_\alpha^p(\mathbb{D}_e)) \subset H^\infty(\mathbb{D}_e)$. On the other hand the inclusion $H^\infty(\mathbb{D}_e) \subset M(\mathcal{A}_\alpha^p(\mathbb{D}_e))$ is obvious. □

By identifying the dual of $\mathcal{D}_\alpha^p$ with $\mathcal{A}^q_{-\alpha q/p}(\mathbb{D}_e)$, we have for $f \in \mathcal{D}_\alpha^p$ and $\varphi \in \mathcal{A}^q_{-\alpha q/p}(\mathbb{D}_e)$,

$$\varphi \in \left([f]_{\mathbb{N}}^{\mathcal{D}_\alpha^p}\right)^\perp \iff \langle z^n f, \varphi \rangle_e = 0, \quad \forall n \in \mathbb{N}.$$

LEMMA 3.6. *Let* $p > 1$*and* $p > \alpha + 1$. *Let* $E \subset \mathbb{T}$ *a closed set of Lebesgue measure,* $\varphi \in \mathscr{H}_E(\mathcal{N}^+, \mathcal{D}_\alpha^{p*})$ *and* $f \in \mathcal{D}_\alpha^p$. *If the family of functions*

$$z \in \mathbb{T} \mapsto f(rz)\varphi(z/r), \quad 1/2 < r < 1,$$

is uniformly integrable on $\mathbb{T}$, *then* $\varphi \in \left([f]_{\mathbb{N}}^{\mathcal{D}_\alpha^p}\right)^\perp$.

PROOF. This result holds by using the analogue arguments like those in [**EKR2**, Lemma 3.4] for the classical Dirichlet space. For the sake of completeness, we include it here. Let $f \in \mathcal{D}^p_\alpha$ and $\varphi|_{\mathbb{D}_e} \in \mathcal{D}^{p\,*}_\alpha = \mathcal{A}^q_{-\alpha q/p}(\mathbb{D}_e)$. By Proposition 2.3, we have

$$\langle f, \varphi\rangle = \lim_{r\to 1^-} \frac{1}{2\pi}\int_0^{2\pi} f(re^{i\theta})\varphi(e^{i\theta}/r)e^{i\theta}d\theta.$$

By Proposition 3.2, $\mathcal{D}^p_\alpha \subset H^p(\mathbb{D})$ and so f^*, the radial limit of f, exists a.e. on $\mathbb{T}$. Since $\varphi \in \mathrm{Hol}(\mathbb{C}\setminus E)$ and E is a closed set of Lebesgue measure zero, $\varphi(z/r) \longrightarrow \varphi(z)$ exists a.e on $\mathbb{T}$ when $r \to 1^-$. So the family of the functions $z \mapsto f(rz)\varphi(z/r)$ converges a.e to $f^*\varphi$ when $r \to 1^-$. By uniform integrability, this family of functions converges in $L^1(\mathbb{T})$. Then

$$\langle f, \varphi\rangle = \frac{1}{2\pi}\int_0^{2\pi} f^*(e^{i\theta})\varphi(e^{i\theta})e^{i\theta}d\theta.$$

Futhermore $\varphi \in \mathcal{N}^+$ and $f \in H^p(\mathbb{D}) \subset \mathcal{N}^+$, so then $f\varphi \in \mathcal{N}^+$. Since the radial limit $(f\varphi)^* = f^*\varphi \in L^1(\mathbb{T})$, by Smirnov's generalized maximum principal [**D**, Theorem 2.11], $f\phi \in H^1(\mathbb{D})$ and so $\widehat{f^*\varphi}(n) = 0$:

$$\widehat{f^*\varphi}(n) = \langle f, \varphi\rangle = \frac{1}{2\pi}\int_0^{2\pi} f^*(e^{i\theta})\varphi(e^{i\theta})e^{i\theta}d\theta = 0.$$

Repeating the same argument with f replaced by $z^n f$, we get $\langle z^n f, \varphi\rangle = 0$ for all $n \in \mathbb{N}$. □

We have the following classical Lemma

LEMMA 3.7. *Let $p > 1$ and $p > \alpha + 1$. Let $E \subset \mathbb{T}$ be a closed set of Lebesgue measure zero and $\varphi \in \mathscr{H}_E(\mathcal{N}^+, \mathcal{D}^{p\,*}_\alpha)$. Then there exists a constant $C > 0$ such that*

$$|\varphi(z)| \le \frac{C}{\mathrm{dist}(z,E)^4}, \qquad 1 < |z| < 2.$$

PROOF. Let $\varphi \in \mathscr{H}_E(\mathcal{N}^+, \mathcal{D}^{p\,*}_\alpha)$. Since $\varphi|_{\mathbb{D}} \in \mathcal{N}^+$, $\varphi|_{\mathbb{D}} = \varphi_i\varphi_o$, where φ_i is an inner function and φ_o is an outer function in $\mathcal{N}$ (see [**D**, p. 25]). Futhermore, since E has Lebesgue measure zero , $\varphi(z) = \varphi^*(z) = \lim_{r\to 1^-}\varphi(rz)$ exists a.e on $\mathbb{T}$. The function $\log|\varphi|$ being in $L^1(\mathbb{T})$, we get

$$\begin{aligned}
|\varphi(z)| &\le |\varphi_o(z)| = \left|\exp\left(\frac{1}{2\pi}\int_0^{2\pi}\frac{e^{it}+z}{e^{it}-z}\log|\varphi(e^{it})|dt\right)\right| \\
&\le \exp\left(\frac{1}{2\pi}\int_0^{2\pi}\frac{1-|z|^2}{|e^{it}-z|^2}\log|\varphi(e^{it})|dt\right) \\
&\le \exp\left(\frac{1-|z|^2}{(1-|z|)^2}\int_0^{2\pi}\left|\log|\varphi(e^{it})|\right|dt\right) \\
&\le \exp\left(\frac{2}{1-|z|}\|\log|\varphi|\|_{L^1(\mathbb{T})}\right). \\
&\le \exp\left(\frac{C_1}{1-|z|}\right),
\end{aligned}$$

for some constant $C_1 > 0$. Let $z \in \mathbb{D}_e$ with $|z| \le 2$. The disc of radius $(|z|-1)/2$ centered at z, $D(z, (|z|-1)/2)$ is contained in $\mathbb{D}_e$. Since $\varphi|_{\mathbb{D}_e} \in \mathcal{A}^q_{-\alpha q/p}(\mathbb{D}_e)$, by subharmonicity of $|\varphi|$ and for $q = p/(p-1) \geq 1$, we obtain

$$\begin{aligned}\frac{(|z|-1)^2}{4}|\varphi(z)|^q &\leq \frac{1}{\pi}\int_{D(z,(|z|-1)/2)} |\varphi(w)|^q dA(w) \\ &\leq \frac{1}{\pi}\int_{D(z,(|z|-1)/2)} |\varphi(w)|^q \frac{(|w|^2-1)^{-\alpha q/p}}{|w|^{4-q-2\alpha q/p}} \frac{|w|^{4-q-2\alpha q/p}}{(|w|^2-1)^{-\alpha q/p}} dA(w) \\ &\leq \max(2^{2\alpha q/p}, 2^{4-q}) \int_{\mathbb{D}_e} |\varphi(w)|^q \frac{(|w|^2-1)^{-\alpha q/p}}{|w|^{4-q-2\alpha q/p}} dA(w) \\ &\leq \max(2^{2\alpha q/p}, 2^{4-q}) \|\varphi|_{\mathbb{D}_e}\|_{\mathcal{B}^q_{-\alpha q/p}}.\end{aligned}$$

So

$$|\varphi(z)| \leq \frac{C_2}{(|z|-1)^2}, \qquad 1 < |z| \leq 2,$$

for some constant $C_2 > 0$. Since $\log|\varphi|$ is subharmonic function, by Taylor-Williams estimates [**RW**, lemma 5.8 and 5.9] and [**EKMR**, Lemma 9.6.5], we get the lemma. □

The following result allows us to reduce the study of cyclic vectors vanishing on a closed set E to the study of cyclicity of particular functions. More precisely we have

THEOREM 3.8. *Let $p > 1$ and $p > \alpha + 1$. Let $f \in \mathcal{D}^p_\alpha$ and $E \subset \mathbb{T}$ be a closed set of Lebesgue measure zero. If there exists a constant $C_1 > 0$ such that,*

$$|f(z)| \leq C_1 \operatorname{dist}(z, E)^4, \qquad z \in \mathbb{D},$$

then

$$\mathscr{H}_E(\mathscr{N}^+, \mathcal{D}^{p*}_\alpha) \subset \left([f]^{\mathcal{D}^p_\alpha}_{\mathbb{N}}\right)^{\perp}.$$

This means that for all $g \in \mathscr{H}_E(\mathscr{N}^+, \mathcal{D}^{p*}_\alpha)$, $g|_{\mathbb{D}_e} \in \left([f]^{\mathcal{D}^p_\alpha}_{\mathbb{N}}\right)^{\perp}$ i.e.

$$\langle z^n f, g|_{\mathbb{D}_e}\rangle_e = 0, \quad \forall n \in \mathbb{N}.$$

PROOF. Let $\varphi \in \mathscr{H}_E(\mathscr{N}^+, \mathcal{D}^{p*}_\alpha)$. By Lemma 3.7, there exists a constant $C_2 > 0$ such that

$$|\varphi(z)| \leq \frac{C_2}{\operatorname{dist}(z,E)^4}, \quad 1 < |z| < 2.$$

So for $1/2 < r < 1$ and $z \in \mathbb{T}$, we have

$$|f(rz)\varphi(z/r)| \leq C_1 C_2 \frac{\operatorname{dist}(rz,E)^4}{\operatorname{dist}(z/r,E)^4} \leq C_1 C_2.$$

The family of the functions $z \mapsto f(rz)\varphi(z/r)$ is uniformly integrable on $\mathbb{T}$ for $1/2 < r < 1$, thus by Lemma 3.6, $\varphi \in \left([f]^{\mathcal{D}^p_\alpha}_{\mathbb{N}}\right)^{\perp}$, which finishes the proof. □

COROLLARY 3.9. *Let $p > 1$ such that $\alpha + 1 < p \leq \alpha + 2$. We have*

$$\mathscr{H}_{\{1\}}(\mathscr{N}^+, \mathcal{D}^{p*}_\alpha) = \{0\}.$$

PROOF. Let $f(z) := (z-1)^4$. We have $f \in \mathcal{D}^p_\alpha$ and $|f(z)| \leq |z-1|^4$. By Theorem 3.8 ,

$$\mathscr{H}_{\{1\}}(\mathscr{N}^+, \mathcal{D}^{p*}_\alpha) \subset \left([f]^{\mathcal{D}^p_\alpha}_{\mathbb{N}}\right)^{\perp}.$$

It suffices to prove that f is cyclic. Let $\varphi \in \mathcal{A}^q_{-\alpha q/p}(\mathbb{D})$ such that

$$\langle z^n(z-1), \varphi\rangle = 0, \quad \forall n \in \mathbb{N}.$$

Write $\varphi(z) = \sum_{n\geq 0} a_n z^n$, we get by (2.4),

$$\overline{a_n} = \langle z^n, \varphi\rangle = \langle z^{n+1}, \varphi\rangle = \overline{a_{n+1}}.$$

Then

$$\varphi(z) = \sum_{n=0}^{\infty} a_n z^n = \frac{a_0}{1-z}, \quad z \in \mathbb{D}.$$

Suppose that $\varphi \neq 0$. Since $\varphi \in \mathcal{A}^q_{-\alpha q/p}(\mathbb{D})$, we have

$$\int_{\mathbb{D}} \frac{(1-|z|^2)^{-\alpha q/p}}{|1-z|^q} \mathrm{d}A(z) < \infty, \tag{3.1}$$

and so $q + \alpha q/p < 2$ (see [**HKZ**, Theorem 1.7]), which contradicts the assumptions on p and α. So $\varphi = 0$ and $[z-1]_{\mathbb{N}}^{\mathcal{D}^p_\alpha} = \mathcal{D}^p_\alpha$. In particular $z-1 \in [(z-1)^2]_{\mathbb{N}}^{\mathcal{D}^p_\alpha}$ and then

$$[(z-1)^2]_{\mathbb{N}}^{\mathcal{D}^p_\alpha} = [z-1]_{\mathbb{N}}^{\mathcal{D}^p_\alpha} = \mathcal{D}^p_\alpha.$$

With the same argument we obtain

$$[(z-1)^4]_{\mathbb{N}}^{\mathcal{D}^p_\alpha} = \mathcal{D}^p_\alpha,$$

and $f(z) = (z-1)^4$ is cyclic in $\mathcal{D}^p_\alpha$. □

Remark. The proof of the previous result also gives us that for $p > \alpha + 2$, the function $f(z) = z - 1$ is not cyclic in $\mathcal{D}^p_\alpha$. Indeed by (3.1), $\varphi(z) = 1/(1-z) \in \mathcal{A}^q_{-\alpha q/p}(\mathbb{D})$ and $\varphi \perp z^n f$, $n \in \mathbb{N}$. More generally if $f \in A(\mathbb{D}) \cap \mathcal{D}^p_\alpha$ with $f(1) = 0$, then f is not cyclic in $\mathcal{D}^p_\alpha$. Indeed for $p > \alpha + 2$, we have $\mathcal{D}^p_\alpha \subset H^\infty(\mathbb{D})$ with $\|\cdot\|_{H^\infty} \lesssim \|\cdot\|_{\mathcal{D}^p_\alpha}$ which implies

$$[f]_{\mathbb{N}}^{\mathcal{D}^p_\alpha} \subset \{g \in A(\mathbb{D}),\ g(1) = 0\}.$$

THEOREM 3.10. *Let $p > 1$ such that $\alpha + 1 < p \leq \alpha + 2$ and let $f \in A(\mathbb{D}) \cap \mathcal{D}^p_\alpha$. If f is an outer function and $\mathcal{Z}(f)$ is countable then f is cyclic in $\mathcal{D}^p_\alpha$.*

PROOF. Since $\mathcal{Z}(f)$ is countable, by [**BS**, Theorem 3] it suffices to prove the theorem when the zero set is reduced to a single point. The result now follows by Theorem 3.4 and Corollary 3.9. □

Acknowledgement. The authors thank Javad Mashreghi and Université Laval for its support and hospitality. We also thank the referee for their comments and suggestions.

References

[B] A. Beurling, *Ensembles exceptionnels* (French), Acta Math. **72** (1940), 1–13, DOI 10.1007/BF02546325. MR0001370

[BS] L. Brown and A. L. Shields, *Cyclic vectors in the Dirichlet space*, Trans. Amer. Math. Soc. **285** (1984), no. 1, 269–303, DOI 10.2307/1999483. MR748841

[C] L. Carleson, *Sets of uniqueness for functions regular in the unit circle*, Acta Math. **87** (1952), 325–345, DOI 10.1007/BF02392289. MR0050011

[D] P. L. Duren, *Theory of H^p spaces*, Pure and Applied Mathematics, Vol. 38, Academic Press, New York-London, 1970. MR0268655

[DRS] P. L. Duren, B. W. Romberg, and A. L. Shields, *Linear functionals on H^p spaces with $0 < p < 1$*, J. Reine Angew. Math. **238** (1969), 32–60. MR0259579

[EKMR] O. El-Fallah, K. Kellay, J. Mashreghi, and T. Ransford, *A primer on the Dirichlet space*, Cambridge Tracts in Mathematics, vol. 203, Cambridge University Press, Cambridge, 2014. MR3185375

[EKR1] O. El-Fallah, K. Kellay, and T. Ransford, *Cyclicity in the Dirichlet space*, Ark. Mat. **44** (2006), no. 1, 61–86, DOI 10.1007/s11512-005-0008-z. MR2237211

[EKR2] O. El-Fallah, K. Kellay, and T. Ransford, *On the Brown-Shields conjecture for cyclicity in the Dirichlet space*, Adv. Math. **222** (2009), no. 6, 2196–2214, DOI 10.1016/j.aim.2009.07.011. MR2562781

[EKR3] O. El-Fallah, K. Kellay, and T. Ransford, *Cantor sets and cyclicity in weighted Dirichlet spaces*, J. Math. Anal. Appl. **372** (2010), no. 2, 565–573, DOI 10.1016/j.jmaa.2010.07.047. MR2678884

[G] J. B. Garnett, *Bounded analytic functions*, 1st ed., Graduate Texts in Mathematics, vol. 236, Springer, New York, 2007. MR2261424

[HS] H. Hedenmalm and A. Shields, *Invariant subspaces in Banach spaces of analytic functions*, Michigan Math. J. **37** (1990), no. 1, 91–104, DOI 10.1307/mmj/1029004068. MR1042516

[HKZ] H. Hedenmalm, B. Korenblum, and K. Zhu, *Theory of Bergman spaces*, Graduate Texts in Mathematics, vol. 199, Springer-Verlag, New York, 2000. MR1758653

[N] N. K. Nikolski, *Operators, functions, and systems: an easy reading. Vol. 1*, Mathematical Surveys and Monographs, vol. 92, American Mathematical Society, Providence, RI, 2002. Hardy, Hankel, and Toeplitz; Translated from the French by Andreas Hartmann. MR1864396

[RS1] S. Richter and C. Sundberg, *Invariant subspaces of the Dirichlet shift and pseudocontinuations*, Trans. Amer. Math. Soc. **341** (1994), no. 2, 863–879, DOI 10.2307/2154587. MR1145733

[Ro] J. W. Roberts, *Cyclic inner functions in the Bergman spaces and weak outer functions in H^p, $0 < p < 1$*, Illinois J. Math. **29** (1985), no. 1, 25–38. MR769756

[RW] B. A. Taylor and D. L. Williams, *Ideals in rings of analytic functions with smooth boundary values*, Canadian J. Math. **22** (1970), 1266–1283, DOI 10.4153/CJM-1970-143-x. MR273024

[Wu] Z. Wu, *Carleson measures and multipliers for Dirichlet spaces*, J. Funct. Anal. **169** (1999), no. 1, 148–163, DOI 10.1006/jfan.1999.3490. MR1726750

Univ. Bordeaux, CNRS, Bordeaux INP, IMB, UMR 5251, F–33405 Talence, France.
Email address: `kkellay@math.u-bordeaux.fr`
Email address: `florian.le-manach@math.u-bordeaux.fr`
Email address: `mohamed.zarrabi@math.u-bordeaux.fr`

Contemporary Mathematics
Volume **743**, 2020
https://doi.org/10.1090/conm/743/14961

Inner vectors for Toeplitz operators

Raymond Cheng, Javad Mashreghi, and William T. Ross

Abstract. In this paper we survey and bring together several approaches to obtaining inner functions for Toeplitz operators. These approaches include the classical definition, the Wold decomposition, the operator-valued Poisson Integral, and Clark measures. We then extend these notions somewhat to inner functions on model spaces. Along the way we present some novel examples.

1. Introduction

For $\varphi \in H^\infty$, the bounded analytic functions on the open unit disk $\mathbb{D}$, let

$$T_\varphi : H^2 \to H^2, \quad T_\varphi f = \varphi f, \tag{1.1}$$

denote the analytic *Toeplitz operator* on the classical Hardy space H^2. In this paper we survey, continue, and synthesize some discussions begun in [**4, 10, 11**] dealing with the notion of an "inner vector" for T_φ along with the general notion of an inner vector for a contraction on a Hilbert space. We connect these results with the operator-valued Poisson kernel and some work from [**2, 3**] concerning "factoring an L^1 function through a contraction". Along the way we also produce some interesting examples and reformulations of these connections.

2. Basic definitions and facts

We begin with the definition of an inner vector for a Toeplitz operator from [**10**]. Recall that the inner product on the Hardy space H^2 is

$$\langle f, g\rangle := \int_{\mathbb{T}} f\overline{g}\, dm, \tag{2.1}$$

where m is normalized Lebesgue measure on the unit circle $\mathbb{T}$. As is tradition, we equate an $f \in H^2$ with its $L^2 = L^2(\mathbb{T}, m)$ radial boundary function, i.e.,

$$f(\zeta) = \lim_{r \to 1^-} f(r\zeta)$$

for almost every $\zeta \in \mathbb{T}$. We will also use the term *inner* function (without any qualifiers like in Definition 2.2 below) to denote an H^∞ function that has unimodular boundary values almost everywhere. Classical theory [**6**] says that an inner function I can be factored uniquely as $I = \xi B S_\mu$, where ξ is a unimodular constant, B is a Blaschke product, and S_μ is a singular inner function associated with a positive

2010 *Mathematics Subject Classification.* Primary: 47B35, Secondary: 30H10.
Key words and phrases. Inner, outer, model space, Toeplitz.
This work was supported by NSERC (Canada).

measure μ on $\mathbb{T}$ that is singular with respect to m. We say the *degree* of I is equal to d if I is a finite Blaschke product of order d, and equal to infinity otherwise. Furthermore, any function $f \in H^2$ can be factored, uniquely up to multiplicative constants, as $f = IG$, where I is an inner function and $G \in H^2$ is an outer function.

For $\varphi \in H^\infty$ the analytic Toeplitz operator T_φ from (1.1) is a bounded operator on H^2 whose norm $\|T_\varphi\|$ satisfies

$$\|T_\varphi\| = \|\varphi\|_\infty := \text{ess-sup}\{\,|\varphi(\xi)| : \xi \in \mathbb{T}\}.$$

Also recall that the adjoint T_φ^* of T_φ satisfies $T_\varphi^* = T_{\overline{\varphi}}$, where $T_{\overline{\varphi}} f = P(\overline{\varphi} f)$ and P is the Riesz projection of L^2 onto H^2. When φ is an inner function, observe from (2.1) that T_φ is an isometry. See [**8**, Ch. 4] for the details of these basic Toeplitz operator facts and [**1**] for a definitive treatise.

Definition 2.2. For $\varphi \in H^\infty$ we say a unit vector $f \in H^2$ is *T_φ-inner* if $\langle T_\varphi^n f, f\rangle = 0$ for all $n \geqslant 1$.

When $\varphi(z) = z$, one can see from Fourier analysis that the T_z-inner vectors are precisely the inner functions. Also observe that replacing φ with $c\varphi$, where $c > 0$, in Definition 2.2 does not change whether or not a function f is T_φ-inner. Thus we can always assume, by scaling φ, that

$$\varphi \in b(H^\infty) := \{g \in H^\infty : \|g\|_\infty \leqslant 1\},$$

the closed unit ball of H^∞. This normalization will be important when we need T_φ to be a contraction operator since in this case $\|T_\varphi\| = \|\varphi\|_\infty \leqslant 1$. Immediate from Definition 2.2 and the inner product formula from (2.1) are the following facts.

Proposition 2.3. *Let $\varphi \in b(H^\infty)$.*

(1) *If $f \in H^2$ is T_φ-inner and I is any inner function, then If is T_φ-inner.*
(2) *If $f \in H^2$ is T_φ-inner and Θ is any inner divisor of f, i.e., $f/\Theta \in H^2$, then f/Θ is T_φ-inner.*
(3) *Any unit vector belonging to* $\ker T_{\overline{\varphi}}$ *is T_φ-inner.*

If u denotes the inner factor of φ, it is known [**8**, p. 108] that

$$\ker T_{\overline{\varphi}} = \mathcal{K}_u := (uH^2)^\perp,$$

the *model space* corresponding to u. Thus we have the simple corollary.

Corollary 2.4. *If I is any inner function and u is the inner factor of $\varphi \in b(H^\infty)$, then any unit vector from $I\mathcal{K}_u$ is T_φ-inner.*

This corollary gives us many specific examples of T_φ-inner vectors. For example, if $\lambda \in \mathbb{D}$, the reproducing kernel functions

$$k_\lambda(z) := \frac{1 - \overline{u(\lambda)}u(z)}{1 - \overline{\lambda}z}$$

belong to $\mathcal{K}_u$. In fact, finite linear combinations of these functions are dense in $\mathcal{K}_u$ [**8**, Ch. 5]. Since

$$\|k_\lambda\| = \sqrt{k_\lambda(\lambda)} = \sqrt{\frac{1 - |u(\lambda)|^2}{1 - |\lambda|^2}},$$

then

$$I\sqrt{\frac{1-|\lambda|^2}{1-|u(\lambda)|^2}}\,\frac{1-\overline{u(\lambda)}u(z)}{1-\overline{\lambda}z}, \quad \lambda \in \mathbb{D}, \ I \text{ inner},$$

are T_φ-inner functions.

When $\varphi = u$ is a finite Blaschke product, then the model space $\mathcal{K}_u$ is a certain finite dimensional space of rational functions that are analytic in a neighborhood of $\overline{\mathbb{D}}$ [**8**, p. 117]. Furthermore, as we will see in a moment in Theorem 3.12, every T_u-inner function is bounded. However, when u is not a finite Blaschke product then $\mathcal{K}_u$ is infinite dimensional [**8**, p. 117] and, since multiplication by an inner function I is an isometry on H^2 (see (2.1)), $I\mathcal{K}_u$ is a closed infinite dimensional subspace of L^2. By a theorem of Grothendieck, it will contain an unbounded function. Putting this all together, we obtain the following.

COROLLARY 2.5. *If the inner factor of $\varphi \in b(H^\infty)$ is not a finite Blaschke product, then there are unbounded T_φ-inner functions.*

A specific version of this was pointed out in [**10**, p. 103].

Of course one needs to discuss the case when φ is an outer function. Since φH^2 is dense in H^2 [**8**, p. 86], we see that $\ker T_{\overline{\varphi}} = \{0\}$. In this case, it is not clear that there are *any* T_φ-inner functions. Indeed, we do not see any obvious ones like $I \ker T_{\overline{\varphi}}$ since, in this case, $\ker T_{\overline{\varphi}} = \{0\}$.

EXAMPLE 2.6. Suppose that φ is the outer function $\varphi(z) = 1 + z$ and that $f \in H^2$ is T_φ-inner, i.e.,

$$\langle T_\varphi^n f, f\rangle = 0, \ \ \forall n \geqslant 1.$$

In other words,

$$\int_{\mathbb{T}} (1+\xi)^n |f(\xi)|^2\, dm(\xi) = 0, \ \ \forall n \geqslant 1. \tag{2.7}$$

Then the L^1 function $|f|^2$ annihilates $(1+z)^n$ for all $n \geqslant 1$, along with all their linear combinations. In particular, $|f|^2$ annihilates

$$(1+z)^2 - (1+z) = 1 + 2z + z^2 - 1 - z = z(1+z).$$

The above observation will be the first step in a proof by induction. Next, suppose that $|f|^2$ annihilates $z^k(1+z)$ for all $1 \leqslant k \leqslant n$. Then

$$z^{n+1}(1+z) = (1+z)^{n+2} - \big[(1+z)^{n+1} - z^{n+1}\big](1+z).$$

By the T_φ-inner property of f notice that $|f|^2$ annihilates the first term on the right. It also annihilates the subtracted expression, by the induction hypothesis (the expression in square brackets is a polynomial of degree n). Thus we have shown by induction that $|f|^2$ annihilates $\{z^n(1+z)\}_{n\geqslant 0}$ (the $n = 0$ case follows from (2.7)). This means that

$$\int_{\mathbb{T}} \xi^n (1+\xi)|f(\xi)|^2 dm(\xi) = 0, \quad n \geqslant 0, \tag{2.8}$$

and by complex conjugation,

$$\int_{\mathbb{T}} \overline{\xi}^n (1+\overline{\xi})|f(\xi)|^2 dm(\xi) = 0, \quad n \geqslant 0.$$

A little algebra yields

$$\int_{\mathbb{T}} \overline{\xi}^{n+1}(1+\xi)|f(\xi)|^2 dm(\xi), \quad n \geqslant 0. \tag{2.9}$$

Equations (2.8) and (2.9) say that all of the Fourier coefficients of $(1+\xi)|f(\xi)|^2$ vanish and so $(1+\xi)|f(\xi)|^2$ is zero. Conclusion: there are no T_φ-inner functions when $\varphi(z) = 1 + z$.

3. Inner vectors via the Wold decomposition

Using some ideas from **[10]**, we can use the Wold decomposition **[9]** to explore the inner vectors for certain Toeplitz operators. Observe that when u is an inner function the Toeplitz operator T_u is an isometry on H^2. Thus the Wold decomposition of H^2 with respect to T_u becomes

$$H^2 = X_0 \oplus X_1 \oplus T_u X_1 \oplus T_u^2 X_1 \oplus \cdots,$$

where

$$X_0 := \bigcap_{n=1}^{\infty} T_u^n H^2 = \{0\}, \quad X_1 := H^2 \ominus T_u H^2 = \mathcal{K}_u.$$

Thus

$$H^2 = \mathcal{K}_u \oplus u\mathcal{K}_u \oplus u^2 \mathcal{K}_u \oplus \cdots.$$

The above decomposition says that every $f \in H^2$ has a unique expansion as

$$f = F_0 + uF_1 + u^2 F_2 + \cdots, \quad F_j \in \mathcal{K}_u. \tag{3.1}$$

Furthermore, for each integer $N \geqslant 1$,

$$\begin{aligned} \langle u^N f, f\rangle &= \Big\langle u^N \sum_{k \geqslant 0} u^k F_k, \sum_{l \geqslant 0} u^l F_l \Big\rangle \\ &= \sum_{k,l \geqslant 0} \langle u^{N+k-l} F_k, F_l\rangle \\ &= \sum_{l-k=N} \langle F_k, F_l\rangle. \end{aligned}$$

This leads us to the following.

PROPOSITION 3.2. *A unit vector $f \in H^2$ with expansion*

$$f = F_0 + uF_1 + u^2 F_2 + \cdots, \quad F_j \in \mathcal{K}_u,$$

as in (3.1) *is T_u-inner if and only if*

$$\sum_{k=0}^{\infty} \langle F_k, F_{N+k}\rangle = 0, \quad N \geqslant 1. \tag{3.3}$$

Though this is just a restatement of the condition for f to be T_u-inner, it is useful for producing more tangible examples of T_u-inner functions.

EXAMPLE 3.4. Choose orthogonal vectors $F_j, j \geqslant 0$ from $\mathcal{K}_u$ so that $\sum_{j \geqslant 0} \|F_j\|^2 = 1$. Then the condition (3.3) is easily satisfied and thus the unit vector $f = \sum_{j \geqslant 0} u^j F_j$ is a T_u-inner function (as is any inner function times this vector).

EXAMPLE 3.5. If $u(z) = z^n$, then $\mathcal{K}_u = \operatorname{span}\{1, z, z^2, \dots z^{n-1}\}$ and the vectors

$$F_j = \frac{z^j}{\sqrt{n}}, \quad 0 \leqslant j \leqslant n-1,$$

satisfy the conditions of the previous example. Thus

$$f = \sum_{j=0}^{n-1} u^j F_j = \frac{1}{\sqrt{n}} + \frac{z^{n+1}}{\sqrt{n}} + \frac{z^{2n+2}}{\sqrt{n}} + \frac{z^{3n+3}}{\sqrt{n}} + \cdots + \frac{z^{(n-1)(n+1)}}{\sqrt{n}}$$

is a T_{z^n}-inner vector.

EXAMPLE 3.6. The previous example can be generalized to a finite Blaschke product

$$u(z) = \prod_{j=1}^{n} \frac{z - a_j}{1 - \overline{a_j} z}, \quad a_j \in \mathbb{D}.$$

If we define

$$F_0(z) = \frac{\sqrt{1 - |a_1|^2}}{1 - \overline{a_1} z},$$

$$F_1(z) = \frac{\sqrt{1 - |a_2|^2}}{1 - \overline{a_2} z} \frac{z - a_1}{1 - \overline{a_1} z},$$

$$F_2(z) = \frac{\sqrt{1 - |a_3|^2}}{1 - \overline{a_3} z} \frac{z - a_1}{1 - \overline{a_1} z} \frac{z - a_2}{1 - \overline{a_2} z},$$

$$\vdots$$

$$F_{n-1}(z) = \frac{\sqrt{1 - |a_n|^2}}{1 - \overline{a_n} z} \prod_{j=1}^{n-1} \frac{z - a_j}{1 - \overline{a_j} z},$$

one can show that $\{F_0, \dots, F_{n-1}\}$ is an orthonormal basis for $\mathcal{K}_u$. Now choose $\alpha_0, \dots, \alpha_{n-1} \in \mathbb{C}$ such that $\sum_{j=0}^{n=1} |\alpha_j|^2 = 1$. Then

$$f = \sum_{j=0}^{n-1} \alpha_j u^j F_j$$

is T_u-inner.

From Corollary 2.4 we know, for an inner function I, that any unit vector from the set $\{I \ker T_{\overline{u}} : I \text{ is inner}\}$ is a $T_{\overline{u}}$-inner vector. Perhaps one might think we have equality here. Indeed, sometimes we do. For example, if $u(z) = z$, then $\ker T_{\overline{z}} = \mathbb{C}$ and, as discussed earlier, the T_z-inner vectors are precisely the inner functions. Here is another positive example of when the unit vectors from $\{I \ker T_{\overline{u}} : I \text{ is inner}\}$ constitute the complete set of T_u-inner vectors.

EXAMPLE 3.7. If the inner function u is the single Blaschke factor

$$u(z) = \frac{z - a}{1 - \overline{a} z}, \quad a \in \mathbb{D},$$

one can show [**8**, Ch. 5] that

$$\ker T_{\overline{u}} = \mathcal{K}_u = \mathbb{C} \frac{1}{1 - \overline{a} z}.$$

As shown in [**4**], the T_u-inner vectors are

$$I\frac{\sqrt{1-|a|^2}}{1-\overline{a}z}, \quad I \text{ inner}.$$

However, in general, the unit vectors from $\{I \ker T_{\overline{u}} : I \text{ is inner}\}$ form a proper subset of the T_u-inner vectors. One can see this with the following example.

EXAMPLE 3.8. Using the technique from Example 3.5, we see that when $u(z) = z^n$ the vector

$$f = \frac{1}{\sqrt{2}} + \frac{z^{n+1}}{\sqrt{2}}$$

is T_u-inner. However, f is not of the form Ig, where I is inner and $g \in \mathcal{K}_u$. This follows from the fact that f is outer and does not belong to

$$\mathcal{K}_u = \operatorname{span}\{1, z, z^2, \ldots, z^{n-1}\}.$$

The papers [**10, 11**] yield a description of the T_u-inner vectors. From the Wold decomposition (3.1) we see that any $f \in H^2$ can be written as

$$f = \sum_{k=0}^{\infty} F_k u^k.$$

If $\{v_j\}_{j\geqslant 1}$ is an orthonormal basis for $\mathcal{K}_u$, then we can expand things a bit further and write

$$\begin{aligned} f &= \sum_{k=0}^{\infty} F_k u^k \\ &= \sum_{k=0}^{\infty} u^k \Big(\sum_{j\geqslant 1} c_{j,k} v_j\Big) \\ &= \sum_{j\geqslant 1} v_j \Big(\sum_{k=0}^{\infty} c_{j,k} u^k\Big). \end{aligned}$$

Observe that

$$\sum_{j\geqslant 1} |c_{j,k}|^2 = \|F_k\|^2$$

and that

$$\begin{aligned} \|f\|^2 &= \sum_{k=0}^{\infty} \|F_k\|^2 \\ &= \sum_{k=0}^{\infty} \sum_{j\geqslant 1} |c_{j,k}|^2 \\ &= \sum_{j\geqslant 1} \sum_{k=0}^{\infty} |c_{j,k}|^2. \end{aligned}$$

Thus for each j, $\sum_{k\geqslant 0} |c_{j,k}|^2 < \infty$ and so

$$f_j(z) = \sum_{k=0}^{\infty} c_{j,k} z^k$$

defines a function in H^2 (square summable power series). By the Littlewood subordination principle [**8**, p. 126], $f_j \circ u$ also belongs to H^2.

Thus every unit vector $f \in H^2$ has the unique representation

$$f(z) = \sum_{j \geqslant 1} v_j(z) f_j(u(z)), \tag{3.9}$$

where $f_j \in H^2$ with $\sum_{j \geqslant 1} \|f_j\|^2 < \infty$, and $\{v_j\}_{j \geqslant 1}$ is an orthonormal basis for $\mathcal{K}_u$. Furthermore, as observed in [**10**, Prop. 1] (and can be proved using the above calculation), if

$$f = \sum_{j \geqslant 1} v_j f_j(u), \quad g = \sum_{j \geqslant 1} v_j g_j(u), \tag{3.10}$$

as in (3.9), then

$$\langle f, g \rangle = \sum_{j \geqslant 1} \langle f_j, g_j \rangle. \tag{3.11}$$

THEOREM 3.12. *A unit vector f written as in* (3.9) *is T_u-inner if and only if*

$$\sum_{j \geqslant 1}^{\infty} |f_j(\xi)|^2 = 1$$

for almost every $\xi \in \mathbb{T}$.

PROOF. Here is the original proof from [**10**]. With

$$f = \sum_{j \geqslant 1} v_j f_j(u),$$

and $n \geqslant 1$, (3.11) yields

$$\begin{aligned}
\langle T_u^n f, f \rangle &= \langle f u^n, f \rangle \\
&= \Big\langle \sum_j v_j u^n f_j(u), \sum_k v_k f_k(u) \Big\rangle \\
&= \sum_{j \geqslant 1} \langle z^n f_j, f_j \rangle \\
&= \sum_{j \geqslant 1} \int_{\mathbb{T}} \xi^n |f_j(\xi)|^2 dm(\xi) \\
&= \int_{\mathbb{T}} \xi^n \Big(\sum_{j \geqslant 1} |f_j(\xi)|^2 \Big) dm(\xi). \qquad (3.13)
\end{aligned}$$

Then $\langle T_u^n f, f \rangle = 0$ for all $n = 1, 2, \ldots$ if and only if, by Fourier analysis, $\sum_{j \geqslant 1} |f_j|^2$ is constant almost everywhere. But since we assuming that f is a unit vector, we see, by putting $n = 0$ in (3.13), that $\sum_{j \geqslant 1} |f_j|^2 = 1$ almost everywhere. □

When u is a finite Blaschke product, then $\mathcal{K}_u$ is finite dimensional. In this case (3.9) is finite and each basis vector v_j is a rational function that is analytic in a neighborhood of $\overline{\mathbb{D}}$ [**8**, Ch. 5]. From here it follows that every T_u-inner vector is a bounded function. Contrast this with Corollary 2.5 which says that when u is not a finite Blaschke product there are always T_u-inner vectors that are unbounded functions.

The two papers [**10, 11**] go further and discuss an "inner-outer" factorization of any $f \in H^2$ in terms of T_u-inner and T_u-outer vectors. They also discuss the concept of T_u-inner in H^p, for $p > 1$, along with some properties of the norms of T_u-inner vectors as well as their growth near $\mathbb{T}$.

4. Inner vectors via the operator-valued Poisson kernel

We can rephrase the language of inner vectors for Toeplitz operators in terms of operator-valued Poisson kernels [**2**]. Moreover, using this new language, we can extend our discussion to inner vectors for contractions on Hilbert spaces. For $\lambda \in \mathbb{D}$ and $\xi \in \mathbb{T}$, define

$$P_\lambda(\xi) := \frac{1}{1 - \overline{\lambda}\xi} + \frac{1}{1 - \lambda\overline{\xi}} - 1 \tag{4.1}$$

and observe that this can be written as

$$P_\lambda(\xi) = \frac{1 - |\lambda|^2}{|\xi - \lambda|^2},$$

which is the standard Poisson kernel. Classical theory says that for any $g \in L^1 = L^1(\mathbb{T}, m)$ the function

$$\int_{\mathbb{T}} P_\lambda(\xi) f(\xi) dm(\xi)$$

is harmonic on $\mathbb{D}$ with

$$\lim_{r \to 1^-} \int_{\mathbb{T}} P_{r\zeta}(\xi) f(\xi) dm(\xi) = f(\zeta) \tag{4.2}$$

for almost every $\zeta \in \mathbb{T}$. Furthermore, if μ is a finite complex measure on $\mathbb{T}$, we have

$$\int_{\mathbb{T}} P_\lambda(\xi) d\mu(\xi) = \widehat{\mu}(0) + \sum_{n \geqslant 1} \widehat{\mu}(n)\lambda^n + \sum_{n \geqslant 1} \widehat{\mu}(-n)\overline{\lambda}^n, \tag{4.3}$$

where

$$\widehat{\mu}(n) := \int_{\mathbb{T}} \overline{\xi}^n d\mu(\xi), \quad n \in \mathbb{Z},$$

are the Fourier coefficients of μ. We will now discuss an operator version of the Poisson kernel.

For a contraction T on a Hilbert space $\mathcal{H}$, we imitate the formula in (4.1) and define, for $\lambda \in \mathbb{D}$, the *operator-valued Poisson kernel* $K_\lambda(T)$ as

$$K_\lambda(T) := (I - \lambda T^*)^{-1} + (I - \overline{\lambda} T)^{-1} - I.$$

By the spectral radius formula, notice how $\sigma(T) \subseteq \overline{\mathbb{D}}$ and thus the formula for $K_\lambda(T)$ above makes sense. A computation with Neumann series will show that for $r \in [0, 1)$ and $\theta \in [0, 2\pi)$

$$K_{re^{i\theta}}(T) = \sum_{n=0}^{\infty} r^n e^{in\theta} T^{*n} + \sum_{n=0}^{\infty} r^n e^{-in\theta} T^n - I. \tag{4.4}$$

The operator identity

$$K_\lambda(T) = (I - \overline{\lambda} T)^{-1}(I - |\lambda|^2 TT^*)(I - \lambda T^*)^{-1}$$

from [**2**, Lemma 2.4] shows that for each $\mathbf{x} \in \mathcal{H}$

$$\langle K_\lambda(T)\mathbf{x}, \mathbf{x} \rangle \geqslant 0, \quad \lambda \in \mathbb{D}.$$

Moreover, the function

$$\lambda \mapsto \langle K_\lambda(T)\mathbf{x}, \mathbf{x}\rangle$$

is harmonic on $\mathbb{D}$. Hence, a classical harmonic analysis result of Herglotz ([**6**, p. 10] or [**8**, p. 17]) produces a unique positive finite Borel measure $\mu_{T,\mathbf{x}}$ on $\mathbb{T}$ such that

$$\langle K_\lambda(T)\mathbf{x}, \mathbf{x}\rangle = \int_{\mathbb{T}} P_\lambda(\zeta) d\mu_{T,\mathbf{x}}(\zeta). \tag{4.5}$$

Since $K_0(T) = I$ we have

$$1 = \langle \mathbf{x}, \mathbf{x}\rangle = \langle K_0(T)\mathbf{x}, \mathbf{x}\rangle = \int_{\mathbb{T}} d\mu_{T,\mathbf{x}}$$

and so $\mu_{T,\mathbf{x}}$ is a probability measure.

As we defined for Toeplitz operators earlier in Definition 2.2, we say that a unit vector $\mathbf{x}$ is T-*inner* if

$$\langle T^n \mathbf{x}, \mathbf{x}\rangle = 0, \quad n \geqslant 1.$$

Note that $\mathbf{x}$ is T-inner if and only if $\mathbf{x}$ is T^*-inner. From (4.4) we see that $\mathbf{x}$ is T-inner if and only if $\langle K_\lambda(T)\mathbf{x}, \mathbf{x}\rangle = 1$ for all $\lambda \in \mathbb{D}$, or equivalently,

$$1 = \int_{\mathbb{T}} P_\lambda(\zeta) d\mu_{T,\mathbf{x}}(\zeta), \quad \lambda \in \mathbb{D}.$$

By (4.3) this is equivalent to the condition $\mu_{T,\mathbf{x}} = m$. This gives us the following.

PROPOSITION 4.6. *Suppose that T is a contraction on a Hilbert space $\mathcal{H}$ and $\mathbf{x}$ is unit vector in $\mathcal{H}$. Then $\mathbf{x}$ is T-inner if and only if $\mu_{T,\mathbf{x}} = m$, where $\mu_{T,\mathbf{x}}$ is defined as in* (4.5).

For an inner function u, note that T_u is an isometry, hence a contraction. Thus we can apply the above analysis to $\mu_{T_u,f}$.

PROPOSITION 4.7. *If*

$$f = \sum_{j \geqslant 1} v_j f_j(u)$$

is a vector from H^2 as in (3.9), *then*

$$d\mu_{T_u,f} = \sum_{j \geqslant 1} |f_j|^2 \, dm. \tag{4.8}$$

PROOF. If

$$f = \sum_{j \geqslant 1} v_j f_j(u),$$

then

$$\|f\|^2 = \sum_{j \geqslant 1} \|f_j\|^2 = \sum_{j \geqslant 1} \int_{\mathbb{T}} |f_j|^2 dm = \int_{\mathbb{T}} \sum_{j \geqslant 1} |f_j|^2 dm$$

and the calculation used to prove Theorem 3.12 yields

$$\langle T_u^n f, f\rangle = \int_{\mathbb{T}} \xi^n \Big(\sum_{j \geqslant 1} |f_j(\xi)|^2\Big) dm(\xi),$$

$$\langle T_u^{*n} f, f\rangle = \int_{\mathbb{T}} \overline{\xi}^n \Big(\sum_{j \geqslant 1} |f_j(\xi)|^2\Big) dm(\xi).$$

From here we observe

$$\begin{aligned}
\int_{\mathbb{T}} P_\lambda(\xi) d\mu_{T_u,f}(\xi) &= \langle K_\lambda(T_u)f, f\rangle \\
&= \sum_{n\geqslant 0} \lambda^n \langle T_u^{*n} f, f\rangle + \sum_{n\geqslant 0} \overline{\lambda}^n \langle T_u^n f, f\rangle - \langle f, f\rangle. \\
&= \sum_{n\geqslant 0} \lambda^n \int_{\mathbb{T}} \overline{\xi}^n \Big(\sum_{j\geqslant 1} |f_j(\xi)|^2\Big) dm(\xi) \\
&\quad + \sum_{n\geqslant 0} \overline{\lambda}^n \int_{\mathbb{T}} \xi^n \Big(\sum_{j\geqslant 1} |f_j(\xi)|^2\Big) dm(\xi) \\
&\quad - \sum_{j\geqslant 1} \int_{\mathbb{T}} |f_j(\xi)|^2 dm \\
&= \int_{\mathbb{T}} \Big(\frac{1}{1-\lambda\overline{\xi}} + \frac{1}{1-\overline{\lambda}\xi} - 1\Big) \sum_{j\geqslant 1} |f_j(\xi)|^2 dm(\xi) \\
&= \int_{\mathbb{T}} P_\lambda(\xi) \sum_{j\geqslant 1} |f_j(\xi)|^2 dm(\xi)
\end{aligned}$$

Now use the uniqueness of the Fourier coefficients of a measure along with (4.3) to obtain (4.8). □

Notice how this gives us another way of thinking about Theorem 3.12: a unit vector $f \in H^2$ is T_u-inner if and only if $\mu_{T_u,f} = m$.

This brings us to an interesting related question. One can also show that for any $f, g \in H^2$, we can define the harmonic function $\langle K_\lambda(T_u)f, g\rangle$ on $\mathbb{D}$ and prove this function also has bounded integral means. This yields, via Herglotz's theorem, a complex valued measure $\mu_{T_u,f,g}$ on $\mathbb{T}$ for which

$$\langle K_\lambda(T_u)f, g\rangle = \int_{\mathbb{T}} P_\lambda(\xi) d\mu_{T,f,g}(\xi), \quad \lambda \in \mathbb{D}. \tag{4.9}$$

See [**2**, Prop. 2.6] for details. A similar calculation used to prove Proposition 4.6 shows that

$$d\mu_{T_u,f,g} = \sum_{j\geqslant 1} f_j \overline{g_j}\, dm. \tag{4.10}$$

In the above formula, f_j and g_j come from the representations of f and g from (3.10). A general result from [**3**] says that given any $F \in L^1$ and a non-constant inner function u that is not an automorphism, there are $f, g \in H^2$ for which

$$F(\zeta) = \frac{d\mu_{T_u,f,g}}{dm}(\zeta) \tag{4.11}$$

m-almost everywhere. In the language of [**3**] this says that any $F \in L^1$ can be "factored through T_u". Equivalently stated, using (4.10) and (4.11), we have

$$F(\zeta) = \sum_{j\geqslant 1} f_j(\zeta)\overline{g_j(\zeta)}.$$

This is an interesting representation for L^1 functions and a refinement of the one from [**3**].

QUESTION 4.12. Proposition 4.7 shows that when φ is an inner function and $f, g \in H^2$, then $d\mu_{T_\varphi, f, g}$ is absolutely continuous with respect to m. When $\varphi \in b(H^\infty)$ is this still the case? For this to be true we would need to know that $\langle \varphi^n f, g\rangle, n \geqslant 1$, are the Fourier coefficients of an L^1 function.

5. Inner vectors via Clark measures

For any fixed $\alpha \in \mathbb{T}$ and inner function u, the function

$$z \mapsto \frac{1 - |u(z)|^2}{|\alpha - u(z)|^2} = \Re\Big(\frac{\alpha + u(z)}{\alpha - u(z)}\Big)$$

is a positive harmonic function on $\mathbb{D}$. Thus by Herglotz's theorem, there is a unique positive measure σ_α on $\mathbb{T}$ for which

$$\frac{1 - |u(z)|^2}{|\alpha - u(z)|^2} = \int_{\mathbb{T}} P_\lambda(\xi) d\sigma_\alpha(\xi).$$

The family of measures $\{\sigma_\alpha : \alpha \in \mathbb{T}\}$ is called the family of *Clark measures* corresponding to u. Let us record some important facts about this family of measures. Proofs can be found in [**5**].

First, one can use the fact that u is an inner function, along with standard harmonic analysis, to prove that each σ_α is singular with respect to m. Second, if E_α is defined to be the set of $\xi \in \mathbb{T}$ for which

$$\lim_{r \to 1^-} u(r\xi) = \alpha,$$

then E_α is a Borel subset of $\mathbb{T}$ with

$$\sigma_\alpha(\mathbb{T} \setminus E_\alpha) = 0. \tag{5.1}$$

In other words, σ_α is "carried" by E_α. From this we also see that the measures $\{\sigma_\alpha : \alpha \in \mathbb{T}\}$ are singular with respect to each other. Third, a beautiful disintegration theorem of Aleksandrov says that if $g \in L^1$ then for m-almost every $\alpha \in \mathbb{T}$, integral

$$\int_{\mathbb{T}} g(\xi) d\sigma_\alpha(\xi)$$

is well defined. Moreover this almost everywhere defined function

$$\alpha \mapsto \int_{\mathbb{T}} g(\xi) d\sigma_\alpha(\xi)$$

is integrable with respect to m and

$$\int_{\mathbb{T}} \Big(\int_{\mathbb{T}} g(\xi) d\sigma_\alpha(\xi) \Big) dm(\alpha) = \int_{\mathbb{T}} g(\zeta) dm(\zeta). \tag{5.2}$$

Using Clark measures, we can use a technique from [**11**] to compute a formula for $\langle K_\lambda(T_u) f, f\rangle$ along with the measure $d\mu_{T_u, f}/dm$. This gives us another way to think about the formula (4.11). The result here is the following.

THEOREM 5.3. *For an inner function u and $f \in H^2$ we have*

$$d\mu_{T_u, f}(\alpha) = \Big(\int_{\mathbb{T}} |f(\xi)|^2 d\sigma_\alpha(\xi) \Big)\, dm(\alpha).$$

PROOF. For any $f \in H^2$ use the formulas from (5.1) and (5.2) to obtain

$$\begin{aligned}\langle T_u^n f, f\rangle &= \int_{\mathbb{T}} |f(\xi)|^2 u(\xi)^n dm(\xi)\\ &= \int_{\mathbb{T}} \Big(\int_{\mathbb{T}} |f(\xi)|^2 u(\xi)^n d\sigma_\alpha(\xi)\Big) dm(\alpha)\\ &= \int_{\mathbb{T}} \Big(\int_{\mathbb{T}} |f(\xi)|^2 \alpha^n d\sigma_\alpha(\xi)\Big) dm(\alpha)\\ &= \int_{\mathbb{T}} \alpha^n \Big(\int_{\mathbb{T}} |f(\xi)|^2 d\sigma_\alpha(\xi)\Big) dm(\alpha).\end{aligned}$$

In a similar way

$$\langle T_u^{*n} f, f\rangle = \int_{\mathbb{T}} \overline{\alpha}^n \Big(\int_{\mathbb{T}} |f(\xi)|^2 d\sigma_\alpha(\xi)\Big) dm(\alpha).$$

Now follow the proof of Proposition 4.7 to get

$$\begin{aligned}&\int_{\mathbb{T}} P_\lambda(\xi) d\mu_{T_u,f}(\xi)\\ =\;& \langle K_\lambda(T_u) f, f\rangle\\ =\;& \sum_{n \geqslant 0} \lambda^n \langle T_u^{*n} f, f\rangle + \sum_{n \geqslant 0} \overline{\lambda}^n \langle T_u^n f, f\rangle - \langle f, f\rangle\\ =\;& \int_{\mathbb{T}} \Big(\Big(\frac{1}{1-\lambda\overline{\alpha}} + \frac{1}{1-\overline{\lambda}\alpha} - 1\Big)\Big(\int_{\mathbb{T}} |f(\xi)|^2 d\sigma_\alpha(\xi)\Big)\Big) dm(\alpha)\\ =\;& \int_{\mathbb{T}} P_\lambda(\alpha)\Big(\int_{\mathbb{T}} |f(\xi)|^2 d\sigma_\alpha(\xi)\Big) dm(\alpha).\end{aligned}$$

Use (4.3) along with the uniqueness of Fourier coefficients of a measure to compute the proof. □

Combing Theorem 5.3 and Proposition 4.6 yields the following result from [**11**].

COROLLARY 5.4. *A unit vector $f \in H^2$ is T_u-inner if and only if*

$$\int_{\mathbb{T}} |f(\xi)|^2 d\sigma_\alpha(\xi) = 1$$

for m-almost every $\alpha \in \mathbb{T}$.

Recall the notation from (4.9) that for a given inner function u and $f, g \in H^2$

$$\langle K_\lambda(T_u) f, g\rangle = \int_{\mathbb{T}} P_\lambda(\xi) d\mu_{T_u,f,g}(\xi).$$

Moreover, if $\deg(u) \geqslant 2$, any $F \in L^1$ can be written as $d\mu_{T_u,f,g}(\xi)/dm$ for some $f, g \in H^2$. Here is another way of thinking about this via Clark measures. The same argument used to prove Theorem 5.3 shows that

$$d\mu_{T_u,f,g} = \int_{\mathbb{T}} f(\xi)\overline{g(\xi)} d\sigma_\alpha(\xi)\, dm \tag{5.5}$$

Since any $F \in L^1$ is equal to $d\mu_{T_u,f,g}/dm$ for some $f, g \in H^2$ [**3**], we see that any $F \in L^1$ can be written as

$$F(\alpha) = \int_{\mathbb{T}} f(\xi)\overline{g(\xi)} d\sigma_\alpha(\xi).$$

This Clark measure viewpoint has the additional feature, via Aleksandrov's theorem, that

$$\begin{aligned}\int_{\mathbb{T}} F(\alpha)dm(\alpha) &= \int_{\mathbb{T}} \Big(\int_{\mathbb{T}} f(\xi)\overline{g(\xi)}d\sigma_\alpha(\xi)\Big)dm(\alpha)\\ &= \int_{\mathbb{T}} f(\zeta)g(\zeta)dm(\zeta).\end{aligned}$$

EXAMPLE 5.6. If u is a finite Blaschke product of degree d and $\alpha \in \mathbb{T}$, then one can compute (see [**5**, p. 209] for the details) the Clark measure to be

$$d\sigma_\alpha = \sum_{j=1}^{d} \frac{1}{|u'(\zeta_j)|}\delta_{\zeta_j},$$

where $\zeta_1, \ldots, \zeta_d$ are the d distinct solutions to the equation $u(z) = \alpha$ and δ_{ζ_j} is the unit point pass as ζ_j. The denominators in the above expression may look troublesome but at the end of the day we have $u' \neq 0$ on $\mathbb{T}$. By Theorem 5.3 we see that

$$\frac{d\mu_{T_u,f}}{dm}(\alpha) = \int_{\mathbb{T}} |f(\xi)|^2 d\sigma_\alpha(\xi) = \sum_{j=1}^{d} \frac{|f(\zeta_j)|^2}{|u'(\zeta_j)|}.$$

Thus the criterion for a unit vector $f \in H^2$ to be a T_u-inner vector is that the above sum is equal to 1 for m-almost every $\alpha \in \mathbb{T}$.

Furthermore, by (5.5), given $F \in L^1$, there are $f, g \in H^2$ so that

$$F(\alpha) = \sum_{j=1}^{d} \frac{f(\zeta_j)\overline{g(\zeta_j)}}{|u'(\zeta_j)|}$$

for m-almost every $\alpha \in \mathbb{T}$. This formula appears in [**3**].

EXAMPLE 5.7. Let us apply this to the simple case where $u(z) = z^2$. Given any $\alpha \in \mathbb{T}$, the two solutions ζ_1, ζ_2 to the equation $z^2 = \alpha$ are

$$\zeta_1 = e^{i \arg \alpha/2}, \quad \zeta_2 = -e^{i \arg \alpha/2}.$$

Thus the condition that a unit f is a T_{z^2}-inner vector becomes

$$|f(e^{i \arg \alpha/2})|^2 + |f(-e^{i \arg \alpha/2})|^2 = 2, \quad m\text{-a.e. } \alpha \in \mathbb{T}.$$

Furthermore, given any $F \in L^1$, there are $f, g \in H^2$ for which

$$F(\alpha) = \tfrac{1}{2} f(e^{i \arg \alpha/2})\overline{g(e^{i \arg \alpha/2})} + \tfrac{1}{2} f(-e^{i \arg \alpha/2})\overline{g(-e^{i \arg \alpha/2})}.$$

This second fact was first observed in [**3**].

EXAMPLE 5.8. Consider the atomic inner function

$$u(z) = \exp\Big(\frac{z+1}{z-1}\Big).$$

For a fixed $t \in [0, 2\pi)$, the solutions to $u(z) = e^{it}$ are

$$\zeta_k = \frac{i(t + 2\pi k) + 1}{i(t + 2\pi k) - 1}, \quad k \in \mathbb{Z}.$$

Noting that

$$|u'(\zeta_k)| = \frac{2}{|\zeta_k - 1|^2},$$

a similar computation as in Example 5.6 shows that

$$d\sigma_{e^{it}} = \frac{1}{2}\sum_{k\in\mathbb{Z}} \delta_{\zeta_k}|\zeta_k - 1|^2.$$

Thus

$$\begin{aligned}\frac{d\mu_{T_u,f}}{dm}(e^{it}) &= \int_{\mathbb{T}} |f(\xi)|^2 d\sigma_{e^{it}}(\xi)\\ &= \frac{1}{2}\sum_{k\in\mathbb{Z}} |f(\zeta_k)|^2|\zeta_k - 1|^2\\ &= \sum_{k\in\mathbb{Z}} \Big|f\Big(\frac{i[t+2\pi k]+1}{i[t+2\pi k]-1}\Big)\Big|^2 \frac{2}{|i(t+2\pi k)-1|^2}.\end{aligned}$$

To create a T_u-inner function, we need to find a unit vector $f \in H^2$ so that the above expression is equal to one for almost every t. Let us create a specific example of when this happens. In fact we can even make f unbounded. We already knew we could do this from Corollary 2.5 but our example below will be explicit, while the proof of Corollary 2.5 needed Grothendieck's theorem and is not an explicit construction.

To see how to do this, fix $\beta \in (\frac{1}{2}, 1)$, and let a_k, $k \in \mathbb{Z}$, be the collection of coefficients

$$a_k = \frac{1}{1+|k|^{\beta}}. \tag{5.9}$$

Note that $\sum_{k\in\mathbb{Z}} |a_k|^2 < \infty$.

Let I_k be the indicator function of the interval $[-\pi + 2\pi k, \pi + 2\pi k)$, $k \in \mathbb{Z}$. Now define F on $\mathbb{T}$ by

$$F(e^{i\theta}) := \sqrt{2}\sum_{k\in\mathbb{Z}} \frac{a_k}{e^{i\theta}-1}\, I_k\Big(i\frac{1+e^{i\theta}}{1-e^{i\theta}}\Big).$$

Then

$$\begin{aligned}\int_{\mathbb{T}} |F|^2\, dm &= 2\int_{-\pi}^{\pi} |F(e^{i\theta})|^2\, \frac{d\theta}{2\pi}\\ &= 2\sum_{k\in\mathbb{Z}} \int_{-\pi}^{\pi} \frac{|a_k|^2}{|e^{i\theta}-1|^2}\, I_k\Big(i\frac{1+e^{i\theta}}{1-e^{i\theta}}\Big)\, \frac{d\theta}{2\pi}\\ &= 2\sum_{k\in\mathbb{Z}} \int_{-\infty}^{\infty} |a_k|^2 \frac{|it-1|^2}{2^2}\, I_k(t)\, \frac{2\,dt}{2\pi|it-1|^2}\\ &= \sum_{k\in\mathbb{Z}} \int_{-\pi}^{\pi} |a_k|^2 \big|i[t+2\pi k]-1\big|^2 \frac{dt}{2\pi|i[t+2\pi k]-1|^2}\\ &= \sum_{k\in\mathbb{Z}} |a_k|^2 < \infty,\end{aligned}$$

i.e., F is square integrable on $\mathbb{T}$ with

$$\|F\|^2 = \sum_{k\in\mathbb{Z}} |a_k|^2. \tag{5.10}$$

Next we establish that $\log|F|$ is integrable. We'll need the following estimates, which hold for all $k \neq 0$. First note that for $k \neq 0$,

$$\begin{aligned}|a_k|\big|i(t+2\pi k)-1\big| &= |a_k|\big([\pi+2\pi|k|]^2+1\big)^{1/2}\\ &\geqslant |a_k|\cdot 2\pi|k|\\ &\geqslant \frac{2\pi|k|}{1+|k|^\beta}\\ &\geqslant 1.\end{aligned}$$

Consequently, for $k \neq 0$ and $t \in [-\pi,\pi)$,

$$\begin{aligned}\Big|\log\Big(|a_k|\big|i(t+2\pi k)-1\big)\Big| &= \log|a_k||i(t+2\pi k)-1|\\ &\leqslant \log\frac{|i(\pi+2\pi|k|)-1|}{1+|k|^\beta}\\ &\leqslant \log\frac{([2\pi(|k|+1/2)]^2+1)^{1/2}}{1+|k|^\beta}\\ &\leqslant \log\frac{([2\pi(|k|+|k|/2)]^2+|k|^2)^{1/2}}{|k|^\beta}\\ &\leqslant \log(|k|^{1-\beta}\sqrt{9\pi^2+1}).\end{aligned}$$

We now have

$$\begin{aligned}&\int_{\mathbb{T}}|\log|F||\,dm\\ =\quad&\int_{-\pi}^{\pi}\Big|\log|F(e^{i\theta})|\Big|\,\frac{d\theta}{2\pi}\\ =\quad&\sum_{k\in\mathbb{Z}}\int_{-\pi}^{\pi}\Big|\log\frac{|a_k|\sqrt{2}}{|e^{i\theta}-1|}\Big|\,I_k\Big(i\frac{1+e^{i\theta}}{1-e^{i\theta}}\Big)\,\frac{d\theta}{2\pi}\\ =\quad&\sum_{k\in\mathbb{Z}}\int_{-\infty}^{\infty}\Big|\log\big(|a_k||it-1|\sqrt{2}/2\big)\Big|\,I_k(t)\,\frac{dt}{2\pi|it-1|^2}\\ =\quad&\sum_{k\in\mathbb{Z}}\int_{-\pi}^{\pi}\Big|\log\big(6\pi|k|^{1-\beta}|i[t+2\pi k]-1|/\sqrt{2}\big)\Big|\,\frac{dt}{2\pi|i[t+2\pi k]-1|^2}.\end{aligned}$$

The series is summable, because the terms behave like $(\log|k|)/|k|^2$.

It follows that there exists an outer function $g \in H^2$ with radial limit function satisfying $|g| = |F|$ almost everywhere on $\mathbb{T}$, namely

$$g(z) := \exp\Big(\int_{\mathbb{T}}\frac{e^{i\theta}+z}{e^{i\theta}-z}\log|F(e^{i\theta})|\,dm(e^{i\theta})\Big).$$

Finally, let J be any classical inner function, and define $f = gJ$. Then

$$\begin{aligned}\frac{d\mu_{T_u,f}}{dm}(e^{it}) &= \sum_{k\in\mathbb{Z}} \Big|f\Big(\frac{i[t+2\pi k]+1}{i[t+2\pi k]-1}\Big)\Big|^2 \frac{2}{|i(t+2\pi k)-1|^2} \\ &= \sum_{k\in\mathbb{Z}} \Big|F\Big(\frac{i[t+2\pi k]+1}{i[t+2\pi k]-1}\Big)\Big|^2 \frac{2}{|i(t+2\pi k)-1|^2} \\ &= \sum_{k\in\mathbb{Z}} \frac{|a_k(i[t+2\pi k]-1)\sqrt{2}|^2}{2^2} \frac{2}{|i(t+2\pi k)-1|^2} \\ &= \sum_{k\in\mathbb{Z}} |a_k|^2.\end{aligned}$$

Notice from (5.10) that

$$\frac{d\mu_{T_u,f}}{dm}(e^{it}) = \|F\|^2$$

and so one can scale F so that it (and hence f) is a unit vector This also gives us $d\mu_{T_u,f}/dm(e^{it}) = 1$ for almost every t. Any such f will be a T_u-inner function.

As a bonus, we get that the f we just constructed is unbounded. To see this, note that F is unbounded, since for θ approaching zero, $F(e^{i\theta})$ takes values

$$F\Big(\frac{i[t+2\pi k]+1}{i[t+2\pi k]-1}\Big) = \frac{a_k}{1-\frac{i[t+2\pi k]+1}{i[t+2\pi k]-1}} = \frac{-i[t+2\pi k]+1}{2+2|k|^\beta}$$

where $t \in [-\pi, \pi)$. Since $\beta < 1$, this expression is unbounded as $|k| \to \infty$.

6. Inner vectors in model spaces

In this section we depart slightly from Toeplitz operators on H^2 to the related topic of compressions of Toeplitz operators on model spaces. For an inner function Θ, recall the model space $\mathcal{K}_\Theta = (\Theta H^2)^\perp$. An important operator to study here is the *compressed shift operator*

$$S_\Theta : \mathcal{K}_u \to \mathcal{K}_u, \quad S_\Theta f = P_\Theta(zf),$$

where P_Θ is the orthogonal projection of L^2 onto $\mathcal{K}_u$. This operator is used to model a certain class of contraction operators on Hilbert space [**8**, Ch. 9] – hence the use of the phrase "model space."

As a generalization of our discussion of classifying the T_z-inner vectors in H^2, one can ask for a description of the S_Θ-inner vectors in $\mathcal{K}_\Theta$, i.e., those unit vectors $f \in \mathcal{K}_\Theta$ for which

$$\langle S_\Theta^n f, f\rangle = 0, \quad n \geqslant 1.$$

Before continuing, let us make a few comments about S_Θ. For the proofs, see [**8**, Ch. 9]. First note that since S_Θ is a compression of T_z to $\mathcal{K}_\Theta$ we have the identity

$$S_\Theta^n = P_\Theta T_{z^n}|_{\mathcal{K}_\Theta}.$$

Furthermore, we have the adjoint formula

$$S_\Theta^* = T_{\overline{z}}|_{\mathcal{K}_\Theta}.$$

For any $\varphi \in H^\infty$ there is the functional calculus for S_Θ which allows us to define

$$\varphi(S_\Theta) = P_\Theta T_\varphi|_{\mathcal{K}_\Theta}$$

along with the adjoint formula

$$\varphi(S_\Theta)^* = P_\Theta T_{\overline{\varphi}}|_{\mathcal{K}_\Theta}.$$

One can actually compute the S_Θ-inner vectors with the following result from [**8**, p. 177].

THEOREM 6.1. *Any S_Θ-inner function is an inner function. Moreover, $\mathcal{K}_\Theta$ contains an inner function if and only if $u(0) = 0$ and the inner functions belonging to $\mathcal{K}_\Theta$ are precisely the inner divisors of $\Theta(z)/z$.*

So now the question becomes the following.

QUESTION 6.2. What are the $\varphi(S_\Theta)$-inner functions?

As we did before with Toeplitz operators, we focus our attention on the case where φ is inner. It is clear that the inner vectors for $\varphi(S_\Theta)$ are the same as those for $\varphi(S_\Theta)^*$. As observed with an analogous result in Proposition 2.3, we see that any (unit) vector in $\ker \varphi(S_\Theta)^*$ is a $\varphi(S_\Theta)^*$-inner vector. It is well-known [**8**] that (assuming φ is an inner function)

$$\ker \varphi(S_\Theta)^* = \mathcal{K}_\Theta \cap \mathcal{K}_\varphi = \mathcal{K}_{\gcd(\Theta,\varphi)},$$

where $\gcd(\Theta, \varphi)$ is the greatest common inner divisor of the inner functions Θ and φ.

At this point, it might the case that $\gcd(\Theta, \varphi)$ is a unimodular constant function whence $\mathcal{K}_{\gcd(\Theta,\varphi)} = \{0\}$ and it is not clear as to whether or not there are any $\varphi(S_\Theta)$-inner vectors.

QUESTION 6.3. We know that if $\gcd(\Theta, \varphi)$ is non-constant, then there are $\varphi(S_\Theta)$-inner vectors. Is the converse true?

For the special case where $\varphi|\Theta$, let us find a class of $\varphi(S_\Theta)$-inner vectors. Define

$$I := \frac{\Theta}{\varphi}$$

and observe from a result in [**7**] that an analytic function g on $\mathbb{D}$ multiplies $\mathcal{K}_\varphi$ to $\mathcal{K}_\Theta$ if and only $g \in \mathcal{K}_{zI}$. Recall from Theorem 6.1 that the inner functions in $\mathcal{K}_{zI}$ are precisely the inner divisors of I. Here is our result about some of the $\varphi(S_\Theta)$-inner vectors.

THEOREM 6.4. *With the notation above, any unit vector from*

$$\{v\mathcal{K}_\varphi : v|I\}$$

is a $\varphi(S_\Theta)$-inner vector.

PROOF. Let f be a unit vector from $\mathcal{K}_\varphi$ and note that $vf \in \mathcal{K}_\Theta$ and hence $P_\Theta(vf) = vf$. Thus for all $n \geqslant 1$ we have

$$\begin{aligned}
\langle (\varphi(S_\Theta))^n(vf), vf\rangle &= \langle P_\Theta(\varphi^n fv), vf\rangle \\
&= \langle \varphi^n vf, P_\Theta(vf)\rangle \\
&= \langle \varphi^n vf, vf\rangle \\
&= \langle \varphi^n f, f\rangle \\
&= \langle f, T_{\overline{\varphi}}^n f\rangle.
\end{aligned}$$

But since $f \in \mathcal{K}_\varphi = \ker T_{\overline{\varphi}}$, this last quantity is equal to zero. This shows that vf is a $\varphi(S_\Theta)$-inner vector. □

When $\Theta(0) = 0$ and $\varphi(z) = z$, notice how this recovers Theorem 6.1. At the other extreme, notice that when $\varphi = \Theta$ then I is a unimodular constant inner function and the theorem above yields $\mathcal{K}_\Theta$ as the complete set of T_Θ-inner functions. Of course this result is obvious once one realizes that $\langle T_\Theta f, f\rangle = 0$ for any $f \in \mathcal{K}_\Theta$ by the definition of the model space $\mathcal{K}_\Theta = (\Theta H^2)^\perp$.

Also observe that one can relax the assumption that $\varphi|\Theta$ and set $I = u/\gcd(\Theta, \varphi)$ and give a more general version of the theorem above.

References

[1] Albrecht Böttcher and Bernd Silbermann, *Analysis of Toeplitz operators*, 2nd ed., Springer Monographs in Mathematics, Springer-Verlag, Berlin, 2006. Prepared jointly with Alexei Karlovich. MR2223704

[2] Isabelle Chalendar, *The operator-valued Poisson kernel and its applications*, Irish Math. Soc. Bull. **51** (2003), 21–44. MR2018850

[3] Isabelle Chalendar and Jonathan R. Partington, *Spectral density for multiplication operators with applications to factorization of L^1 functions*, J. Operator Theory **50** (2003), no. 2, 411–422. MR2050137

[4] R. Cheng, J. Mashreghi, and William T. Ross. Inner functions in reproducing kernel spaces. to appear, Trends in Mathematics (S. Shimorin Memorial Volume), Birkhauser.

[5] Joseph A. Cima, Alec L. Matheson, and William T. Ross, *The Cauchy transform*, Mathematical Surveys and Monographs, vol. 125, American Mathematical Society, Providence, RI, 2006. MR2215991

[6] Peter L. Duren, *Theory of H^p spaces*, Pure and Applied Mathematics, Vol. 38, Academic Press, New York-London, 1970. MR0268655

[7] Emmanuel Fricain, Andreas Hartmann, and William T. Ross, *Multipliers between model spaces*, Studia Math. **240** (2018), no. 2, 177–191, DOI 10.4064/sm8782-4-2017. MR3720929

[8] Stephan Ramon Garcia, Javad Mashreghi, and William T. Ross, *Introduction to model spaces and their operators*, Cambridge Studies in Advanced Mathematics, vol. 148, Cambridge University Press, Cambridge, 2016. MR3526203

[9] Paul R. Halmos, *Shifts on Hilbert spaces*, J. Reine Angew. Math. **208** (1961), 102–112, DOI 10.1515/crll.1961.208.102. MR0152896

[10] T. L. Lance and M. I. Stessin, *Multiplication invariant subspaces of Hardy spaces*, Canad. J. Math. **49** (1997), no. 1, 100–118, DOI 10.4153/CJM-1997-005-9. MR1437202

[11] Michael Stessin and Kehe Zhu, *Generalized factorization in Hardy spaces and the commutant of Toeplitz operators*, Canad. J. Math. **55** (2003), no. 2, 379–400, DOI 10.4153/CJM-2003-017-1. MR1969797

Department of Mathematics and Statistics, Old Dominion University, Norfolk, VA 23529

Email address: rcheng@odu.edu

Département de mathématiques et de statistique, Université Laval, Québec, QC, Canada, G1V 0A6

Email address: javad.mashreghi@mat.ulaval.ca

Department of Mathematics and Computer Science, University of Richmond, Richmond, VA 23173, USA

Email address: wross@richmond.edu

Contemporary Mathematics
Volume **743**, 2020
https://doi.org/10.1090/conm/743/14962

Jack and Julia

Richard Fournier and Oliver Roth

ABSTRACT. We state and prove a multi-point version of Jack's Lemma for functions which are analytic on the open unit disc, but are not necessarily analytic at boundary points. Our proof, in particular, does not rely on Julia's lemma.

1. Introduction and statement of the main result

Let $\mathbb{D}$ denote the open unit disc $\{z : |z| < 1\}$ of the complex plane $\mathbb{C}$ and $\mathcal{H}(\mathbb{D})$ the class of functions analytic in $\mathbb{D}$; $\overline{\mathbb{D}}$ is the closed disc $\{z : |z| \leq 1\}$ and $\mathcal{H}(\overline{\mathbb{D}})$ is the class of functions analytic on some open set containing $\overline{\mathbb{D}}$. The following result was first stated by Jack [**9**] who attributed its proof to Clunie.

JACK'S LEMMA (smooth version). *Let* $F \in \mathcal{H}(\overline{\mathbb{D}})$ *with*

$$|F(\zeta)| = \max_{|z|\leq 1}|F(z)| = \max_{|z|=1}|F(z)| > 0,$$

where $|\zeta| = 1$. *Then*

$$\zeta\frac{F'(\zeta)}{F(\zeta)} \geq 0$$

and in fact $\operatorname{Re}\left(1 + \zeta\frac{F''(\zeta)}{F'(\zeta)}\right) \geq \zeta\frac{F'(\zeta)}{F(\zeta)} > 0$ *if* F *is non-constant.*

After its publication by Jack, it has been observed that the lemma is indeed valid for functions F in $\mathcal{H}(\mathbb{D})$, analytic in a neighbourhood of $\zeta \in \partial\mathbb{D}$, and this result goes back to Loewner at least in the 1930's (see [**14**, p. 162]). Under the milder hypothesis, the result has been rediscovered, improved and applied by a number of mathematicians (see for example [**7**], the book of Miller and Mocanu [**12**] or the interesting survey by Boas [**4**]. The survey by Elin et al. [**6**] also contains relevant information).

The following result is indeed valid (we still call it Jack's lemma in what follows).

LEMMA (less smooth Jack's lemma). *Let* $F \in \mathcal{H}(\mathbb{D})$ *with* $F(\mathbb{D}) \subseteq \mathbb{D}$ *and* $\zeta \in \partial\mathbb{D}$. *The following statements are equivalent*:

(i) $\liminf_{z\to\zeta} \frac{1-|F(z)|}{1-|z|} < \infty$.

2010 *Mathematics Subject Classification.* 30C80.
Key words and phrases. Jack's lemma, Blaschke products.

(ii) *The radial limits* $F(\zeta) := \lim_{r\to 1} F(r\zeta)$ *and* $F'(\zeta) := \lim_{r\to 1} F'(r\zeta)$ *exist with* $|F(\zeta)| = 1$ *and* $|F'(\zeta)| < \infty$.

Moreover, under (i) *or* (ii),

$$\zeta\frac{F'(\zeta)}{F(\zeta)} = \lim_{r\to 1}\frac{1-|F(r\zeta)|}{1-r} = \lim_{r\to 1}\frac{1-F(r\zeta)/F(\zeta)}{1-r} > 0.$$

The strict positivity of $\zeta\frac{F'(\zeta)}{F(\zeta)}$ above also follows, as observed for example by Tom Ransford [**15**, p. 33], from the lemma of Hopf. We stress the fact that a proof of the less smooth Jack's lemma can be obtained, as in [**1**] or [**8**], from the properties of the measure in the representation

$$\frac{1+F(\zeta)}{1-F(\zeta)} = \int_0^{2\pi}\frac{1+\mathrm{e}^{-i\theta}\zeta}{1-\mathrm{e}^{-i\theta}\zeta}\,\mathrm{d}\mu(\theta).$$

Such a proof is in particular "horocycle free" and does not rely on Julia's lemma [**10**] which may be conveniently stated as follows:

Julia's lemma. *Let* $F \in \mathcal{H}(\mathbb{D})$ *with* $F(\mathbb{D}) \subseteq \mathbb{D}$, *and* $\lim_{z\to\zeta}\frac{1-|F(z)|}{1-|z|} < \infty$ *for some* $\zeta \in \partial\mathbb{D}$. *Then*

$$\zeta\frac{F'(\zeta)}{F(\zeta)} \geq \frac{|1-F(z)\overline{F(\zeta)}|^2}{1-|F(z)|^2}\,\frac{1-|z|^2}{|1-z\overline{\zeta}|^2}, \qquad z \in \mathbb{D}.$$

To the best of our knowledge, the relation between Jack's lemma and Julia's lemma was first made explicit by Ruscheweyh [**5**].We shall prove that a multi-point version of Julia's lemma can be obtained from the apparently weaker less smooth Jack's lemma. Our main result is the following:

Theorem A. *Let* $f \in \mathcal{H}(\mathbb{D})$, $f(\mathbb{D}) \subseteq \mathbb{D}$ *and* $\zeta \in \partial\mathbb{D}$ *such that*

$$\liminf_{z\to\zeta}\frac{1-|f(z)|}{1-|z|} < \infty. \tag{1}$$

Let also $\{z_k\} \subset \mathbb{D}$ *and define a (possibly finite) sequence* $\{f_k\} \subset \mathcal{H}(\mathbb{D})$ *by* $f_0 = f$ *and*

$$f_{k+1}(z) = \frac{1-\overline{z}_k z}{z-z_k}\,\frac{f_k(z)-f_k(z_k)}{1-\overline{f_k(z_k)}f_k(z)}, \qquad k \geq 0,$$

provided that f_k *is not a unimodular constant. Then*

$$\zeta\frac{f'(\zeta)}{f(\zeta)} \geq \sum_{j=0}^{n}\left(\prod_{k=0}^{j}\frac{\left|1-\overline{f_k(z_k)}f_k(\zeta)\right|^2}{1-|f_k(z_k)|^2}\right)\frac{1-|z_j|^2}{|1-\overline{z}_j\zeta|^2}.$$

Equality holds if and only if f *is a Blaschke product of degree* $n+1$.

Remark that here the functions f_k are the hyperbolic divided differences of the initial function f at the points $\{z_k\}$, cf. the work of Beardon and Minda [**3**] and Baribeau, Rivard and Wegert [**2**] amongst others.

2. Proof of our main result

Each function f_k belongs to $\mathcal{H}(\mathbb{D})$, $f_k(\mathbb{D}) \subseteq \mathbb{D}$ and satisfies (1) together with

$$\zeta\frac{f'_{k+1}(\zeta)}{f_{k+1}(\zeta)} = \zeta\frac{f'_k(\zeta)}{f_k(\zeta)}\,\frac{1-|f_k(z_k)|^2}{\left|1-\overline{f_k(z_k)}f_k(\zeta)\right|^2} - \frac{1-|z_k|^2}{|1-\overline{z}_k\zeta|^2}$$

for all $k \geq 0$. In particular Jack's lemma yields

$$0 \leq \zeta \frac{f_1'(\zeta)}{f_1(\zeta)} = \zeta \frac{f'(\zeta)}{f(\zeta)} \frac{1 - |f(z_0)|^2}{\left|1 - \overline{f(z_0)} f(\zeta)\right|^2} - \frac{1 - |z_0|^2}{|1 - \overline{z}_0 \zeta|^2}$$

and

$$\zeta \frac{f'(\zeta)}{f(\zeta)} \geq \frac{\left|1 - \overline{f(z_0)} f(\zeta)\right|^2}{1 - |f(z_0)|^2} \frac{1 - |z_0|^2}{|1 - \overline{z}_0 \zeta|^2}. \tag{2}$$

This is Julia's lemma and clearly equality shall hold in (2) if and only if the function f_1 is constant and therefore if f is a Blaschke product of order 1. An iteration of this procedure shall lead to, for each $n \geq 0$,

$$\zeta \frac{f'(\zeta)}{f(\zeta)} \geq \sum_{j=0}^{n} \left(\prod_{k=0}^{j} \frac{\left|1 - \overline{f_k(z_k)} f_k(\zeta)\right|^2}{1 - |f_k(z_k)|^2} \right) \frac{1 - |z_j|^2}{|1 - \overline{z}_j \zeta|^2} \tag{3}$$

and equality holds in (3) if and only if f is a Blaschke product of order $n + 1$.

3. Two special cases

Case 1. Let us take $z_k = 0$ for $k \geq 0$. Then

$$\zeta \frac{f'(\zeta)}{f(\zeta)} \geq \sum_{j=0}^{n} \left(\prod_{k=0}^{j} \frac{\left|1 - \overline{f_k(0)} f_k(\zeta)\right|^2}{1 - |f_k(0)|^2} \right) \geq \sum_{j=0}^{n} \prod_{k=0}^{j} \frac{1 - |f_k(0)|}{1 + |f_k(0)|}. \tag{4}$$

It is not difficult to see that the righthand side of (4) is a quantity depending on the $n+1$ first Taylor coefficients $\{\alpha_k\}_{k=0}^{n}$ of $f(\zeta) =: \sum_{j=0}^{\infty} \alpha_j z^j$. The case $n = 0$ is due to Osserman [**13**] and the case $n = 1$ is due to Lecko and Uzar [**11**]. The series $\sum_{j=0}^{\infty} \prod_{k=0}^{j} \frac{1-|f_k(0|}{1+|f_k(0)|}$ is convergent with $\left\{\prod_{k=0}^{j} \frac{1-|f_k(0)|}{1+|f_k(0)|}\right\}$ monotonically decreasing, and hence

$$\lim_{j \to \infty} j \prod_{k=0}^{j} \frac{1 - |f_k(0)|}{1 + |f_k(0)|} = 0.$$

We recall that, according to a result of Boyd [**8**, p. 175],

$$\prod_{k=0}^{\infty} (1 - |f_k(0)|) = \exp\left(\frac{1}{2\pi} \int_0^{2\pi} \ln\left(1 - |f(\mathrm{e}^{\mathrm{i}\theta})|^2\right) \mathrm{d}\theta \right).$$

Case 2. We apply our idea to a function f in $\mathcal{H}(\mathbb{D})$ with $f(\mathbb{D}) \subseteq \mathbb{D}$ satisfying (1) with zeros $\{z_k\}$ in $\mathbb{D}$ and to the sequence defined by $f_{j+1}(z) = f(z) \prod_{k=0}^{j} \frac{1-\overline{z}_k z}{z - z_k}$. We get

$$\zeta \frac{f_{j+1}'(\zeta)}{f_j(\zeta)} = \zeta \frac{f'(\zeta)}{f(\zeta)} - \sum_{k=0}^{j} \frac{1 - |z_k|^2}{|1 - \overline{z}_k \zeta!|^2} \geq 0,$$

and the series $\sum_{k=0}^{\infty} \frac{1-|z_k|^2}{|1-\overline{z}_k\zeta|^2}$ is convergent. We finally recall that in the case where f is a Blaschke product, the convergence of $\sum_{k=0}^{\infty} \frac{1-|z_k|^2}{|1-\overline{z}_k\zeta|^2}$ is necessary and sufficient for the existence of $F(\zeta)$ and $F'(\zeta)$ and in fact $\zeta \frac{f'(\zeta)}{f(\zeta)} = \sum_{k=0}^{\infty} \frac{1-|z_k|^2}{|1-\overline{z}_k\zeta|^2}$. This is a result of Frostman (see [**5**, p. 15]).

References

[1] L. V. Ahlfors, *Conformal invariants*, AMS Chelsea Publishing, Providence, RI, 2010. Topics in geometric function theory; Reprint of the 1973 original; With a foreword by Peter Duren, F. W. Gehring and Brad Osgood. MR2730573

[2] L. Baribeau, P. Rivard, and E. Wegert, *On hyperbolic divided differences and the Nevanlinna-Pick problem*, Comput. Methods Funct. Theory **9** (2009), no. 2, 391–405, DOI 10.1007/BF03321735. MR2572646

[3] A. F. Beardon and D. Minda, *A multi-point Schwarz-Pick lemma*, J. Anal. Math. **92** (2004), 81–104, DOI 10.1007/BF02787757. MR2072742

[4] H. P. Boas, *Julius and Julia: mastering the art of the Schwarz lemma*, Amer. Math. Monthly **117** (2010), no. 9, 770–785, DOI 10.4169/000298910X521643. MR2760379

[5] P. Colwell, *Blaschke products*, University of Michigan Press, Ann Arbor, MI, 1985. Bounded analytic functions. MR779463

[6] M. Elin, F. Jacobzon, M. Levenshtein, and D. Shoikhet, *The Schwarz lemma: rigidity and dynamics*, Harmonic and complex analysis and its applications, Trends Math., Birkhäuser/Springer, Cham, 2014, pp. 135–230, DOI 10.1007/978-3-319-01806-5_3. MR3203101

[7] R. Fournier, *On a new proof and an extension of Jack's lemma*, J. Appl. Anal. **23** (2017), no. 1, 21–24, DOI 10.1515/jaa-2017-0003. MR3657924

[8] J. B. Garnett, *Bounded analytic functions*, 1st ed., Graduate Texts in Mathematics, vol. 236, Springer, New York, 2007. MR2261424

[9] I. S. Jack, *Functions starlike and convex of order* α, J. London Math. Soc. (2) **3** (1971), 469–474, DOI 10.1112/jlms/s2-3.3.469. MR0281897

[10] G. Julia, *Extension nouvelle d'un lemme de Schwarz* (French), Acta Math. **42** (1920), no. 1, 349–355, DOI 10.1007/BF02404416. MR1555173

[11] A. Lecko and B. Uzar, *A note on Julia-Carathéodory theorem for functions with fixed initial coefficients*, Proc. Japan Acad. Ser. A Math. Sci. **89** (2013), no. 10, 133–137, DOI 10.3792/pjaa.89.133. MR3161534

[12] S. S. Miller and P. T. Mocanu, *Differential subordinations*, Monographs and Textbooks in Pure and Applied Mathematics, vol. 225, Marcel Dekker, Inc., New York, 2000. Theory and applications. MR1760285

[13] R. Osserman, *A sharp Schwarz inequality on the boundary*, Proc. Amer. Math. Soc. **128** (2000), no. 12, 3513–3517, DOI 10.1090/S0002-9939-00-05463-0. MR1691000

[14] G. Pólya and G. Szegő, *Problems and theorems in analysis. I*, Classics in Mathematics, Springer-Verlag, Berlin, 1998.

[15] T. Ransford, *Potential theory in the complex plane*, London Mathematical Society Student Texts, vol. 28, Cambridge University Press, Cambridge, 1995. MR1334766

[16] S. Ruscheweyh, *Neighborhoods of univalent functions*, Proc. Amer. Math. Soc. **81** (1981), no. 4, 521–527, DOI 10.2307/2044151. MR601721

Dép. de mathématiques et de statistique, Université de Montréal, C.P. 6128, Succ. Centre-ville, Montréal, QC H3C 3J7, Canada

Email address: `fournier@dms.umontreal.ca`

Mathematisches Institut, Universität Würzburg, 97074 Würzburg, Germany

Email address: `roth@mathematik.uni-wuerzburg.de`

Contemporary Mathematics
Volume **743**, 2020
https://doi.org/10.1090/conm/743/14963

Spectrum and local spectrum preservers of skew Lie products of matrices

Z. Abdelali, A. Bourhim, and M. Mabrouk

ABSTRACT. Let $\mathcal{M}_n(\mathbb{C})$ be the algebra of all $n \times n$ complex matrices, and fix a nonzero vector $x_0 \in \mathbb{C}^n$. For any matrix $T \in \mathcal{M}_n(\mathbb{C})$, let $\sigma(T)$ be its spectrum and $\sigma_T(x_0)$ be its local spectrum at x_0. We show that a map φ on $\mathcal{M}_n(\mathbb{C})$ satisfies

$$\sigma_{\varphi(T)\varphi(S)-\varphi(S)\varphi(T)^*}(x_0) = \sigma_{TS-ST^*}(x_0), \ (T,\ S \in \mathcal{M}_n(\mathbb{C}))$$

if and only if there exists a unitary matrix $U \in \mathcal{M}_n(\mathbb{C})$ and a nonzero scalar α such that $Ux_0 = \alpha x_0$ and $\varphi(T) = \pm UTU^*$ for all $T \in \mathcal{M}_n(\mathbb{C})$. To prove this result, we also describe the form of all maps φ on $\mathcal{M}_n(\mathbb{C})$ satisfying

$$\sigma\left(\varphi(T)\varphi(S) - \varphi(S)\varphi(T)^*\right) = \sigma\left(TS - ST^*\right), \ (T,\ S \in \mathcal{M}_n(\mathbb{C})).$$

As immediate consequences, we characterize all maps on $\mathcal{M}_n(\mathbb{C})$ preserving the local spectrum and spectrum of the skew Jordan product of matrices.

1. Introduction

In recent decades, numerous authors studied *nonlinear preserver problems.* These problems demand the characterization of maps on algebras that preserve various spectral quantities or subsets or relations but without assuming any algebraic condition like linearity or additivity or multiplicativity. The first nonlinear preserver problem was considered by Kowalski and Słodkowski who proved in [**62**] that a complex-valued function f on a Banach algebra $\mathcal{A}$ is linear and multiplicative provided that $f(0) = 0$ and $f(x) - f(y)$ lies in the spectrum of $x - y$ for all x and y in $\mathcal{A}$, and thus generalized the well-known theorem of Gleason–Kahane–Żelazko in the theory of Banach algebra [**54, 61**]. Since then, a number of techniques have been developed to treat nonlinear preserver problems and many results have been obtained mainly in matrix theory and in operator theory; see for instance [**8, 9, 17, 26, 27, 31, 37, 38, 40, 43, 48, 52, 56–59, 64, 67, 69, 70, 73–75**]. In [**9**], Bhatia, Šemrl and Sourour described the form of all surjective maps on the algebra $M_n(\mathbb{C})$ of all complex $n \times n$-matrices preserving the spectral radius of the difference of matrices, and thus, in particular, they provided an extension of Marcus and Moyls' result [**66**] in the absence of the linearity. In [**69**], Molnár studied maps preserving the spectrum of operator or matrix products and showed, in particular, that a surjective map φ on $\mathcal{L}(\mathcal{H})$, the algebra of all bounded linear operators on

2010 *Mathematics Subject Classification.* Primary 47B49; Secondary 47A10, 47A11.
Key words and phrases. Nonlinear preservers, Spectrum, Local spectrum, Skew Lie Product.

an infinite-dimensional complex Hilbert space $\mathcal{H}$, preserves the spectrum of operator products if and only if φ is an automorphism or an automorphism multiplied by -1. His results have been extended in several directions, and a number of results were obtained on maps preserving several spectral quantities of operator or matrix product, or Jordan product, or Jordan triple product, etc; see for example[**10, 11, 26, 45, 57–59, 64**] and the references therein.

In recent years, there has been an upsurge of interest for preservers of the skew Lie product, which is defined on any $*$-ring $\mathcal{R}$ by

$$[x, y]_* := xy - yx^*, \ (x, \ y \in \mathcal{R}).$$

This product has been implicitly or explicitly studied by several authors and in various contexts; see for instance [**7, 25, 27, 44, 53, 65, 71, 72, 76, 77**] and the references therein. Particularly, a number of authors described maps on algebras preserving a spectral set or a quantity of the skew Lie product of matrices or operators. Maps on factor Von Neumann algebras preserving skew Lie product, strong skew Lie product and zero skew Lie product are consider by several authors; see [**28, 42, 44, 46, 52, 78**]. In [**42**], Cui and Li proved that, if $\mathcal{A}$ and $\mathcal{B}$ are factor Von Neumann algebras and $\Phi : \mathcal{A} \to \mathcal{B}$ is a bijective map preserving the skew Lie product (i.e., $\Phi([S, T]_*) = [\Phi(S), \Phi(T)]_*$ for all $S, \ T \in \mathcal{A}$), then Φ is a $*$-ring isomorphism. In [**41**], Cui and Hou characterized, in particular, all linear bijective maps preserving zero skew Lie product of operators. In [**28**], the authors characterized all maps on $\mathcal{L}(\mathcal{H})$ preserving numerical range and the maps preserving pseudo-spectrum of skew Lie product of operators.

The topic of this paper belongs to the subject of linear and nonlinear local spectra preserver problems that attracted in recent years the attention of numerous researchers; see [**1–5, 12–16, 18, 19, 22, 24**] and the references therein. In [**16, 19, 22, 24, 29, 30, 35, 55**], many results on linear maps on matrices or Banach space operators preserving the local spectrum have been obtained. Linear maps on matrices or Banach space operators preserving the inner and outer local spectral radii have been obtained in [**16, 19, 21, 24, 36, 39, 49, 50**]. While, nonlinear maps preserving local spectra of matrices and operators have been discussed by various authors; see for instance [**1–5, 10, 12–16, 18, 32–35, 37, 50, 51, 60**] and the references therein. In particular, nonlinear maps on matrices or Banach space operators preserving the local spectrum of different products of matrices and operators has been investigated in [**1–5, 10, 12–16**]. In this paper, we describe maps φ on the algebra $\mathcal{M}_n(\mathbb{C})$ of all $n \times n$ complex matrices preserving the local spectrum at a fixed nonzero vector $x_0 \in \mathbb{C}^n$ of the skew Lie product of matrices. We show that such a map φ is a self-adjoint automorphism multiplied by either 1 or -1 and the intertwining matrix sends x_0 to a nonzero multiple of itself. Besides some arguments quoted from [**13**], the proof of this result uses new techniques and intermediate results. Among these results, one characterizes all maps on $\mathcal{M}_n(\mathbb{C})$ preserving the spectrum of the skew Lie product of matrices. Another one provides the local spectra of the skew Lie product of any rank one operator on a complex Hilbert space $\mathcal{H}$ by an arbitrary bounded linear operator on $\mathcal{H}$. We also use a local spectral identity principal that tells us that if Ω is a dense subset of $\mathcal{M}_n(\mathbb{C})$ then two matrices A and B in $\mathcal{M}_n(\mathbb{C})$ coincides if and only if the local spectra at x_0 of $[T, A]_*$ and $[T, B]_*$ are the same for all $T \in \Omega$.

2. Main results

Throughout this paper, let $\mathcal{L}(\mathcal{H})$ denote the algebra of all bounded linear operators on a complex Hilbert space $\mathcal{H}$, and $\mathcal{M}_n(\mathbb{C})$ be the algebra of all $n \times n$ complex matrices. Let $\mathbf{1}$ stand for the identity operator of $\mathcal{L}(\mathcal{H})$ and the identity matrix of $\mathcal{M}_n(\mathbb{C})$, and denote by Tr the usual trace functional on $\mathcal{M}_n(\mathbb{C})$. An operator $T \in \mathcal{L}(\mathcal{H})$ is said to have the single-valued extension property (SVEP) provided that for every open subset U of $\mathbb{C}$, the equation $(T-\lambda)f(\lambda) = 0$, $(\lambda \in U)$, has no nontrivial analytic solution f. Every operator $T \in \mathcal{L}(\mathcal{H})$ for which the interior of its point spectrum, $\sigma_p(T)$, is empty enjoys this property. The *local resolvent set*, $\rho_T(x)$, of an operator $T \in \mathcal{L}(\mathcal{H})$ at a point $x \in \mathcal{H}$ is the union of all open subsets U of $\mathbb{C}$ for which there is an analytic function $\zeta : U \to \mathcal{H}$ such that $(T-\lambda)\zeta(\lambda) = x$, $(\lambda \in U)$. The *local spectrum* of T at x is

$$\sigma_T(x) := \mathbb{C} \backslash \rho_T(x),$$

and is obviously a closed subset (possibly empty) of $\sigma(T)$, the spectrum of T. In fact, $\sigma_T(x) \neq \emptyset$ for all nonzero vectors x in $\mathcal{H}$ precisely when T has SVEP. In this case, for every $x \in \mathcal{H}$, there is a maximal analytic function, denoted by $\widetilde{x}_T$, on $\rho_T(x)$ such that $(T-\lambda)\widetilde{x}_T(\lambda) = x$ for all $\lambda \in \rho_T(x)$. It is worth mentioning that, as demonstrated by weighted shift operators, sometimes the description of the local spectra of an operator is difficult; see [**20, 23, 68**]. However, the local spectra of matrices is well understood and can be found for instance in [**21, 55, 79**].

The first main result of this paper is the following theorem. It describes the form of all maps φ on $\mathcal{M}_n(\mathbb{C})$ preserving the local spectrum at a fixed nonzero vector $x_0 \in \mathbb{C}^n$ of the skew Lie product of matrices.

Theorem 2.1. *If $x_0 \in \mathbb{C}^n$ is a nonzero vector, then a map φ on $\mathcal{M}_n(\mathbb{C})$ satisfies*

$$\sigma_{\varphi(T)\varphi(S)-\varphi(S)\varphi(T)^*}(x_0) = \sigma_{TS-ST^*}(x_0), \quad (T,\ S \in \mathcal{M}_n(\mathbb{C})), \tag{2.1}$$

if and only if there exists a unitary matrix $U \in \mathcal{M}_n(\mathbb{C})$ and a nonzero scalar $\alpha \in \mathbb{C}$ such that $Ux_0 = \alpha x_0$ and either $\varphi(T) = UTU^*$ for all $T \in \mathcal{M}_n(\mathbb{C})$, or $\varphi(T) = -UTU^*$ for all $T \in \mathcal{M}_n(\mathbb{C})$.

An immediate consequence of the above theorem is the following corollary. It shows that Theorem 2.1 remains valid if the subtraction in (2.1) is replaced by the sum. It suffices to observe that if x_0 is a nonzero fixed vector in $\mathbb{C}^n$ and φ is a map on $\mathcal{M}_n(\mathbb{C})$ satisfying

$$\sigma_{\varphi(T)\varphi(S)+\varphi(S)\varphi(T)^*}(x_0) = \sigma_{TS+ST^*}(x_0), \quad (T,\ S \in \mathcal{M}_n(\mathbb{C})), \tag{2.2}$$

then the map $T \mapsto \phi(T) := i\varphi(iT)$ satisfies (2.1), and thus Theorem 2.1 applies.

Corollary 2.2. *Let x_0 be a fixed nonzero vector in $\mathbb{C}^n$. A map φ on $\mathcal{M}_n(\mathbb{C})$ satisfies* (2.2) *if and only if there exists a unitary matrix $U \in \mathcal{M}_n(\mathbb{C})$ and a nonzero scalar $\alpha \in \mathbb{C}$ such that $Ux_0 = \alpha x_0$ and either $\varphi(T) = UTU^*$ for all $T \in \mathcal{M}_n(\mathbb{C})$, or $\varphi(T) = -UTU^*$ for all $T \in \mathcal{M}_n(\mathbb{C})$.*

The proof of Theorem 2.1 is presented in Section 7.1, and uses a series of auxiliary results that are of independent interest. One of them is the following theorem that completely characterizes all maps on $\mathcal{M}_n(\mathbb{C})$ preserving the spectrum of the skew Lie product of matrices.

Theorem 2.3. *A map φ from $\mathcal{M}_n(\mathbb{C})$ into itself satisfies*

$$\sigma\left(\varphi(T)\varphi(S) - \varphi(S)\varphi(T)^*\right) = \sigma\left(TS - ST^*\right), \; (T, \; S \in \mathcal{M}_n(\mathbb{C})), \tag{2.3}$$

if and only if there exists a unitary matrix $U \in \mathcal{M}_n(\mathbb{C})$ such that either $\varphi(T) = UTU^*$ for all $T \in \mathcal{M}_n(\mathbb{C})$, or $\varphi(T) = -UTU^*$ for all $T \in \mathcal{M}_n(\mathbb{C})$.

This theorem seems to be a natural result but we couldn't find it in the literature. Its proof uses some ideas from the proof of [**26**, Theorem 2.1] together with certain auxiliary lemmas established in Section 3. The first one describes the spectrum of the skew Lie product of any rank one operator by an arbitrary operator in $\mathcal{L}(\mathcal{H})$. The second lemma tells us that if $\mathcal{O}$ is a dense subset of $\mathcal{M}_n(\mathbb{C})$ then two matrices A and B in $\mathcal{M}_n(\mathbb{C})$ coincide if and only if $\sigma([T, A]_*) = \sigma([T, B]_*)$ for all $T \in \mathcal{O}$.

With no extra efforts, the same proof of Corollary 2.2 yields the following consequence of Theorem 2.3.

Corollary 2.4. *A map φ from $\mathcal{M}_n(\mathbb{C})$ into itself satisfies*

$$\sigma\left(\varphi(T)\varphi(S) + \varphi(S)\varphi(T)^*\right) = \sigma\left(TS + ST^*\right), \; (T, \; S \in \mathcal{M}_n(\mathbb{C})), \tag{2.4}$$

if and only if there exists a unitary matrix $U \in \mathcal{M}_n(\mathbb{C})$ such that either $\varphi(T) = UTU^*$ for all $T \in \mathcal{M}_n(\mathbb{C})$, or $\varphi(T) = -UTU^*$ for all $T \in \mathcal{M}_n(\mathbb{C})$.

Throughout the rest of this paper, we may and shall assume for the sake of simplicity that $n \geq 3$. If $n = 1$, then the proof of our main results is an easy exercise. If $n = 2$, then the proof of our results remain valid but the statements of Lemma 3.1 should be adjusted. Because if A and R are matrices in $\mathcal{M}_2(\mathbb{C})$ such that R has rank one, then $[R, A]_*$ is a matrix of rank at most 2 and thus it can be invertible and 0 may or may not belong to the spectrum of $[R, A]_*$. Finally, note that, in the above results, no linearity or surjectivity condition is imposed on the maps φ. But these conditions are parts of the conclusion of these results.

3. Spectra and skew Lie product

For two nonzero vectors x and y in $\mathcal{H}$, denote by $x \otimes y$ the operator of rank one defined by

$$(x \otimes y)(z) := \langle z, y \rangle x.$$

Note that $(x \otimes y)^* = y \otimes x$ and that every rank one operator in $\mathcal{L}(\mathcal{H})$ can be written as $x \otimes y$. Given an operator $A \in \mathcal{L}(\mathcal{H})$, we also note that

$$[x \otimes y, A]_* = (x \otimes y)A - A(x \otimes y)^* = x \otimes (A^* y) \; - \; (Ay) \otimes x.$$

It is an operator of rank at most two, and its spectrum is described by the following result. To state it, set

$$\Delta_A(x, y) := \left(\langle Ax, y\rangle + \langle Ay, x\rangle\right)^2 - 4\|x\|^2 \langle A^2 y, y\rangle$$

for all $x, y \in \mathcal{H}$ and $A \in \mathcal{L}(\mathcal{H})$.

Lemma 3.1. *For any nonzero vectors $x, y \in \mathcal{H}$ and $A \in \mathcal{L}(\mathcal{H})$, we have*

$$\sigma\left([x \otimes y, A]_*\right) = \frac{1}{2}\left\{0, \langle Ax, y\rangle - \langle Ay, x\rangle \pm \sqrt{\Delta_A(x, y)}\right\}. \tag{3.5}$$

PROOF. Assume that there is a nonzero scalar α in $\sigma\left([x \otimes y, A]_*\right)$ and let h be a nonzero vector in $\mathcal{H}$ such that $[x \otimes y, A]_* h = \alpha h$. It follows that

$$\langle Ah, y\rangle x - \langle h, x\rangle Ay = \alpha h, \tag{3.6}$$

and consequently,

$$\langle Ah, y\rangle \|x\|^2 - \langle h, x\rangle \langle Ay, x\rangle = \alpha \langle h, x\rangle, \tag{3.7}$$

and

$$\langle Ah, y\rangle \langle Ax, y\rangle - \langle h, x\rangle \langle A^2 y, y\rangle = \alpha \langle Ah, y\rangle. \tag{3.8}$$

If $\langle h, x\rangle = 0$, then from (3.7) it follows that $\langle Ah, y\rangle = 0$ and (3.6) implies that either $\alpha = 0$ or $h = 0$. This contradiction shows that $\langle h, x\rangle \neq 0$.

If $\langle A^2 y, y\rangle \neq 0$, then $\langle Ah, y\rangle \neq 0$ by (3.8). Therefore, after rewriting (3.7) and (3.8) as

$$\langle Ah, y\rangle \|x\|^2 - \langle h, x\rangle \left(\langle Ay, x\rangle + \alpha\right) = 0 \tag{3.9}$$

and

$$\langle Ah, y\rangle \left(\langle Ax, y\rangle - \alpha\right) - \langle h, x\rangle \langle A^2 y, y\rangle = 0, \tag{3.10}$$

we see that

$$-\alpha^2 + \left(\langle Ax, y\rangle - \langle Ay, x\rangle\right)\alpha + \langle Ay, x\rangle \langle Ax, y\rangle - \|x\|^2 \langle A^2 y, y\rangle = 0. \tag{3.11}$$

Accordingly, $\alpha = \dfrac{1}{2}\left(\langle Ax, y\rangle - \langle Ay, x\rangle \pm \sqrt{\Delta_A(x, y)}\right)$ and thus (3.5) holds.

If, however, $\langle A^2 y, y\rangle = 0$, then

$$[x \otimes y, A]_* Ay = \left((x \otimes y)A - A(y \otimes x)\right) Ay = -\langle Ay, x\rangle Ay, \tag{3.12}$$

and $-\langle Ay, x\rangle \in \sigma\left([x \otimes y, A]_*\right)$. Moreover, if $\langle Ax, y\rangle + \langle Ay, x\rangle \neq 0$, set $z := x - \frac{\|x\|^2}{\langle Ax,y\rangle + \langle Ay,x\rangle} Ay$. It then follows that $z \neq 0$ and

$$[x \otimes y, A]_* z = \left((x \otimes y)A - A(y \otimes x)\right) z = \langle Ax, y\rangle z.$$

This shows that $\langle Ax, y\rangle \in \sigma\left([x \otimes y, A]_*\right)$ and (3.5) holds in this case. If $\langle Ax, y\rangle + \langle Ay, x\rangle = 0$, then (3.12) together with (3.6) ,(3.7) and (3.8) imply that $\alpha = \langle Ax, y\rangle = -\langle Ay, x\rangle$, and $\sigma\left([x \otimes y, A]_*\right) = \{0, \langle Ax, y\rangle\}$. Thus (3.5) is established in this case too, and the proof is now complete. □

As a consequence of the above lemma, we obtain the following corollary which gives necessary and sufficient conditions for two matrices to be the same.

Corollary 3.2. *If $\mathcal{O}$ is a dense subset of $\mathcal{M}_n(\mathbb{C})$, then two matrices A and B in $\mathcal{M}_n(\mathbb{C})$ coincide if and only if $\sigma([T, A]_*) = \sigma([T, B]_*)$ for every $X \in \mathcal{O}$.*

PROOF. Assume that $\sigma([T, A]_*) = \sigma([T, B]_*)$ for all $T \in \mathcal{O}$, and note that the continuity of the spectrum and the involution on $\mathcal{M}_n(\mathbb{C})$ together with the density of $\mathcal{O}$ in $\mathcal{M}_n(\mathbb{C})$ imply that

$$\sigma([T, A]_*) = \sigma([T, B]_*) \tag{3.13}$$

for every $T \in \mathcal{M}_n(\mathbb{C})$. Fix a unit vector $x \in \mathbb{C}^n$, and note that (3.13) and Lemma 3.1 entail that

$$\begin{aligned}\left\{0, i\langle Ax, x\rangle \pm i\sqrt{\langle A^2 x, x\rangle}\right\} &= \sigma\left([(ix) \otimes x, A]_*\right) = \sigma\left([(ix) \otimes x, B]_*\right) \\ &= \left\{0, i\langle Bx, x\rangle \pm i\sqrt{\langle B^2 x, x\rangle}\right\}\end{aligned} \tag{3.14}$$

for all unit vectors $x \in \mathbb{C}^n$. Therefore, either

$$\langle Ax, x\rangle + \sqrt{\langle A^2x, x\rangle} = \langle Bx, x\rangle + \sqrt{\langle B^2x, x\rangle}$$

and

$$\langle Ax, x\rangle - \sqrt{\langle A^2x, x\rangle} = \langle Bx, x\rangle - \sqrt{\langle B^2x, x\rangle},$$

or

$$\langle Ax, x\rangle + \sqrt{\langle A^2x, x\rangle} = \langle Bx, x\rangle - \sqrt{\langle B^2x, x\rangle}$$

and

$$\langle Ax, x\rangle - \sqrt{\langle A^2x, x\rangle} = \langle Bx, x\rangle + \sqrt{\langle B^2x, x\rangle}.$$

Combining the two equations in either case, we clearly get that $\langle Ax, x\rangle = \langle Bx, x\rangle$. Since x is an arbitrary unit vector, we conclude that $A = B$; as desired. □

4. Local spectra and skew Lie product

In this section, we collect and provide several lemmas needed for the proof of Theorem 2.1. The first one summarizes some known basic properties of the local spectrum that will be used frequently through this paper. Among the rest of these lemmas, one of them describes the local spectra of the skew Lie product of any rank one operator by an arbitrary operator in $\mathcal{L}(\mathcal{H})$. Another one is a local spectral identity principal that tells us that if $x_0 \in \mathbb{C}^n$ is a nonzero vector and Ω is a dense subset of $\mathcal{M}_n(\mathbb{C})$ then two matrices A and B in $\mathcal{M}_n(\mathbb{C})$ coincide if and only if $\sigma_{[T,A]_*}(x_0) = \sigma_{[T,B]_*}(x_0)$ for all $T \in \Omega$.

Lemma 4.1. *Let x and y be two vectors in $\mathcal{H}$, and α a nonzero scalar in $\mathbb{C}$. For every operator $T \in \mathcal{L}(\mathcal{H})$, the following statements hold.*

(1) *$\sigma_T(\alpha x) = \sigma_T(x)$ and $\sigma_{\alpha T}(x) = \alpha\sigma_T(x)$.*
(2) *$\sigma_T(x+y) \subset \sigma_T(x) \cup \sigma_T(y)$. The equality holds if $\sigma_T(x) \cap \sigma_T(y) = \emptyset$.*
(3) *For any $n \geq 1$, $\sigma_{T^n}(x) = \{\sigma_T(x)\}^n$.*
(4) *If T has SVEP, then $\sigma(T) = \bigcup\{\sigma_T(x) : x \in \mathcal{H}\}$.*
(5) *If T has SVEP, $x \neq 0$ and $Tx = \lambda x$ for some $\lambda \in \mathbb{C}$, then $\sigma_T(x) = \{\lambda\}$.*
(6) *If T has SVEP and $(T-\lambda)x = y$ for some $\lambda \in \mathbb{C}$, then $\sigma_T(y) \subset \sigma_T(x) \subset \sigma_T(y) \cup \{\lambda\}$.*
(7) *If T has SVEP, then $\sigma_T(T^k x) \subset \sigma_T(x) \subset \sigma_T(T^k x) \cup \{0\}$ for all positive integers k.*
(8) *If $R \in \mathcal{L}(\mathcal{H})$ commutes with T, then $\sigma_T(Rx) \subset \sigma_T(x)$.*

PROOF. See for instance [**6**] or [**63**]. □

The next lemma characterizes when a finite rank operator in $\mathcal{L}(\mathcal{H})$ has a trivial local spectrum at a nonzero vector. In its proof, we use the analytic spectral subspaces of operators. Recall that for any operator $T \in \mathcal{L}(\mathcal{H})$ and a closed subset F of $\mathbb{C}$, the corresponding analytic spectral subspace is defined by

$$\mathcal{H}_T(F) := \{h \in \mathcal{H} : \sigma_T(h) \subset F\}.$$

It is a T-invariant subspace but it is not necessarily closed. However, if, in particular, T is a finite rank operator, then $\mathcal{H}_T(F)$ is closed for all closed subsets F of $\mathbb{C}$; see [**63**, Propositions 1.4.3 & 1.4.5].

Lemma 4.2. *If $T \in \mathcal{L}(\mathcal{H})$ is an operator of rank n and $x_0 \in \mathcal{H}$ is a nonzero vector, then $\sigma_T(x_0) = \{0\}$ if and only if $T^{n+1}(x_0) = 0$.*

PROOF. If $T^{n+1}(x_0) = 0$, then $\sigma_T(x_0) = \{0\}$ by Lemma 4.1-(6).

Conversely, assume that $\sigma_T(x_0) = \{0\}$ for certain nonzero vector $x_0 \in \mathcal{H}$, and set $\mathcal{H}_1 := \bigvee\{T^k x_0 : k \geq 0\}$. Note that $\mathcal{H}_1$ is a T-invariant subspace and its dimension is at most $n+1$. Then with respect to the space decomposition $\mathcal{H} = \mathcal{H}_1 \oplus \mathcal{H}_1^{\perp}$, the operator T can be written as

$$T = \begin{bmatrix} A & C \\ 0 & B \end{bmatrix}$$

Since T is a finite rank operator, both A and B must be too finite rank operators. Now, fix a nonzero vector $y \in \mathcal{H}_1$ and let us first prove that $\sigma_T(y) = \sigma_A(y) = \{0\}$. Indeed, since $\sigma_T(T^k x_0) \subset \sigma_T(x_0) = \{0\}$ for all $k \geq 0$ and $\mathcal{H}_T(\{0\})$ is a closed linear space, we note that $\mathcal{H}_1 \subset \mathcal{H}_T(\{0\})$. In particular, $y \in \mathcal{H}_T(\{0\})$ and $\sigma_T(y) = \{0\}$. To prove that $\sigma_A(y) = \{0\}$, write $\widetilde{y}_T = f_1 \oplus f_2$ on $\rho_T(y) = \mathbb{C}\backslash\{0\}$ and note that

$$\mathcal{H}_1 \ni y = (T-\lambda)\widetilde{y}_T(\lambda) = [(A-\lambda)f_1(\lambda) + Cf_2(\lambda)] \oplus (B-\lambda)f_2(\lambda)$$

for all $\lambda \in \mathbb{C}\backslash\{0\}$. It then follows that $(B-\lambda)f_2(\lambda) = 0$ for all $\lambda \in \mathbb{C}\backslash\{0\}$ and thus f_2 must be identically zero since B is a finite rank operator. Hence,

$$y = (T-\lambda)\widetilde{y}(\lambda) = (A-\lambda)f_1(\lambda)$$

for all $\lambda \in \mathbb{C}\backslash\{0\}$, and $\sigma_A(y) = \{0\}$.

Now, we show that $T^{n+1}x_0 = 0$. Since A is a finite rank operator and

$$\sigma(A) = \bigcup_{y \in \mathcal{H}_1} \sigma_A(y) = \{0\},$$

we see that A is nilpotent and thus $A^{n+1} = 0$, by Cayley-Hamilton theorem. This implies that $T^{n+1}x_0 = A^{n+1}x_0 = 0$, and the proof is complete. □

The following lemma gives complete description of the local spectra of the skew Lie product of any rank one operator by an arbitrary operator in $\mathcal{L}(\mathcal{H})$. To state it, we need to introduce some notation and the concept of the nonzero local spectrum introduced in [**14, 15**]. Recall that the nonzero local spectrum of an operator $T \in \mathcal{L}(\mathcal{H})$ at a nonzero vector x_0 is defined by

$$\sigma_T^*(x_0) := \begin{cases} \{0\} & \text{if } \sigma_T(x_0) = \{0\} \\ \sigma_T(x_0)\backslash\{0\} & \text{if } \sigma_T(x_0) \neq \{0\}. \end{cases}$$

For any vectors x and y in $\mathcal{H}$, we have

$$\sigma_{x\otimes y}^*(x_0) = \begin{cases} \{0\} & \text{if } \langle x_0, y\rangle = 0, \\ \{\langle x, y\rangle\} & \text{otherwise.} \end{cases} \tag{4.15}$$

In particular, if x and Ay are linearly dependent, then $[x \otimes y, A]_*$ is an operator of rank at most one, and the description of $\sigma^*_{[x\otimes y, A]_*}(x_0)$ can be deduced from (4.15).

For any $A \in \mathcal{L}(\mathcal{H})$ and $x,\ y \in \mathcal{H}$, set

$$\alpha_1(A,x,y) := \frac{1}{2}\left(\langle Ax, y\rangle - \langle Ay, x\rangle + \sqrt{\Delta_A(x,y)}\right),$$

$$\alpha_2(A,x,y) := \frac{1}{2}\left(\langle Ax, y\rangle - \langle Ay, x\rangle - \sqrt{\Delta_A(x,y)}\right),$$

$$\Gamma_0(A,x,y) := \langle Ay, x\rangle + \langle Ax, y\rangle, \tag{4.16}$$

(4.17)
$$\Gamma_1(A,x,y) := \langle Ax_0, y\rangle \frac{\alpha_1(A,x,y) - \langle Ax, y\rangle}{(\alpha_2(A,x,y) - \alpha_1(A,x,y))\langle A^2y, y\rangle} + \frac{\langle x_0, x\rangle}{\alpha_2(A,x,y) - \alpha_1(A,x,y)},$$

(4.18)
$$\Gamma_2(A,x,y) := \langle Ax_0, y\rangle \frac{\alpha_2(A,x,y) - \langle Ax, y\rangle}{(\alpha_2(A,x,y) - \alpha_1(A,x,y))\langle A^2y, y\rangle} + \frac{\langle x_0, x\rangle}{\alpha_2(A,x,y) - \alpha_1(A,x,y)},$$

(4.19)
$$\Gamma_3(A,x,y) = \Big[\langle Ax_0, y\rangle\left(\langle Ax, y\rangle^2 - \langle A^2y, y\rangle\|x\|^2\right) + \langle x_0, x\rangle\langle A^2y, y\rangle\left(\langle Ay, x\rangle - \langle Ax, y\rangle\right)\Big],$$

and
(4.20)
$$\Gamma_4(A,x,y) = \Big[\langle Ax_0, y\rangle\left(\langle A^2y, y\rangle\langle Ay, x\rangle - \langle Ax, y\rangle\|x\|^2\right) + \langle x_0, x\rangle\left(\langle A^2y, y\rangle\|x\|^2 - \langle Ay, x\rangle^2\right)\Big].$$

Lemma 4.3. *Let $x_0 \in \mathcal{H}$ be a nonzero vector. If $A \in \mathcal{L}(\mathcal{H})$ and $x, y \in \mathcal{H}$ such that x and Ay are linearly independent, then the following assertions hold.*

(1) *If $\langle A^2y, y\rangle = 0$ and $\Gamma_0(A,x,y) = 0$, then*
$$\sigma_{[x\otimes y,A]_*}(x_0) = \begin{cases} \{0\} & \text{if } \langle Ax_0, y\rangle = \langle x_0, x\rangle = 0 \\ \{\langle Ax, y\rangle\} & \text{if } \langle Ax_0, y\rangle \neq 0 \text{ or } \langle x_0, x\rangle \neq 0 \end{cases}$$

(2) *If $\langle A^2y, y\rangle = 0$ and $\Gamma_0(A,x,y) \neq 0$, then*
$$\sigma_{[x\otimes y,A]_*}(x_0) = \begin{cases} \{0\} & \text{if } \langle Ax_0, y\rangle = \langle x_0, x\rangle = 0 \\ \{-\langle Ay, x\rangle\} & \text{if } \langle Ax_0, y\rangle = 0 \text{ and } \langle x_0, x\rangle \neq 0 \\ \{\langle Ax, y\rangle\} & \text{if } \langle Ax_0, y\rangle \neq 0 \text{ and } \frac{\langle Ax_0, y\rangle\|x\|^2}{\Gamma_0(A,x,y)} - \langle x_0, x\rangle = 0 \\ \{-\langle Ay, x\rangle, \langle Ax, y\rangle\} & \text{if } \langle Ax_0, y\rangle \neq 0 \text{ and } \frac{\langle Ax_0, y\rangle\|x\|^2}{\Gamma_0(A,x,y)} - \langle x_0, x\rangle \neq 0 \end{cases}$$

(3) *If $\langle A^2y, y\rangle \neq 0$ and $\alpha_1(A,x,y) \neq \alpha_2(A,x,y)$, then*
$$\sigma^*_{[x\otimes y,A]_*}(x_0) = \begin{cases} \{0\} & \text{if } \Gamma_1(A,x,y) = \Gamma_2(A,x,y) = 0 \\ \{\alpha_1(A,x,y)\} & \text{if } \Gamma_1(A,x,y) \neq 0 \text{ and } \Gamma_2(A,x,y) = 0 \\ \{\alpha_2(A,x,y)\} & \text{if } \Gamma_1(A,x,y) = 0 \text{ and } \Gamma_2(A,x,y) \neq 0 \\ \{\alpha_1(A,x,y), \alpha_2(A,x,y)\}\setminus\{0\} & \text{if } \Gamma_1(A,x,y) \neq 0 \text{ and } \Gamma_2(A,x,y) \neq 0 \end{cases}$$

(4) *If $\langle A^2y, y\rangle \neq 0$ and $\alpha := \alpha_1(A,x,y) = \alpha_2(A,x,y)$, then*
$$\sigma^*_{[x\otimes y,A]_*}(x_0) = \begin{cases} \{0\} & \text{if } \Gamma_3(A,x,y) = \Gamma_4(A,x,y) = 0 \\ \{\alpha\} & \text{if } \Gamma_3(A,x,y) \neq 0 \text{ or } \Gamma_4(A,x,y) \neq 0 \end{cases}$$

Proof. Note that

$$[x\otimes y, A]_*(x_0) = \langle Ax_0, y\rangle\, x - \langle x_0, x\rangle\, Ay, \tag{4.21}$$

and let us distinguish three cases.

Case 1. Assume that $\langle A^2y, y\rangle = 0$.

In this case, we note that
$$\sigma([x\otimes y, A]_*) = \{0, -\langle Ax, y\rangle, \langle Ay, x\rangle\},$$

and that $[x \otimes y, A]_* Ay = -\langle Ay, x\rangle Ay$. It then follows that

$$\sigma_{[x\otimes y,A]_*}(Ay) = \{-\langle Ay, x\rangle\}. \tag{4.22}$$

We also note that $([x \otimes y, A]_* - \langle Ax, y\rangle \mathbf{1})\, x = -\|x\|^2 Ay$, and then

$$\{-\langle Ay, x\rangle\} \subseteq \sigma_{[x\otimes y,A]_*}(x) \subseteq \{-\langle Ax, y\rangle, \langle Ay, x\rangle\}, \tag{4.23}$$

by Lemma 4.1 and (4.27). If $\Gamma_0(A, x, y) = 0$, then $\langle Ay, x\rangle = -\langle Ax, y\rangle$ and (4.23) implies that

$$\sigma_{[x\otimes y,A]_*}(x) = \{-\langle Ax, y\rangle\}. \tag{4.24}$$

This together with (4.21) and (4.27) entail that

$$\sigma_{[x\otimes y,A]_*}(x_0) = \begin{cases} \{0\} & \text{if } \langle Ax_0, y\rangle = \langle x_0, x\rangle = 0 \\ \{\langle Ax, y\rangle\} & \text{if } \langle Ax_0, y\rangle \neq 0 \text{ or } \langle x_0, x\rangle \neq 0 \end{cases}$$

and the statement (1) is established.

If, however, $\Gamma_0(A, x, y) \neq 0$, then $\langle Ay, x\rangle \neq -\langle Ax, y\rangle$. Set $z := x - \frac{\|x\|^2}{\Gamma_0(A,x,y)} Ay$, and note that $[x \otimes y, A]_* z = \langle Ax, y\rangle z$. Then

$$\sigma_{[x\otimes y,A]_*}(z) = \{\langle Ax, y\rangle\}. \tag{4.25}$$

Since

$$[x \otimes y, A]_*(x_0) = \langle Ax_0, y\rangle z + \left(\frac{\langle Ax_0, y\rangle \|x\|^2}{\Gamma_0(A, x, y)} - \langle x_0, x\rangle\right) Ay,$$

we infer that

$$\sigma_{[x\otimes y,A]_*}(x_0) = \begin{cases} \{0\} & \text{if } \langle Ax_0, y\rangle = \langle x_0, x\rangle = 0 \\ \{-\langle Ay, x\rangle\} & \text{if } \langle Ax_0, y\rangle = 0 \text{ and } \langle x_0, x\rangle \neq 0 \\ \{\langle Ax, y\rangle\} & \text{if } \langle Ax_0, y\rangle \neq 0 \text{ and } \frac{\langle Ax_0,y\rangle\|x\|^2}{\Gamma_0(A,x,y)} - \langle x_0, x\rangle = 0 \\ \{-\langle Ay, x\rangle, \langle Ax, y\rangle\} & \text{if } \langle Ax_0, y\rangle \neq 0 \text{ and } \frac{\langle Ax_0,y\rangle\|x\|^2}{\Gamma_0(A,x,y)} - \langle x_0, x\rangle \neq 0 \end{cases}$$

CASE 2. Assume that $\langle A^2y, y\rangle \neq 0$ and $\alpha_1(A, x, y) \neq \alpha_2(A, x, y)$.

Set

$$z_i := \langle A^2y, y\rangle\, x \;-\; (\langle Ax, y\rangle \;-\; \alpha_i(A, x, y))\, Ay, \;(i = 1,\ 2),$$

and note that

$$[x \otimes y, A]_*(z_i) = \alpha_i(A, x, y)\; z_i.$$

It then follows that

$$\sigma_{[x\otimes y,A]_*}(z_i) = \{\alpha_i(A, x, y)\},$$

and

$$\sigma_{[x\otimes y,A]_*}([x \otimes y, A]_*(x_0)) = \begin{cases} \{0\} & \text{if } \Gamma_1(A, x, y) = \Gamma_2(A, x, y) = 0 \\ \{\alpha_1(A, x, y)\} & \text{if } \Gamma_1(A, x, y) \neq 0 \text{ and } \Gamma_2(A, x, y) = 0 \\ \{\alpha_2(A, x, y)\} & \text{if } \Gamma_1(A, x, y) = 0 \text{ and } \Gamma_2(A, x, y) \neq 0 \\ \{\alpha_1(A, x, y), \alpha_2(A, x, y)\} & \text{if } \Gamma_1(A, x, y) \neq 0 \text{ and } \Gamma_2(A, x, y) \neq 0 \end{cases}$$

From this, we infer that

$$\sigma^*_{[x\otimes y,A]_*}(x_0)=\begin{cases}\{0\} & \text{if } \Gamma_1(A,x,y)=\Gamma_2(A,x,y)=0\\ \{\alpha_1(A,x,y)\} & \text{if } \Gamma_1(A,x,y)\neq 0 \text{ and } \Gamma_2(A,x,y)=0\\ \{\alpha_2(A,x,y)\} & \text{if } \Gamma_1(A,x,y)=0 \text{ and } \Gamma_2(A,x,y)\neq 0\\ \{\alpha_1(A,x,y),\alpha_2(A,x,y)\}\backslash\{0\} & \text{if } \Gamma_1(A,x,y)\neq 0 \text{ and } \Gamma_2(A,x,y)\neq 0\end{cases}$$

CASE 3. Assume that $\langle A^2y,y\rangle\neq 0$ and $\alpha_1(A,x,y)=\alpha_2(A,x,y)$.

We have

$$[x\otimes y,A]_*(x)=\langle Ax,y\rangle\, x-\|x\|^2\, Ay, \tag{4.26}$$

and

$$[x\otimes y,A]_*(Ay)=\langle A^2y,y\rangle\, x-\langle Ay,x\rangle\, Ay. \tag{4.27}$$

These and (4.21) imply that

$$\begin{aligned}[x\otimes y,A]^2_*(x_0)=&\left[\langle Ax_0,y\rangle\langle Ax,y\rangle - \langle x_0,x\rangle\langle A^2y,y\rangle\right]\, x\\ &+\left[\langle Ax_0,y\rangle\|x\|^2+\langle x_0,x\rangle\langle Ay,x\rangle\right]\, Ay,\end{aligned}$$

and thus

$$\begin{aligned}[x\otimes y,A]^3_*(x_0)=&\left[\langle Ax_0,y\rangle\langle Ax,y\rangle - \langle x_0,x\rangle\langle A^2y,y\rangle\right]\,[x\otimes y,A]_*(x)\\ &+\left[\langle Ax_0,y\rangle\|x\|^2+\langle x_0,x\rangle\langle Ay,x\rangle\right]\,[x\otimes y,A]_*(Ay)\\ =&\Big[\langle Ax_0,y\rangle\left(\langle Ax,y\rangle^2-\langle A^2y,y\rangle\|x\|^2\right)+\langle x_0,x\rangle\langle A^2y,y\rangle\left(\langle Ay,x\rangle-\langle Ax,y\rangle\right)\Big]x\\ &+\Big[\langle Ax_0,y\rangle\left(\langle A^2y,y\rangle\langle Ay,x\rangle-\langle Ax,y\rangle\|x\|^2\right)\\ &\quad+\langle x_0,x\rangle\left(\langle A^2y,y\rangle\|x\|^2-\langle Ay,x\rangle^2\right)\Big]Ay\\ =&\,\Gamma_3(A,x,y)\ x+\ \Gamma_4(A,x,y)\ Ay.\end{aligned}$$

Now, note that, since $\alpha_1(A,x,y)=\alpha_2(A,x,y):=\alpha$, we have $\sigma([x\otimes y,A]_*)=\{0,\alpha\}$ and then either $\sigma^*_{[x\otimes y,A]_*}(x_0)=\{0\}$ or $\sigma^*_{[x\otimes y,A]_*}(x_0)=\{\alpha\}$. By Lemma 4.2, we note that

$$\begin{aligned}\sigma^*_{[x\otimes y,A]_*}(x_0)=\{0\}\quad&\Longleftrightarrow\quad \sigma^*_{[x\otimes y,A]_*}([x\otimes y,A]_*\,x_0)=\{0\}\\ &\Longleftrightarrow\quad [x\otimes y,A]^3_*\,x_0=0\\ &\Longleftrightarrow\quad \Gamma_3(A,x,y)\ x+\ \Gamma_4(A,x,y)\ Ay=0\\ &\Longleftrightarrow\quad \Gamma_3(A,x,y)=0 \text{ and } \Gamma_4(A,x,y)=0.\end{aligned}$$

Therefore,

$$\sigma^*_{[x\otimes y,A]_*}(x_0)=\begin{cases}\{0\} & \text{if } \Gamma_3(A,x,y)=\Gamma_4(A,x,y)=0\\ \{\alpha\} & \text{if } \Gamma_3(A,x,y)\neq 0 \text{ or } \Gamma_4(A,x,y)\neq 0\end{cases}$$

The proof is then complete. □

5. A local spectral identity principal

In this section, we establish a local spectral identity principal which might be interesting in its own right. It shows that if $x_0 \in \mathcal{H}$ is a fixed vector then two operators A and B in $\mathcal{L}(\mathcal{H})$ coincide precisely when the local spectrum of $[T, A]_*$ and $[T, B]_*$ at x_0 are the same whenever $T \in \mathcal{L}(\mathcal{H})$ is a rank one operator. Its proof uses a density argument together with the following lemma that shows, with the help of Lemma 4.3-(3), that if $x_0 \in \mathcal{H}$ is a nonzero vector and $A \in \mathcal{L}(\mathcal{H})$ is an operator with a nonzero square then the set of all $(x, y) \in \mathcal{H}^2$ for which $\sigma_{[x\otimes y, A]_*}(x_0)$ contains two nonzero elements is dense in $\mathcal{H}^2$.

Lemma 5.1. *Let $x_0 \in \mathcal{H}$ be a nonzero vector and $A \in \mathcal{L}(\mathcal{H})$ be an operator. If $A^2 \neq 0$, then the following assertions hold.*

(1) *The set*

$$\mathcal{W} := \left\{(x, y) \in \mathcal{H}^2 \ : \ \langle A^2y, y\rangle \neq 0,\ \Delta_A(x, y) \neq 0 \text{ and } \Gamma_1(A, x, y)\Gamma_2(A, x, y) \neq 0\right\}$$

is a dense open subset of $\mathcal{H}^2$.

(2) *The set*

$$\mathcal{O} := \left\{(x, y) \in \mathcal{H}^2 \ : \ \alpha_1 (A, x, y)\, \alpha_2 (A, x, y) \neq 0\right\}$$

is a dense open subset of $\mathcal{H}^2$, and x and Ay are linearly independent for all $(x, y) \in \mathcal{O}$.

PROOF. (1) For every $(x, y) \in \mathcal{H}^2$, set

$$\Phi(x, y) := \left(-\langle Ax_0, y\rangle\ (\langle Ax, y\rangle + \langle Ay, x\rangle)\ +\ 2\langle x_0, x\rangle\langle A^2y, y\rangle\right)^2 - \langle Ax_0, y\rangle^2\, \Delta_A(x, y).$$

Let us show that Φ is not identically zero, and the set

$$\mathcal{W}_0 := \left\{(x, y) \in \mathcal{H}^2 : \Phi(x, y) \neq 0\right\}$$

is a dense open subset of $\mathcal{H}^2$. Indeed, if $Ax_0 = 0$, then

$$\Phi(x_0, y) = \left(2\|x_0\|^2\langle A^2y, y\rangle\right)^2 \neq 0$$

for all $y \in \mathcal{H}$ for which $\langle A^2y, y\rangle \neq 0$. If, however, $Ax_0 \neq 0$, then there is $y \in \mathcal{H}$ such that $\langle A^2y, y\rangle \neq 0$ and $\langle Ax_0, y\rangle \neq 0$. Therefore,

$$\Phi(x, y) = 4\langle Ax_0, y\rangle^2\ \|x\|^2\langle A^2y, y\rangle \neq 0$$

for all nonzero $x \in \mathcal{H}$ for which $\langle x_0, x\rangle = 0$. Moreover, since Φ is continuous, the set $\mathcal{W}_0$ is open. Now, assume that $\Phi(x_1, y_1) = 0$ for some $(x_1, y_1) \in \mathcal{H}^2$, and fix $(x_2, y_2) \in \mathcal{H}^2$ such that $\Phi(x_2, y_2) \neq 0$. Set

$$P(t) := \Phi\left(x_1 + t(x_2 - x_1), y_1 + t(y_2 - y_1)\right),\ (t \in \mathbb{R}),$$

and note that P is a polynomial of degree at most 6. It is nonconstant since $P(0) = 0$ and $P(1) = \Phi(x_2, y_2) \neq 0$. Therefore, $P(t) \neq 0$ for all scalars t except for a finite number of zeros and $(x_1 + t(x_2 - x_1), y_1 + t(y_2 - y_1)) \in \mathcal{W}_0$ for all scalars t except for a finite number of zeros. As $\lim_{t\to 0} (x_1 + t(x_2 - x_1), y_1 + t(y_2 - y_1)) = (x_1, y_1)$, we clearly see that $\mathcal{W}_0$ is dense.

Set

$$\mathcal{W}_1 := \left\{(x, y) \in \mathcal{H}^2 \ : \ \langle A^2y, y\rangle \neq 0\right\},$$

and

$$\mathcal{W}_2 := \left\{(x, y) \in \mathcal{H}^2 \ : \ \Delta_A(x, y) \neq 0\right\}$$

Note that, since $A^2 \neq 0$ and $\Delta_A(iy, y) = -4\|y\|^2\langle A^2y, y\rangle$ for all $y \in \mathcal{H}$, the two sets $\mathcal{W}_1$ and $\mathcal{W}_2$ are nonempty and thus are dense open subsets of $\mathcal{H}^2$. Now, observe that for every $(x, y) \in \mathcal{W}_1 \cap \mathcal{W}_2$, we have

$$\Gamma_1(A, x, y) = \left(\frac{1}{(\alpha_2(A, x, y) - \alpha_1(A, x, y))\langle A^2y, y\rangle}\right) \times \left(\frac{-1}{2}\langle Ax_0, y\rangle\ (\langle Ax, y\rangle + \langle Ay, x\rangle) + \frac{1}{2}\langle Ax_0, y\rangle\ \sqrt{\Delta_A(x, y)}\ +\ \langle x_0, x\rangle\langle A^2y, y\rangle\right),$$

and

$$\Gamma_2(A, x, y) = \left(\frac{1}{(\alpha_2(A, x, y) - \alpha_1(A, x, y))\langle A^2y, y\rangle}\right) \times \left(\frac{-1}{2}\langle Ax_0, y\rangle\ (\langle Ax, y\rangle + \langle Ay, x\rangle) - \frac{1}{2}\langle Ax_0, y\rangle\ \sqrt{\Delta_A(x, y)}\ +\ \langle x_0, x\rangle\langle A^2y, y\rangle\right)$$

Thus,

$$\begin{aligned}&\Gamma_1(A, x, y)\Gamma_2(A, x, y)\\ &= \frac{\left(-\langle Ax_0, y\rangle\ (\langle Ax, y\rangle + \langle Ay, x\rangle)\ +\ 2\langle x_0, x\rangle\langle A^2y, y\rangle\right)^2 - \langle Ax_0, y\rangle^2\ \Delta_A(x, y)}{16\Delta_A(x, y)\langle A^2y, y\rangle^2}\\ &= \frac{\Phi(x, y)}{16\Delta_A(x, y)\langle A^2y, y\rangle^2}\end{aligned}$$

for all $(x, y) \in \mathcal{W}_1 \cap \mathcal{W}_2$. It then follows that

$$\mathcal{W} = \mathcal{W}_0 \cap \mathcal{W}_1 \cap \mathcal{W}_2$$

is a dense open subset of $\mathcal{H}^2$; as desired.

(2) By a simple computation, we obtain that

$$\mathcal{O} = \left\{(x, y) \in \mathcal{H}^2\ :\ \|x\|^2\langle A^2y, y\rangle - \langle Ax, y\rangle\langle Ay, x\rangle \neq 0\right\}.$$

Note that, since $A^2 \neq 0$, there is a nonzero vector $y_0 \in \mathcal{H}$ such that $\langle A^2y_0, y_0\rangle \neq 0$. Now, observe that $(x, y_0) \in O$ for all $x \in \mathcal{H}$ for which $\langle Ay_0, x\rangle = 0$, and thus $\mathcal{O}$ is not trivial. Hence, $\mathcal{O}$ is a dense open subset of $\mathcal{H}^2$.

Finally, observe that if x and Ay are linearly dependent for some vector $(x, y) \in \mathcal{H}^2$, then

$$\|x\|^2\langle A^2y, y\rangle - \langle Ax, y\rangle\langle Ay, x\rangle = 0,$$

and $(x, y) \notin \mathcal{O}$. □

Now, we are in a position to state and prove the promised local spectral identity principal.

Theorem 5.2. *Let $x_0 \in \mathcal{H}$ be a nonzero vector and A and B two operators in $\mathcal{L}(\mathcal{H})$. Then the following assertions hold.*

(1) $A = B$.
(2) $\sigma_{[T,A]_*}(x_0) = \sigma_{[T,B]_*}(x_0)$ *for all rank one operators* $T \in \mathcal{L}(\mathcal{H})$.
(3) $\sigma_{[T,A]_*}(x_0) \cup \sigma_{[T,B]_*}(x_0) \subset \sigma([T, A]_*) \cap \sigma([T, B]_*)$ *for all rank one operators* $T \in \mathcal{L}(\mathcal{H})$.

PROOF. It suffices that establish the implication (3) ⇒ (1). So, assume that

$$\sigma_{[T,A]_*}(x_0) \cup \sigma_{[T,B]_*}(x_0) \subset \sigma([T, A]_*) \cap \sigma([T, B]_*) \tag{5.28}$$

for all rank one operators $T \in \mathcal{L}(\mathcal{H})$. First, assume that $A^2 \neq 0$, and note that Lemma 5.1 implies that the sets

$$\mathcal{W} := \{(x, y) \in \mathcal{H}^2 \ : \ \langle A^2 y, y\rangle \neq 0,\ \Delta_A(x, y) \neq 0 \text{ and } \Gamma_1(A, x, y)\Gamma_2(A, x, y) \neq 0\}$$

and

$$\mathcal{O} := \{(x, y) \in \mathcal{H}^2 \ : \ \alpha_1(A, x, y)\,\alpha_2(A, x, y) \neq 0\}$$

are dense open subsets of $\mathcal{H}^2$, and x and Ay are linearly independent for all $(x, y) \in \mathcal{O}$. By Lemma 3.1 and Lemma 4.3-(3), we have

$$\{\alpha_1(A, x, y), \alpha_2(A, x, y)\} = \sigma^*_{[x\otimes y, A]_*}(x_0) \subset \sigma\left([x \otimes y, B]_*\right) = \{0, \alpha_1(B, x, y), \alpha_2(B, x, y)\}$$

for all $(x, y) \in \mathcal{W} \cap \mathcal{O}$. It then follows that

$$\langle Ax, y\rangle - \langle Ay, x\rangle = \alpha_1(A, x, y) + \alpha_2(A, x, y) = \alpha_1(B, x, y) + \alpha_2(B, x, y) = \langle Bx, y\rangle - \langle By, x\rangle$$

for all $(x, y) \in \mathcal{W} \cap \mathcal{O}$. But, since $\mathcal{W} \cap \mathcal{O}$ is dense in $\mathcal{H}^2$, we have

$$\langle Ax, y\rangle - \langle Ay, x\rangle = \langle Bx, y\rangle - \langle By, x\rangle$$

for all $(x, y) \in \mathcal{H}^2$. Therefore,

$$-2i\langle Ax, x\rangle = \langle Ax, ix\rangle - \langle A(ix), x\rangle = \langle Bx, ix\rangle - \langle B(ix), x\rangle = -2i\langle Bx, x\rangle$$

for all $x \in \mathcal{H}$, and $A = B$.

If, however, $A^2 = 0$, then what has been discussed previously implies that $B^2 = 0$. Fix a nonzero $x \in \mathcal{H}$ for which $\langle x_0, x\rangle \neq 0$ and note that, since $A^2 = 0$, either $Ax = 0$ or the vectors x and Ax are linearly independent. Since $\Gamma_0(A, x, (ix)) = \Gamma_0(B, x, (ix)) = 0$, in both cases (4.15) and Lemma 4.3 imply that

$$\sigma_{[x\otimes(ix), A]_*}(x_0) = \{-i\langle Ax, x\rangle\} \text{ and } \sigma([x \otimes (ix), A]_*) = \{0, -i\langle Ax, x\rangle\}.$$

We also have

$$\sigma_{[x\otimes(ix), B]_*}(x_0) = \{-i\langle Bx, x\rangle\} \text{ and } \sigma([x \otimes (ix), B]_*) = \{0, -i\langle Bx, x\rangle\}.$$

In view or (5.28), we have

$$\langle Ax, x\rangle = \langle Bx, x\rangle.$$

Now, let $x \in \mathcal{H}$ such that $\langle x_0, x\rangle = 0$. We have $\langle x_0, x + tx_0\rangle = t\|x_0\|^2 \neq 0$ for all nonzero real scalars t and then $\langle A(x+tx_0), (x+tx_0)\rangle = \langle B(x+tx_0), (x+tx_0)\rangle$ for all nonzero real scalars t. Now take the limit as t goes to 0 to get that $\langle Ax, x\rangle = \langle Bx, x\rangle$ in this case too. Since x is an arbitrary vector in $\mathcal{H}$, we clearly have $A = B$. $\square$

Finally, we close this section with a local spectral identity principal that gives necessary and sufficient conditions for two matrices to be the same. It is a consequence of Theorem 5.2 and lower semi-continuity of the local spectrum on $\mathcal{M}_n(\mathbb{C})$ at a fixed vector x_0 of $\mathbb{C}^n$; see [**47**]. That is if $(T_k)_k \subset \mathcal{M}_n(\mathbb{C})$ is a sequence of matrices converging to $T \in \mathcal{M}_n(\mathbb{C})$ then $\sigma_T(x_0) \subset \liminf_{k\to\infty} \sigma_{T_k}(x_0)$.

Corollary 5.3. *If $x_0 \in \mathbb{C}^n$ is a nonzero vector and Ω is a dense subset of $\mathcal{M}_n(\mathbb{C})$, then two matrices A and B in $\mathcal{M}_n(\mathbb{C})$ coincide if and only if*

$$\sigma_{[T,A]_*}(x_0) = \sigma_{[T,B]_*}(x_0) \tag{5.29}$$

for all $T \in \Omega$.

Proof. Assume that $\sigma_{[T,A]_*}(x_0) = \sigma_{[T,B]_*}(x_0)$ for all $T \in \Omega$. Take a matrix $T \in \mathcal{M}_n(\mathbb{C})$ and a sequence $(T_k)_k \subset \Omega$ converging to T. We have

$$\sigma_{[T_k,A]_*}(x_0) = \sigma_{[T_k,B]_*}(x_0) \subset \sigma([T_k, A]_*) \cap \sigma([T_k, B]_*)$$

for all k. By the continuity of the spectrum and lower semi-continuity of the local spectrum on $\mathcal{M}_n(\mathbb{C})$, we have

$$\sigma_{[T,A]_*}(x_0) \subset \sigma([T, A]_*) \cap \sigma([T, B]_*)$$

and

$$\sigma_{[T,B]_*}(x_0) \subset \sigma([T, A]_*) \cap \sigma([T, B]_*).$$

Accordingly

$$\sigma_{[T,A]_*}(x_0) \cup \sigma_{[T,B]_*}(x_0) \subset \sigma([T, A]_*) \cap \sigma([T, B]_*) \tag{5.30}$$

for all $T \in \mathcal{M}_n(\mathbb{C})$. By Theorem 5.2, we have $A = B$ and the proof is complete. □

6. Useful dense and spanning subsets of $\mathcal{M}_n(\mathbb{C})$

The proofs of the main results are based on density arguments that use some open dense subsets of $\mathcal{M}_n(\mathbb{C})$. Let $\mathrm{GL_n}(\mathbb{C})$ denote, as usual, the group of all invertible matrices in $\mathcal{M}_n(\mathbb{C})$, and let $\mathcal{D}_n(\mathbb{C})$ be the set of all matrices having n distinct nonzero eigenvalues; i.e.,

$$\mathcal{D}_n(\mathbb{C}) := \{T \in \mathrm{GL_n}(\mathbb{C}) : |\sigma(T)| = n\}.$$

Here, $|\sigma(T)|$ denotes the cardinality of $\sigma(T)$. It is well known that $\mathcal{D}_n(\mathbb{C})$ is an open, arcwise connected and dense subset of $\mathcal{M}_n(\mathbb{C})$. Set

$$\Omega_n(\mathbb{C}) := \{A \in \mathrm{GL_n}(\mathbb{C}) : \sigma(A) \cap \sigma(A^*) = \emptyset\}.$$

With minor changes, the same argument of [**13**, Lemma 3.6] shows that this set is open and dense in $\mathcal{M}_n(\mathbb{C})$. Therefore,

$$\Lambda_n(\mathbb{C}) := \Omega_n(\mathbb{C}) \cap \mathcal{D}_n(\mathbb{C})$$

is also an open and dense subset of $\mathcal{M}_n(\mathbb{C})$. Moreover, reasoning in the same way as the proof of [**13**, Lemma 3.6], we can show the following lemma.

Lemma 6.1. *$\Lambda_n(\mathbb{C})$ is an open dense and arcwise connected subset of $\mathcal{M}_n(\mathbb{C})$.*

Recall that an operator $T \in \mathcal{L}(\mathcal{H})$ is said to be cyclic with a cyclic vector $x_0 \in \mathcal{H}$ provided that the linear span of $\{T^k x_0 : k \geq 0\}$ is dense in $\mathcal{H}$. When $\mathcal{H} = \mathbb{C}^n$ is a finite-dimensional space, the collection of all matrices $T \in \mathcal{M}_n(\mathbb{C})$ with the same cyclic vector $x_0 \in \mathbb{C}^n$ is an open dense subset of $\mathcal{M}_n(\mathbb{C})$. Therefore,

$$\begin{aligned} \mathcal{G}(n, x_0) &:= \{T \in \Omega_n(\mathbb{C}) : |\sigma(T)| = n \text{ and } T \text{ is cyclic with cyclic vector } x_0\} \\ &= \Lambda_n(\mathbb{C}) \cap \{T \in \mathcal{M}_n(\mathbb{C}) : T \text{ is cyclic with cyclic vector } x_0\} \end{aligned}$$

is an open and dense subset of $\mathcal{M}_n(\mathbb{C})$ as it is the intersection of two open dense subsets of $\mathcal{M}_n(\mathbb{C})$. Finally, note that

$$\sigma(T) = \sigma_T(x_0) \tag{6.31}$$

for all cyclic matrices $T \in \mathcal{M}_n(\mathbb{C})$ with cyclic vector x_0.

We close this section with the following lemma that tells us that the set $\{X - X^* : X \in \mathcal{O}\}$ spans $\mathcal{M}_n(\mathbb{C})$ whenever $\mathcal{O}$ is a nonempty open subset of $\mathcal{M}_n(\mathbb{C})$.

Lemma 6.2. *If $\mathcal{O}$ is a nonempty open subset $\mathcal{M}_n(\mathbb{C})$, then the set $\{X - X^* : X \in \mathcal{O}\}$ spans $\mathcal{M}_n(\mathbb{C})$.*

PROOF. Let $R_0 \in \mathcal{M}_n(\mathbb{C})$ such that $\mathrm{Tr}\,(R_0\,(X - X^*)) = 0$ for all $X \in \mathcal{O}$, and let us show that $R_0 = 0$. Let $X_0 \in \mathcal{O}$ and $R \in \mathcal{M}_n(\mathbb{C})$, and note that there exists $\epsilon >$ such that $X_0 + tR \in \mathcal{O}$ for all $t \in (-\epsilon, \epsilon)$. Then $\mathrm{Tr}\,(R_0\,(X_0 - X_0{}^*)) = 0$ and

$$\begin{aligned} 0 &= \mathrm{Tr}\,(R_0\,((X_0 + tR) - (X_0 + tR)^*)) \\ &= 2\,\mathrm{Tr}\,(R_0\,(X_0 - X_0{}^*)) + t\,\mathrm{Tr}\,(R_0(R - R^*)) \\ &= t\,\mathrm{Tr}\,(R_0(R - R^*))\,, \end{aligned}$$

and $\mathrm{Tr}\,(R_0(R - R^*)) = 0$. Replacing R by iR, we also get that $\mathrm{Tr}\,(R_0(R + R^*)) = 0$ and thus $\mathrm{Tr}(R_0 R) = 0$ for all $R \in \mathcal{M}_n(\mathbb{C})$. Accordingly, $R_0 = 0$, and $\{X - X^*, X \in \mathcal{O}\}$ spans $\mathcal{M}_n(\mathbb{C})$. □

7. Proofs of the main results

In this section, by use of the fundamental theorem of projective geometry, we present the proofs of the main results of this paper. The one of Theorem 2.1 uses Theorem 2.3. So, it is more convenient to start first by proving Theorem 2.3 which has interest of its own. As usual, denote by $E_{ij} \in \mathcal{M}_n(\mathbb{C})$ the matrix whose ij entry is 1 and all its other entries are 0, and by $A^\top$ the transpose of any $m \times n$-matrix A. If $a_1, ..., a_n$ are scalars, we denote by $\mathrm{diag}(a_1, a_2, ..., a_n)$ the diagonal matrix with $a_1, ..., a_n$ on the main diagonal in this order. For any matrix $X = (x_{ij}) \in \mathcal{M}_n(\mathbb{C})$, consider the following row and column vectors

$$R_X := (x_{11}, \ldots, x_{1n}, x_{21}, \ldots, x_{2n}, \ldots, x_{n1}, \ldots, x_{nn}),$$

and

$$C_X := (x_{11}, x_{21} \ldots, x_{n1}, x_{12}, \ldots, x_{n2}, \ldots, x_{1n}, \ldots, x_{nn})^\top.$$

7.1. Proof of Theorem 2.3. Checking the "*if*" part is on the straightforward side, and we therefore will only deal with the "*only if*" part. Assume that φ verifies (2.3), and let us break down the proof into five steps to show that φ takes the desired form. The proofs of the first and second steps use similar arguments to those of [**26**, Theorem 2.1]. We include them for the sake of completeness.

STEP 1. For every $A \in \mathcal{D}_n(\mathbb{C})$, there is an open neighborhood N_A of A such that the restriction of φ on N_A equals an invertible linear map L_A.

Fix $A \in \mathcal{D}_n(\mathbb{C})$, and note that, since $[i\mathbf{1}, A]_* = 2iA \in \mathcal{D}_n(\mathbb{C})$, the continuity of the eigenvalues and the map $(X, Y) \mapsto [X, Y]_*$ entail the existence of two open neighborhoods $N_{i\mathbf{1}}$ of $i\mathbf{1}$ and N_A of A such that $[X, Y]_* \in \mathcal{D}_n(\mathbb{C})$ for all $X \in N_{i\mathbf{1}}$ and $Y \in N_A$. By (2.3), the matrices $[X, Y]_*$ and $[\varphi(X), \varphi(Y)]_*$ have the same n distinct eigenvalues whenever $X \in N_{i\mathbf{1}}$ and $Y \in N_A$. Hence,

$$\mathrm{Tr}\,(\varphi(Y)\,(\varphi(X) - \varphi(X)^*)) \;=\; \mathrm{Tr}\,(Y\,(X - X^*)) \tag{7.32}$$

for all $X \in N_{i\mathbf{1}}$ and $Y \in N_A$. It then follows that

$$R_{\varphi(Y)} C_{(\varphi(X) - \varphi(X)^*)} = \mathrm{Tr}(\varphi(X)(\varphi(Y) - \varphi(Y)^*)) = \mathrm{Tr}((X - X^*)Y) = R_Y C_{(X - X^*)} \tag{7.33}$$

for all $X \in N_{i\mathbf{1}}$ and $Y \in N_A$.

Next, we use similar argument as in the proof of [**26**, Assertion 1 of the proof of Theorem 2.1]. In view of Lemma 6.2, there are n^2 matrices $X_1, \ldots, X_{n^2}$ in $N_{i\mathbf{1}}$ such that $\{X_i - X_i{}^* : 1 \le i \le n^2\}$ is a basis of $\mathcal{M}_n(\mathbb{C})$. Let $\mathcal{X}$ and $\mathcal{Z}$ be the $n^2 \times n^2$-matrices with columns

$$C_{X_1 - X_1{}^*}, \ldots, C_{X_{n^2} - X_{n^2}{}^*}$$

and

$$C_{\varphi(X_1)-\varphi(X_1)^*},\ldots,C_{\varphi(X_{n^2})-\varphi(X_{n^2})^*},$$

respectively. By (7.33), we have

$$R_{\varphi(Y)}\mathcal{Z} = R_Y\mathcal{X} \tag{7.34}$$

for all $Y \in N_A$. Now, let us show that $\mathcal{Z}$ is invertible. Indeed, let $Y_1,\ldots,Y_{n^2}$ be n^2 matrices in N_A such that $\{Y_1,\ldots,Y_{n^2}\}$ is a basis of $\mathcal{M}_n(\mathbb{C})$, and let $\mathcal{Y}$ and $\mathcal{W}$ be the $n^2\times n^2$-matrices with rows $R_{Y_1},\ldots,R_{Y_{n^2}}$ and $R_{\varphi(Y_1)},\ldots,R_{\varphi(Y_{n^2})}$, respectively. In view of (7.34), we have $\mathcal{W}\mathcal{Z} = \mathcal{Y}\mathcal{X}$ and then $\mathcal{Z}$ is invertible since both $\mathcal{X}$ and $\mathcal{Y}$ are invertible. Thus, (7.34) implies that

$$R_{\varphi(Y)} = R_Y\mathcal{X}\mathcal{Z}^{-1} \tag{7.35}$$

for all $Y \in N_A$, and the restriction of φ to N_A coincides with an invertible linear map L_A.

STEP 2. When restricted on $\mathcal{D}_n(\mathbb{C})$, the map φ is equal an invertible linear map L.

Let $A, \in \mathcal{D}_n(\mathbb{C})$, and note that, since $\mathcal{D}_n(\mathbb{C})$ is arcwise connected, there is a continuous function f from $[0,1]$ into $\mathcal{D}_n(\mathbb{C})$ such that $f(0) = A$ and $f(1) = B$. Set

$$\mathcal{C} := \{t \in [0,1]\ : \varphi = L_A \text{ on an open neighborhood of } f(t)\}.$$

Just as in the proof of [**26**, Assertion 2 of the proof of Theorem 2.1], one can observe that $\mathcal{C}$ is open and closed in $[0,1]$ to conclude that in fact $\mathcal{C} = [0,1]$ and thus $L_A = L_B$. This shows that on $\mathcal{D}_n(\mathbb{C})$ the map φ coincides with an invertible linear map L.

STEP 3. The mapping L is selfadjoint; i.e., $L(X^*) = L(X)^*$ whenever $X \in \mathcal{M}_n(\mathbb{C})$.

Let $B := \operatorname{diag}(i, 2i,\ldots, ni)$, and note that $B \in \mathcal{D}_n(\mathbb{C})$ and

$$[B,B]_* = \operatorname{diag}(-2,-8,\ldots,-2n^2) \in \mathcal{D}_n(\mathbb{C}).$$

Then the continuity of the maping $(X,Y) \mapsto [X,Y]_*$ on $\mathcal{M}_n(\mathbb{C})$ implies that there is an open neighborhood N_B of B such that $N_B \subseteq \mathcal{D}_n(\mathbb{C})$ and $\{[X,Y]_* : (X,Y) \in N_B^2\} \subseteq \mathcal{D}_n(\mathbb{C})$. We therefore have

$$\operatorname{Tr}(L(Y)(L(X)-L(X)^*) = \operatorname{Tr}(\varphi(Y)(\varphi(X)-\varphi(X)^*) = \operatorname{Tr}(Y(X-X^*)) \tag{7.36}$$

for all $(X,Y) \in N_B^2$. Now, let X and Y be two matrices in $\mathcal{M}_n(\mathbb{C})$, and note that there exists $\epsilon > 0$ such that $(B+tX, B+tY) \in N_B^2$ for all $t \in (-\epsilon,\epsilon)$. Thus, (7.36) gives that

$$\operatorname{Tr}(L(B+tY)(L(B+tX)-L(B+tX)^*) = \operatorname{Tr}((B+tY)(B-B^*+t(X-X^*)))$$

for all $t \in (-\epsilon,\epsilon)$. Use the linearity of the trace, and then expand both sides of this identity and compare the coefficients of t^2 to see that

$$\operatorname{Tr}(L(Y)(L(X)-L(X)^*) = \operatorname{Tr}(Y(X-X^*)).$$

If X is selfadjoint, then this implies that $\operatorname{Tr}(L(Y)(L(X)-L(X)^*) = 0$. Since Y is an arbitrary matrix in $\mathcal{M}_n(\mathbb{C})$ and L is bijective, we conclude that $L(X) = L(X)^*$ for all selfadjoint matrices $X \in \mathcal{M}_n(\mathbb{C})$. Now, let $X \in \mathcal{M}_n(\mathbb{C})$ and X_1 and X_2 be two selfadjoint matrices in $\mathcal{M}_n(\mathbb{C})$ such that $X = X_1 + iX_2$, and note that

$$L(X^*) = L(X_1 - iX_2) = L(X_1) - iL(X_2) = (L(X_1)+iL(X_2))^* = L(X)^*.$$

This shows that L is a selfadjoint mapping and the proof of this step is complete.

STEP 4. There exist a unitary matrix U in $\mathcal{M}_n(\mathbb{C})$ and a scalar $\alpha \in \{-1, 1\}$ such that $L(T) = \alpha UTU^*$ for all $T \in \mathcal{M}_n(\mathbb{C})$.

Since L and φ coincide on the open dense subset $\mathcal{D}_n(\mathbb{C})$, the continuity of the spectrum and L implies that L too satisfies (2.3). It then follows that

$$\begin{aligned}\sigma\left(L(T)L(S) + L(S)L(T)^*\right) &= \sigma\left(L(iT)L(-iS) - L(-iS)L(iT)^*\right)\\ &= \sigma\left((iT)(-iS) - (-iS)(iT)^*\right)\\ &= \sigma\left(TS + ST^*\right)\end{aligned}$$

for all $T,\ S \in \mathcal{M}_n(\mathbb{C})$. In particular, we have

$$\sigma\left(L(T)L(S) + L(S)L(T)\right) = \sigma\left(TS + ST\right)$$

for all $T, S \in \mathcal{H}_n(\mathbb{C})$. As L maps $\mathcal{H}_n(\mathbb{C})$ into itself, [**26**, Theorem 3.2] and the linearity of L on $\mathcal{M}_n(\mathbb{C})$ entail that there exist a unitary matrix $U \in \mathcal{M}_n(\mathbb{C})$ and $\alpha \in \{-1, 1\}$ such that either

$$L(T) = \alpha U^*TU, \ (T \in \mathcal{M}_n(\mathbb{C})) \tag{7.37}$$

or

$$L(T) = \alpha U^*T^\top U, \ (T \in \mathcal{M}_n(\mathbb{C})). \tag{7.38}$$

Assume for the sake of contradiction that L takes the form (7.38) and note that

$$\{-i, i\} \subset \sigma\left(E_{21}E_{11} - E_{11}{E_{21}}^*\right) = \sigma\left(L\left(E_{21}\right)L\left(E_{11}\right) - L\left(E_{11}\right)L\left(E_{21}\right)^*\right) = \sigma(0) = \{0\}.$$

This contradiction shows that (7.38) can not occur and thus L takes only the form (7.37); as desired.

STEP 5. φ has the desired form.

For every $T \in \mathcal{D}_n(\mathbb{C})$ and $S \in \mathcal{M}_n(\mathbb{C})$, we have

$$\begin{aligned}\sigma\left(L(T)L(S) - L(S)L(T)^*\right) &= \sigma\left(TS - ST^*\right)\\ &= \sigma\left(\varphi(T)\varphi(S) - \varphi(S)\varphi(T)^*\right)\\ &= \sigma\left(L(T)\varphi(S) - \varphi(S)L(T)^*\right).\end{aligned}$$

Since L bijectively maps $\mathcal{D}_n(\mathbb{C})$ onto itself, Corollary 3.2 entails that $\varphi(S) = L(S) = \pm U^*SU$ for all $S \in \mathcal{M}_n(\mathbb{C})$.

7.2. Proof of Theorem 2.1. Checking the 'if' part is on the straightforward side, and we therefore will only deal with the 'only if' part. Assume that

$$\sigma_{[\varphi(T),\varphi(S)]_*}(x_0) \ = \ \sigma_{[T,S]_*}(x_0) \tag{7.39}$$

for all T and $S \in \mathcal{M}_n(\mathbb{C})$, and let us break down the proof into six claims to show that φ takes the desired form.

CLAIM 1. Let $A \in \mathcal{G}(n, x_0)$ be an arbitrary matrix. Then there is an open neighborhood $\mathcal{N}_A$ and an invertible linear mapping L_A on $\mathcal{M}_n(\mathbb{C})$ such that $\varphi(X) = L_A(X)$ for all $X \in \mathcal{N}_A$. In particular, φ is continuous on $\mathcal{G}(n, x_0)$.

Fix $A \in \mathcal{G}(n, x_0)$, and note that, since $[(i\mathbf{1}), A]_* = 2iA \in \mathcal{G}(n, x_0)$ and $\mathcal{G}(n, x_0)$ is an open set, we can find an open neighborhood $\mathcal{N}_A \subset \mathcal{G}(n, x_0)$ of A and an open neighborhood $\mathcal{N}_{i\mathbf{1}}$ of $i\mathbf{1}$ such that $[T, S]_* = TS - ST^* \in \mathcal{G}(n, x_0)$ for all $S \in \mathcal{N}_A$ and $T \in \mathcal{N}_{i\mathbf{1}}$. Since $\sigma(T) = \sigma_T(x_0)$, the identity (7.39) entails that $\sigma(TS - ST^*) = \sigma(\varphi(T)\varphi(S) - \varphi(S)\varphi(T)^*)$ for all $T \in \mathcal{N}_A$ and $S \in \mathcal{N}_{i\mathbf{1}}$. As $|\sigma(T)| = n$ for all $T \in \mathcal{G}(n, x_0)$, it then follows that

$$\operatorname{Tr}(S(T - T^*)) = \operatorname{Tr}\left(\varphi(S)(\varphi(T) - \varphi(T)^*)\right)$$

for all $S \in \mathcal{N}_A$ and $T \in \mathcal{N}_{i\mathbf{1}}$. Now, using the same argument as the ones in the proof of Step 1 in the previous proof, one can show that there exists an invertible linear mapping L_A such that $\varphi(X) = L_A(X)$ for all $X \in \mathcal{N}_A$. In particular, φ is continuous at A.

Claim 2. For every $T \in \Omega_n(\mathbb{C})$, the set

$$\Delta_{T,\, x_0} := \{S \in \mathcal{G}(n, x_0) : TS - ST^* \in \mathcal{G}(n, x_0)\}$$

is a nonempty open subset of $\mathcal{M}_n(\mathbb{C})$.

The proof of this claim follows closely the proof of Step 2 of [**13**, Proof of Theorem 2.2]. However, we include it here for the convenience of the reader. Fix $T \in \Omega_n(\mathbb{C})$, and set

$$\Delta'_{T,\, x_0} := \{S \in \mathrm{GL_n}(\mathbb{C}) : TS - ST^* \in \mathcal{G}(n, x_0)\}.$$

Observe that $\Delta_{T,\, x_0} = \Delta'_{T,\, x_0} \cap \mathcal{G}(n, x_0)$, and note that, since $\mathcal{G}(n, x_0)$ is a nonempty open dense subset, it suffices to show that $\Delta'_{T,\, x_0}$ is a nonempty open subset of $\mathcal{M}_n(\mathbb{C})$. Evidently $\Delta_{T,\, x_0}$ is open since the map $g : \mathrm{GL_n}(\mathbb{C}) \longrightarrow \mathcal{M}_n(\mathbb{C})$ defined by $g(T) := TS - ST^*$ is continuous and $\Delta'_{S,\, x_0} = g^{-1}(\mathcal{G}(n, x_0))$. It remains to show that $\Delta'_{T,\, x_0}$ is a nonempty set. To that end, let $A \in \mathcal{G}(n, x_0)$ and note that, since $T \in \Omega_n(\mathbb{C})$, Sylvester's theorem tells us that the equation $TS - ST^* = A$ has a unique solution $S \in \mathcal{M}_n(\mathbb{C})$. Accordingly

$$T(S - \epsilon T^{-1}) - (S - \epsilon T^{-1})T^* = A + \epsilon T^{-1}T^* - \epsilon\mathbf{1},$$

for all $\epsilon > 0$. Note that, since $S - \epsilon T^{-1} = T^{-1}(TS - \epsilon)$ and $\mathcal{G}(n, x_0)$ is a non empty open set, there is $\epsilon > 0$ such that $S - \epsilon T^{-1} \in \mathrm{GL_n}(\mathbb{C})$ and $A + \epsilon T^{-1}T^* - 2\epsilon\mathbf{1} \in \mathcal{G}(n, x_0)$. Thus $S - \epsilon T^{-1} \in \Delta'_{S,\, x_0} \neq \emptyset$.

Claim 3. The map φ is continuous on $\Omega_n(\mathbb{C})$.

Pick up an element $T \in \Omega_n(\mathbb{C})$ and a sequence $(T_k)_k \in \mathcal{M}_n(\mathbb{C})$ converging to T, and let us show that $(\varphi(T_k))_k$ converges towards $\varphi(T)$. To do so, we first show that $(\varphi(T_k))_k$ is a bounded sequence. Fix $A \in \Delta_{T,\, x_0}$, and note that, since $\Delta_{T,\, x_0} \subset \mathcal{G}(n, x_0)$, Claim 1 tells us that there exists an open neighborhood $\mathcal{N}_A \subset \Delta_{T,\, x_0}$ of A and an invertible linear map $L_A : \mathcal{N}_A \longrightarrow L_A(\mathcal{N}_A)$ such that $\varphi(X) = L_A(X)$ for all $X \in \mathcal{N}_A$. Let $S \in \mathcal{N}_A$ and observe that $ST - TS^* \in \mathcal{G}(n, x_0)$ since $\mathcal{N}_A \subset \Delta_{T,\, x_0}$. Since $\lim ST_k - T_kS^* = ST - TS^*$, there is $N > 0$ such that $ST_k - T_kS^* \in \mathcal{G}(n, x_0)$ for all $k \geq N$. By (6.31) and the continuity of the spectrum, we get

$$\begin{aligned}
\sigma(ST - TS^*) &= \lim_{k\to\infty} \sigma(ST_k - T_kS^*) \\
&= \lim_{k\to\infty} \sigma_{ST_k - T_kS^*}(x_0) \\
&= \lim_{k\to\infty} \sigma_{\varphi(S)\varphi(T_k) - \varphi(T_k)\varphi(S)^*}(x_0) \\
&= \lim_{k\to\infty} \sigma(\varphi(S)\varphi(T_k) - \varphi(T_k)\varphi(S)^*).
\end{aligned}$$

Accordingly the consequence $(\mathrm{Tr}(\varphi(T_k)(\varphi(S)-\varphi(S)^*))_k$ is bounded for all $S \in \mathcal{N}_A$. Since $\varphi(\mathcal{N}_A)$ is a nonempty open subset of $\mathcal{M}_n(\mathbb{C})$, Lemma 6.2 tells us that $\{\varphi(S)-\varphi(S)^*) : S \in \mathcal{N}_A\}$ spans $\mathcal{M}_n(\mathbb{C})$. Therefore the linearity of the trace implies that $(\mathrm{Tr}(\varphi(T_k)X))_k$ is bounded for all $X \in \mathcal{M}_n(\mathbb{C})$, and the sequence $(\varphi(T_k))_k$ is itself bounded.

By first choosing a subsequence, if necessary, we may assume that $\lim\limits_{k\to\infty} \varphi(T_k) = T_0$ exists. A similar reasoning as above yields

$$\sigma(ST-TS^*) = \sigma(\varphi(S)\varphi(T)-\varphi(T)\varphi(S)^*) \tag{7.40}$$

and

$$\sigma(ST_k-T_kS^*) = \sigma(\varphi(S)\varphi(T_k)-\varphi(T_k)\varphi(S)^*) \tag{7.41}$$

for all $S \in \mathcal{N}_A$ and k large enough. Take the limit as k goes to infinity and use above the equations to see that

$$\sigma(\ \varphi(S)T_0 - T_0\varphi(S)^*) = \sigma(ST-TS^*) = \sigma(\varphi(S)\varphi(T)-\varphi(T)\varphi(S)^*)$$

for all $S \in \mathcal{N}_A$. Thus $\mathrm{Tr}(T_0(\varphi(S)-\varphi(S)^*)) = \mathrm{Tr}(\varphi(T)(\varphi(S)-\varphi(S)^*))$ for all $S \in \mathcal{N}_A$. By [**13**, Lemma 3.5], we conclude that $\varphi(T) = T_0$ and φ is continuous at T.

CLAIM 4. The restriction of φ on $\Omega_n(\mathbb{C})$ equals to a bijective linear map L.

Firstly, we show that

$$\sigma(TS-ST^*) = \sigma(\varphi(T)\varphi(S)-\varphi(S)\varphi(T)^*) \tag{7.42}$$

for all S and T in $\Omega_n(\mathbb{C})$. Let S and T in $\Omega_n(\mathbb{C})$ and take a sequence $(V_k)_k \subset \mathcal{G}(n,x_0)$ so that $\lim V_k = TS-ST^*$. As $T \in \Omega_n(\mathbb{C})$, by Sylvester's theorem, for each k, there exists a unique matrix $W_k \in \mathcal{M}_n(\mathbb{C})$ such that $TW_k - W_kT^* = V_k$. Note that the sequence $(W_k)_k$ is bounded since the mapping $X \mapsto TX - XT^*$ is bijective. Thus, we may assume without loss of generality (i.e.; after an eventual extraction of a subsequence) that $\lim W_k = W$ for some $W \in \mathcal{M}_n(\mathbb{C})$. So, we get $TS-ST^* = TW-WT^*$. Which implies by the uniqueness of the solution of Sylvester's equation that $\lim W_k = W = S$. Using the continuity properties for the spectrum and the continuity of φ on $\Omega_n(\mathbb{C})$, we get

$$\begin{aligned}
\sigma([T,S]_*) &= \lim_{k\to\infty} \sigma(V_k) \\
&= \lim_{k\to\infty} \sigma_{V_k}(x_0) \\
&= \lim_{k\to\infty} \sigma_{[T,W_k]^*}(x_0) \\
&= \lim_{k\to\infty} \sigma_{[\varphi(T),\varphi(W_k)]_*}(x_0) \\
&= \lim_{k\to\infty} \sigma([\varphi(T),\varphi(W_k)]_*) \\
&= \sigma([\varphi(T),\varphi(S)]_*);
\end{aligned}$$

as desired.

By Lemma 6.1, the set $\Lambda_n(\mathbb{C})$ is an open dense and path-connected subset of $\mathcal{M}_n(\mathbb{C})$, and obviously contains $\mathcal{G}(n,x_0)$. Thus, upon replacing the set $\mathcal{G}(n,x_0)$ by $\Lambda_n(\mathbb{C})$, similar argument to the one used in the proof of the Claim 1, allows us to conclude that for every $A \in \Lambda_n(\mathbb{C})$, there is an open neighborhood $\mathcal{N}_A$ of A such that the restriction of φ on $\mathcal{N}_A$ equals an invertible linear map L_A. As $\Lambda_n(\mathbb{C})$ is a arcwise connected subset of $\mathcal{G}(n,x_0)$, then just as in Step 2 of the previous proof one argues as in the proof of Assertion 2 of [**26**, Proof of Theorem 2.1] to see that all linear maps L_A are the same, and then conclude that the restriction of φ on $\Lambda_n(\mathbb{C})$ is equal to a bijective linear map L. Furthermore, from the continuity of the maps φ and L on $\Omega_n(\mathbb{C})$ and the density of the set $\Lambda_n(\mathbb{C})$, we infer that the map φ coincides with L on $\Omega_n(\mathbb{C})$; which yields the desired conclusion.

CLAIM 5. There exist a unitary matrix $U \in \mathcal{M}_n(\mathbb{C})$ and a nonzero scalar $\alpha \in \mathbb{C}$ such that $Ux_0 = \alpha x_0$ and the mapping L in Claim 4 has the form $L(T) = \pm UTU^*$.

In view of the density of $\Omega_n(\mathbb{C})$ and the continuity of L and the spectrum, (7.42) yields that

$$\sigma([T,S]_*) = \sigma([L(T),L(S)]_*) \tag{7.43}$$

for any S and T in $\mathcal{M}_n(\mathbb{C})$. Therefore, Theorem 2.1 entails that there is a unitary matrix $U \in \mathcal{M}_n(\mathbb{C})$ such that

$$L(T) = \pm UTU^{-1}, \ (T \in \mathcal{M}_n(\mathbb{C})). \tag{7.44}$$

Let $T \in \Omega_n(\mathbb{C})$, and note that this together with Claim 4 and (7.39) imply that

$$2i\sigma_T(x_0) = \sigma_{[(i\mathbf{1}),T]_*}(x_0) = \sigma_{[\varphi(i\mathbf{1}),\varphi(T)]_*}(x_0) = \sigma_{[L(i\mathbf{1}),L(T)]_*}(x_0) = 2i\sigma_{UTU^*}(x_0).$$

Now, [**13**, Lemma 3.9] ensures that there is a nonzero scalar $\alpha \in \mathbb{C}$ such that $Ux_0 = \alpha x_0$.

CLAIM 6. φ has the asserted form

By Claim 3, we have $L(T) = \pm UTU^*$ for all T in $\mathcal{M}_n(\mathbb{C})$. Further, $\varphi(T) = L(T) = \pm UTU^*$ for all $T \in \Omega_n(\mathbb{C})$. With these, for every T in $\Omega_n(\mathbb{C})$ and $S \in \mathcal{M}_n(\mathbb{C})$, we have

$$\begin{aligned}\sigma_{[T,\varphi(S)]_*}(x_0) &= \sigma_{[\varphi(U^*TU),\varphi(S)]_*}(x_0)\\ &= \sigma_{[U^*TU,S]_*}(x_0)\\ &= \sigma_{[L(U^*TA),L(S)]_*}(x_0)\\ &= \sigma_{[T,L(S)]_*}(x_0).\end{aligned}$$

Whence Corollary 5.3 entails that $\varphi(S) = L(S) = \pm USU^*$ for all $S \in \mathcal{M}_n(\mathbb{C})$, and the proof is thus complete.

References

[1] Z. E. A. Abdelali and A. Bourhim, *Maps preserving the local spectrum of quadratic products of matrices*, Acta Sci. Math. (Szeged) **84** (2018), no. 1-2, 49–64. MR3792765

[2] Z. Abdelali, A. Bourhim, and M. Mabrouk, *Lie product and local spectrum preservers*, Linear Algebra Appl. **553** (2018), 328–361, DOI 10.1016/j.laa.2018.05.013. MR3809383

[3] Z. E. A. Abdelali, A. Achchi, and R. Marzouki, *Maps preserving the local spectrum of some matrix products*, Oper. Matrices **12** (2018), no. 2, 549–562, DOI 10.7153/oam-2018-12-34. MR3812190

[4] A. Achchi, M. Mabrouk, and R. Marzouki, *Maps preserving the local spectrum of the skew Jordan product of operators*, Oper. Matrices **11** (2017), no. 1, 133–146, DOI 10.7153/oam-11-10. MR3602635

[5] Z. Abdelali, A. Achchi, and R. Marzouki, *Maps preserving the local spectrum of skew-product of operators*, Linear Algebra Appl. **485** (2015), 58–71, DOI 10.1016/j.laa.2015.07.019. MR3394138

[6] P. Aiena, *Fredholm and local spectral theory, with applications to multipliers*, Kluwer Academic Publishers, Dordrecht, 2004. MR2070395

[7] Z. Bai and S. Du, *Maps preserving products $XY - YX^*$ on von Neumann algebras*, J. Math. Anal. Appl. **386** (2012), no. 1, 103–109, DOI 10.1016/j.jmaa.2011.07.052. MR2834869

[8] L. Baribeau and T. Ransford, *Non-linear spectrum-preserving maps*, Bull. London Math. Soc. **32** (2000), no. 1, 8–14, DOI 10.1112/S0024609399006426. MR1718765

[9] R. Bhatia, P. Šemrl, and A. R. Sourour, *Maps on matrices that preserve the spectral radius distance*, Studia Math. **134** (1999), no. 2, 99–110. MR1688218

[10] A. Bourhim and J. E. Lee, *Multiplicatively local spectrum-preserving maps*, Linear Algebra Appl. **549** (2018), 291–308, DOI 10.1016/j.laa.2018.03.042. MR3784351

[11] A. Bourhim, J. Mashreghi, and A. Stepanyan, *Maps between Banach algebras preserving the spectrum*, Arch. Math. (Basel) **107** (2016), no. 6, 609–621, DOI 10.1007/s00013-016-0960-9. MR3571153

[12] A. Bourhim and M. Mabrouk, *Jordan product and local spectrum preservers*, Studia Math. **234** (2016), no. 2, 97–120. MR3549185

[13] A. Bourhim and M. Mabrouk, *Maps preserving the local spectrum of Jordan product of matrices*, Linear Algebra Appl. **484** (2015), 379–395, DOI 10.1016/j.laa.2015.06.034. MR3385068

[14] A. Bourhim and J. Mashreghi, *Maps preserving the local spectrum of product of operators*, Glasg. Math. J. **57** (2015), no. 3, 709–718, DOI 10.1017/S0017089514000585. MR3395343

[15] A. Bourhim and J. Mashreghi, *Maps preserving the local spectrum of triple product of operators*, Linear Multilinear Algebra **63** (2015), no. 4, 765–773, DOI 10.1080/03081087.2014.898299. MR3291562

[16] A. Bourhim and J. Mashreghi, *A survey on preservers of spectra and local spectra*, Invariant subspaces of the shift operator, Contemp. Math., vol. 638, Amer. Math. Soc., Providence, RI, 2015, pp. 45–98, DOI 10.1090/conm/638/12810. MR3309349

[17] A. Bourhim, J. Mashreghi, and A. Stepanyan, *Nonlinear maps preserving the minimum and surjectivity moduli*, Linear Algebra Appl. **463** (2014), 171–189, DOI 10.1016/j.laa.2014.09.002. MR3262395

[18] A. Bourhim and J. Mashreghi, *Local spectral radius preservers*, Integral Equations Operator Theory **76** (2013), no. 1, 95–104, DOI 10.1007/s00020-013-2041-9. MR3041723

[19] A. Bourhim, *Surjective linear maps preserving local spectra*, Linear Algebra Appl. **432** (2010), no. 1, 383–393, DOI 10.1016/j.laa.2009.08.020. MR2566487

[20] A. Bourhim, *Spectral pictures of operator-valued weighted bi-shifts*, J. Operator Theory **59** (2008), no. 1, 193–210. MR2404470

[21] A. Bourhim and V. G. Miller, *Linear maps on $M_n(\mathbb{C})$ preserving the local spectral radius*, Studia Math. **188** (2008), no. 1, 67–75, DOI 10.4064/sm188-1-4. MR2430550

[22] A. Bourhim and T. Ransford, *Additive maps preserving local spectrum*, Integral Equations Operator Theory **55** (2006), no. 3, 377–385, DOI 10.1007/s00020-005-1392-2. MR2244195

[23] A. Bourhim, *On the local spectral properties of weighted shift operators*, Studia Math. **163** (2004), no. 1, 41–69, DOI 10.4064/sm163-1-3. MR2047464

[24] J. Bračič and V. Müller, *Local spectrum and local spectral radius of an operator at a fixed vector*, Studia Math. **194** (2009), no. 2, 155–162, DOI 10.4064/sm194-2-3. MR2534182

[25] M. Brešar and M. Fošner, *On rings with involution equipped with some new product*, Publ. Math. Debrecen **57** (2000), no. 1-2, 121–134. MR1771679

[26] J.-T. Chan, C.-K. Li, and N.-S. Sze, *Mappings preserving spectra of products of matrices*, Proc. Amer. Math. Soc. **135** (2007), no. 4, 977–986, DOI 10.1090/S0002-9939-06-08568-6. MR2262897

[27] M. A. Chebotar, Y. Fong, and P.-H. Lee, *On maps preserving zeros of the polynomial $xy-yx$*, Linear Algebra Appl. **408** (2005), 230–243, DOI 10.1016/j.laa.2005.06.015. MR2166866

[28] C. Chen and F. Lu, *Nonlinear maps preserving higher-dimensional numerical range of skew Lie product of operators*, Oper. Matrices **10** (2016), no. 2, 335–344, DOI 10.7153/oam-10-18. MR3517730

[29] C. Costara, *Linear surjective maps preserving at least one element from the local spectrum*, Proc. Edinb. Math. Soc. (2) **61** (2018), no. 1, 169–175, DOI 10.1017/S001309151700027X. MR3774850

[30] C. Costara, *Automatic continuity for linear surjective maps compressing the local spectrum at fixed vectors*, Proc. Amer. Math. Soc. **145** (2017), no. 5, 2081–2087, DOI 10.1090/proc/13364. MR3611322

[31] C. Costara, *On nonlinear maps preserving the reduced minimum modulus on differences of matrices*, Linear Algebra Appl. **507** (2016), 288–299, DOI 10.1016/j.laa.2016.06.016. MR3536959

[32] C. Costara, *Continuous maps preserving local spectra of matrices*, Linear Algebra Appl. **492** (2016), 1–8, DOI 10.1016/j.laa.2015.11.017. MR3440142

[33] C. Costara, *Holomorphic maps preserving parts of the local spectrum*, Arch. Math. (Basel) **105** (2015), no. 2, 153–162, DOI 10.1007/s00013-015-0792-z. MR3372599

[34] C. Costara, *Surjective maps on matrices preserving the local spectral radius distance*, Linear Multilinear Algebra **62** (2014), no. 7, 988–994, DOI 10.1080/03081087.2013.801967. MR3232673

[35] C. Costara, *Local spectrum linear preservers at non-fixed vectors*, Linear Algebra Appl. **457** (2014), 154–161, DOI 10.1016/j.laa.2014.05.031. MR3230437

[36] C. Costara, *Linear maps preserving operators of local spectral radius zero*, Integral Equations Operator Theory **73** (2012), no. 1, 7–16, DOI 10.1007/s00020-012-1953-0. MR2913657

[37] C. Costara, *Maps on matrices that preserve the spectrum*, Linear Algebra Appl. **435** (2011), no. 11, 2674–2680, DOI 10.1016/j.laa.2011.04.026. MR2825274

[38] C. Costara and D. Repovš, *Nonlinear mappings preserving at least one eigenvalue*, Studia Math. **200** (2010), no. 1, 79–89, DOI 10.4064/sm200-1-5. MR2720208

[39] C. Costara, *Automatic continuity for linear surjective mappings decreasing the local spectral radius at some fixed vector*, Arch. Math. (Basel) **95** (2010), no. 6, 567–573, DOI 10.1007/s00013-010-0191-4. MR2745466

[40] C. Costara and T. Ransford, *On local irreducibility of the spectrum*, Proc. Amer. Math. Soc. **135** (2007), no. 9, 2779–2784, DOI 10.1090/S0002-9939-07-08779-5. MR2317952

[41] J. Cui and J. Hou, *Linear maps preserving elements annihilated by the polynomial $XY-YX$*, Studia Math. **174** (2006), no. 2, 183–199, DOI 10.4064/sm174-2-5. MR2238461

[42] J. Cui and C.-K. Li, *Maps preserving product $XY - YX^*$ on factor von Neumann algebras*, Linear Algebra Appl. **431** (2009), no. 5-7, 833–842, DOI 10.1016/j.laa.2009.03.036. MR2535555

[43] J. Cui and J. Hou, *Maps leaving functional values of operator products invariant*, Linear Algebra Appl. **428** (2008), no. 7, 1649–1663, DOI 10.1016/j.laa.2007.10.010. MR2388647

[44] J. Cui, Q. Li, J. Hou, and X. Qi, *Some unitary similarity invariant sets preservers of skew Lie products*, Linear Algebra Appl. **457** (2014), 76–92, DOI 10.1016/j.laa.2014.05.009. MR3230434

[45] J. Cui and C.-K. Li, *Maps preserving peripheral spectrum of Jordan products of operators*, Oper. Matrices **6** (2012), no. 1, 129–146, DOI 10.7153/oam-06-09. MR2952440

[46] J. Cui and C. Park, *Maps preserving strong skew Lie product on factor von Neumann algebras*, Acta Math. Sci. Ser. B (Engl. Ed.) **32** (2012), no. 2, 531–538, DOI 10.1016/S0252-9602(12)60035-6. MR2921895

[47] M. B. Dollinger and K. K. Oberai, *Variation of local spectra*, J. Math. Anal. Appl. **39** (1972), 324–337, DOI 10.1016/0022-247X(72)90205-3. MR0318950

[48] G. Dolinar and B. Kuzma, *General preservers of quasi-commutativity*, Canad. J. Math. **62** (2010), no. 4, 758–786, DOI 10.4153/CJM-2010-041-x. MR2674700

[49] M. Ech-Cherif El Kettani and H. Benbouziane, *Additive maps preserving operators of inner local spectral radius zero*, Rend. Circ. Mat. Palermo (2) **63** (2014), no. 2, 311–316, DOI 10.1007/s12215-014-0160-z. MR3274066

[50] M. Elhodaibi and A. Jaatit, *Inner local spectral radius preservers*, Rend. Circ. Mat. Palermo (2) **67** (2018), no. 2, 215–225, DOI 10.1007/s12215-017-0308-8. MR3833003

[51] M. Elhodaibi and A. Jaatit, *On maps preserving operators of local spectral radius zero*, Linear Algebra Appl. **512** (2017), 191–201, DOI 10.1016/j.laa.2016.10.001. MR3567521

[52] A. Fošner and B. Kuzma, *Preserving zeros of Lie product on alternate matrices*, Spec. Matrices **4** (2016), 80–100, DOI 10.1515/spma-2016-0009. MR3451272

[53] M. Fošner, *Prime rings with involution equipped with some new product*, Southeast Asian Bull. Math. **26**, no. 1, (2003) 27–31.

[54] A. M. Gleason, *A characterization of maximal ideals*, J. Analyse Math. **19** (1967), 171–172, DOI 10.1007/BF02788714. MR0213878

[55] M. González and M. Mbekhta, *Linear maps on $M_n(\mathbb{C})$ preserving the local spectrum*, Linear Algebra Appl. **427** (2007), no. 2-3, 176–182, DOI 10.1016/j.laa.2007.07.005. MR2351350

[56] A. E. Guterman and B. Kuzma, *Preserving zeros of a polynomial*, Comm. Algebra **37** (2009), no. 11, 4038–4064, DOI 10.1080/00927870802545687. MR2573234

[57] J. Hou and Q. Di, *Maps preserving numerical ranges of operator products*, Proc. Amer. Math. Soc. **134** (2006), no. 5, 1435–1446, DOI 10.1090/S0002-9939-05-08101-3. MR2199190

[58] J. Hou, C.-K. Li, and N.-C. Wong, *Jordan isomorphisms and maps preserving spectra of certain operator products*, Studia Math. **184** (2008), no. 1, 31–47, DOI 10.4064/sm184-1-2. MR2365474

[59] J. Hou, C.-K. Li, and N.-C. Wong, *Maps preserving the spectrum of generalized Jordan product of operators*, Linear Algebra Appl. **432** (2010), no. 4, 1049–1069, DOI 10.1016/j.laa.2009.10.018. MR2577648

[60] T. Jari, *Nonlinear maps preserving the inner local spectral radius*, Rend. Circ. Mat. Palermo (2) **64** (2015), no. 1, 67–76, DOI 10.1007/s12215-014-0181-7. MR3324374

[61] J.-P. Kahane and W. Żelazko, *A characterization of maximal ideals in commutative Banach algebras*, Studia Math. **29** (1968), 339–343, DOI 10.4064/sm-29-3-339-343. MR0226408

[62] S. Kowalski and Z. Słodkowski, *A characterization of multiplicative linear functionals in Banach algebras*, Studia Math. **67** (1980), no. 3, 215–223, DOI 10.4064/sm-67-3-215-223. MR592387

[63] K. B. Laursen and M. M. Neumann, *An introduction to local spectral theory*, London Mathematical Society Monographs. New Series, vol. 20, The Clarendon Press, Oxford University Press, New York, 2000. MR1747914

[64] C.-K. Li, P. Šemrl, and N.-S. Sze, *Maps preserving the nilpotency of products of operators*, Linear Algebra Appl. **424** (2007), no. 1, 222–239, DOI 10.1016/j.laa.2006.11.013. MR2324385

[65] W.-H. Lin, *Nonlinear $*$-Lie-type derivations on von Neumann algebras*, Acta Math. Hungar. **156** (2018), no. 1, 112–131, DOI 10.1007/s10474-018-0803-1. MR3856906

[66] M. Marcus and B. N. Moyls, *Linear transformations on algebras of matrices*, Canadian J. Math. **11** (1959), 61–66, DOI 10.4153/CJM-1959-008-0. MR99996

[67] J. Mashreghi and A. Stepanyan, *Nonlinear maps preserving the reduced minimum modulus of operators*, Linear Algebra Appl. **493** (2016), 426–432, DOI 10.1016/j.laa.2015.12.010. MR3452747

[68] T. L. Miller, V. G. Miller, and M. M. Neumann, *Local spectral properties of weighted shifts*, J. Operator Theory **51** (2004), no. 1, 71–88. MR2055805

[69] L. Molnár, *Some characterizations of the automorphisms of $B(H)$ and $C(X)$*, Proc. Amer. Math. Soc. **130** (2002), no. 1, 111–120, DOI 10.1090/S0002-9939-01-06172-X. MR1855627

[70] L. Molnár, *Selected preserver problems on algebraic structures of linear operators and on function spaces*, Lecture Notes in Mathematics, vol. 1895, Springer-Verlag, Berlin, 2007. MR2267033

[71] L. Molnár, *Jordan $*$-derivation pairs on a complex $*$-algebra*, Aequationes Math. **54** (1997), no. 1-2, 44–55, DOI 10.1007/BF02755445. MR1466293

[72] L. Molnár, *A condition for a subspace of $\mathcal{B}(H)$ to be an ideal*, Linear Algebra Appl. **235** (1996), 229–234, DOI 10.1016/0024-3795(94)00143-X. MR1374262

[73] P. Šemrl, *Order and spectrum preserving maps on positive operators*, Canad. J. Math. **69** (2017), no. 6, 1422–1435, DOI 10.4153/CJM-2016-039-0. MR3715017

[74] P. Šemrl, *Maps on matrix and operator algebras*, Jahresber. Deutsch. Math.-Verein. **108** (2006), no. 2, 91–103. MR2248755

[75] P. Šemrl, *Maps on matrix spaces*, Linear Algebra Appl. **413** (2006), no. 2-3, 364–393, DOI 10.1016/j.laa.2005.03.011. MR2198941

[76] P. Šemrl, *Jordan $*$-derivations of standard operator algebras*, Proc. Amer. Math. Soc. **120** (1994), no. 2, 515–518, DOI 10.2307/2159889. MR1186136

[77] P. Šemrl, *Quadratic functionals and Jordan $*$-derivations*, Studia Math. **97** (1991), no. 3, 157–165, DOI 10.4064/sm-97-3-157-165. MR1100685

[78] A. Taghavi, F. Kolivand, and E. Tavakoli, *A note on strong (skew) η-Lie products preserving maps on some algebras*, Linear Multilinear Algebra **67** (2019), no. 5, 886–895, DOI 10.1080/03081087.2018.1435624. MR3923033

[79] A. Torgašev, *On operators with the same local spectra*, Czechoslovak Math. J. **48**, no. 1, (1998) 77–83.

Department of Mathematics, Faculty of sciences, Mohammed-V University, Rabat, Morocco

Email address: zineelabidineabdelali@gmail.com

Syracuse University, Department of Mathematics, 215 Carnegie Building, Syracuse, NY 13244, USA

Email address: abourhim@syr.edu

Departement of Mathematics, Faculty of Applied Sciences, Umm Al-Qura University 21955 Makkah, Saudi Arabia & D Faculty of Sciences of Sfax, Department of Mathematics University of Sfax Tunisia

Email address: msmabrouk@uqu.edu.sa

Contemporary Mathematics
Volume **743**, 2020
https://doi.org/10.1090/conm/743/14964

Numerical range and compressions of the shift

Kelly Bickel and Pamela Gorkin

ABSTRACT. The numerical range of a bounded, linear operator on a Hilbert space is a set in $\mathbb{C}$ that encodes important information about the operator. In this survey paper, we first consider numerical ranges of matrices and discuss several connections with envelopes of families of curves. We then turn to the shift operator, perhaps the most important operator on the Hardy space $H^2(\mathbb{D})$, and compressions of the shift operator to model spaces, i.e. spaces of the form $H^2 \ominus \theta H^2$ where θ is inner. For these compressions of the shift operator, we provide a survey of results on the connection between their numerical ranges and the numerical ranges of their unitary dilations. We also discuss related results for compressed shift operators on the bidisk associated to rational inner functions and conclude the paper with a brief discussion of the Crouzeix conjecture.

1. Introduction

Let $B(\mathcal{H})$ denote the set of bounded, linear operators on a Hilbert space $\mathcal{H}$. Then for $A \in B(\mathcal{H})$, its *numerical range* or field of values is the subset of $\mathbb{C}$ defined as follows:

$$W(A) = \{\langle Ax, x\rangle : x \in \mathcal{H}, \|x\| = 1\}.$$

This crucial object encodes many properties of the operator A and is closely related to the spectrum of A, denoted $\sigma(A)$. Indeed, $\sigma(A)$ is always contained in $\overline{W(A)}$ and the convex hull of $\sigma(A)$ can be recovered from the numerical ranges of operators similar to A [**45**]. Typically, $W(A)$ encodes significantly more information about A than the spectrum does. For instance, if $W(A)$ is contained in $\mathbb{R}$, then A must be Hermitian. Similarly, if $\mathcal{H}$ is finite dimensional, then $W(A)$ is compact and the maximal elements of $W(A)$ are related to the combinatorial structure of A [**54**].

Due to these and many other such properties, numerical ranges and related objects have found numerous applications in diverse areas including differential equations, numerical analysis, and quantum computing, see for example [**5, 28, 34, 36, 50, 51, 60**]. As the topic of numerical ranges is both natural and useful, it has been extensively studied and the current body of research is quite vast. Thus this

---

2010 Mathematics Subject Classification. Primary 47A12; Secondary 47A13, 30C15

Key words and phrases. Numerical Range, Envelopes, Compressions of Shifts, Unitary Dilations.

The first author's research was supported in part by National Science Foundation DMS grant #1448846.

The second author's research was supported in part by Simons Foundation Grant #243653.

survey is not meant to be in any way exhaustive. Instead we refer the interested readers to the books [**39, 42, 46**].

This survey primarily covers two topics: connections between numerical ranges and envelopes and the numerical ranges of compressions of the shift. Section 2 presents several relationships between envelopes of families of curves $\mathcal{F}$ and numerical ranges of matrices $W(A)$. Specifically, let F be continuously differentiable and let $\mathcal{F}$ denote the family of curves of (x, y) points that satisfy $F(x, y, t) = 0$ for different values of t ranging over an interval. Then, intuitively, an envelope of $\mathcal{F}$ is a curve that, at each of its points, is tangent to a member of the family. Envelopes have a number of applications and appear, for example, in both economics and in robotics and gear construction, [**53, 59**]. They are also connected to numerical ranges in several ways. First, in [**48**], Kippenhahn showed that for any matrix A, the boundary $\partial W(A)$ of the numerical range is–after removing a finite number of corners–an envelope of the family of support lines of $W(A)$. Similarly in [**27**], Donoghue outlined a proof of the elliptical range theorem, which characterizes $W(A)$ for 2×2 matrices, that constructs $\partial W(A)$ as an envelope of a family of circles.

Sections 3-5 concern compressions of the shift and their numerical ranges. To define these classical operators, recall that the Hardy space on the unit disk $H^2(\mathbb{D})$ consists of functions of the form

$$f(z) = \sum_{n=0}^{\infty} a_n z^n, \quad \text{where} \quad \sum_{n=0}^{\infty} |a_n|^2 < \infty. \tag{1}$$

A particularly important operator acting on H^2 is S, the (forward) shift operator defined by $[S(f)](z) = zf(z)$. To compress S, first recall that an inner function θ is a bounded analytic function on $\mathbb{D}$ whose radial limits have modulus one almost everywhere. Then for each inner θ, the space θH^2 is a subspace of H^2 and the *model space* K_θ and the *compression of the shift* S_θ can be defined as follows:

$$K_\theta := H^2 \ominus \theta H^2 \quad \text{and} \quad S_\theta = P_\theta S|_{K_\theta},$$

where P_θ is the orthogonal projection onto K_θ. The study of such spaces and operators has been extensive and forms a key subarea of both classic operator theory and complex analysis; for the main theory, we direct the readers to [**30, 63**]. Indeed, compressed shifts represent a large class of operators. For a contraction $T \in B(\mathcal{H})$, define the defect operator $D_T = \sqrt{1 - T^\star T}$ and the defect space $\mathcal{D}_T = \overline{D_T \mathcal{H}}$. Then if T is a completely non-unitary contraction with defect indices $\dim \mathcal{D}_T = \dim \mathcal{D}_{T^\star} = 1$, the Nagy-Foias functional model says that T is unitarily equivalent to a compressed shift, see [**64**].

Arguably the prettiest results concern finite Blaschke products

$$B(z) = \lambda \prod_{j=1}^{n} \frac{z - a_j}{1 - \overline{a_j} z}$$

with $|\lambda| = 1$, which we discuss in Section 3. In this case, K_B is finite dimensional and S_B has a nice (upper-triangular) matrix representation in terms of the zeros of B. This allowed Gau and Wu to obtain a simple characterization of the unitary 1-dilations of S_B and show that

$$W(S_B) = \cap\{W(U) : U \text{ is a unitary 1-dilation of } S_B\}, \tag{2}$$

see **[31, 32]**. Their work–and that of Mirman in **[55]**–shows that each $\partial W(S_B)$ also satisfies an elegant geometric condition called the *Poncelet property.*

There are numerous ways to generalize or extend these investigations of $W(S_B)$. For example, researchers have studied S_θ for infinite Blaschke products and general inner functions, considered operators with higher defect indices, and studied compressions of shifts in the bivariate setting **[8, 12, 15]**. While versions of (2) are true in some settings, many open questions remain. For details about such generalizations, see Sections 4-5.

As shown by two previously-discussed topics, numerical ranges are at the heart of many beautiful results and open questions in both operator and function theory. Perhaps the most famous open question concerning numerical ranges is Crouzeix's conjecture **[21]**, which states:

Conjecture (2004): There is a constant C such that for any polynomial $p \in \mathbb{C}[z]$ and $n \times n$ matrix A, the following inequality holds:

$$\|p(A)\| \leq C \max |p(z)|_{z \in W(A)}.$$

The best constant should be $C = 2$.

Initially in **[23]**, Crouzeix showed that $2 \leq C \leq 11.08$. However, significant recent progress has been made on improving C, proving special cases, and identifying other questions that imply the conjecture, see **[17, 22, 35, 61]**. We include the details in Section 6.

2. Numerical Ranges and Envelopes

2.1. Preliminaries. To examine the connections between numerical ranges of matrices and envelopes of families of curves, we need some well-known results about numerical ranges. We state these for matrices, but many results have generalizations to bounded linear operators on a Hilbert space. First, it is easy to show that numerical ranges are well behaved with respect to operations like unitary conjugation and affine transformation:

THEOREM 2.1. *If A is an $n \times n$ matrix, then*

a. *For U an $n \times n$ unitary matrix, $W(U^\star AU) = W(A)$.*
b. *For $\alpha, \beta \in \mathbb{C}$, $W(\alpha A + \beta I) = \alpha W(A) + \beta := \{\alpha z + \beta : z \in W(A)\}$.*

It is also easy to see that $W(A)$ contains the eigenvalues of A; indeed if λ is an eigenvalue of A with normalized eigenvector x, then

$$\langle Ax, x\rangle = \langle \lambda x, x\rangle = \lambda\langle x, x\rangle = \lambda.$$

One of the deepest results about numerical ranges follows from theorems of Toeplitz and Hausdorff in **[44, 67]** and states:

THEOREM 2.2 (Toeplitz-Hausdorff theorem). *If A is an $n \times n$ matrix, then $W(A)$ is convex.*

If A is normal, then this is the entire story. Indeed, the numerical range of a normal matrix A is the convex hull of its eigenvalues. To see this, recall that A must be unitarily equivalent to some diagonal matrix

$$D = \begin{bmatrix} \lambda_1 & & \\ & \ddots & \\ & & \lambda_n \end{bmatrix},$$

where $\lambda_1, \ldots, \lambda_n$ are the eigenvalues of A. For each $x \in \mathbb{C}^n$, we have $\langle Dx, x \rangle = \sum_{j=1}^n \lambda_j |x_j|^2$. This implies that $W(D)$, and hence $W(A)$, is the convex hull of the eigenvalues of A. More generally, the closure of the numerical range of a (bounded) normal operator is the convex hull of its spectrum, see [**42**, pp. 112].

In contrast, non-normal matrices typically have more points in their numerical ranges. Consider A_1 and A_2 given below:

$$A_1 = \begin{bmatrix} 0 & 0 \\ 0 & 0 \end{bmatrix} \quad \text{and} \quad A_2 = \begin{bmatrix} 0 & 1 \\ 0 & 0 \end{bmatrix}. \tag{3}$$

They have the same eigenvalues with the same multiplicity, so the spectrum does not distinguish between them. However, the numerical range does. Indeed, the elliptical range theorem says

THEOREM 2.3. *Let A be a 2×2 matrix with eigenvalues a and b. Then $W(A)$ is an elliptical disk with foci at a and b and minor axis given by $(tr(A^\star A) - |a|^2 - |b|^2)^{1/2}$.*

This implies $W(A_1) = \{0\}$, but $W(A_2) = \{z : |z| \leq 1/2\}$. More generally, the elliptical range theorem is a key tool in several proofs of the Toeplitz-Hausdorff theorem, see for example [**39**, pp. 4].

2.2. Envelopes. In what follows, we study numerical ranges of matrices using *envelopes* of families of curves. To make this precise, let $\mathcal{F}$ be the family of curves given by $F(x, y, t) = 0$, for some continuously differentiable function F. For each t in some interval, let Γ_t denote the curve of (x, y) points satisfying $F(x, y, t) = 0$.

There is some historic vagueness concerning the definition of the envelope of a family of curves, and many sources indicate at least three different ways to define it [**14**, **20**, **47**, **65**]. We discuss two of these below. For more specific details, see [**13**].

Arguably, the most natural definition is the following:

DEFINITION 2.4. *A* geometric envelope *E_1 of $\mathcal{F}$ is a curve so that each point on E_1 is a point of tangency to some member of the family Γ_t (and often, each Γ_t is touched by E_1).*

In practice, it can be hard to use Definition 2.4 to find exact formulas for a geometric envelope. In contrast, one can often compute the following set explicitly:

DEFINITION 2.5. *The* discriminant envelope *E_2 of $\mathcal{F}$ is the set of points (x, y) for which there is a value of t so that both $F(x, y, t) = 0$ and $F_t(x, y, t) = 0$.*

In general, these definitions do not yield the same set of points. However it is known that E_1 is contained in E_2, see [**14**, Propositions 1 and 2]. Moreover, if the curves in E_2 can be parameterized as $(x(t), y(t))$ and the relevant derivatives are nonvanishing in the following sense:

$$F_x^2(x, y, t) + F_y^2(x, y, t) \neq 0 \quad \text{and} \quad x'(t)^2 + y'(t)^2 \neq 0,$$

then $E_1 = E_2$, see [**20**, pp. 173].

Often, it is easy to compute the discriminant envelope E_2. For simple $\mathcal{F}$, one can find E_2 by setting $F(x, y, t) = 0$ and $F_t(x, y, t) = 0$ and then eliminating the parameter t; this process is called the *envelope algorithm.* There are also connections between the boundary of $\mathcal{F}$ and its envelope(s) and often, the boundary (or a piece of the boundary) of $\mathcal{F}$ will correspond to an envelope. For example in [**47**], Kalman observes that if the boundary is smoothly parameterized by t, then it is part of the geometric envelope. However, such a condition is difficult to check. For more information about these envelopes, additional definitions, and connections to boundaries, see [**14**, **20**, **47**, **65**] and the references therein.

Let us now consider two connections between numerical ranges and envelopes.

2.3. Finding the numerical range via Kippenhahn. Let A be an $n \times n$ matrix. Then A can be decomposed as

$$A = \Re(A) + i\Im(A), \quad \text{where } \Re(A) = \tfrac{A+A^\star}{2} \text{ and } \Im(A) = \tfrac{A-A^\star}{2i}.$$

Using this decomposition, Kippenhahn developed a method that produces $\partial W(A)$ as the geometric envelope of a family of lines. Specifically, we say that a line is a *support line* of $W(A)$ if it touches $\partial W(A)$ in either one point or along a line segment. The following theorem, which can be found in Hochstenbach and Zachlin's translation of Kippenhahn's paper [**48**, Theorem 9], allows us to identify support lines of $W(A)$:

THEOREM 2.6. *If $A = \Re(A) + i\Im(A)$ with $\alpha_1 \leq \alpha_2 \leq \cdots \leq \alpha_n$ the eigenvalues of $\Re(A)$ and $\beta_1 \leq \beta_2 \leq \cdots \leq \beta_n$ the eigenvalues of $\Im(A)$, then the points of $W(A)$ lie in the interior or on the boundary of the rectangle constructed by the line $x = \alpha_1, x = \alpha_n; y = \beta_1, y = \beta_n$ positioned parallel to the axes. The sides of the rectangle share either one point (possibly with multiplicity > 1) or one closed interval with the boundary of $W(A)$.*

For a matrix A, let $M_e(A)$ denote the maximum eigenvalue of $\Re(A)$. Then Theorem 2.6 says that the vertical line

$$x = M_e(A)$$

is a support line for $W(A)$. To identify other support lines of $W(A)$, fix $\gamma \in (0, 2\pi)$ and consider the rotated matrix $e^{-i\gamma}A$. As shown in the accompanying figure, the numerical range $W(e^{-i\gamma}A)$ is exactly the numerical range of $W(A)$ rotated by the angle $-\gamma$. By Theorem 2.6, it has vertical support line

$$x = M_e(e^{-i\gamma}A).$$

Rotating this line by an angle γ gives the new line

$$x\cos\gamma + y\sin\gamma = M_e(e^{-i\gamma}A), \tag{4}$$

which is a support line of $W(A)$.

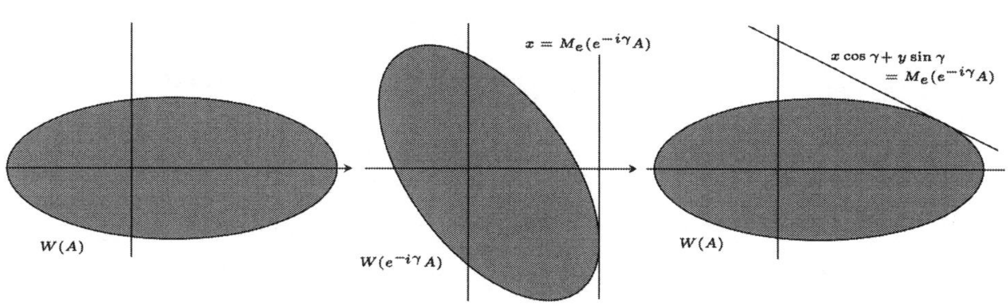

FIGURE 2.1. The Kippenhahn construction giving support lines of $W(A)$.

Letting γ vary over $[0, 2\pi]$ gives a family of support lines of $W(A)$. Then the convexity of $W(A)$ implies that the intersections of $\partial W(A)$ with these support lines must give the entire boundary of $W(A)$.

To connect this to envelopes, consider the family of lines $\mathcal{F}$ given in (4) for $\gamma \in [0, 2\pi]$. Then the differentiable components of $\partial W(A)$ are geometric envelopes for $\mathcal{F}$; this is easy to see because each point of $\partial W(A)$ lies on a line in (4) and as long as $\partial W(A)$ is differentiable at that point, it must be tangent to the line. Restricting to differentiable components of $\partial W(A)$ is reasonable because, as also proved by Kippenhahn, there are at most a finite number of places where $\partial W(A)$ is not differentiable. Moreover these singular points must occur at the eigenvalues of A, see [**48**, Theorem 13].

Kippenhahn actually proved much more than this. In particular, he completely analyzed the boundary of the numerical range in the 3×3 setting. For more information on this, see [**48**, Section 7]. For results about $\partial W(A)$ for general $A \in B(\mathcal{H})$, see Agler's paper [**1**].

2.4. The elliptical range theorem. Recall that the elliptical range theorem, given in Theorem 2.3, characterizes the numerical ranges of 2×2 matrices. Indeed, if A is a 2×2 matrix with eigenvalues a and b, then $W(A)$ is an elliptical disk with foci a and b and minor axis $(\mathrm{tr}(A^\star A) - |a|^2 - |b|^2)^{1/2}$. C.-K. Li gave a simple computational proof of this in [**52**] and other proofs can be found in [**46, 56**].

One can also use envelopes of families of circles to prove the elliptical range theorem. The proof idea described here is due to Donoghue [**27**], but many of the details appear in [**13**]. First observe that each A is unitarily equivalent to

$$B = \begin{bmatrix} a & p \\ 0 & b \end{bmatrix}, \text{ where } p = (\mathrm{tr}(A^\star A) - |a|^2 - |b|^2)^{1/2}.$$

If A has a repeated eigenvalue a and $J = \begin{bmatrix} 0 & 1 \\ 0 & 0 \end{bmatrix}$, then Theorem 2.1 gives $W(A) = W(B) = pW(J) + a$. A simple computation shows $W(J)$ is the disk of radius $\frac{1}{2}$ centered at $(0, 0)$, which gives the result.

If A has distinct eigenvalues, we can define

$$T = \begin{bmatrix} 0 & m \\ 0 & 1 \end{bmatrix}$$

where $m = \frac{p}{|b-a|}$ and apply Theorem 2.1 to show $W(A) = W(B) = (b-a)W(T)+a$. Thus it suffices to study $W(T)$, which can be realized as a family of circles. Indeed each normalized $z \in \mathbb{C}^2$ can be written as $z = \begin{bmatrix} te^{i\theta_1} \\ \sqrt{1-t^2}e^{i\theta_2} \end{bmatrix}$ for some $t \in [0, 1]$ and $\theta_1, \theta_2 \in [0, 2\pi)$, which gives

$$\langle Tz, z \rangle = (1 - t^2) + me^{i(\theta_2 - \theta_1)}(t\sqrt{1 - t^2}).$$

This implies $W(T)$ is the union of circles $\bigcup_{t \in [0,1]} \mathcal{C}_t$, where $\mathcal{C}_t$ is the circle with center $(1 - t^2, 0)$ and radius $mt\sqrt{1 - t^2}$. Equivalently, $W(T)$ is the family of curves satisfying $F(x, y, t) = 0$ for

$$F(x, y, t) := (x - (1 - t^2))^2 + y^2 - m^2t^2(1 - t^2)$$

and $t \in [0, 1]$. To find the discriminant envelope, we apply the envelope algorithm. Taking the equations $F_t(x, y, t) = 0$ and $F(x, y, t) = 0$ and solving for x and y gives the curves

$$x(t) = (1-t^2) + \tfrac{m^2}{2}(1-2t^2) \text{ and } y(t) = \pm\sqrt{m^2(t^2-t^4) - \tfrac{m^4}{4}(1-2t^2)^2} \tag{5}$$

and the point $(1,0)$. It is easy to check that the curves in (5) give exactly the ellipse

$$\frac{(x-\frac{1}{2})^2}{1+m^2} + \frac{y^2}{m^2} = \frac{1}{4}. \tag{6}$$

This leads to the question:

Do the envelope curves in (5) *give the boundary of the union* $\bigcup_{t\in[0,1]} \mathcal{C}_t$*?*

In general, the relationship between the boundary of a family of curves and its envelope is murky. However in this case, the answer is yes. For the details about that and the fact that $W(T)$ is the closed elliptical disk with boundary (5), see [**13**]. Then the elliptical range theorem follows immediately from this result about $W(T)$.

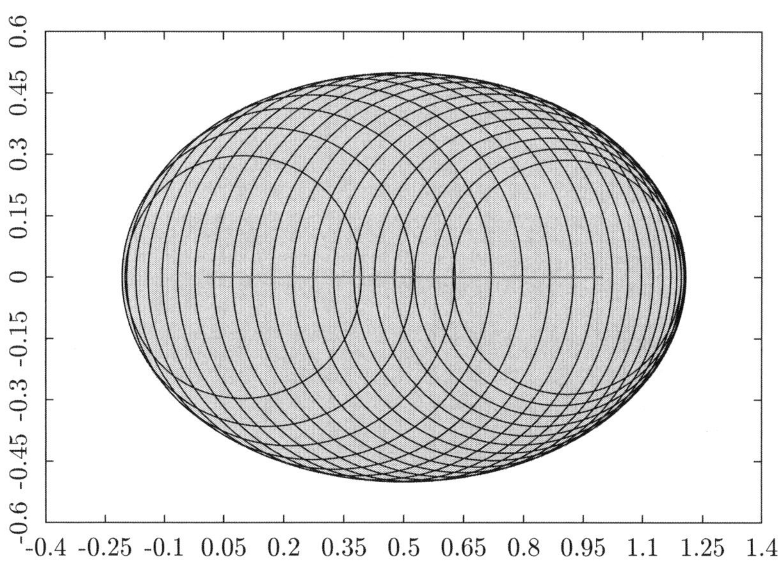

FIGURE 2.2. $W(T)$ as a union of circles $\mathcal{C}_t$ with elliptical boundary from (6).[1]

3. The numerical range of a compressed shift operator (single variable)

Recall that H^2 is the Hardy space on the unit circle $\mathbb{T}$ consisting of functions of the form $f(z) = \sum_{n=0}^{\infty} a_n z^n$ where $\sum_{n=0}^{\infty} |a_n|^2 < \infty$ and an inner function is a bounded analytic function on $\mathbb{D}$ with radial limits of modulus one almost everywhere. Perhaps the most important operator acting on this space is S, the (forward) shift operator on H^2 defined by $[S(f)](z) = zf(z)$; the adjoint of S is the backward shift $[S^\star(f)](z) = (f(z) - f(0))/z$. In 1949, Beurling [**9**] proved the following theorem about (closed) nontrivial invariant subspaces of the shift S, which has implications for the invariant subspaces of the backward shift operator.

[1]This figure was created by Trung Tran. We thank him for his permission to use the figure in this paper.

THEOREM 3.1 (Beurling's theorem). *The nontrivial invariant subspaces under S are*

$$\theta H^2 = \{\theta h : h \in H^2\},$$

where θ is a (nonconstant) inner function.

Thus we see that the invariant subspaces for the adjoint $S^\star$ are $K_\theta := H^2 \ominus \theta H^2$. These subspaces are called *model spaces*. The following description of K_θ is often helpful and it is not difficult to prove.

THEOREM 3.2. *Let θ be inner. Then $K_\theta = H^2 \cap \theta\, \overline{zH^2}$.*

Let $|a_j| < 1$ for $j = 1, \ldots, n$ and consider K_B where $B(z) = \prod_{j=1}^n \frac{z-a_j}{1-\overline{a_j}z}$ is a finite Blaschke product. The reproducing kernel corresponding to the point $a \in \mathbb{D}$ is defined by $g_a(z) = \dfrac{1}{1-\overline{a}z}$ and it has the property that $\langle f, g_a\rangle = f(a)$ for all $f \in H^2$. As a consequence we see that $\langle Bh, g_{a_j}\rangle = B(a_j)h(a_j) = 0$ for all $h \in H^2$. So if B is a Blaschke product with zeros $a_1, \ldots, a_n$, then $g_{a_j} \in K_B$ for $j = 1, 2, \ldots, n$. In fact, if the points a_j are distinct, $K_B = \text{span}\{g_{a_j} : j = 1, \ldots, n\}$ and the reproducing kernels $g_{a_1}, \ldots, g_{a_n}$ will be linearly independent.

It is not really essential that the points be distinct, but certain adjustments must be made if they are not. However, the representations for our matrices will not change; we refer the reader to **[31]**.

The operators that we are interested in here are *compressions of the shift operator*: For θ an inner function, we define $S_\theta : K_\theta \to K_\theta$ by

$$S_\theta(f) = P_\theta(S(f))$$

where P_θ is the orthogonal projection from H^2 onto K_θ. In this section, we are particularly interested in the case in which $\theta = B$ is a finite Blaschke product. In this case (and precisely in this case) K_B is finite dimensional and with the appropriate choice of an orthonormal basis, we can analyze this operator. The basis that we choose is obtained from applying the Gram-Schmidt process to the basis we obtained from the reproducing kernels. In finite dimensions, this basis is called the *Takenaka-Malmquist basis*: Letting $b_a(z) = \frac{z-a}{1-\overline{a}z}$, we take it to be the following ordered basis:

$$\left(\frac{\sqrt{1-|a_1|^2}}{1-\overline{a_1}z}\prod_{j=2}^n b_{a_j}, \frac{\sqrt{1-|a_2|^2}}{1-\overline{a_2}z}\prod_{j=3}^n b_{a_j}, \ldots, \frac{\sqrt{1-|a_{n-1}|^2}}{1-\overline{a_{n-1}}z} b_{a_n}, \frac{\sqrt{1-|a_n|^2}}{1-\overline{a_n}z}\right).$$

We have chosen this ordered basis to yield an upper triangular matrix for S_B. For example, for two zeros a and b we obtain

$$A = \begin{bmatrix} a & \sqrt{1-|a|^2}\sqrt{1-|b|^2} \\ 0 & b \end{bmatrix}.$$

So A is the matrix representing S_B when B has two zeros a and b. By the elliptical range theorem, the numerical range is an elliptical disk with foci at a and b and minor axis of length $\sqrt{1-|a|^2}\sqrt{1-|b|^2}$. When $a = b$ we see that the numerical range is a circular disk with center at a and radius $(1-|a|^2)/2$. Thus, as we mentioned earlier, for the 2×2 Jordan block A_2 that we met in (3), it follows that the numerical range is the closed disk centered at the origin of radius $1/2$. What about the $n \times n$ case?

Here things are more complicated, but we can still obtain a matrix representing S_B: A computation shows that the $n \times n$ matrix representing S_B with respect to the Takenaka-Malmquist basis is

$$(7)\quad A = \begin{bmatrix} a_1 & \sqrt{1-|a_1|^2}\sqrt{1-|a_2|^2} & \dots & (\prod_{k=2}^{n-1}(-\overline{a_k}))\sqrt{1-|a_1|^2}\sqrt{1-|a_n|^2} \\ 0 & a_2 & \dots & (\prod_{k=3}^{n-1}(-\overline{a_k}))\sqrt{1-|a_2|^2}\sqrt{1-|a_n|^2} \\ \dots & \dots & \dots & \dots \\ 0 & 0 & 0 & a_n \end{bmatrix}.$$

Note that for each $\lambda \in \mathbb{T}$, by adding only one row and one column, we can put A "inside" a unitary matrix

$$U_{ij}^{\lambda} = \begin{cases} A_{ij} & \text{if } 1 \le i,j \le n, \\ \lambda\left(\prod_{k=1}^{j-1}(-\overline{a_k})\right)\sqrt{1-|a_j|^2} & \text{if } i = n+1 \text{ and } 1 \le j \le n, \\ \left(\prod_{k=i+1}^{n}(-\overline{a_k})\right)\sqrt{1-|a_i|^2} & \text{if } j = n+1 \text{ and } 1 \le i \le n, \\ \lambda\prod_{k=1}^{n}(-\overline{a_k}) & \text{if } i = j = n+1. \end{cases}$$

EXAMPLE 3.3. Let $B(z) = z^n$. Then

$$K_B = \operatorname{span}(1, z, z^2, \dots, z^{n-1})$$

and S_B is represented (with respect to the Takenaka-Malmquist basis) by

$$\begin{pmatrix} 0 & 1 & 0\cdots & 0 & 0 \\ 0 & 0 & 1\cdots & 0 & 0 \\ \cdots & \cdots & \cdots & \cdots & 1 \\ 0 & 0 & 0 & 0 & 0 \end{pmatrix}.$$

An example of a unitary 1-dilation of this matrix (the matrix with entries from S_B in bold) is

$$\begin{pmatrix} \mathbf{0} & \mathbf{1} & \mathbf{0}\cdots & \mathbf{0} & \mathbf{0} & 0 \\ \mathbf{0} & \mathbf{0} & 1\cdots & \mathbf{0} & \mathbf{0} & 0 \\ \cdots & \cdots & \cdots & \cdots & \mathbf{1} & 0 \\ \mathbf{0} & \mathbf{0} & \mathbf{0} & \mathbf{0} & \mathbf{0} & 1 \\ 1 & 0 & 0 & 0 & 0 & 0 \end{pmatrix}.$$

3.1. Unitary dilations. Let A be a matrix with $\|A\| \le 1$. Then, following Halmos, we may consider $D_A = \sqrt{1 - AA^\star}$ and $D_{A^\star} = \sqrt{1 - A^\star A}$. Halmos noted that

$$U = \begin{pmatrix} A & D_A \\ D_{A^\star} & -A^\star \end{pmatrix}$$

is a unitary dilation of A to a space twice as large as the original and he posed the following question **[43]**: Is

$$\overline{W(A)} = \bigcap\{\overline{W(U)} : U \text{ a unitary dilation of } A\}?$$

We note that Halmos considered operators on a Hilbert space $\mathcal{H}$ and posed this question for operators on (possibly) infinite dimensional spaces. In particular, the closures of the numerical ranges are required in this general setting. For finite Blaschke products, all numerical ranges in question will be closed.

In this paper, we focus on unitary dilations of a compressed shift operator S_B. We begin with the case in which B is a finite Blaschke product (and we will see later that a similar result holds for S_θ when θ is a general inner function). For these, we know that the unitary dilations can be parametrized as a family $\{U_\lambda\}$ for each $\lambda \in \mathbb{T}$, where

$$U_\lambda = \begin{bmatrix} S_B & *_\lambda \\ *_\lambda & *_\lambda \end{bmatrix}, \text{ where we have added one row and one column.}$$

(In this representation, each $*_\lambda$ can be determined once we have S_B. This will be clear for S_B when B is finite and we discuss this parametrization later briefly for arbitrary inner functions.)

In fact, up to unitary equivalence, these are all the unitary 1-dilations of S_B. Note also that it makes sense that there is a unitary 1-dilation: Looking at the Halmos dilation, recalling that $\operatorname{rank}(I - S_B^\star S_B) = 1 = \operatorname{rank}(I - S_B S_B^\star)$ and that this implies $\operatorname{rank}(I - S_B^\star S_B)^{1/2} = \operatorname{rank}(I - S_B S_B^\star)^{1/2}$, we might expect that we need only add one row and one column to get to the unitary dilation. To investigate this further, we mention some important connections between the Blaschke product B and the unitary dilations of S_B. Before we do so, however, we note that one half of Halmos's conjecture is easy: We show that

$$W(S_B) \subseteq \bigcap \{W(U_\lambda) : \lambda \in \mathbb{T}\}.$$

To see this, let $V = [I_n, 0]$ be $n \times (n+1)$ and $\lambda \in \mathbb{T}$. Then $S_B = VU_\lambda V^t$ and we see that for x with $\|x\| = 1$ we have $V^t x = \begin{bmatrix} x \\ 0 \end{bmatrix}$, so $\|V^t x\| = 1$. Suppose $\beta \in W(S_B)$. Then there exists a unit vector x such that $\beta = \langle S_B x, x\rangle$. Thus,

$$\beta = \langle S_B x, x\rangle = \langle VU_\lambda V^t x, x\rangle = \langle U_\lambda V^t x, V^t x\rangle.$$

Consequently, $\beta \in W(U_\lambda)$. Since this holds for each $\lambda \in \mathbb{T}$, we see that β lies in $\cap_{\lambda \in \mathbb{T}} W(U_\lambda)$. Thus containment holds because S_B is a compression of U_λ and the same argument works in greater generality. From this it is not difficult to see that it is the other containment that is interesting.

Because we know that the numerical range of a unitary matrix is the convex hull of its eigenvalues, we first consider the eigenvalues of the unitary 1-dilation of S_B where B is a finite Blaschke product. In this case, it can be shown that the eigenvalues of U_λ are the values $\hat{B}(z) := zB(z)$ maps to λ **[31, 33]**. Recalling that a finite Blaschke product is continuous on an open subset of $\mathbb{C}$ containing the closed unit disk, maps the unit circle to itself, the open unit disk to itself, and the complement of the closed unit disk to itself, we see that for $\lambda \in \mathbb{T}$ the only possible solutions to this equation will lie on the unit circle. It is also well known (see, for example, **[24]**) that the argument of a finite Blaschke product increases on the unit circle. As a consequence, the solutions to $\hat{B} = \lambda$ will be distinct. If B has degree n, there will precisely $n+1$ distinct solutions to $\hat{B} = \lambda$. Thus $W(U_\lambda)$ is the convex hull of these $n+1$ distinct points.

We are now ready to put all this together using a result of Gau and Wu **[32]** who studied the class $\mathcal{S}_n$ of compressions of the shift to an n-dimensional space: These are operators that have no eigenvalues of modulus 1, are contractions (completely non-unitary contractions) with $\operatorname{rank}(I - T^\star T) = \operatorname{rank}(I - TT^\star) = 1$ and they are compressions of the shift operator with finite Blaschke product symbol:

$$S_B(f) = P_B(S(f)) \text{ where } f \in K_B, P_B : H^2 \to K_B, \tag{8}$$

where P_B is defined by

$$P_B(g) = BP_-(\overline{B}g) = B(I - P_+)(\overline{B}g)$$

and P_- the orthogonal projection for L^2 onto $L^2 \ominus H^2$. Their result is the following (see Figure 3.1):

THEOREM 3.4. **[32]** *For $T \in \mathcal{S}_n$ and any point $\lambda \in \mathbb{T}$, there is an $(n+1)$-gon inscribed in $\mathbb{T}$ that circumscribes the boundary of $W(T)$ and has λ as a vertex.*

As a consequence of this result, Gau and Wu were able to prove the following:

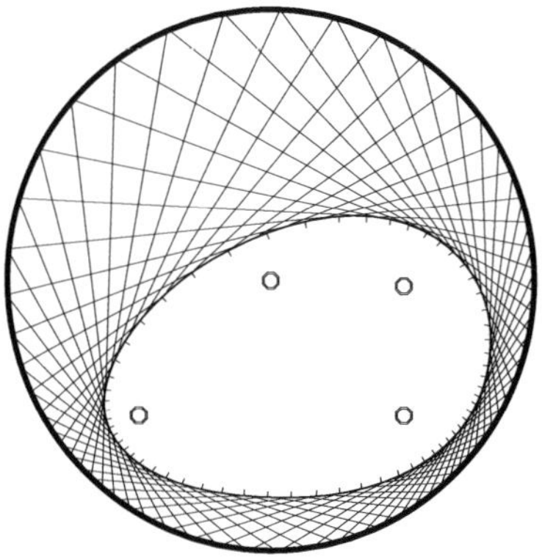

FIGURE 3.1. Polygons intersecting to yield numerical range

THEOREM 3.5. **[32]** *Let B be a finite Blaschke product. Then*

$$W(S_B) = \bigcap \{W(U) : U \text{ a unitary 1-dilation of } S_B\}.$$

The authors note that, for compressions of the shift operator, this is the "most economical" intersection; that is, we need only consider dilations of our operators to a space one dimension larger. For general operators, Choi and Li answered Halmos's question in 2001:

THEOREM 3.6 (General theorem, **[16]**). *Let T be a contraction on a Hilbert space $\mathcal{H}$. Then*

$$\overline{W(T)} = \bigcap \{\overline{W(U)} : U \text{ a unitary dilation of } T \text{ on } \mathcal{H} \oplus \mathcal{H}\}.$$

In addition to answering Halmos's theorem for these operators, Gau and Wu's result has a very beautiful geometric consequence that we investigate in the next subsection.

3.2. The Poncelet property. In this section, we discuss the connection to a famous theorem from projective geometry known as Poncelet's theorem. Poncelet was born in Metz, France in 1788. He joined Napoleon's army as it was approaching Russia. On October 19 Napoleon ordered the army to withdraw. The Russians then attacked the retreating French army and sources say that Poncelet was left for dead on the battlefield. Poncelet was held as a prisoner in Saratov and it was during this time that he discovered the following theorem, now bearing his name.

THEOREM 3.7. *(Poncelet's Theorem, 1813, ellipse version) Given one ellipse contained entirely inside another, if there exists one circuminscribed (simultaneously inscribed in the outer and circumscribed around the inner) n-gon, then every point on the boundary of the outer ellipse is the vertex of some circuminscribed n-gon.*

Poncelet's theorem says that if you shoot a ball starting at a point on the exterior ellipse, shooting tangent to the smaller ellipse, and the path closes in n steps, then *no matter where you begin* the path will close in n steps. There are now many proofs of Poncelet's theorem – of course there is one due to Poncelet **[58]**, one due to Griffiths and Harris **[38]**, and in 2015 for the bicentennial of Poncelet's theorem, a proof due to Halbeisen and Hungerbühler appeared **[41]**.

Though this version of Poncelet's theorem is about two ellipses, using an affine transformation does not change the Poncelet nature of an inner ellipse. Thus we may assume that the outer ellipse is the unit circle. Returning to Gau and Wu's theorem, we see that it says that the boundary of the numerical range of a compression of the shift operator is a Poncelet curve; that is, it is the case that given any point λ on the unit circle we can find a polygon with all vertices on the unit circle that circumscribes the bounding curve. All of the polygons will have the same number of sides. While the curves have this beautiful property, they are not usually ellipses and we therefore call them *Poncelet curves.* For further study in this regard, we note that work of Mirman [**55**] looks at this same property as well as *packages of Poncelet curves*; that is, what if instead of connecting successive points, we connect every other point? What happens if we connect every third point? Examples of higher-degree cases in which the numerical range is elliptical can be found in [**25, 29, 37**].

EXAMPLE 3.8. Consider the special case in which $B(z) = z^n$, and the matrix representing S_B is the $n \times n$ Jordan block. Then the unitary 1-dilations are parametrized by the unit circle and for each $\lambda \in \mathbb{T}$ the numerical range of U_λ is the convex hull of the points for which $\hat{B}(z) = zB(z) = z^{n+1} = \lambda$. By Theorem 3.4, the intersection of all $W(U_\lambda)$ over $\lambda \in \mathbb{T}$ is the numerical range of S_B. It is now easy to see that the numerical range must be bounded by a circle. Looking at the points on the unit circle that give rise to the real eigenvalues, we see that the radius of the bounding circle must be $\cos(\pi/(n+1))$. (See also [**40, 69**] for this and related results.)

Thus, we have the following result.

THEOREM 3.9. *The numerical range of the $n \times n$ Jordan block is a circular disk of radius* $\cos(\pi/(n+1))$.

We note that this circular disk of radius $\cos(\pi/(n+1))$ is inscribed in the convex $(n+1)$-gon with vertices equally spaced on the unit circle; in other words, the boundary is a Poncelet circle.

For Poncelet ellipses inscribed in triangles, we refer to the paper [**24**]. For more on a Blaschke product perspective of Poncelet's theorem, see also [**25, 26**].

4. Extensions: General inner functions and other defect indices

Now we consider infinite Blaschke products as well as general inner functions. The compression of the shift is defined as in (8).

For an infinite Blaschke product, we require that the zeros $a_n \in \mathbb{D}$ satisfy the Blaschke condition $\sum_{n=1}^{\infty}(1-|a_n|) < \infty$. We recall ([**8**]) that if an operator T is a completely nonunitary contraction with a unitary 1-dilation, then

(1) every eigenvalue of T is in the interior of $W(T)$;
(2) $\overline{W(T)}$ has no corners in $\mathbb{D}$.

4.1. Compressions of the shift with inner functions as symbol. To obtain the matrix representation for our operators with inner function θ as symbol, we consider two orthogonal decompositions of K_θ: The first decomposition will be

$$\mathcal{M}_1 = \mathbb{C}(S^\star\theta) = \{x(\theta(z)-\theta(0))/z : x \in \mathbb{C}\} \text{ and } \mathcal{N}_1 = K_\theta \ominus \mathcal{M}_1 \tag{9}$$

while the second will be

$$\mathcal{M}_2 = \mathbb{C}(\theta\,\overline{\theta(0)} - 1) \text{ and } \mathcal{N}_2 = K_\theta \ominus \mathcal{M}_2. \tag{10}$$

Computations then show that

$$S_\theta(xS^\star\theta + w) = x((\theta\,\overline{\theta(0)} - 1)\theta(0) + Sw$$

for $x \in \mathbb{C}$ and $w \in \mathcal{N}_1$. Thus, we get this matrix representation for unitary 1-dilations on $K = K_\theta \oplus \mathbb{C}$:

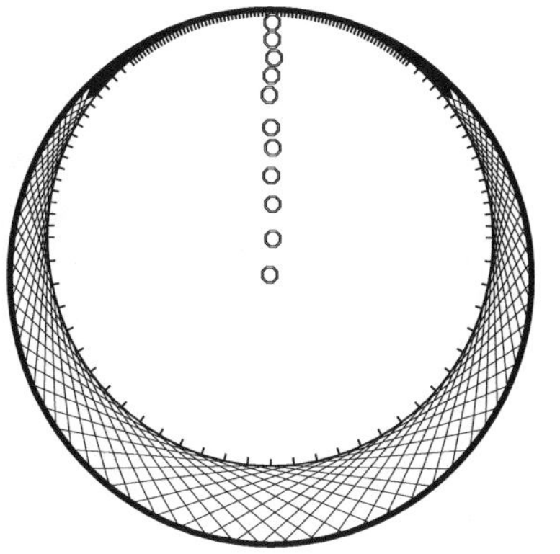

FIGURE 4.1. An approximation of an infinite Blaschke product with one singularity

$$S_\theta = \begin{bmatrix} \lambda & 0 \\ 0 & S \end{bmatrix} \text{ and } U_{\alpha\beta} = \begin{bmatrix} \lambda & 0 & \alpha\sqrt{1-|\lambda|^2} \\ 0 & S & 0 \\ \beta\sqrt{1-|\lambda|^2} & 0 & -\alpha\beta\overline{\lambda} \end{bmatrix}.$$

Here α, β have modulus 1 and $|\lambda| < 1$. If $\theta(0) = 0$, then $\lambda = 0$. It appears that there are several free variables, but up to unitary equivalence there is only one free parameter and that is the value of $\alpha\beta$. Thus, the unitary dilations may be parametrized by $\gamma \in \mathbb{T}$. We have the celebrated theorem of D. Clark.

THEOREM 4.1. **[18]** *If θ is inner and $\theta(0) = 0$, then all unitary 1-dilations of S_θ are equivalent to rank-1 perturbations of $S_{z\theta}$.*

For compressions of the shift with a Blaschke product as symbol, we obtain the following:

THEOREM 4.2. **[15]** *Let B be an infinite Blaschke product. Then the closure of the numerical range of S_B satisfies*

$$\overline{W(S_B)} = \bigcap_{\gamma \in \mathbb{T}} \overline{W(U_\gamma)},$$

where the U_γ are the unitary 1-dilations of S_B (or, equivalently, the rank-1 Clark perturbations of $S_{\hat{B}}$).

For some functions, we get an infinite version of Poncelet's theorem, see Figure 4.1.

To extend this theorem to arbitrary inner functions, the following well-known result of Frostman is useful.

THEOREM 4.3 (Frostman's Theorem). *Let I be an inner function. Let $a \in \mathbb{D}$ and $\varphi_a(z) = \frac{z-a}{1-\bar{a}z}$. Then $\varphi_a \circ I$ is a Blaschke product for almost all $a \in \mathbb{D}$.*

Every inner function is, therefore, a uniform limit of Blaschke products. So, if θ is an arbitrary inner function, we may find a sequence (B_n) of Blaschke products convering uniformly to θ. Since

$$P_\theta f = \theta P_-(\overline{\theta} f) \text{ for } f \in H^2,$$

where $P_- : L^2(\mathbb{T}) \to L^2(\mathbb{T}) \ominus H^2$ is the orthogonal projection, $\|B_n - \theta\|_\infty \to 0$ implies that $\|P_{B_n} - P_\theta\| \to 0$. We may use this to obtain $W(S_\theta)$ from $W(S_{B_n})$ where B_n is a Blaschke product. For more details, see **[15]**.

Combining results in **[8]** with Frostman's theorem tells us that Theorem 4.2 holds for arbitrary inner functions.

4.2. Higher defect index. Thus far, our operators have defect index equal to 1. However, the situation for more general defect index and an operator on a complex separable Hilbert space $\mathcal{H}$ was studied by Bercovici and Timotin **[8]**. They considered *n-dilations* of contractions, which are unitary dilations of T that act on $\mathcal{H} \oplus \mathbb{C}^n$. In general, a contraction can be written as a direct sum of a unitary operator and a completely nonunitary contraction. Since we understand the numerical range of the unitary piece of the operator, most of the work in a proof focuses on the completely nonunitary part of the contraction. To state the next result, we let $D_T = (I - T^\star T)^{1/2}$ denote *the defect operator* and $\mathcal{D}_T = \overline{D_T \mathcal{H}}$ denote *the defect space.*

THEOREM 4.4. *Let T be a contraction with dim $\mathcal{D}_T$ = dim $\mathcal{D}_{T^\star} = n < \infty$. Then*

$$\overline{W(T)} = \bigcap \{\overline{W(U)} : U \text{ a unitary } n - \text{dilation of } T\}.$$

Once again, we see that we can use the "most economical" dilations of T. Bercovici and Timotin also show that if ℓ is a support line for the closure of the numerical range of T, then there is a unitary n-dilation of T for which ℓ is a support line for $\overline{W(U)}$. In some sense, then, the geometry extends to this situation as well.

5. Compressed shifts on the bidisk

5.1. Two-variable Setup. While the earlier sections discuss many results on $\mathbb{D}$, there is another direction to pursue, namely compressions of shifts in several variables. To that end, let $\mathbb{D}^2$ denote the unit bidisk and $\mathbb{T}^2$ its distinguished boundary

$$\mathbb{D}^2 = \{(z_1, z_2) : |z_1|, |z_2| < 1\} \quad \text{and} \quad \mathbb{T}^2 = \{(\tau_1, \tau_2) : |\tau_1|, |\tau_2| = 1\}.$$

In this setting, many one-variable objects generalize easily. Indeed, the Hardy space $H^2(\mathbb{D}^2)$ consists of functions of the form

$$f(z) = \sum_{m,n=0}^{\infty} a_{m,n} z_1^m z_2^n, \quad \text{where} \quad \sum_{m,n=0}^{\infty} |a_{m,n}|^2 < \infty.$$

Then there are two natural shift operators, S_{z_1} and S_{z_2}, on $H^2(\mathbb{D}^2)$, defined by $[S_{z_j}(f)](z) = z_j f(z)$ for $j = 1, 2$. Similarly, a function Θ is inner if $\Theta \in \text{Hol}(\mathbb{D}^2)$ and $\lim_{r \nearrow 1} |\Theta(r\tau)| = 1$ for a.e. $\tau \in \mathbb{T}^2$. Then $\Theta H^2(\mathbb{D}^2)$ is a shift-invariant subspace of the two-variable Hardy space and in analogy with the one-variable setting, one can define *two-variable model spaces* as

$$K_\Theta = H^2(\mathbb{D}^2) \ominus \Theta H^2(\mathbb{D}^2)$$

and the associated compressed shifts

$$\tilde{S}^1_\Theta = P_\Theta S_{z_1}|_{K_\Theta} \quad \text{and} \quad \tilde{S}^2_\Theta = P_\Theta S_{z_2}|_{K_\Theta},$$

where P_Θ is the orthogonal projection onto K_Θ. The study of such compressed shifts is aided by the existence of nice decompositions of K_Θ into shift-invariant subspaces. Indeed, Ball, Sadosky, and Vinnikov **[7]** showed that there are subspaces $\mathcal{S}_1$ and $\mathcal{S}_2$ such that

$$K_\Theta = \mathcal{S}_1 \oplus \mathcal{S}_2 \quad \text{and} \quad S_{z_1}\mathcal{S}_1 \subseteq \mathcal{S}_1, \; S_{z_2}\mathcal{S}_2 \subseteq \mathcal{S}_2. \tag{11}$$

Such decompositions are called *Agler decompositions* and were introduced by J. Agler in a different form in **[2]**. For more information about Agler decompositions and their properties see **[3,10,11,19,49,68]** and the references therein. Then using basic properties of multiplication operators, one can show (as in **[12]**):

LEMMA 5.1. *Assume $K_\Theta = \mathcal{S}_1 \oplus \mathcal{S}_2$ gives an Agler decomposition as in* (11). *Then*

a. *If $\mathcal{S}_1 \neq \{0\}$, then $\overline{W(\tilde{S}^1_\Theta)} = \overline{W(S_{z_1}|\mathcal{S}_1)} = \overline{\mathbb{D}}$.*

b. *If $\mathcal{S}_1 = \{0\}$, then $W(\tilde{S}^1_\Theta|\mathcal{S}_1) = \emptyset$.*

Typically $\mathcal{S}_1 \neq \{0\}$ and so Lemma 5.1 says that *most* compressed shifts have maximal numerical ranges. This renders the standard numerical range questions trivial. To obtain interesting questions, one can further compress the multiplication operators and define

$$S^1_\Theta := P_{\mathcal{S}_2} \tilde{S}^1_\Theta|_{\mathcal{S}_2} = P_{\mathcal{S}_2} S_{z_1}|_{\mathcal{S}_2},$$

where $\mathcal{S}_2$ is any subspace arising from an Agler decomposition of K_Θ as in (11). The numerical ranges of such S^1_Θ have quite interesting properties, which will be discussed later. In what follows, we always assume $\mathcal{S}_2 \neq \{0\}$.

5.2. Rational Inner Functions. First consider the two-variable analogues of finite Blaschke products, called *rational inner functions.* Although more complicated than finite Blaschke products, rational inner functions still have fairly nice structures. Indeed, as shown in **[4, 62]**, every rational inner function Θ with $\deg \Theta = (m, n)$ can be written as

$$\Theta = \lambda \frac{\tilde{p}}{p}, \text{ where } \tilde{p}(z) = z_1^m z_2^n \overline{p\left(\tfrac{1}{\bar{z}_1}, \tfrac{1}{\bar{z}_2}\right)},$$

and $\lambda \in \mathbb{T}$. Here $\deg \Theta = (m, n)$ means that, after canceling any common factors of the numerator and denominator, m is the largest power of z_1 appearing in Θ and n is the largest power of z_2 appearing in Θ. Furthermore, one can choose p so that it has at most a finite number of zeros on $\mathbb{T}^2$, is nonvanishing on $\mathbb{D}^2 \cup (\mathbb{D} \times \mathbb{T}) \cup (\mathbb{T} \times \mathbb{D})$, and shares no common factors with $\tilde{p}$. For example, up to a unimodular constant, a general degree $(1, 1)$ rational inner function has the form

$$\Theta(z) = \frac{\tilde{p}(z)}{p(z)} = \frac{\bar{a} z_1 z_2 + \bar{b} z_2 + \bar{c} z_1 + \bar{d}}{a + b z_1 + c z_2 + d z_1 z_2},$$

where $p = a + bz_1 + cz_2 + dz_1z_2$ satisfies the stated conditions on its zero set and shares no common factors with $\tilde{p}$. For rational inner functions Θ, the associated model spaces K_Θ have particularly nice Agler decompositions. For example, the following result describes properties of $\mathcal{S}_2$ from (11):

LEMMA 5.2. *Let $\Theta = \frac{\tilde{p}}{p}$ be rational inner with degree (m, n) and let $H = \mathcal{S}_2 \ominus S_{z_2}\mathcal{S}_2$. Then* $\dim H = m$ *and if $g \in H$, then $g = \frac{q}{p}$ where q is a polynomial with* $\deg q \leq (m-1, n)$.

The complete result appears in **[12]**, but related results and ideas appeared earlier in **[11, 68]**. Rational inner functions also have close connections to one-variable finite Blaschke products. For Θ a rational inner function with $\deg \Theta = (m, n)$, define the exceptional set

$$E_\Theta = \{\tau \in \mathbb{T} : p(\tau_1, \tau) = 0 \text{ for some } \tau_1 \in \mathbb{T}\}.$$

Then if $\tau \in \mathbb{T} \setminus E_\Theta$, the function $\theta_\tau(z) := \Theta(z, \tau)$ is a finite Blaschke product with $\deg \theta_\tau = m$. In what follows, we let K_{θ_τ} denote the one-variable model space associated to each θ_τ.

5.3. Results. Let us restrict to rational inner Θ and consider the compressed shift S^1_Θ and its numerical range.

Stating the main results requires some notation. Let $H^2_2(\mathbb{D})$ denote the one-variable Hardy space with variable z_2 and $H^2_2(\mathbb{D})^m := \oplus_{j=1}^m H^2_2(\mathbb{D})$ denote the space of vector-valued functions $\vec{f} = (f_1, \ldots, f_m)$ with $f_j \in H^2_2(\mathbb{D})$. If M is a bounded, $m \times m$ matrix-valued function, let T_M denote the matrix-valued Toeplitz operator defined by

$$T_M \vec{f} = P_{H^2_2(\mathbb{D})^m}(M \vec{f}). \tag{12}$$

Then the following results and their corollaries appear in **[12]**. The proofs rely heavily on the structure of Agler decompositions, as given in Lemma 5.2 and other results.

THEOREM 5.3. *Let $\Theta = \frac{\tilde{p}}{p}$ be rational inner of degree (m,n) and let $\mathcal{S}_2$ be as in* (11). *There is an $m \times m$ matrix-valued function M_Θ with entries continuous on $\overline{\mathbb{D}}$ and rational in $\overline{z_2}$ such that S^1_Θ is unitarily equivalent to T_{M_Θ}, the matrix-valued Toeplitz operator with symbol M_Θ, defined as in* (12).

The symbol M_Θ generalizes the classical matrix associated to a compressed shift. Indeed, if $\Theta(z) = B(z_1)$ is a one-variable finite Blaschke product, then

$$M_\Theta = \text{ the constant matrix of } S_B \text{ on } K_B \text{ given in (7).}$$

In some more complicated situations, we can still compute M_Θ. For example, let $\Theta(z) = \left(\frac{2z_1z_2 - z_1 - z_2}{2 - z_1 - z_2}\right)\left(\frac{3z_1z_2 - z_1 - 2z_2}{3 - 2z_1 - z_2}\right)$. Then an application of [**12**, Theorem 4.4] gives

$$M_\Theta(z_2) = \begin{bmatrix} \dfrac{1}{2 - \bar{z}_2} & 0 \\ \\ \dfrac{-\sqrt{6}(1 - \bar{z}_2)^2}{(2 - \bar{z}_2)(3 - \bar{z}_2)} & \dfrac{2}{3 - \bar{z}_2} \end{bmatrix}.$$

As in this example, Theorem 4.4 from [**12**] gives lower triangular matrices, rather than upper triangular ones like in (7), because the proof orders the basis elements differently than in the classical one-variable setup. This relationship between S^1_Θ and the Toeplitz operator with symbol M_Θ gives information about the numerical range. Specifically,

THEOREM 5.4. *Let $\Theta = \frac{\tilde{p}}{p}$ be rational inner of degree (m,n), let $\mathcal{S}_2$ be as in* (11), *and let M_Θ be as in Theorem* 5.3. *Then $\overline{W(S^1_\Theta)}$ = the convex hull of $\left(\bigcup_{\tau \in \mathbb{T}} W(M_\Theta(\tau))\right)$.*

For example, this means that if $\Theta(z) = \left(\frac{2z_1z_2 - z_1 - z_2}{2 - z_1 - z_2}\right)\left(\frac{3z_1z_2 - z_1 - 2z_2}{3 - 2z_1 - z_2}\right)$, then by the elliptical range theorem, the closure of $W(S^1_\Theta)$ is the convex hull of a union of elliptical disks. Moreover, Theorem 5.4 allows us to connect the study of $W(S^1_\Theta)$ to the one-variable setting as follows. Recall that the notation in the following theorem was defined after Lemma 5.2.

THEOREM 5.5. *Let $\Theta = \frac{\tilde{p}}{p}$ be rational inner of degree (m,n) and let $\mathcal{S}_2$ be as in* (11). *Then $\overline{W(S^1_\Theta)}$ = the closure of the convex hull of $\left(\bigcup_{\tau \in \mathbb{T}\setminus E_\Theta} W\left(S_{\theta_\tau} \text{ on } K_{\theta_\tau}\right)\right)$.*

Thus, $\overline{W(S^1_\Theta)}$ can also be obtained using the numerical ranges of one-variable compressions of the shift associated to Θ. As this formula no longer involves $\mathcal{S}_2$, it implies the following:

COROLLARY 5.6. *Let $\Theta = \frac{\tilde{p}}{p}$ be rational inner of degree (m,n) and let $\mathcal{S}_2, \tilde{\mathcal{S}}_2$ be as in* (11). *Then $\overline{W(P_{\mathcal{S}_2} S_{z_1}|_{\mathcal{S}_2})} = \overline{W(P_{\tilde{\mathcal{S}}_2} S_{z_1}|_{\tilde{\mathcal{S}}_2})}$.*

This is important because, in general, Agler decompositions are not unique and one would expect that $W(S^1_\Theta)$ would depend heavily on the choice of $\mathcal{S}_2$. Finally, one can use the connection to one-variable compressions of the shift to characterize when S^1_Θ has maximal numerical radius $w(S^1_\Theta)$, where the numerical radius is $\sup\{|z| : z \in W(S^1_\Theta)\}$. Then:

COROLLARY 5.7. *Let $\Theta = \frac{\tilde{p}}{p}$ be rational inner of degree (m,n) and let $\mathcal{S}_2$ be as in* (11). *Then $w\left(S^1_\Theta\right) = 1$ if and only if Θ has a singularity on $\mathbb{T}^2$.*

This indicates, for example, that many $\partial W(S^1_\Theta)$ *cannot* satisfy a Poncelet property because they touch $\mathbb{T}$.

To say more about the geometry of $W(S^1_\Theta)$, we restrict to very simple rational inner functions, namely $\Theta = \theta_1^2$ where $\theta_1 = \frac{\tilde{p}}{p}$ with $p(z) = a - z_1 + cz_2$ with $a, c > 0$ and

$p(1,-1)=0$. Then $\deg\Theta=(2,2)$ and so each $M_\Theta(\tau)$ is a 2×2 matrix. Indeed, each $W(M_\Theta(\tau))$ is a circular disk and so the numerical range looks like the convex hull of this:

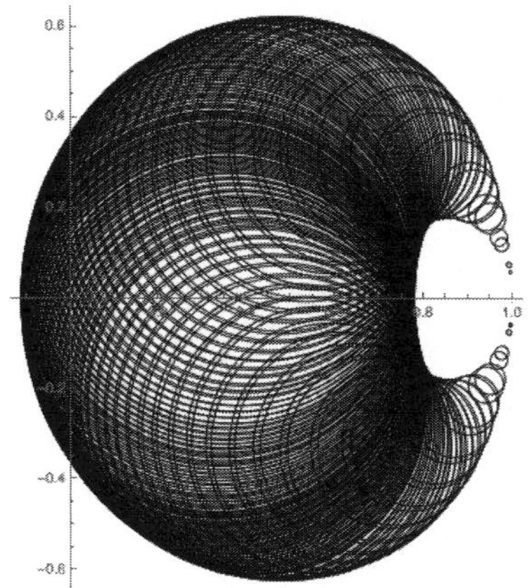

FIGURE 5.1. A collection of boundaries of $W(M_\Theta(\tau))$ associated to $\Theta(z)=\left(\frac{2z_1z_2+z_1-z_2}{2-z_1+z_2}\right)^2$.

In this case, taking the convex hull merely fills in the hole in this set. So, the boundary of $W(S^1_\Theta)$ is precisely the outer boundary of the family of disks $W(M_\Theta(\tau))$. This should bring to mind envelopes. Indeed, in this case, one can use the discriminant envelope from Definition 2.5 to obtain formulas for the boundaries of these numerical ranges:

THEOREM 5.8. *For $\Theta=\theta_1^2$ given above, the boundary of $W(S^1_\Theta)$ is the curve $E=(x(t),y(t))$ for $t\in[0,2\pi)$, where*

$$x(t)=\frac{a+c\cos t}{a+c}+\frac{ac(1-\cos t)}{(a+c)^2}\cos\left(t-\arcsin\left(\frac{a}{a+c}\sin t\right)\right);$$
$$y(t)=\frac{c\sin t}{a+c}+\frac{ac(1-\cos t)}{(a+c)^2}\sin\left(t-\arcsin\left(\frac{a}{a+c}\sin t\right)\right).$$

For more details and additional geometric results about $W(S^1_\Theta)$, see **[12]**.

6. Open Questions

In **[21]** M. Crouzeix stated the following conjecture:

Conjecture (2004): There exists a constant C such that for any polynomial $p\in\mathbb{C}[z]$ and A an $n\times n$ matrix, the inequality holds:

$$\|p(A)\|\le C\max|p(z)|_{z\in W(A)}.$$

The best constant should be $C=2$.

First let us see why the constant must be at least 2. Let $p(z)=z$ and $A=\begin{bmatrix}0&1\\0&0\end{bmatrix}$. Then $\|p(A)\|=\|A\|=1$ and $\max|p(z)|_{z\in W(A)}=\max_{\{z:|z|\le 1/2\}}|z|=1/2$. So $C\ge 2$.

Even though it is unclear that such a constant exists let alone is equal to 2, there is reason to believe the conjecture is true. Crouzeix showed that in fact such a constant does exist and is between 2 and 11.08. Okubo and Ando **[57]** showed that the conjecture is true if the numerical range is a disk. Badea, Crouzeix and Delyon presented several approaches to this problem and others in **[6]**. Recently, Glader, Kurula and Lindström **[35]** considered the problem for tridiagonal 3×3 matrices with constant diagonal; this includes 3×3 matrices with elliptical numerical range and one eigenvalue at the center of the ellipse. Choi **[17]** showed the conjecture holds for 3×3 matrices that are "nearly" Jordan blocks. And recently, Crouzeix and Palencia **[22]** showed that the best constant is between 2 and $1 + \sqrt{2}$.

Crouzeix and Palencia's proof relies on a crucial lemma, which we reproduce below. In what follows, we let Ω be a bounded open convex set with smooth boundary and let $A(\Omega)$ denote the algebra of functions continuous on $\overline{\Omega}$ and holomorphic on Ω.

LEMMA 6.1 (Crouzeix and Palencia). *Let T be a bounded Hilbert space operator and let Ω be a bounded open set containing the spectrum of T. Suppose that for each $f \in A(\Omega)$ there exists $g \in A(\Omega)$ such that*

$$\|g\|_\Omega \leq \|f\|_\Omega \text{ and } \|f(T) + g(T)^\star\| \leq 2\|f\|_\Omega.$$

Then

$$\|f(T)\| \leq (1 + \sqrt{2})\|f\|_\Omega, \ f \in A(\Omega).$$

In proving their theorem, Crouzeix and Palencia apply the lemma with

$$g = C\overline{f} \text{ where } (C\overline{f})(z) = \frac{1}{2\pi i}\int_{\partial\Omega} \frac{\overline{f(\zeta)}}{\zeta - z} d\zeta, z \in \Omega.$$

That is, g is the Cauchy transform of $\overline{f}$.

Ransford and Schwenninger **[61]** give a short proof of this and show that in this lemma, the constant $(1 + \sqrt{2})$ is sharp. However, as they point out, this is not a counterexample to the theorem; it just shows that the theorem will not be established "merely by adjusting the proof" of the lemma. Taking g to be the Cauchy transform of $\overline{f}$ and considering the map from f to g, we see that this is an antilinear map and it maps 1 to 1; we say that it is antilinear and *unital.* The example appearing in **[61]** is antilinear, but sends 1 to -1. The authors of **[61]** suggest considering the following question, as a positive answer to this would establish the Crouzeix conjecture.

QUESTION. **[61]** *Let T be a bounded Hilbert space operator and let Ω be a bounded open set containing the spectrum of T. Suppose that there exists a unital antilinear map $\alpha : A(\Omega) \to A(\Omega)$ such that for all $f \in A(\Omega)$*

$$\|\alpha(f)\|_\Omega \leq \|f\|_\Omega \text{ and } \|f(T) + (\alpha(f))(T)^\star\| \leq 2\|f\|_\Omega.$$

Does it follow that

$$\|f(T)\| \leq 2\|f\|_\Omega?$$

References

[1] J. Agler, *Geometric and topological properties of the numerical range*, Indiana Univ. Math. J. **31** (1982), no. 6, 767–777, DOI 10.1512/iumj.1982.31.31053. MR674866

[2] J. Agler, *On the representation of certain holomorphic functions defined on a polydisc*, Topics in operator theory: Ernst D. Hellinger memorial volume, Oper. Theory Adv. Appl., vol. 48, Birkhäuser, Basel, 1990, pp. 47–66. MR1207393

[3] J. Agler and J. E. McCarthy, *Pick interpolation and Hilbert function spaces*, Graduate Studies in Mathematics, vol. 44, American Mathematical Society, Providence, RI, 2002. MR1882259

[4] J. Agler, J. E. McCarthy, and M. Stankus, *Toral algebraic sets and function theory on polydisks*, J. Geom. Anal. **16** (2006), no. 4, 551–562, DOI 10.1007/BF02922130. MR2271943

[5] O. Axelsson, H. Lu, and B. Polman, *On the numerical radius of matrices and its application to iterative solution methods*, Linear and Multilinear Algebra **37** (1994), no. 1-3, 225–238, DOI 10.1080/03081089408818325. Special Issue: The numerical range and numerical radius. MR1313771

[6] C. Badea, M. Crouzeix, and B. Delyon, *Convex domains and K-spectral sets*, Math. Z. **252** (2006), no. 2, 345–365, DOI 10.1007/s00209-005-0857-y. MR2207801

[7] J. A. Ball, C. Sadosky, and V. Vinnikov, *Scattering systems with several evolutions and multidimensional input/state/output systems*, Integral Equations Operator Theory **52** (2005), no. 3, 323–393, DOI 10.1007/s00020-005-1351-y. MR2184571

[8] H. Bercovici and D. Timotin, *The numerical range of a contraction with finite defect numbers*, J. Math. Anal. Appl. **417** (2014), no. 1, 42–56, DOI 10.1016/j.jmaa.2014.02.037. MR3191411
[9] A. Beurling, *On two problems concerning linear transformations in Hilbert space*, Acta Math. **81** (1948), 17, DOI 10.1007/BF02395019. MR0027954
[10] K. Bickel, *Fundamental Agler decompositions*, Integral Equations Operator Theory **74** (2012), no. 2, 233–257, DOI 10.1007/s00020-012-1992-6. MR2983064
[11] K. Bickel and G. Knese, *Inner functions on the bidisk and associated Hilbert spaces*, J. Funct. Anal. **265** (2013), no. 11, 2753–2790, DOI 10.1016/j.jfa.2013.08.002. MR3096989
[12] K. Bickel and P. Gorkin, *Compressions of the shift on the bidisk and their numerical ranges*, J. Operator Theory **79** (2018), no. 1, 225–265. MR3764150
[13] K. Bickel, P. Gorkin, and T. Trung. *Applications of Envelopes.* In preparation.
[14] J. W. Bruce and P. J. Giblin, *What is an envelope?*, Math. Gaz. **65** (1981), no. 433, 186–192, DOI 10.2307/3617131. MR634468
[15] I. Chalendar, P. Gorkin, and J. R. Partington, *Numerical ranges of restricted shifts and unitary dilations*, Oper. Matrices **3** (2009), no. 2, 271–281, DOI 10.7153/oam-03-17. MR2522781
[16] M.-D. Choi and C.-K. Li, *Constrained unitary dilations and numerical ranges*, J. Operator Theory **46** (2001), no. 2, 435–447. MR1870416
[17] D. Choi, *A proof of Crouzeix's conjecture for a class of matrices*, Linear Algebra Appl. **438** (2013), no. 8, 3247–3257, DOI 10.1016/j.laa.2012.12.045. MR3023274
[18] D. N. Clark, *One dimensional perturbations of restricted shifts*, J. Analyse Math. **25** (1972), 169–191, DOI 10.1007/BF02790036. MR0301534
[19] B. J. Cole and J. Wermer, *Ando's theorem and sums of squares*, Indiana Univ. Math. J. **48** (1999), no. 3, 767–791, DOI 10.1512/iumj.1999.48.1716. MR1736979
[20] R. Courant, *Differential and integral calculus. Vol. II*, Wiley Classics Library, John Wiley & Sons, Inc., New York, 1988. Translated from the German by E. J. McShane; Reprint of the 1936 original; A Wiley-Interscience Publication. MR1009559
[21] M. Crouzeix, *Bounds for analytical functions of matrices*, Integral Equations Operator Theory **48** (2004), no. 4, 461–477, DOI 10.1007/s00020-002-1188-6. MR2047592
[22] M. Crouzeix and C. Palencia, *The numerical range is a* $(1+\sqrt{2})$*-spectral set*, SIAM J. Matrix Anal. Appl. **38** (2017), no. 2, 649–655, DOI 10.1137/17M1116672. MR3666309
[23] M. Crouzeix, *Numerical range and functional calculus in Hilbert space*, J. Funct. Anal. **244** (2007), no. 2, 668–690, DOI 10.1016/j.jfa.2006.10.013. MR2297040
[24] U. Daepp, P. Gorkin, and R. Mortini, *Ellipses and finite Blaschke products*, Amer. Math. Monthly **109** (2002), no. 9, 785–795, DOI 10.2307/3072367. MR1933701
[25] U. Daepp, P. Gorkin, A. Shaffer, and K. Voss, *Möbius transformations and Blaschke products: the geometric connection*, Linear Algebra Appl. **516** (2017), 186–211, DOI 10.1016/j.laa.2016.11.032. MR3589712
[26] U. Daepp, P. Gorkin, A. Shaffer, K Voss, *Finding Ellipses: What Blaschke Products, Poncelet's Theorem, and the Numerical Range Know about Each Other*, The Carus Mathematical Monographs, AMS/MAA press, **34**, 2018.
[27] W. F. Donoghue Jr., *On the numerical range of a bounded operator*, Michigan Math. J. **4** (1957), 261–263. MR0096127
[28] M. Eiermann, *Fields of values and iterative methods*, Linear Algebra Appl. **180** (1993), 167–197, DOI 10.1016/0024-3795(93)90530-2. MR1206415
[29] M. Fujimura, *Inscribed ellipses and Blaschke products*, Comput. Methods Funct. Theory **13** (2013), no. 4, 557–573, DOI 10.1007/s40315-013-0037-8. MR3138353
[30] S. R. Garcia and W. T. Ross, *Model spaces: a survey*, Invariant subspaces of the shift operator, Contemp. Math., vol. 638, Amer. Math. Soc., Providence, RI, 2015, pp. 197–245, DOI 10.1090/conm/638/12811. MR3309355
[31] H.-L. Gau and P. Y. Wu, *Numerical range and Poncelet property*, Taiwanese J. Math. **7** (2003), no. 2, 173–193, DOI 10.11650/twjm/1500575056. MR1978008
[32] H.-L. Gau and P. Y. Wu, *Numerical range of* $S(\phi)$, Linear and Multilinear Algebra **45** (1998), no. 1, 49–73, DOI 10.1080/03081089808818577. MR1665615
[33] H.-L. Gau and P. Y. Wu, *Numerical range circumscribed by two polygons*, Linear Algebra Appl. **382** (2004), 155–170, DOI 10.1016/j.laa.2003.12.003. MR2050103
[34] P. Gawron, Z. Puchała, J. A. Miszczak, Ł. Skowronek, and K. Życzkowski, *Restricted numerical range: a versatile tool in the theory of quantum information*, J. Math. Phys. **51** (2010), no. 10, 102204, 24, DOI 10.1063/1.3496901. MR2761297

[35] C. Glader, M. Kurula, and M. Lindström, *Crouzeix's conjecture holds for tridiagonal* 3×3 *matrices with elliptic numerical range centered at an eigenvalue*, SIAM J. Matrix Anal. Appl. **39** (2018), no. 1, 346–364, DOI 10.1137/17M1110663. MR3769703

[36] M. Goldberg and E. Tadmor, *On the numerical radius and its applications*, Linear Algebra Appl. **42** (1982), 263–284, DOI 10.1016/0024-3795(82)90155-0. MR656430

[37] P. Gorkin and N. Wagner, *Ellipses and compositions of finite Blaschke products*, J. Math. Anal. Appl. **445** (2017), no. 2, 1354–1366, DOI 10.1016/j.jmaa.2016.01.067. MR3545246

[38] P. Griffiths and J. Harris, *On Cayley's explicit solution to Poncelet's porism*, Enseign. Math. (2) **24** (1978), no. 1-2, 31–40. MR497281

[39] K. E. Gustafson and D. K. M. Rao, *Numerical range*, Universitext, Springer-Verlag, New York, 1997. The field of values of linear operators and matrices. MR1417493

[40] U. Haagerup and P. de la Harpe, *The numerical radius of a nilpotent operator on a Hilbert space*, Proc. Amer. Math. Soc. **115** (1992), no. 2, 371–379, DOI 10.2307/2159255. MR1072339

[41] L. Halbeisen and N. Hungerbühler, *A simple proof of Poncelet's theorem (on the occasion of its bicentennial)*, Amer. Math. Monthly **122** (2015), no. 6, 537–551, DOI 10.4169/amer.math.monthly.122.6.537. MR3361732

[42] P. R. Halmos, *A Hilbert space problem book*, 2nd ed., Graduate Texts in Mathematics, vol. 19, Springer-Verlag, New York-Berlin, 1982. Encyclopedia of Mathematics and its Applications, 17. MR675952

[43] P. R. Halmos, *Numerical ranges and normal dilations*, Acta Sci. Math. (Szeged) **25** (1964), 1–5. MR0171168

[44] F. Hausdorff, *Der Wertvorrat einer Bilinearform* (German), Math. Z. **3** (1919), no. 1, 314–316, DOI 10.1007/BF01292610. MR1544350

[45] S. Hildebrandt, *Über den numerischen Wertebereich eines Operators* (German), Math. Ann. **163** (1966), 230–247, DOI 10.1007/BF02052287. MR0200725

[46] R. A. Horn and C. R. Johnson, *Topics in matrix analysis*, Cambridge University Press, Cambridge, 1994. Corrected reprint of the 1991 original. MR1288752

[47] D. Kalman, *Solving the ladder problem on the back of an envelope*, Math. Mag. **80** (2007), no. 3, 163–182, DOI 10.1080/0025570x.2007.11953477. MR2322082

[48] R. Kippenhahn, *On the numerical range of a matrix*, Linear Multilinear Algebra **56** (2008), no. 1-2, 185–225, DOI 10.1080/03081080701553768. Translated from the German by Paul F. Zachlin and Michiel E. Hochstenbach [MR0059242]. MR2378310

[49] G. Knese, *Rational inner functions in the Schur-Agler class of the polydisk*, Publ. Mat. **55** (2011), no. 2, 343–357, DOI 10.5565/PUBLMAT_55211_04. MR2839446

[50] D. W. Kribs, A. Pasieka, M. Laforest, C. Ryan, and M. P. da Silva, *Research problems on numerical ranges in quantum computing*, Linear Multilinear Algebra **57** (2009), no. 5, 491–502, DOI 10.1080/03081080802677441. MR2532418

[51] C.-K. Li and Y.-T. Poon, *Generalized numerical ranges and quantum error correction*, J. Operator Theory **66** (2011), no. 2, 335–351. MR2844468

[52] C.-K. Li, *A simple proof of the elliptical range theorem*, Proc. Amer. Math. Soc. **124** (1996), no. 7, 1985–1986, DOI 10.1090/S0002-9939-96-03307-2. MR1322932

[53] F. Maclachlan, Long-Run and Short-Run Cost Curves,*Famous figures and diagrams in economics.* In Blaug, M., Lloyd, P. eds., Edward Elgar Publishing, (2010).

[54] J. Maroulas, P. J. Psarrakos, and M. J. Tsatsomeros, *Perron-Frobenius type results on the numerical range*, Linear Algebra Appl. **348** (2002), 49–62, DOI 10.1016/S0024-3795(01)00574-2. MR1902115

[55] B. Mirman, *Numerical ranges and Poncelet curves*, Linear Algebra Appl. **281** (1998), no. 1-3, 59–85, DOI 10.1016/S0024-3795(98)10037-X. MR1645335

[56] F.D. Murnaghan, *On the field of values of a square matrix.* Proc. Natl. Acad. Sci. 18 (1932), no. 3, 246–248.

[57] K. Okubo and T. Ando, *Constants related to operators of class* C_ρ, Manuscripta Math. **16** (1975), no. 4, 385–394, DOI 10.1007/BF01323467. MR0377562

[58] J.-V. Poncelet, *Traité des propriétés projectives des figures. Tome I* (French), Les Grands Classiques Gauthier-Villars. [Gauthier-Villars Great Classics], Éditions Jacques Gabay, Sceaux, 1995. Reprint of the second (1865) edition. MR1399774

[59] H. Pottman, M. Peternell, *Envelopes—computational theory and applications.* Proceedings of Spring Conference in Computer Graphics, (2000) Budmerice, Slovakia, pp. 3–23.

[60] P. J. Psarrakos and M. J. Tsatsomeros, *On the stability radius of matrix polynomials*, Linear Multilinear Algebra **50** (2002), no. 2, 151–165, DOI 10.1080/03081080290019577. MR1892837

[61] T. Ransford and F. L. Schwenninger, *Remarks on the Crouzeix-Palencia proof that the numerical range is a* $(1+\sqrt{2})$*-spectral set*, SIAM J. Matrix Anal. Appl. **39** (2018), no. 1, 342–345, DOI 10.1137/17M1143757. MR3769702

[62] W. Rudin, *Function theory in polydiscs*, W. A. Benjamin, Inc., New York-Amsterdam, 1969. MR0255841

[63] D. Sarason, *Sub-Hardy Hilbert spaces in the unit disk*, University of Arkansas Lecture Notes in the Mathematical Sciences, vol. 10, John Wiley & Sons, Inc., New York, 1994. A Wiley-Interscience Publication. MR1289670

[64] B. Sz.-Nagy, C. Foias, H. Bercovici, and L. Kérchy, *Harmonic analysis of operators on Hilbert space*, Revised and enlarged edition, Universitext, Springer, New York, 2010. MR2760647

[65] J. W. Rutter, *Geometry of curves*, Chapman & Hall/CRC Mathematics, Chapman & Hall/CRC, Boca Raton, FL, 2000. MR1750083

[66] M. H. Stone, *Linear transformations in Hilbert space*, American Mathematical Society Colloquium Publications, vol. 15, American Mathematical Society, Providence, RI, 1990. Reprint of the 1932 original. MR1451877

[67] O. Toeplitz, *Das algebraische Analogon zu einem Satze von Fejér* (German), Math. Z. **2** (1918), no. 1-2, 187–197, DOI 10.1007/BF01212904. MR1544315

[68] H. J. Woerdeman, *A general Christoffel-Darboux type formula*, Integral Equations Operator Theory **67** (2010), no. 2, 203–213, DOI 10.1007/s00020-010-1776-9. MR2650771

[69] P. Y. Wu, *A numerical range characterization of Jordan blocks*, Linear and Multilinear Algebra **43** (1998), no. 4, 351–361, DOI 10.1080/03081089808818536. MR1616464

Kelly Bickel, Department of Mathematics, Bucknell University, 380 Olin Science Building, Lewisburg, PA 17837, USA.
Email address: kelly.bickel@bucknell.edu

Pamela Gorkin, Department of Mathematics, Bucknell University, 380 Olin Science Building, Lewisburg, PA 17837, USA.
Email address: pgorkin@bucknell.edu

Contemporary Mathematics
Volume **743**, 2020
https://doi.org/10.1090/conm/743/14965

On the asymptotics of n-times integrated semigroups

José E. Galé, María M. Martínez, and Pedro J. Miana

ABSTRACT. We discuss the behaviour at infinity of n-times integrated semigroups with nonquasianalytic growth. The results obtained provide in this setting extensions of the Arendt-Batty-Lyubich-Vũ theorem on stability of C_0-semigroups and of a theorem of El Mennaoui on stability of bounded once integrated semigroups.

1. Introduction

The study of the asymptotic behaviour at infinity of discrete and continuous semigroups is a well established topic in operator theory. In this setting, peripheral spectral conditions on the semigroup are of significant importance. As a matter of fact, the following results hold true. Assume that a bounded operator T on a Banach space X, with spectrum $\sigma(T)$, is power-bounded, which is to say $\sup_{n\geq 1}\|T^n\| < \infty$. Gelfand proved that if T is invertible, T and T^{-1} are power-bounded, and $\sigma(T) = \{1\}$ then T is the identity operator I; see [**12**]. This was extended by Esterle, who showed that for every power-bounded T such that $\sigma(T) = \{1\}$ one has $\|T^n(I-T)\| \to 0$ as $n \to \infty$; see [**8**]. Further, Katznelson and Tzafriri proved that in order to arrive at the property $\lim_{n\to\infty}\|T^n(I-T)\| = 0$ is enough to assume $\sigma(T)\cap\mathbb{T} \subset \{1\}$ where $\mathbb{T}$ is the unit circle of complex numbers; see [**13**]. Other proofs of the Katznelson-Tzafriri theorem were given by Allan, O'Farrell and Ransford in [**1**], Allan and Ransford in [**2**], and also in [**16**] where, to complete the picture, Vũ did notice that the Katznelson-Tzafriri theorem can be reduced to the Esterle's theorem. Of course, it is interesting to relax the growth conditions on T. In this direction, T is said to be power-dominated if there exists a sequence of positive numbers μ_n $(n \geq 0)$, with $\lim_{n\to\infty}\mu_n^{-1}\mu_{n+1} = 1$, such that $\|T^n\| \leq \mu_n$ $(n \geq 0)$. In [**2**], Allan and Ransford revealed the interest of that growth condition by proving its equivalence with the spectral condition $\sigma(T) \subset \overline{\mathbb{D}}$, where $\mathbb{D}$ is the (open) unit disc. They also established some quantitative generalizations of the Katznelson-Tzafriri theorem for power-dominated operators.

The above results apply to monothetic semigroups. As regards the continuous semigroup case, let us first remark that Katznelson and Tzafriri proved in fact the

---

2010 *Mathematics Subject Classification.* Primary 47D62; Secondary 47A35.

Key words and phrases. n-times integrated semigroup; asymptotic behaviour, stability, nonquasianalytic weight.

The research has been partially supported by Project MTM2013-42105-P, Project MTM2016-7710-P (DGI-FEDER) of the MEYC and Project E26_17R, fondos FEDER, D.G. Aragón.

fairly general result that if T is a power-bounded operator and f is an analytic function of spectral synthesis with respect to its peripheral spectrum, lying in the Wiener algebra on the unit disc, then $\lim_{n\to\infty} \|T^n f(T)\| = 0$. Here $f(T)$ is given by the usual functional calculus associated with T, see [**13**]. The analogous version of the above theorem for bounded C_0-semigroups of operators was obtained by J. Esterle, E. Strouse and F. Zouakia in [**9**, Théorème 3.4] and independently by Q. P. Vũ in [**17**, Theorem 3.2]. In such a continuous version T^n is replaced by $T(t)$, $t > 0$, and $f(T)$ is given by the corresponding calculus for $T(t)$ associated with f in $L^1(\mathbb{R}^+)$. This result provides convergence in the norm topology, as in the discrete semigroup case quoted above. There are other significant results about the asymptotic behaviour of C_0-semigroups that are referred to the strong topology, which is to say that they are given in terms of orbits.

Recall that if A is a general closed operator on a Banach space X with dense domain $D(A)$ the abstract Cauchy problem of first order for A is the differential equation with initial value x

$$u'(t) = Au(t),\ t \geq 0; \quad u(0) = x \in X. \tag{1.1}$$

When A satisfies the Hille-Yosida condition, then A is the infinitesimal generator of a C_0-semigroup $T_0(t) = e^{tA}$ and the solution $u\colon [0,\infty) \to D(A) \subseteq X$ of (1) is given by $u(t) = T_0(t)x$. Thus the study of the limit $\lim_{t\to\infty} T_0(t)x$ reflects the asymptotic behavior of the solution u of (1.1) at infinity. In this respect, a bounded C_0-semigroup $(T_0(t))_{t\geq 0}$ is said to be stable if $\lim_{t\to\infty} T_0(t)x = 0$, $x \in X$. The stability of a bounded C_0-semigroup $T(t) = e^{tA}$ on a Banach space was established in [**3**] and [**14**] -under certain peripheral spectral assumptions on A- with different proofs and independently one from each other. We refer to this stability result as the Arendt-Batty-Lyubich-Vũ theorem.

Similarly to the discrete case, the boundedness of the semigroup can also be replaced with a less restrictive condition. Namely, the Arendt-Batty-Lyubich-Vũ theorem was extended by Vũ to semigroups of nonquasianalytic growth, in due form; see [**18**] and the next section. For more information on asymptotic behaviour and stability of operator semigroups we refer the reader to [**4**], [**6**] and [**15**].

Not all closed operators A generate C_0-semigroups, but there is a class of them which is still important in the study of abstract Cauchy equations. It is formed by the generators of families -of bounded operators- so called *n-times integrated semigroups*, for $n \geq 0$ (see next section for the definition and some properties). Let $T_n(t)$ denote such a family, with generator A, where t runs over $[0,\infty)$. Then, as before, the asymptotic behaviour of the solution u of the Cauchy equation for A should be described in terms of the limit $\lim_{t\to\infty} T_n(t)x$, or ergodic versions of it. However, it is still not entirely clear what should be an accurate version of the Arendt-Batty-Lyubich-Vũ theorem in the setting of integrated semigroups. In [**7**], El Mennaoui establishes a partial result like the above one, for uniformly bounded once integrated semigroups with *invertible* generators.

In the present note, we extend El Mennaoui's theorem to integrated semigroups with non-quasianalytic growth and generator not necessarily invertible. For this, we combine results and arguments of [**7**] and [**18**]. Doing so, we in fact find an extension of the Arendt-Batty-Lyubich-Vũ theorem -and its extension by Vũ in [**17**]- to integrated semigroups.

The organization of the paper is simple. After the introduction just written above, Section 2 is devoted to definitions and a minimum of properties which seem necessary in order to understand the subject. We also state in Section 2 the main theorem and its corollaries. Section 3 contains the proofs.

2. Preliminaries and statement of the main theorem

Let X be a Banach space and let $\mathcal{B}(X)$ be the Banach algebra of bounded operators on X. Let n be a nonnegative integer. A closed operator on X with domain $D(A)$ is said to be the generator of an exponentially bounded *n-times integrated semigroup* if there exist a family $(T_n(t))_{t\geq 0}$ in $\mathcal{B}(X)$ and constants $w \in \mathbb{R}$, $C \geq 0$ for which $\|T_n(t)\| \leq Ce^{wt}$ $(t \geq 0)$ such that

$$(\lambda - A)^{-1}x = \lambda^n \int_0^\infty e^{-t\lambda} T_n(t)x\, dt, \quad \Re\lambda > w, \quad x \in X.$$

When $n = 0$ the family $T_0(t)$ is a C_0-semigroup and A is its infinitesimal generator; see [**4**, Sections 3.2 and 8.3], [**5**] and references therein. Let $\sigma(A)$, $\sigma_P(A^*)$ denote the spectrum of A and the point spectrum of the adjoint operator of A, respectively. Also, we denote by $\rho(A)$ the resolvent set of A.

A positive measurable locally bounded function $\omega(t)$ with domain $\mathbb{R}$ or $[0, \infty)$ is called a weight if $\omega(t) \geq 1$ and $\omega(s+t) \leq \omega(s)\omega(t)$ for all s, t in its domain. A weight ω on $[0, \infty)$ is called nonquasianalytic if

$$\int_0^\infty \frac{\log \omega(t)}{t^2+1} dt < \infty.$$

We assume additionally, as in [**18**], that $\liminf_{t\to\infty} \omega(t)^{-1}\omega(s+t) \geq 1$ for all $s > 0$. Then one can define the function $\widetilde{\omega}$ on $\mathbb{R}$ given by

$$\widetilde{\omega}(s) := \limsup_{t\to\infty} \frac{\omega(t+s)}{\omega(t)}, \text{ if } s \geq 0; \quad \widetilde{\omega}(s) := 1, \text{ if } s < 0.$$

Clearly, $\widetilde{\omega}$ is a weight function and $\widetilde{\omega}(t) \leq \omega(t)$ for every $t \geq 0$.

Recall that a bounded C_0-semigroup $(T_0(t))_{t\geq 0}$ is said to be stable if $\lim_{t\to\infty} T_0(t)x = 0$ for all $x \in X$. The following is Vũ's extension in [**18**] of the original Arendt-Batty-Lyubich-Vũ theorem on stability (for $\omega(t) \equiv 1$) in [**3**], [**14**].

THEOREM 2.1. *Let $T_0(t)$ be a C_0-semigroup on a Banach space X with generator A such that $\sup_{t\geq 1} \omega(t)^{-1}\|T_0(t)\| < \infty$ for a nonquasianalytic weight $\omega\colon [0, \infty) \to [1, \infty)$ for which $\widetilde{\omega}(t) = O(t^k)$ as $t \to \infty$, for some $k \geq 0$. Assume that $\sigma(A) \cap i\mathbb{R}$ is countable and $\sigma_P(A^*) \cap i\mathbb{R} = \emptyset$. Then*

$$\lim_{t\to\infty} \omega(t)^{-1}T_0(t)x = 0 \quad (x \in X).$$

It seems interesting to obtain stability type results for n-times integrated semigroups. In this direction, a general once integrated semigroup $(T_1(t))_{t\geq 0}$ is called stable in [**7**, p. 363] when $\lim_{t\to\infty} T_1(t)x$ exists for every $x \in \overline{D(A)}$. Then it is shown in [**7**, Prop. 5.1] that if a once integrated semigroup $T_1(t)$ is stable in the above sense then A must be invertible. The following result is proved by O. El Mennaoui in [**7**, Theorem 5.6]. It gives a version of the Arendt-Batty-Lyubich-Vũ theorem for once integrated semigroups.

THEOREM 2.2. *Let A be the generator of a once integrated semigroup $(T_1(t))_{t\geq 0}$ such that $\sup_{t>0} \|T_1(t)\| < +\infty$. Assume that A is invertible, $\sigma(A) \cap i\mathbb{R}$ is countable and $\sigma_P(A^*) \cap i\mathbb{R} = \emptyset$. Then there exists $\lim_{t\to\infty} T_1(t)x$ for every $x \in \overline{D(A)}$, with*

$$\lim_{t\to\infty} T_1(t)x = -A^{-1}x \quad (x \in \overline{D(A)}).$$

The purpose of this note is to extend Theorem 2.1 and Theorem 2.2 to n-times integrated semigroups with nonquasianalytic growth, for every $n \in \mathbb{N}$, in a suitable way. We do not discuss here what could be the most accurate notion of stability for integrated semigroups, but wish to remark the following fact. For a general bounded C_0-semigroup $(T_0(t))_{t\geq 0}$ its induced n-times integrated semigroup is defined by

$$T_n(t)x = \frac{1}{(n-1)!}\int_0^t (t-s)^{n-1} T_0(s)x\, ds, \qquad t > 0,\, x \in X.$$

Then the derived estimate on $(T_n(t))_{t\geq 0}$ which is to be expected from the integral expression is $\sup_{t>0} t^{-n}\|T_n(t)\| < +\infty$ (so that for $n = 1$ it is $\sup_{t>0} t^{-1}\|T_1(t)\| < +\infty$ instead of boundedness). Under such a condition a suitable behaviour of a n-integrated semigroup $T_n(t)$ at infinity would be $\lim_{t\to\infty} t^{-n}T_n(t)x = 0$, $x \in X$.

Our main result is as follows. In the statement, and throughout the paper, the symbol " $\sim$ " in $a(t) \sim b(t)$ as $t \to \infty$ means that $\lim_{t\to\infty} b(t)^{-1}a(t) = c > 0$ as $t \to \infty$.

THEOREM 2.3. *Let ω be a nonquasianalytic weight on $[0,\infty)$ such that $\widetilde{\omega}$ is of polynomial growth at infinity. For a fixed natural number n let $(T_n(t))_{t\geq 0}$ be a n-times integrated semigroup on a Banach space X, with generator A, such that $\sup\{\omega(t)^{-1}\|T_n(t)\| : t \geq 1\} < +\infty$.*

Assume in addition that $\sigma(A) \cap i\mathbb{R}$ is countable, $\sigma_P(A^*) \cap i\mathbb{R} = \emptyset$. For every $\mu > 0$, we have:

(i) *If $\omega(t)^{-1} = o(t^{-n+1})$ as $t \to \infty$, then*

$$\lim_{t\to\infty} \omega(t)^{-1}T_n(t)A^n(\mu - A)^{-2n}x = 0, \quad x \in X.$$

(ii) *If $\omega(t) \sim t^{n-1}$ as $t \to \infty$, then for all $x \in X$,*

$$\lim_{t\to\infty} t^{-n+1}T_n(t)A^n(\mu - A)^{-2n}x = -\frac{1}{(n-1)!}A^{n-1}(\mu - A)^{-2n}x.$$

From the above result, we will obtain extensions of Theorem 2.1 and Theorem 2.2; namely, see Corollary 3.2 and Corollary 3.3.

3. Proofs

In order to establish Theorem 2.3 one needs to extend [**18**, Theorem 7] and [**7**, Theorem 5.6]. Thus Proposition 3.1 below is a slight improvement of [**18**, Theorem 7], which is in turn an extension of the Arendt-Batty-Lyubich-Vũ theorem. The case $\beta(t) \equiv 1$ in Proposition 3.1 appears in [**3**, Remark 3.3].

PROPOSITION 3.1. *Let Y be a Banach space and let $(U(t))_{t\geq 0} \subset \mathcal{B}(Y)$ be a C_0-semigroup with generator L. Let β be a nonquasianalytic weight on $[0,\infty)$ such that $\widetilde{\beta}(t) = O(t^k)$ as $t \to \infty$, for some $k \geq 0$. Assume that there exists $W \in \mathcal{B}(Y)$ such that $U(t)W = WU(t)$ for all $t \geq 0$ and $\|U(t)W\| \leq \beta(t)$ for $t \geq 0$. If $\sigma(L) \cap i\mathbb{R}$ is countable and $\sigma_P(L^*) \cap i\mathbb{R} = \emptyset$ then*

$$\lim_{t\to\infty} \frac{1}{\beta(t)}U(t)Wy = 0, \quad y \in Y.$$

PROOF. The overall argument goes along similar lines as in [**18**, Theorem 7], lemmata included, suitably adapted to the present situation.

Put $q(y) := \limsup_{t\to\infty} \beta(t)^{-1}\|U(t)Wy\|$, $y \in Y$. Then q is a seminorm on Y such that $q(y) \le \|y\|$ for all $y \in Y$. Moreover, $q(U(s)y) \le \widetilde{\beta}(s)q(y)$ for every $s \ge 0$, $y \in Y$, and so $N := \{y \in Y : q(y) = 0\}$ is a $U(t)$-invariant closed subspace of Y. Hence one can define a norm $\widehat{q}$ and an operator $\widehat{U}(t)$ on Y/N given for $y \in Y, t \ge 0$ by $\widehat{q}(\pi(y)) := q(y)$ and $\widehat{U}(t)(\pi(y)) := \pi(U(t)y)$, respectively, where π is the projection $Y \to Y/N$. It is readily seen that $(\widehat{U}(t))_{t\ge 0}$ is a strongly continuous semigroup in the norm $\widehat{q}$ on Y/N, such that $\widehat{q}(\widehat{U}(t)\pi(y)) \ge \widehat{q}(\pi(y))$ for every $y \in Y$ and $t \ge 0$. Let $(Z, \|\cdot\|_Z)$ be the $\widehat{q}$-completion of Y/N, and let $V(t)$ be the continuous extension on Z of $\widehat{U}(t)$ for all $t > 0$. Then:

(a) $\|\pi(y)\|_Z = \limsup_{t\to\infty} \frac{1}{\beta(t)}\|U(t)Wy\|$ for $y \in Y$. This is obvious.

(b) $\|V(t)\|_{Z\to Z} \le \widetilde{\beta}(t)$, $t \ge 0$. This follows by continuity and density from the estimate $\widehat{q}(\widehat{U}(t)\pi(y)) \le \widetilde{\beta}(t)q(y) \le \widetilde{\beta}(t)\widehat{q}(\pi(y))$, for every $y \in Y$, $t \ge 0$. Then one easily obtains that $(V(t))_{t>0}$ is a C_0-semigroup in $\mathcal{B}(Z)$.

(c) $\|V(t)z\|_Z \ge \|z\|_Z$ for all $z \in Z$: It follows also by continuity and density since for $y \in Y$ and $t \ge 0$,
$$\widehat{q}(\widehat{U}(t)\pi(y)) = \limsup_{s\to\infty} \frac{\beta(t+s)}{\beta(t)} \frac{\|U(t+s)Wy\|_Y}{\beta(t+s)} \ge \widehat{q}(\pi(y)).$$

(d) $V(t)\circ\pi = \pi\circ U(t)$ $(t \ge 0)$ and then one obtains that $\pi(D(L)) \subseteq D(H)$ and $H\circ\pi = \pi\circ L$ on $D(L)$, where H is the infinitesimal generator of $(V(t))_{t\ge 0}$.

(e) $\sigma(H) \subseteq \sigma(L)$: By hypothesis, $(U(t))_{t\ge 0}$ is of finite exponential type τ whence
$$R(\lambda, L)y := -(\lambda - L)^{-1}y = -\int_0^\infty e^{-\lambda t}U(t)y\,dt,$$
for $y \in Y$ and $\lambda \in \mathbb{C}$, $\Re\lambda > \delta > \min\{\tau, 0\}$.

Similarly, $(V(t))_{t\ge 0}$ is of exponential type 0 by (b), and therefore we have for $z \in Z$ and $\lambda \in \mathbb{C}$, $\Re\lambda > 0$,
$$R(\lambda, H)z := -(\lambda - H)^{-1}z = -\int_0^\infty e^{-\lambda t}V(t)z\,dt.$$
On the other hand since W commutes with $U(t)$, $t \ge 0$, one has that W commutes with $R(\lambda, L)$ for $\Re\lambda > \delta$. Then $q(R(\lambda, L)y) \le \|R(\lambda, L)\|q(y)$ for all $y \in Y$, which implies that N is $R(\lambda, L)$-invariant. Hence one can define the bounded operator $\widehat{R}(\lambda, L)$ on Z given by $\widehat{R}(\lambda, L)(\pi(y)) := \pi\left(R(\lambda, L)y\right)$, $y \in Y$. Thus,
$$\begin{aligned}\widehat{R}(\lambda, L)\pi(y) &= \pi\left(R(\lambda, L)y\right)\\ &= -\int_0^\infty e^{-\lambda t}\pi\left(U(t)y\right)dt\\ &= -\int_0^\infty e^{-\lambda t}V(t)\pi(y)dt = R(\lambda, H)\pi(y)\end{aligned}$$
where (d) has been applied in the last but one equality. Hence $\widehat{R}(\lambda, L) = R(\lambda, H)$, for $\Re\lambda > \delta$. Now, for $\Re\lambda > \delta$ and any $\mu \in \rho(L)$, by using the resolvent identity $R(\lambda, L) - R(\mu, L) = (\lambda - \mu)R(\lambda, L)R(\mu, L)$ on Y and

its corresponding identity for $\widehat{R}(\lambda, L)$ and $\widehat{R}(\mu, L)$ on Z, one readily finds that there exists $R(\mu, H)$ with $R(\mu, H) = \widehat{R}(\mu, L)$, see [**18**, p. 234]. Thus $\mu \in \rho(H)$. Hence $\rho(L) \subseteq \rho(H)$ as we claimed.

(f) $\sigma_P(H^*) \subseteq \sigma_P(L^*)$. This is straightforward to see, using restrictions of functionals; see [**18**, p. 234].

Now, suppose, if possible, that $Z \neq \{0\}$. By (e), we have that $\sigma(H) \cap i\mathbb{R}$ is countable and then $i\mathbb{R} \setminus \sigma(H) \neq \emptyset$. So, by (c) and [**18**, Lemma 2], the C_0-semigroup $(V(t))_{t \geq 0}$ can be extended to a C_0-group $(\widetilde{V}(t))_{t \in \mathbb{R}}$ such that $\|\widetilde{V}(-t)\|_{Z \to Z} \leq 1$ $(t > 0)$ and $\|\widetilde{V}(t)\|_{Z \to Z} = O(t^k)$, as $t \to +\infty$. Also, $\sigma(H)$ is nonempty by (b) and [**17**, Lemma 5].

Then reasoning as in [**18**, Theorem 7] one gets $\sigma_P(H^*) \cap i\mathbb{R} \neq \emptyset$ whence $\sigma_P(L^*) \cap i\mathbb{R} \neq \emptyset$ by (f) above. This is a contradiction and so we have proved that $Z = \{0\}$. By (a) we get the statement. □

Proof of Theorem 2.3. Since ω is nonquasianalytic one has $(0, \infty) \subseteq \rho(A)$. Take $\delta \in \mathbb{R}$ such that $\mu > \delta > 0$ and for $x \in X$ define

$$\|x\|_Y := \sup_{t \geq 0} \left\| e^{-\delta t} (T_n(t) A^n (\mu - A)^{-n} x + \sum_{j=0}^{n-1} \frac{t^j}{j!} A^j (\mu - A)^{-n} x) \right\|. \tag{3.1}$$

Note that $A(\mu - A)^{-1} = -I + \mu(\mu - A)^{-1}$ is a bounded operator on X and $T_n(0) = 0$, so $\| \cdot \|_Y$ is a norm on X and there exists a constant $M_\delta > 0$ such that

$$\|(\mu - A)^{-n} x\| \leq \|x\|_Y \leq M_\delta \|x\|, \quad x \in X. \tag{3.2}$$

Let Y be the Banach space obtained as the completion of X in the norm $\| \cdot \|_Y$. As in the proof of the Extrapolation Theorem [**5**, Theorem 0.2], there exists a closed operator B on Y which generates a C_0-semigroup $(S(t))_{t \geq 0} \subset \mathcal{B}(Y)$ such that $D(B^n) \hookrightarrow X \hookrightarrow Y$ and $A = B_X$ where the operator B_X is given by the conditions $D(B_X) := \{x \in D(B) \cap X : Bx \in X\}$, $B_X(x) := B(x)$ $(x \in X)$. Moreover, $\sigma_P(B^*) \subseteq \sigma_P(A^*)$, and also $\rho(A) = \rho(B)$ with

$$(\lambda - A)^{-1} x = (\lambda - B)^{-1} x, \quad \lambda \in \rho(A) = \rho(B), \quad x \in X; \tag{3.3}$$

see [**5**, Remark 3.1].

Let $(S_n(t))_{t \geq 0}$ be the n-times integrated semigroup generated by B on Y, given by

$$S_n(t) y := \frac{1}{(n-1)!} \int_0^t (t - s)^{n-1} S(s) y \, ds, \quad y \in Y.$$

Then $S_n(t)x = T_n(t)x$ for all $x \in X$ and $t \geq 0$. To see this, note that $(T_n(t))_{t \geq 0}$ and $(S_n(t))_{t \geq 0}$ are of exponential type so one can rewrite (3.3) above in terms of the Laplace transforms of $(T_n(t))_{t \geq 0}$ and $(S_n(t))_{t \geq 0}$ respectively, for $\Re \lambda$ large enough. Then it suffices to apply the uniqueness of the Laplace transform.

From the above identification between $T_n(t)$ and $S_n(t)$, it follows by application of (3.1) to $S_n(t)x$, $x \in X$, that

$$\begin{aligned}\|S_n(t)x\|_Y &= \sup_{s\geq 0}\left\| e^{-\delta s}\left(T_n(s)\frac{A^n}{(\mu-A)^n} + \sum_{j=0}^{n-1}\frac{s^j}{j!}\frac{A^j}{(\mu-A)^n}\right)S_n(t)x\right\|_X \\ &= \sup_{s\geq 0}\left\| T_n(t)e^{-\delta s}\left(T_n(s)\frac{A^n}{(\mu-A)^n} + \sum_{j=0}^{n-1}\frac{s^j}{j!}\frac{A^j}{(\mu-A)^n}\right)x\right\| \\ &\leq \|T_n(t)\|_{Y\to Y}\|x\|_Y\end{aligned}$$

which is to say, by density, that $\|S_n(t)\|_{Y\to Y} \leq C\omega(t)$, for all $t \geq 0$ and some constant C.

Now, for $t \geq 0$, by reiteration of the well known equality

$$S(t)y - y = \int_0^t BS(s)y\,ds, \quad y \in D(B),$$

we have

$$S(t)y = S_n(t)B^n y + \sum_{j=0}^{n-1}\frac{t^j}{j!}B^j y, \quad y \in D(B^n).$$

Hence, for every $t \geq 0$ and $y \in Y$,

$$S(t)(\mu-B)^{-n}y = S_n(t)\left(\frac{B}{\mu-B}\right)^n y + \sum_{j=0}^{n-1}\frac{t^j}{j!}\left(\frac{B}{\mu-B}\right)^j(\mu-B)^{-(n-j)}y \tag{3.4}$$

and therefore there exists a constant $C_\mu > 0$ such that

$$\|S(t)(\mu-B)^{-n}\|_{Y\to Y} \leq C_\mu\omega(t), \quad t \geq 0.$$

Then, by applying Proposition 3.1 with $U(t) = S(t)$, $B = L$ and $W = (\mu - A)^{-n}$, we obtain

$$\lim_{t\to\infty}\frac{1}{\omega(t)}\|S(t)(\mu-B)^{-n}y\|_Y = 0, \quad y \in Y,$$

whence, by (3.2), (3.3) and (3.4),

$$\begin{aligned}0 &= \lim_{t\to\infty}\frac{1}{\omega(t)}\left\| T_n(t)A^n(\mu-A)^{-n}x + \sum_{j=0}^{n-1}\frac{t^j}{j!}A^j(\mu-A)^{-n}x\right\|_Y \\ &\geq \limsup_{t\to\infty}\frac{1}{\omega(t)}\left\| T_n(t)A^n(\mu-A)^{-2n}x + \sum_{j=0}^{n-1}\frac{t^j}{j!}A^j(\mu-A)^{-2n}x\right\|_X,\end{aligned}$$

for every $x \in X$.

To finish the proof, notice that the assumptions in (i) imply

$$\lim_{t\to\infty}\frac{1}{\omega(t)}\sum_{j=0}^{n-1}\frac{t^j}{j!}A^j(\mu-A)^{-2n}x = 0$$

whereas assumptions in (ii) entails

$$\lim_{t\to\infty}\frac{1}{\omega(t)}\sum_{j=0}^{n-1}\frac{t^j}{j!}A^j(\mu-A)^{-2n}x=\frac{1}{(n-1)!}A^{n-1}(\mu-A)^{-2n}x.$$

□

Let us consider explicitly the case when the generator A is invertible, which provides us with a generalization of the El Mennaoui result in [**7**, Theorem 5.6].

COROLLARY 3.2. *In the setting of Theorem* 2.3, *assume also that A is invertible. We have:*

(i) *If* $\omega(t)^{-1}=o(t^{-n+1})$ *as* $t\to\infty$, *then*

$$\lim_{t\to\infty}\omega(t)^{-1}T_n(t)z=0,\quad z\in\overline{D(A^n)}.$$

(ii) *If* $\omega(t)\sim t^{n-1}$ *as* $t\to\infty$, *then*

$$\lim_{t\to\infty}t^{-n+1}T_n(t)z=-\frac{1}{(n-1)!}A^{-1}z,\quad x\in\overline{D(A^n)}.$$

PROOF. Let $y\in D(A^n)=D(\mu-A)^n$ where $\mu\in(0,\infty)\subseteq\rho(A)$. Then there exists $x\in X$ given by $x:=(\mu A^{-1}-I)^n(\mu-A)^ny=A^{-n}(\mu-A)^{2n}y$. Thus we have that for every $y\in D(A^n)$ there is $x\in X$ such that $y=A^n(\mu-A)^{-2n}x$. So it suffices to apply Theorem 2.3 to prove (i) and (ii) of this corollary for $y\in D(A^n)$, and then, for $z\in\overline{D(A^n)}$, that $\sup_{t>0}\omega(t)^{-1}\|T_n(t)\|<\infty$. □

Note that for $n=1$, Corollary 3.2 (ii) is [**7**, Theorem 5.6].

Now we give a generalization of the Arendt-Batty-Lyubich-Vũ theorem for integrated semigroups without assuming that A is invertible.

COROLLARY 3.3. *In the setting of Theorem* 2.3, *assume part (i), that is,* $\omega(t)^{-1}=o(t^{-n+1})$ *as* $t\to\infty$, *and also that A is densely defined. Then*

$$\lim_{t\to\infty}\frac{1}{\omega(t)}T_n(t)x=0,\ \textit{for all } x\in X.$$

PROOF. Fix $\mu\in(0,\infty)\subseteq\rho(A)$. For every $j\geq 1$, one has $D(A^j)=D((\mu-A)^j)=R((\mu-A)^{-j})$, the range of $(\mu-A)^{-j}$. Hence since $D(A)$ is dense in X we have that $(\mu-A)^{-1}$ has dense image, so $(\mu-A)^{-j}$ has dense image; that is, $D(A^j)$ is also dense in X.

Take $y\in D(A^n)$. It follows from the above remark that there is $(z_j)\subset D(A^n)$ such that $\lim_{j\to\infty}z_j=(\mu-A)^ny$. Put $y_j:=(\mu-A)^{-n}z_j$. Since $A^n(\mu-A)^{-n}=(-I+\mu(\mu-A)^{-1})^n\in\mathcal{B}(X)$ one has moreover that $y_j\in D(A^{2n})$, so that $y_j=(\mu-A)^{-2n}\xi_j$ with ξ_j in X, for each j. Hence $\lim\limits_{t\to\infty}\omega(t)^{-1}T_n(t)A^ny_j=0$, for every j, by Theorem 2.3(i). On the other hand,

$$\begin{aligned}\|A^ny_j-A^ny\|&=\|A^n(\mu-A)^{-n}(\mu-A)^ny_j-A^n(\mu-A)^{-n}(\mu-A)^ny\|\\&\leq\|A^n(\mu-A)^{-n}\|\|z_j-(\mu-A)^ny\|\to 0,\ \text{as } j\to\infty,\end{aligned}$$

Altogether, we have $\lim\limits_{t\to\infty}\omega(t)^{-1}T_n(t)A^ny=0$ for every $y\in D(A^n)$. Finally, it is enough to use once again the density of $D(A^n)$ in X and the fact that $\sup_{t>0}\omega(t)^{-1}\|T_n(t)\|<\infty$ to obtain the statement. □

REMARK 3.4. Let $L^1(\mathbb{R})$, $L^1(\mathbb{R}^+)$ be the usual convolution Banach algebras on the real line $\mathbb{R}$ and the positive half-line $\mathbb{R}^+$, respectively. Suppose A is the infinitesimal generator of a uniformly bounded C_0-semigroup $(T(t))_{t\geq 0}$ on X with spectrum $\sigma(A)$. Define $\mathfrak{S}_0$ as the subspace of functions of $L^1(\mathbb{R}^+)$ which are of spectral synthesis in $L^1(\mathbb{R})$ with respect to the subset $i\sigma(A)\cap\mathbb{R}$.

Let $\Theta_0\colon L^1(\mathbb{R}^+)\to\mathfrak{B}(X)$ be the bounded Banach algebra homomorphism defined by

$$\Theta_0(f)x := \int_0^\infty f(t)\,T(t)x\,dt \quad (x\in X,\ f\in L^1(\mathbb{R}^+)).$$

Then Esterle, Strouse and Zouakia in [**9**] and Vũ in [**17**] prove, with different methods, that $\lim_{t\to\infty}\|T(t)\Theta_0(f)\|=0$ for every $f\in\mathfrak{S}_0$. Moreover, it is shown in [**9**] that, under the assumptions that $\sigma(A)\cap i\mathbb{R}$ is countable and $\sigma_P(A^*)\cap i\mathbb{R}=\emptyset$, the subspace $\pi_0(\mathfrak{S}_0)X$ is dense in X so one gets another different proof of the Arendt-Batty-Lyubich-Vũ theorem.

We now consider a n-times integrated semigroup $T_n(t)$ in $\mathcal{B}(X)$, with generator A, such that

$$\sup_{t>0} t^{-n}\|T_n(t)\|<\infty \text{ and } \lim_{t\to 0^+} n!t^{-n}T_n(t)x=x \quad (x\in X).$$

Let $\mathcal{T}^{(n)}(|t|^n)$ be the Banach space obtained as the completion of the Schwarz class in the norm $\|f\|_{(n)}:=\int_{-\infty}^{\infty}|f^{(n)}(t)|\,|t|^n\,dt$. In fact, $\mathcal{T}^{(n)}(|t|^n)$ is contained in $L^1(\mathbb{R})$ and is a convolution Banach algebra. Its character space is $\mathbb{R}$ with Gelfand transform equal to the Fourier transform $f\mapsto\widehat{f}$. Moreover, $\mathcal{T}^{(n)}(|t|^n)$ is regular on $\mathbb{R}$. An element f of $\mathcal{T}^{(n)}(|t|^n)$ is said to be of spectral synthesis with respect to a closed subset F of $\mathbb{R}$ if there exists $(f_j)\subset\mathcal{T}^{(n)}(|t|^n)$ such that $\widehat{f_j}$ vanishes in a neighbourhood of F for each j and $\lim_{j\to\infty}\|f-f_j\|_{(n)}=0$ (see [**10**]). Let $\mathcal{T}_+^{(n)}(t^n)$ be the restriction of $\mathcal{T}^{(n)}(|t|^n)$ on $(0,\infty)$, and let $\mathfrak{S}_n$ denote the subspace of functions of $\mathcal{T}_+^{(n)}(t^n)$ which are of spectral synthesis in $\mathcal{T}^{(n)}(|t|^n)$ with respect to the subset $i\sigma(A)\cap\mathbb{R}$. Define the Banach algebra homorphism $\Theta_n\colon\mathcal{T}_+^{(n)}(t^n)\to\mathcal{B}(X)$ given by

$$\Theta_n(f):=(-1)^n\int_0^\infty f^{(n)}(t)T_n(t)dt,\quad f\in\mathcal{T}_+^{(n)}(t^n).$$

Then, as an extension of the Esterle-Strouse-Vũ-Zouakia theorem, it is proven in [**10**] that $\lim_{t\to\infty}t^{-n}\|T_n(t)\Theta_n(f)\|=0$ for every $f\in\mathfrak{S}_n$.

One can ask if the argument dealt with in [**9**] to deduce the Arendt-Batty-Lyubich-Vũ theorem works for a n-times integrated semigroup $T_n(t)$ as above; that is, if $\Theta_n(\mathfrak{S}_n)X$ is dense in X whenever $\sigma(A)\cap i\mathbb{R}$ is countable and $\sigma_P(A^*)\cap i\mathbb{R}=\emptyset$. This would give us

$$\lim_{t\to\infty}t^{-n}T_n(t)x=0,\quad x\in X. \tag{3.5}$$

Unfortunately, the proof of that density is rather difficult and we only know how to obtain (3.5) when $\sigma(A)\cap i\mathbb{R}$ is finite, see [**11**]. In any case, Corollary 3.3 is more general.

Acknowledgements. The authors wish to thank the referee for a careful reading of the initial version of this paper, including a number of accurate comments which have contributed to improve the presentation of this work.

References

[1] G. R. Allan, A. G. O'Farrell, and T. J. Ransford, *A Tauberian theorem arising in operator theory*, Bull. London Math. Soc. **19** (1987), no. 6, 537–545, DOI 10.1112/blms/19.6.537. MR915430

[2] G. R. Allan and T. J. Ransford, *Power-dominated elements in a Banach algebra*, Studia Math. **94** (1989), no. 1, 63–79, DOI 10.4064/sm-94-1-63-79. MR1008239

[3] W. Arendt and C. J. K. Batty, *Tauberian theorems and stability of one-parameter semigroups*, Trans. Amer. Math. Soc. **306** (1988), no. 2, 837–852, DOI 10.2307/2000826. MR933321

[4] W. Arendt, C. J. K. Batty, M. Hieber, and F. Neubrander, *Vector-valued Laplace transforms and Cauchy problems*, Monographs in Mathematics, vol. 96, Birkhäuser Verlag, Basel, 2001. MR1886588

[5] W. Arendt, F. Neubrander, and U. Schlotterbeck, *Interpolation of semigroups and integrated semigroups*, Semigroup Forum **45** (1992), no. 1, 26–37, DOI 10.1007/BF03025746. MR1161416

[6] R. Chill and Y. Tomilov, *Stability of operator semigroups: ideas and results*, Perspectives in operator theory, Banach Center Publ., vol. 75, Polish Acad. Sci. Inst. Math., Warsaw, 2007, pp. 71–109, DOI 10.4064/bc75-0-6. MR2336713

[7] O. El-Mennaoui, *Asymptotic behaviour of integrated semigroups*, J. Comput. Appl. Math. **54** (1994), no. 3, 351–369, DOI 10.1016/0377-0427(94)90256-9. MR1321079

[8] J. Esterle, *Quasimultipliers, representations of H^∞, and the closed ideal problem for commutative Banach algebras*, Radical Banach algebras and automatic continuity (Long Beach, Calif., 1981), Lecture Notes in Math., vol. 975, Springer, Berlin, 1983, pp. 66–162, DOI 10.1007/BFb0064548. MR697579

[9] J. Esterle, E. Strouse, and F. Zouakia, *Stabilité asymptotique de certains semi-groupes d'opérateurs et idéaux primaires de $L^1(\mathbf{R}^+)$* (French), J. Operator Theory **28** (1992), no. 2, 203–227. MR1273043

[10] J. E. Galé, M. M. Martínez, and P. J. Miana, *Katznelson-Tzafriri type theorem for integrated semigroups*, J. Operator Theory **69** (2013), no. 1, 59–85, DOI 10.7900/jot.2010jul02.1960. MR3029488

[11] J. E. Galé, M. Martínez and P. J. Miana, *On spectral synthesis in convolution Sobolev algebras on the real line*, to appear in the Proc. Conf. Banach Algebras, Oulu, Finland, 2017.

[12] I. Gelfand, *Zur Theorie der Charaktere der Abelschen topologischen Gruppen* (German, with Russian summary), Rec. Math. [Mat. Sbornik] N. S. **9 (51)** (1941), 49–50. MR0004635

[13] Y. Katznelson and L. Tzafriri, *On power bounded operators*, J. Funct. Anal. **68** (1986), no. 3, 313–328, DOI 10.1016/0022-1236(86)90101-1. MR859138

[14] Yu. I. Lyubich and Vũ Quốc Phóng, *Asymptotic stability of linear differential equations in Banach spaces*, Studia Math. **88** (1988), no. 1, 37–42, DOI 10.4064/sm-88-1-37-42. MR932004

[15] J. van Neerven, *The asymptotic behaviour of semigroups of linear operators*, Operator Theory: Advances and Applications, vol. 88, Birkhäuser Verlag, Basel, 1996. MR1409370

[16] Vũ Quốc Phóng, *A short proof of the Y. Katznelson's and L. Tzafriri's theorem*, Proc. Amer. Math. Soc. **115** (1992), no. 4, 1023–1024, DOI 10.2307/2159349. MR1087468

[17] Vũ Quốc Phóng, *Theorems of Katznelson-Tzafriri type for semigroups of operators*, J. Funct. Anal. **103** (1992), no. 1, 74–84, DOI 10.1016/0022-1236(92)90135-6. MR1144683

[18] Vũ Quốc Phóng, *Semigroups with nonquasianalytic growth*, Studia Math. **104** (1993), no. 3, 229–241, DOI 10.4064/sm-104-3-229-241. MR1220663

Departamento de Matemáticas e IUMA, Universidad de Zaragoza, 50009, Zaragoza, Spain

Email address: `gale@unizar.es`

Departamento de Matemáticas e IUMA, Universidad de Zaragoza, 50009, Zaragoza, Spain

Email address: `martinezmartinezm@gmail.com`

Departamento de Matemáticas e IUMA, Universidad de Zaragoza, 50009, Zaragoza, Spain

Email address: `pjmiana@unizar.es`

Contemporary Mathematics
Volume **743**, 2020
https://doi.org/10.1090/conm/743/14966

Powers of operators: convergence and decomposition

W. Arendt and I. Chalendar

ABSTRACT. We study the asymptotic behaviour of the powers of an operator or a semigroup with respect to a triangular decomposition.

1. Powers of composition operators

In recent papers [**2, 3**], we investigated convergence of powers of composition operators.

What was surprising is the following result: whenever a composition operator T on one of those Banach spaces $X \hookrightarrow \mathrm{Hol}(\mathbb{D})$, that we investigated, has the property that $\lim_{n\to\infty} T^n f$ exists for all $f \in X$, then $(T^n)_n$ converges already *uniformly* in $\mathcal{L}(X)$. More precisely we proved the following result.

THEOREM 1.1. *Let X be one of the Banach spaces:*

- *Hardy space $H^p(\mathbb{D})$ with $1 \le p < \infty$*
- *weighted Bergman spaces A^p_β with $\beta > -1$ and $1 \le p < \infty$*
- *little Bloch space $\mathcal{B}_0$*
- *Bloch-type space $\mathcal{B}^\alpha$ with $\alpha > 0$*
- *standard weighted Bergman spaces of infinite order $H^\infty_{\nu_p}$ with $0 < p < \infty$.*

Let $\varphi : \mathbb{D} \to \mathbb{D}$ be holomorphic such that $f \circ \varphi$ for all $f \in X$. Consider the operator $T \in \mathcal{L}(X)$ given by $Tf = f \circ \varphi$. If $Pf := \lim_{n\to\infty} T^n f$ exists in X for all $f \in X$, then $\lim_{n\to\infty} \|T^n - P\| = 0$.

Of course, in general uniform convergence is much stronger than strong convergence (think of a self-adjoint operator). Theorem 1.1 expresses a special and remarkable property of composition operators.
We found a curious exception to this

"strong convergence implies uniform convergence".

However it takes place on a quotient space and not exactly on a space which is continuously embedded in $\mathrm{Hol}(\mathbb{D})$. Let us make things more precise.

Let $\widetilde{\mathcal{D}} := \mathcal{D}/\mathbb{C}\mathbf{1}_{\mathbb{D}}$ be the Dirichlet space $\mathcal{D}$ modulo the constant functions. It is a Banach space for the norm

$$\|[f]\|_{\widetilde{\mathcal{D}}} = \int_{\mathbb{D}} |f'(z)|dA(z),$$

2010 *Mathematics Subject Classification.* 30D05, 47D03, 47B33.

Key words and phrases. Asymptotic behaviour, composition operator, Banach space of holomorphic functions, triangular decomposition, C_0-semigroup of composition operators.

where $[\ ] : \mathcal{D} \to \mathcal{D}/\mathbf{1}_{\mathbb{D}}$ denotes the quotient map.

Let $\varphi \in \mathcal{D}$ be a univalent map. Then it is not difficult to check that $f \circ \varphi \in \mathcal{D}$ if $f \in \mathcal{D}$. Moreover, if $f, g \in \mathcal{D}$ and $f - g = c$ is a constant, then $f \circ \varphi - g \circ \varphi = c$. It follows that $[f \circ \varphi] = [g \circ \varphi]$ and thus

$$\widetilde{C}_\varphi([f]) = [f \circ \varphi]$$

defines a linear operator on $\widetilde{\mathcal{D}}$.

This operator may also be obtained in a different way.
Let $\mathcal{D}_0 := \{f \in \mathcal{D} : f(0) = 0\}$. Then $\mathcal{D}_0$ is a closed subspace of $\mathcal{D}$. The mapping

$$S : \mathcal{D}_0 \to \widetilde{\mathcal{D}}, \quad Sf = [f]$$

is an isometric isomorphism. Then $\widehat{C}_\varphi := S^{-1}\widetilde{C}_\varphi S$ is given by

$$(\widehat{C}_\varphi f)(z) = f(\varphi(z)) - f(\varphi(0)).$$

Now let $\varphi(z) = \frac{az+b}{cz+d}$ be a linear fractional map with $ad - bc \neq 0$. Then φ has two fixed points in $\mathbb{C} \cup \{\infty\}$. One calls φ *parabolic* if they coincide. If in addition $\varphi(\mathbb{D}) \subset \mathbb{D}$, the unique fixed point lies on the unit circle.

The spectrum of $\widetilde{C}_\varphi$ on $\mathcal{D}_0$ has been investigated in [**6**] and [**7**]. Using these results we obtained the following in [**3**].

THEOREM 1.2. *Let φ be a parabolic linear fractional self-map of $\mathbb{D}$ which is not an automorphism of $\mathbb{D}$. Then for all $f \in \mathcal{D}$,*

$$\lim_{n \to \infty} \widetilde{C}^n_\varphi([f]) = 0,$$

but $(\widetilde{C}^n_\varphi)_n$ does not converge uniformly. Moreover, the operator C_φ on $\mathcal{D}$ is not power-bounded.

The space $\widetilde{\mathcal{D}}$ is not embedded into $\mathrm{Hol}(\mathbb{D})$, so Theorem 1.2 does not give a counterexample against the phenomenon of Theorem 1.1. If we replace $\widetilde{\mathcal{D}}$ by the isometric space $\mathcal{D}_0$, then $\mathcal{D}_0$ is embedded into $\mathrm{Hol}(\mathbb{D})$, but the operator is no longer a composition operator. We continue to consider the situation of the quotient. This gives rise to the following investigation.

2. Abstract decomposition

Theorem 1.2 is interesting from an abstract point of view. Let E be a Banach space, $T \in \mathcal{L}(E)$ and $F \subset E$ a closed subspace such that $TF \subset F$. Assume that F is complemented, i.e., there exists a closed subspace G of E such that $E = F \oplus G$. Then there exist unique operators $T_2 \in \mathcal{L}(G)$ and $S \in \mathcal{L}(G, F)$ such that

$$T(y + w) = (T_1 y + Sw) + T_2 w \in F \oplus G,$$

for $y + w \in F \oplus G$, where $T_1 = T_{|F}$. In other words, we represent T by a triangular 2×2-matrix:

$$\begin{pmatrix} T_1 & S \\ 0 & T_2 \end{pmatrix} \tag{1}$$

with respect to the decomposition $E = F \oplus G$.

It is interesting to compare the asymptotic behaviour of the powers of T with the one of the powers of T_1 and T_2. Obviously if T^n converges uniformly (or strongly) as $n \to \infty$, the same is true for T_1^n and T_2^n as $n \to \infty$.

Recall that $T^n \to 0$ uniformly as $n \to \infty$ if and only if $r(T) < 1$, where $r(T)$ denotes the spectral radius of T. This in turn implies that $\|T^n\| \leq cr^n$ $(n \in \mathbb{N})$ for some $c \geq 0$, $r \in (0,1)$ as can be seen by the formula for the spectral radius.

Since $\sigma(T) = \sigma(T_1) \cup \sigma(T_2)$, it follows that $r(T) < 1$ if and only if $r(T_j) < 1$ for $j = 1, 2$. Thus $\|T^n\| \to 0$ as $n \to \infty$ if and only if $\|T_1^n\| \to 0$ and $\|T_2^n\| \to 0$ as $n \to \infty$.

If we are interested in strong convergence, the following theorem describes another situation which is less obvious.

THEOREM 2.1. *Let $T \in \mathcal{L}(E)$, F, G, T_1, T_2 and S be defined as in (1).*

(a) *If $r(T_1) < 1$ and $T_2^n \to P_2$ strongly as $n \to \infty$, then T^n converges strongly to P as $n \to \infty$, where P is given by*

$$P(y + w) = \sum_{k=0}^{\infty} T_1^k S P_2 w, \quad y \in F, w \in G.$$

(b) *If $r(T_2) < 1$ and $T_1^n \to P_1$ strongly as $n \to \infty$, then T^n converges strongly to Q as $n \to \infty$, where Q is defined by*

$$Q(y + w) = P_1 y + \sum_{k=0}^{\infty} P_1 S T_2^k w, \quad y \in F, w \in G.$$

PROOF. By (1), it follows that

$$T^n = \begin{pmatrix} T_1^n & S_n \\ 0 & T_2^n \end{pmatrix},$$

where $S_n = \sum_{k=0}^{n-1} T_1^{n-1-k} S T_2^k$. Thus we have to show that S_n converges strongly as $n \to \infty$ in the two cases.
(a) Assume that $r(T_1) < 1$ and $T_2^n \to P_2$ strongly as $n \to \infty$. Then there exist $c_1 \geq 0$ and $r \in (0,1)$ such that $\|T_1^n\| \leq c_1 r^n$. Let $w \in G$.
First case: Assume that $P_2 w = w$. Then $T_2^k w = w$ for all $k \geq 0$ and thus

$$S_n w = \sum_{k=0}^{n-1} T_1^{n-1-k} S w = \sum_{k=0}^{n-1} T_1^k S w \to \sum_{k=0}^{\infty} T_1^k S w,$$

as $n \to \infty$, since the F is a Banach space and $\sum_{k=0}^{\infty} \|T_1^k\| \|Sw\| < \infty$.
Second case: Assume that $w \in \ker P_2$. Then $T_2^k w \to 0$ as $k \to \infty$. We show that $S_n w \to 0$ as $n \to \infty$.
Let $\varepsilon > 0$. Choose $m \in \mathbb{N}$ such that $\|T_2^k w\| < \varepsilon$ for all $k \geq m$. Now choose

$n_0 \geq m+1$ such that $r^{n_0-m} \leq \varepsilon$. Letting $c_2 := \sup_{k\in\mathbb{N}} \|T_2^k\|$ we obtain, for $n \geq n_0$,

$$\begin{aligned}
\|S_n w\| &\leq \left\|\sum_{k=0}^{m-1} T_1^{n-1-k} S T_2^k w\right\| + \left\|\sum_{k=m}^{n-1} T_1^{n-1-k} S T_2^k w\right\| \\
&\leq \sum_{k=0}^{m-1} c_1 r^{n-1-k}\|S\| c_2 \|w\| + \sum_{k=m}^{n-1} r^{n-1-k}\|S\|\varepsilon \\
&= r^{n-m} c_1 c_2 \|S\|\|w\| \sum_{k=0}^{m-1} r^{m-1-k} + \|S\|\varepsilon \sum_{k=0}^{n-1-m} r^k \\
&\leq \varepsilon \frac{1}{1-r}\|S\|(c_1 c_2 \|w\| + 1).
\end{aligned}$$

This proves the claim.
(b) Assume that $r(T_2) < 1$ and that $T_1^n \to P_1$ strongly as $n \to \infty$. Then there exist $c_2 \geq 0$ and $r \in (0,1)$ such that $\|T_2^n\| \leq c_2 r^n$ for all $n \in \mathbb{N}$. We show that

$$S_n w \to \sum_{k=0}^{\infty} P_1 S T_2^k w$$

as $n \to \infty$ and for all $w \in G$.
Write $S_n = U_n + V_n$ with

$$U_n := \sum_{k=0}^{n-1} T_1^{n-1-k} P_1 S T_2^k = \sum_{k=0}^{n-1} P_1 S T_2^k \text{ and } V_n := \sum_{k=0}^{n-1} T_1^{n-1-k}(Id - P_1) S T_2^k.$$

Then $U_n \to \sum_{k=0}^{\infty} P_1 S T_2^k$ as $n \to \infty$.
Let $w \in G$. We show that $V_n w \to 0$ as $n \to \infty$. Let $c_1 := \sup_{n\in\mathbb{N}} \|T_1^n\|$. Let $\varepsilon > 0$. Choose $m \in \mathbb{N}$ such that

$$c_1 c_2 (1 + \|P_1\|)\|S\|\|w\| \frac{r^m}{1-r} \leq \frac{\varepsilon}{2}.$$

Since $\|T_1^n(Id - P_1)z\| \to 0$ as $n \to \infty$ for all $z \in F$, there exists $n_0 \geq m+1$ such that

$$\|T_1^{n-1-k}(Id - P_1) S T_2^k w\| \leq \frac{\varepsilon}{2m}$$

for all $k = 1, \cdots, m-1$ whenever $n \geq n_0$.
Let $n \geq n_0$. Then

$$\begin{aligned}
\|V_n w\| &\leq \left\|\sum_{k=0}^{m-1} T_1^{n-1-k}(Id - P_1) S T_2^k w\right\| + \left\|\sum_{k=m}^{n-1} T_1^{n-1-k}(Id - P_1) S T_2^k w\right\| \\
&\leq \frac{\varepsilon}{2} + \sum_{k=m}^{n-1} c_1 r^k (1 + \|P_1\|)\|S\| c_2 \|w\| \\
&\leq \frac{\varepsilon}{2} + \frac{r^m}{1-r} c_1 c_2 (1 + \|P_1\|)\|S\|\|w\| \leq \varepsilon.
\end{aligned}$$

This proves the claim. □

In Theorem 2.1 it is important to suppose that T_1^n or T_2^n converge uniformly to 0 as $n \to \infty$. It does not suffice to suppose that one of the two operators converges uniformly to an arbitrary limit. Indeed, in Theorem 1.2 we have $T_1 = P_1$ (so $T_1^n = P_1$ converges uniformly as $n \to \infty$) and T_2^n converges strongly as $n \to \infty$.

Nevertheless T^n is not power-bounded.

One can reformulate Theorem 2.1 in the following way.

COROLLARY 2.2. *Let E be a Banach space and F a closed subspace. Let $T \in \mathcal{L}(E)$ be such that $TF \subset F$. Let $T_1 = T_{|F}$ and define $\widetilde{T} \in \mathcal{L}(E/F)$ by $\widetilde{T}([x]) = [T(x)]$ $(x \in E)$. Suppose that F is complemented and that*

(a) *$r(T_1) < 1$ and $\widetilde{T}^n$ converges strongly as $n \to \infty$,*
or that
(b) *$r(\widetilde{T}) < 1$ and T_1^n converges strongly as $n \to \infty$.*

Then T^n converges strongly as $n \to \infty$.

PROOF. By assumption, there exists a closed subspace G such that $E = F \oplus G$. Thus there exist $S \in \mathcal{L}(G, F)$ and $T_2 \in \mathcal{L}(G)$ such that

$$T(y + w) = (T_1 y + Sw) + T_2 w \in F \oplus G,$$

for all $y \in F$, $w \in G$. The operator T_2 is similar to $\widetilde{T}$. Thus the claim follows from Theorem 2.1. □

The operators T_1 and $\widetilde{T}$ play an important role in the theory of semigroups, and in particular for studying the spectrum and asymptotic behaviour (see [**10**]).

Note that we have always

$$\sigma(T) \subset \sigma(T_1) \cup \sigma(\widetilde{T}),$$

even if F is not complemented. It follows that if $T_1^n \to 0$ uniformly and if $\widetilde{T}^n \to 0$ uniformly, then $T^n \to 0$ uniformly.

Another corollary of Theorem 2.1 describes the asymptotic behaviour of an operator with respect to an upper-triangular decomposition with n closed subspaces.

COROLLARY 2.3. *Assume that $T \in \mathcal{L}(E)$ with $E = F_1 \oplus F_2 \oplus \cdots \oplus F_n$, where $n \geq 2$ and F_k are closed subspaces of E such $T(F_k) \subset F_1 \oplus \cdots \oplus F_k$ for $1 \leq k \leq n$. In other words, with respect to the above decomposition of E, T has the form*

$$T = \begin{pmatrix} T_1 & A_1^2 & \cdots & \cdots & A_1^n \\ 0 & T_2 & A_2^3 & \cdots & A_2^n \\ 0 & 0 & T_3 & \cdots & \cdots \\ \vdots & \vdots & \ddots & \cdots & \cdots \\ 0 & 0 & 0 & 0 & T_n \end{pmatrix}$$

where T_k is linear and bounded on F_k for $1 \leq k \leq n$.
If there exists $k_0 \in \{1, \ldots, n\}$ such that $(T_{k_0}^n)_n$ converges strongly to P_{k_0} and if $r(T_k) < 1$ for all $k \neq k_0$, $1 \leq k \leq n$, then $(T^n)_n$ converges strongly as $n \to \infty$.

Of course we can have a cheaper example showing that in Theorem 2.1 uniform convergence of $(T_1^n)_n$ and $(T_2^n)_n$ to a limit different from 0 does not suffice to ensure strong convergence of T. Take $E = \mathbb{R}$ and $T = \begin{pmatrix} 1 & 1 \\ 0 & 1 \end{pmatrix}$. Then $T^n = \begin{pmatrix} 1 & n \\ 0 & 1 \end{pmatrix}$ is not power-bounded as for our composition operator on $\mathcal{D}$. But we may ask whether power-boundedness of T suffices for strong convergence.

Question: Let $T \in \mathcal{L}(E)$ be power-bounded, E a Banach space and F and G closed subspaces of E such that $TF \subset F$ and $E = F \oplus G$. Consider the matrix of T corresponding to this decomposition

$$\begin{pmatrix} T_1 & S \\ 0 & T_2 \end{pmatrix}.$$

If $(T_j^n)_n$ converges strongly as $n \to \infty$ for $j = 1, 2$, does $(T^n)_n$ converges strongly as $n \to \infty$?

We describe a situation where this is true. For this we recall the countable-spectrum theorem.

THEOREM 2.4 (ABLV-Theorem [**4**, **9**]). *Let E be a reflexive Banach space and $T \in \mathcal{L}(E)$ be a power-bounded operator. Suppose that*

$$\sigma(T) \cap \mathbb{T} \textit{ is countable and } \sigma_p(T) \cap \mathbb{T} \subset \{1\}. \tag{2}$$

Then $(T^n)_n$ converges strongly as $n \to \infty$.

We mention that this theorem should be seen in the context of a movement in the eighties starting with the Katznelson–Tzafriri theorem [**8**] and a beautiful contribution of Allan, O'Farrell and Ransford [**1**].

Now suppose that we are in the situation where

$$T = \begin{pmatrix} T_1 & S \\ 0 & T_2 \end{pmatrix}$$

on a reflexive Banach space where T is power-bounded and T_j ($j = 1, 2$) satisfies (2). Then also T does so since

$$\sigma(T) \subset \sigma(T_1) \cup \sigma(T_2) \text{ and } \sigma_p(T) \subset \sigma_p(T_1) \cup \sigma_p(T_2).$$

So the answer is "yes" in this case.

3. Decomposition of semigroups of operators

The aim of this section is to study the asymptotic behaviour of semigroups with respect to a given triangular decomposition.

Let E be a Banach space and $T = (T(t))_{t>0}$ be a C_0-semigroup on E. Its *growth bound* is defined by

$$w(T) := \inf\{w \in \mathbb{R} : \exists M > 0, \|T(t)\| \leq Me^{wt} \text{ for all } t > 0\}.$$

By [**5**, Prop. 5.1.1] one has:

$$w(T) = \lim_{t\to\infty} \log \|T(t)\| = \inf_{t>0} \log \|T(t)\| \text{ and} \tag{3}$$

in particular, if $\|T(t_0)\| < 1$ for some $t_0 > 0$, then T is *exponentially stable*, i.e.

$$\|T(t)\| \leq Me^{-\varepsilon t}$$

for some $\varepsilon > 0$ and $M \geq 0$.

Suppose that T is a C_0-semigroup on E. Then it is well-known that

$$r(T(t)) = e^{w(T)t}, \, t \geq 0,$$

where $r(T(t))$ denotes the spectral radius of $T(t)$. Now suppose that F is a closed subspace of E such that $T(t)F \subset F$ for all $t \geq 0$.
Suppose that F is completed, i.e., there exists a closed subspace G such that $E = F \oplus G$. Then $T(t)$ has the following matrix representation:

$$\begin{pmatrix} T_1(t) & S(t) \\ 0 & T_2(t) \end{pmatrix}.$$

Here $T_1(t) = T(t)_{|F}$ defines a C_0-semigroup on F and it is easy to see that T_2 is a C_0-semigroup on G.

THEOREM 3.1. *If*

(a) $w(T_1) < 0$ *and* $\lim_{t\to\infty} T_2(t)w$ *exists for all* $w \in G$ *or*
(b) $w(T_2) < 0$ *and* $\lim_{t\to\infty} T_1(t)y$ *exists for all* $y \in F$,

then $\lim_{t\to\infty} T(t)x$ *exists for all* $x \in E$.

We shall prove the theorem using the results of Section 2 via the following proposition which we owe to Jochen Glück with many thanks.

PROPOSITION 3.2. *Let T be a C_0-semigroup on E. If* $\lim_{n\to\infty} T(n\tau) =: P_\tau x$ *exists for all* $x \in E$ *and* $\tau \in \mathbb{Q} \cap (0, \infty)$, *then* $\lim_{t\to\infty} T(t)x$ *exists for all* $x \in E$.

PROOF. It follows from the definition that $P_{t+s} = P_t P_s$ and

$$P_t^2 = P_t = P_t T(t) = T(t) P_t$$

for all $t, s \in \mathbb{Q} \cap (0, \infty)$. It follows that $P_{nt} = P_t$ for all $t \in \mathbb{Q} \cap (0, \infty)$.
Let $M = \sup_{s\in[0,1]} \|T(s)\|$, $x \in E$ and $\varepsilon > 0$. There exists $n_0 \in \mathbb{N}$ such that

$$\|T(n)x - P_1 x\| < \varepsilon \text{ for all positive integers } n \geq n_0.$$

Now, for $t \geq n_0$, there exist $n \in \mathbb{N}$ and $s \in [0, 1)$ such that $t = n + s$.
Obviously if $t \geq n_0$ and $t \in \mathbb{Q} \cap (0, \infty)$ it follows that $s \in \mathbb{Q} \cap (0, \infty)$ and

$$\begin{aligned} \|T(t)x - P_1 x\| &= \|T(s)T(n)x - P_1 x\| \\ &= \|T(s)(T(n)x - P_1 x)\| \\ &\leq M\|T(n)x - P_1 x\| \leq M\varepsilon. \end{aligned}$$

If $t \notin \mathbb{Q} \cap (0, \infty)$, choose $r > t$ with $r \in \mathbb{Q} \cap (0, \infty)$. Then

$$\begin{aligned} \|T(t)x - P_1 x\| &\leq \|T(t)x - T(r)x\| + \|T(r)x - P_1 x\| \\ &\leq \|T(t)x - T(r)x\| + M\varepsilon. \end{aligned}$$

Letting $r \downarrow t$, we find that $\|T(t)x - P_1 x\| \leq M\varepsilon$ whenever $t \geq n_0$. □

PROOF OF THEOREM 3.1. (*i*): Suppose that $\lim_{t\to\infty} T_2(t)w =: P_2 w$ exists for all $w \in G$ and that $w(T_1) < 0$. Then $r(T_1(t)) < 1$ for all $t > 0$. It follows from Theorem 2.1 that $\lim_{n\to\infty} T(nt)x$ exists for all $x \in E$ and $t \neq 0$. Now the claim follows from Proposition 3.2.
(*ii*): The proof is similar to the proof of (*i*). □

References

[1] G. R. Allan, A. G. O'Farrell, and T. J. Ransford, *A Tauberian theorem arising in operator theory*, Bull. London Math. Soc. **19** (1987), no. 6, 537–545, DOI 10.1112/blms/19.6.537. MR915430

[2] W. Arendt, I. Chalendar, M. Kumar, and S. Srivastava, *Asymptotic behaviour of the powers of composition operators on Banach spaces of holomorphic functions*, Indiana Univ. Math. J. **67** (2018), no. 4, 1571–1595, DOI 10.1512/iumj.2018.67.7389. MR3853919

[3] W. Arendt, I. Chalendar, M. Kumar and S. Srivastava, *Powers of composition operators: asymptotic behaviour on Bergman, Dirichlet and Bloch spaces*, Journal of the Australian Mathematical Society, 1–32. doi:10.1017/S1446788719000235

[4] W. Arendt and C. J. K. Batty, *Tauberian theorems and stability of one-parameter semigroups*, Trans. Amer. Math. Soc. **306** (1988), no. 2, 837–852, DOI 10.2307/2000826. MR933321

[5] W. Arendt, C. J. K. Batty, M. Hieber, and F. Neubrander, *Vector-valued Laplace transforms and Cauchy problems*, 2nd ed., Monographs in Mathematics, vol. 96, Birkhäuser/Springer Basel AG, Basel, 2011. MR2798103

[6] E. A. Gallardo-Gutiérrez and A. Montes-Rodríguez, *Adjoints of linear fractional composition operators on the Dirichlet space*, Math. Ann. **327** (2003), no. 1, 117–134, DOI 10.1007/s00208-003-0442-9. MR2005124

[7] W. M. Higdon, *The spectra of composition operators from linear fractional maps acting upon the Dirichlet space*, J. Funct. Anal. **220** (2005), no. 1, 55–75, DOI 10.1016/j.jfa.2004.10.015. MR2114698

[8] Y. Katznelson and L. Tzafriri, *On power bounded operators*, J. Funct. Anal. **68** (1986), no. 3, 313–328, DOI 10.1016/0022-1236(86)90101-1. MR859138

[9] Yu. I. Lyubich and Vũ Quốc Phóng, *Asymptotic stability of linear differential equations in Banach spaces*, Studia Math. **88** (1988), no. 1, 37–42, DOI 10.4064/sm-88-1-37-42. MR932004

[10] W. Arendt, A. Grabosch, G. Greiner, U. Groh, H. P. Lotz, U. Moustakas, R. Nagel, F. Neubrander, and U. Schlotterbeck, *One-parameter semigroups of positive operators*, Lecture Notes in Mathematics, vol. 1184, Springer-Verlag, Berlin, 1986. MR839450

Wolfgang Arendt, Institute of Applied Analysis, University of Ulm. Helmholtzstr. 18, D-89069 Ulm (Germany)

Email address: wolfgang.arendt@uni-ulm.de

Isabelle Chalendar, Université Paris-Est, LAMA, (UMR 8050), UPEM, UPEC, CNRS, F-77454, Marne-la-Vallée (France)

Email address: isabelle.chalendar@u-pem.fr

Selected Published Titles in This Series

743 **H. Garth Dales, Dmitry Khavinson, and Javad Mashreghi, Editors,** Complex Analysis and Spectral Theory, 2020

742 **Francisco-Jesús Castro-Jiménez, David Bradley Massey, Bernard Teissier, and Meral Tosun, Editors,** A Panorama of Singularities, 2020

741 **Houssam Abdul-Rahman, Robert Sims, and Amanda Young, Editors,** Analytic Trends in Mathematical Physics, 2020

740 **Alina Bucur and David Zureick-Brown, Editors,** Analytic Methods in Arithmetic Geometry, 2019

739 **Yaiza Canzani, Linan Chen, and Dmitry Jakobson, Editors,** Probabilistic Methods in Geometry, Topology and Spectral Theory, 2019

738 **Shrikrishna G. Dani, Surender K. Jain, Jugal K. Verma, and Meenakshi P. Wasadikar, Editors,** Contributions in Algebra and Algebraic Geometry, 2019

737 **Fernanda Botelho, Editor,** Recent Trends in Operator Theory and Applications, 2019

736 **Jane Hawkins, Rachel L. Rossetti, and Jim Wiseman, Editors,** Dynamical Systems and Random Processes, 2019

735 **Yanir A. Rubinstein and Bernard Shiffman, Editors,** Advances in Complex Geometry, 2019

734 **Peter Kuchment and Evgeny Semenov, Editors,** Differential Equations, Mathematical Physics, and Applications, 2019

733 **Peter Kuchment and Evgeny Semenov, Editors,** Functional Analysis and Geometry, 2019

732 **Samuele Anni, Jay Jorgenson, Lejla Smajlović, and Lynne Walling, Editors,** Automorphic Forms and Related Topics, 2019

731 **Robert G. Niemeyer, Erin P. J. Pearse, John A. Rock, and Tony Samuel, Editors,** Horizons of Fractal Geometry and Complex Dimensions, 2019

730 **Alberto Facchini, Lorna Gregory, Sonia L'Innocente, and Marcus Tressl, Editors,** Model Theory of Modules, Algebras and Categories, 2019

729 **Daniel G. Davis, Hans-Werner Henn, J. F. Jardine, Mark W. Johnson, and Charles Rezk, Editors,** Homotopy Theory: Tools and Applications, 2019

728 **Nicolás Andruskiewitsch and Dmitri Nikshych, Editors,** Tensor Categories and Hopf Algebras, 2019

727 **André Leroy, Christian Lomp, Sergio López-Permouth, and Frédérique Oggier, Editors,** Rings, Modules and Codes, 2019

726 **Eugene Plotkin, Editor,** Groups, Algebras and Identities, 2019

725 **Shijun Zheng, Marius Beceanu, Jerry Bona, Geng Chen, Tuoc Van Phan, and Avy Soffer, Editors,** Nonlinear Dispersive Waves and Fluids, 2019

724 **Lubjana Beshaj and Tony Shaska, Editors,** Algebraic Curves and Their Applications, 2019

723 **Donatella Danielli, Arshak Petrosyan, and Camelia A. Pop, Editors,** New Developments in the Analysis of Nonlocal Operators, 2019

722 **Yves Aubry, Everett W. Howe, and Christophe Ritzenthaler, Editors,** Arithmetic Geometry: Computation and Applications, 2019

721 **Petr Vojtěchovský, Murray R. Bremner, J. Scott Carter, Anthony B. Evans, John Huerta, Michael K. Kinyon, G. Eric Moorhouse, and Jonathan D. H. Smith, Editors,** Nonassociative Mathematics and its Applications, 2019

720 **Alexandre Girouard, Editor,** Spectral Theory and Applications, 2018

719 **Florian Sobieczky, Editor,** Unimodularity in Randomly Generated Graphs, 2018